Fractal Worlds

Grown,

Built,

and Imagined

Fractal Worlds
Grown, Built, and Imagined

Michael Frame and Amelia Urry

Foreword by Steven Strogatz

Yale UNIVERSITY PRESS

New Haven and London

Copyright © 2016 by Michael Frame and Amelia Urry. Foreword copyright © 2016 by Steven Strogatz.

All rights reserved.

This book may not be reproduced, in whole or in part, including illustrations, in any form (beyond that copying permitted by Sections 107 and 108 of the U.S. Copyright Law and except by reviewers for the public press), without written permission from the publishers.

Yale University Press books may be purchased in quantity for educational, business, or promotional use. For information, please e-mail sales.press@yale.edu (U.S. office) or sales@yaleup.co.uk (U.K. office).

Printed in the United States of America.

Library of Congress Control Number: 2015955525
ISBN 978-0-300-19787-7 (pbk. : alk. paper)

A catalogue record for this book is available from the British Library.

This paper meets the requirements of ANSI/NISO Z39.48-1992 (Permanence of Paper).

10 9 8 7 6 5 4 3 2 1

Frontispiece: Great Barrier Reef coral photographed by Amelia Urry; branching fractal antenna built by Nathan Cohen; and Julia set fragment coded by Michael Frame.

Contents

Foreword by Steven Strogatz — ix

Introduction — xi

1 What is the shape of the world around us? — 1
  1.1 Types of symmetries 1
  1.2 Symmetry under magnification 2
  1.3 The Sierpinski gasket 4
  1.4 Fractals with memory 5
  1.5 Self-affine fractals 6
  1.6 Nonlinear fractals 6
  1.7 Random fractals 7
  1.8 Some non-fractals 8
  1.9 Physical world examples 10

2 Self-similarity in geometry — 15
  2.1 A simple way to grow fractals 15
  2.2 Some classical fractals 21
  2.3 Fractal trees and ferns 29
  2.4 Fractal spirals 33
  2.5 IFS with memory 34
  2.6 Random rendering of fractal images 42
  2.7 Circle inversion fractals 46
  2.8 Random fractals 51
  2.9 And flavors stranger still 53

3 Self-similarity in the wild — 55
  3.1 MathWorld vs. PhysicalWorld 56
  3.2 A foreshadowing of fractal dimension 56
  3.3 Coastlines, mountains, and rivers 58
  3.4 How (and why) the lungs are fractal 64
  3.5 Power laws 67
  3.6 Forests and trees 70
  3.7 Our complex hearts 71

3.8	Metabolic rates	73
3.9	Fractals and DNA	75
3.10	How planets grow	80
3.11	Reversals of the Earth's magnetic field	82
3.12	The distribution of galaxies	83
3.13	Is this all?	85

4 Manufactured fractals — 89

4.1	Chemical mixers	90
4.2	Capacitors	91
4.3	Wide-band antennas	92
4.4	Invisibility cloaks	94
4.5	Fractal metamaterials	95
4.6	Internet infrastructure	97
4.7	Music	99
4.8	Literature	105
4.9	Visual art	119
4.10	Building fractals	128

5 The Mandelbrot set: infinite complexity from a simple formula — 129

5.1	Some pictures	130
5.2	The algorithm	131
5.3	Julia sets	134
5.4	The Mandelbrot set	138
5.5	Other Mandelbrot sets	147
5.6	The universality of the Mandelbrot set	149
5.7	The Mandelbrot set in four dimensions	153
5.8	Unanswered questions	156

6 Quantifying fractals: What is fractal dimension? — 159

6.1	Similarity dimension	160
6.2	Box-counting dimension	167
6.3	Mass dimension	177
6.4	Random, with memory, and nonlinear	179
6.5	Dimension rules	189
6.6	Are unearthly dimensions of earthly use?	196
6.7	A speculation about dimensions	201

7 Further developments — 205

7.1	Driven IFS	205
7.2	Driven IFS and synchronization	216
7.3	Multifractals from IFS	226
7.4	Applications of multifractals	233
7.5	Fractals and stories, again	242

8 Valediction	**249**
A A look under the hood: Some technical notes	**253**
B Solutions to the problems	**407**
References	**439**
Acknowledgments	**491**
Figure credits	**493**
Index	**495**

Foreword

Every college has its legendary professors.
At Yale, one of them is Michael Frame.
Students have been flocking to his course on fractal geometry for the past twenty years. Its description in the course catalog hints at what makes it (and him) so extraordinary:

> **MATH 190a, Fractal Geometry** Michael Frame
> A visual introduction to the geometry of fractals and the dynamics of chaos, accessible to students not majoring in science. Study of mathematical patterns repeating on many levels and expressions of these patterns in nature, art, music, and literature.

Doesn't that sound appealing? "A visual introduction" – perfect for intuitive, right-brained types. "Accessible to students not majoring in science" – thank you! Here's a teacher who believes that everyone can fall in love with math if it's presented with empathy and humor and clarity and context. And speaking of context, who can resist wanting to learn how "patterns repeating on many levels" express themselves in "nature, art, music, and literature"?

This book is the distillation of Michael Frame's celebrated course. You're about to see why so many of his students thank him for being the best teacher they ever had.

Earlier today I read a profile of Michael Frame written by his coauthor on this book, Amelia Urry. It moved me to laughter and tears. And it made me want to meet both of them. Titled "Curious Geometries," it was published in the *Yale Daily News Magazine* in 2011 and is available online at
 http://yaledailynews.com/magazine/2011/11/08/curious-geometries/
With great lyricism and poignance, Amelia Urry tells the story of how Michael Frame came to be at Yale and describes his decades-long friendship and collaboration with Benoit Mandelbrot, the genius who made fractals a household word. The voices of all three of them – Frame, Urry, and Mandelbrot – sing through the book before you, and they harmonize perfectly.

With so much hand wringing these days about STEM and innumeracy and the importance of math for our nation's global competitiveness, a book like *Fractal Worlds* could be just what's needed. It portrays math as math lovers know it: a beautiful garden, a place of curiosity and delight, a tribute to human creativity and the wonders of nature.

Steven Strogatz
Cornell University

Introduction

Curiosity about the world is one of the joys of childhood. Discovering everything for the first, a child never stops asking, "Why?" Why this and not some other world? Why blue, why green, why thunder, why snow, why? If we are very fortunate, this curiosity stays with us throughout our lives. Wondering about the world and trying to understand how it works and why is one of the finest things we do as a species.

To understand the world, we seek simple descriptions of what we see. What do we notice, which features are important, which can be ignored? A honeycomb appears complicated when we focus on the variations in the structure of each cell, but reveals a simpler aspect when we focus on its repeating pattern of hexagons. The seeds of a sunflower or of a pinecone are not packed together randomly, but arranged in two families of interwoven spirals, one bending left, one right.

Figure 1: A tree silhouette; a dog lung cast.

Classical geometry is sufficient to understand these and many other shapes of nature. But what about a fern, the silhouette of a tree in winter, the veins in a leaf, the intricate branching of your lungs, or the roughness of a coastline? Euclidean geometry, familiar from school, cannot provide simple descriptions of these shapes, but that does not mean they are without order. They all share a common property called self-similarity: every small piece contains a copy of the whole shape, or at least something that looks like it, and every smaller piece contains still smaller copies of the whole, and on and on.

Being clear about what we mean by "looks like" will take some effort, but the idea is simple: if you magnify a picture of a fern, you will see more fern-like shapes making up the whole.

In fractal geometry, shapes can be understood by their self-similarity. When it can be applied, self-similarity is a tool of great power and simplicity: obvious once understood, but hidden in plain sight for centuries.

This book explores examples of mathematical and physical fractals, some assembled by people, some grown or sculpted by nature. Using high school geometry, we will learn how to grow basic fractals and how to understand the surprisingly simple rules that define the infinitely complex Mandelbrot set (right). For those familiar with logarithms, we will learn how to quantify the roughness of fractals and to find in what sense an object can be said to have a dimension other than the usual one, two, or three.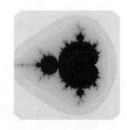

We can appreciate fractal geometry purely as one geometry among many, but it is so much more impressive to consider the range of natural structures that can be described by self-similarity. A new way to understand the shapes of the world, fractal geometry is elegant and spare in its concept, drop-dead beautiful in its simple categorization of shapes, world-shaking in its applications.

Much of our intellectual development is the story of how we learn to sort impressions: self or environment, rocks, trees, clouds, books, cats. It is a story of how we learn to judge and recognize colors, numbers, shapes, and abstract concepts. When we learn a new category, our internal model of the world rotates – often slightly, occasionally more. If we are very fortunate, a few times this change of perspective spins much of our world in a new direction altogether. Vertiginous, wonderful, exhilarating, these shifts echo through our lives for years. Some shifts are emotional: holding a newborn in your hands and understanding just what a rich and varied life will come to this tiny seed of an individual, looking into the eyes of an animal and recognizing a kinship despite our having traveled very different evolutionary paths. Some shifts are abstract: learning the crystalline pure beauty of a geometric proof's logic. Fractal geometry also represents a shift, both emotional and abstract, and convincing you of that is one of the goals of this book.

There is another goal, on a different level altogether. Having seen parts of a pattern for years, how can we fill in the gaps? How did Benoit Mandelbrot, the brilliant scientist who invented fractal geometry, knit together the observations of hundreds of artists and scientists into a unified field? Is fractal geometry the last such surprise to be discovered? Almost surely not. Perhaps in learning about the unfolding of fractal geometry, you will find your own way to make sense of our world. We shall see.

How the older author came to know Benoit and how your authors came to work together

This is the older author writing. My graduate training is in differential topology and in mathematical physics; fractal geometry was not yet a well-defined field when I was a student. In the mid-1980s, fairly early in my teaching career, a student asked if I knew anything about fractals. Though I'd seen Philip Morrison's enthusiastic *Scientific American* review of Benoit Mandelbrot's book *The Fractal Geometry of Nature*, I had not yet read it myself. What happened next is among my clearest memories. I got a copy of *The Fractal Geometry of Nature*, and because the day was pleasant, I sat under a tree and began reading. Reading, and looking at the pictures.

The picture of the Koch curve on page 45 caught my eye with its complicated articulations. To how many levels was it resolved? I held the book close to my eyes, and the trajectory of my life changed abruptly and significantly. Focusing ever  more closely on that picture of the curve, the physical world disappeared, and I had my first view of FractalWorld. Faster and faster came the images – roots of a tree, veins in a leaf, veins on my hand, mountain ranges, coastlines, clouds. So much of the world looks about the same over many scales. How had I not noticed this? It was so very clear. I had to learn more.

At Union College a few years later, the physicist David Peak and I developed a course on fractals as an introduction to scientific thinking for students not pursuing science degrees. We thought these students would be interested in learning how to understand some of nature's messiness, rather than the smooth simplifications common in many math and physics courses.

The next year, Benoit was to receive an honorary degree from SUNY Albany. James Corbett, Benoit's host in Albany, organized a mini symposium on fractals for the morning before Benoit's afternoon lecture. Jim had arranged several talks on interesting applications of fractals but needed someone to give a general introduction. He invited Dave, but a schedule conflict kept Dave from attending. This is how on a clear April morning I found myself walking to the front of a crowded room, clutching a folder of overhead transparencies. I turned to face the room and saw ... Benoit sitting in the middle of the front row. I had not expected him to attend the morning session, certainly not my simple introduction to the field he had invented. I felt that I was summarizing the Ten Commandments with Moses sitting in the middle of the front row. What could I do?

The only sensible thing: give the talk as planned. The other talks were more technical. These guys knew what they were saying; I was a beginner. Why in the world was Benoit wasting his time listening to me? I'd heard stories about his arrogance, but nothing Benoit did that day – or any other day, at least as far as I ever saw – suggested arrogance. In the break between the morning and afternoon sessions, Benoit made a point of finding me to compliment my exposition. "We must work together some day," he told me. Well, OK, he was more polite, gentler, than I'd expected. But me work with him? Not likely.

So three years later I was very surprised when Benoit invited me to spend the next year working with him at Yale. Our first long conversation there was free-ranging, about some historical precursors of fractals. Benoit dredged facts from his amazing memory, and I flexed my web-search muscles. That afternoon brought a vivid image to my mind, in many ways, my clearest of Benoit. We were two little kids, running around in a big grassy field under a sunny sky, finding all sorts of interesting things, excited to share what we found with one another. With me, Benoit was a little kid, too. I have had no other relation of that kind with anyone, though some conversations with Dave and some with my coauthor have been pretty close. I am still amazed when I realize this happened to me. We worked together for two decades, had hundreds and hundreds of conversations about mathematics, physics, art, literature, music, pedagogy; about our families, his dogs, my cats; about life. I've seen a bit of how Benoit thought. I have written code for him and watched him stare at the output, bothered by a pattern he didn't think was right and then say something like, "Check the order of the i and j loops. I believe you've reversed them." Time and again I was amazed at how conversations would unfold in directions unexpected by me, yet remarkably (almost) always finish with an answer to my question asked minutes, sometimes hours, earlier. This was so much fun.

In the spring semester of 1993, I introduced at Yale a version of the fractal geometry course Dave and I had developed at Union College. That first year about 180 students took the course and enrollment has remained high, with over 100 taking the course in the spring semester of 2016.

The level of student interest energized Benoit's ideas about pedagogy. In 1997 we ran a small meeting at Yale of people who had used fractals in teaching, and Benoit and I edited a book, *Fractals, Graphics, and Mathematics Education*. With help from high school math teacher Nial Neger and middle school math and music teacher Harlan Brothers, Benoit and I ran summer workshops on fractals and education for middle school, high school, and college teachers from 2000 to 2006. All together, several hundred teachers participated, most from the northeast, but one from California and one from Brazil.

Understanding Benoit's extensive obligations – juggling many research projects, always flying off to give talks, preparing the volumes of his *Selecta*, and working on his memoirs, *The Fractalist* – many of the teachers thought Benoit's involvement would begin and end with adding his name to the NSF pro-posal that funded the workshops. So when Benoit walked into the room with a new group of teachers, the pleasant pre-workshop chatter stopped. The teachers were accustomed to Nial, Harlan, and me; we were comfortable and decidedly not famous. But here was a world-class scientist talking directly to them. Benoit began with a joke or self-deprecating comment, for example, "Sometimes mathematicians ask me what is the hardest theorem I've proved. Please don't ask that. I prove only easy theorems. However, I reserve the right to ask very, very hard questions." The teachers relaxed and had a fascinating discussion with Benoit .

These workshops and my classes used a website which I'd been building since 2000. But Benoit's death in 2010 reminded me that I won't be able to maintain the site forever. Teaching fractal geometry for so many years, I've learned a few useful tricks that I'd hate to see lost. I thought I could extract key ideas from the website and write a sketch of the parts of the subject I enjoy the most.

On my own, I'd never finish the book. Sorting the ideas would be too complicated, and I wanted the prose to fly. Words are important, but not just as signifiers of ideas; they have their own rhythm, their own layering; pages sprinkled with hundreds of tiny mirrors, reflecting one another and bits of the world. But my writing is mechanical, not lyrical. I needed a partner, but who?

I needed someone who knows fractal geometry, and lots of science in general, and who has an energetic imagination. Someone who writes wonderfully, and has poetry in her bones. Because there were bound to be so many, conversations with this person must be an unalloyed joy. And most importantly, I needed someone who would be critical of my attempts at explanation, who would point out lack of clarity on scales small and large. How could I possibly find such a person? But in fact, I knew this person, already had had many conversations with her, often with my cat Bopper sleeping on her lap. So Amelia and I wrote the book you're reading. She and I offer this addition, a kind of companion to the memorial volume Nathan Cohen and I edited, to the legacy of Benoit's vision.

I cannot understand why Benoit brought me into his world. Yet for year after year we worked together. Through the challenges of aging, and the illness and death of dear family members, those glimpses of Benoit's nimble mind at work have been a constant of my life. Was I in awe of his intellect? You bet. Did I love him? Yes. Yes, indeed.

This is the younger author writing. I met Michael in the fall of 2011. At the time, I was a junior newly committed to the English major. Elbow-deep in the abstractions of Wordsworth, I found I missed the compelling riddle-like quality of especially tricky math problems, that tight flourish of logic unfolding step by step. This is probably the same quality that also attracts me to particularly tricky poems. In both cases, a sense of wonder animates the premise: how can these constructed symbols mean something true about the world? That semester, Michael was teaching his class on introductory fractal geometry for the 20th year in a row. The class was immensely popular and therefore immensely crowded. Though he would resist the designation, Michael was something of a campus celebrity. This was probably a result of an incredibly approachable teaching style, rare in the department: His classes punctuated with jokes stolen from Prairie Home Companion, his extreme generosity toward his students, the care and time he took to make sure every question is answered, and, more importantly, that the answer is understood – and his sincere obsession with cats.

I visited Michael in his office one afternoon in September of that year, to ask an unusual favor. I had pitched an idea for an article to the Yale Daily News Magazine, a profile of Michael ("Curious Geometries," published in November, 2011) both as an influential teacher and as a colleague of Mandelbrot, who had died very suddenly just one year earlier. As I had been warned, Michael was quickly dismissive of this idea. He was (he claimed) too boring to be the subject of a profile. I countered with a compromise. I offered to write not only Michael's story but also the many Mandelbrot stories which interwove it. To this, I believe, he could not have said no.

Over the next month, we met every week. I even managed to finagle an invitation to dinner, where I met his wife (Jean Maatta) and their seven cats (Chessie, Dusty, Fuzzy, Dinky, Leo, Crumples, and Bopper). I felt at ease almost immediately, as our conversations stretched longer and longer, reaching out from fractals to literature to the history of science and back again. Of course, we were both hypnotized by the way in which fractals seemed to be constantly reaching out of the world around us to dazzle with their complex simplicity. As my vision adapted to this way of seeing the world, it seemed that fractals could be found almost anywhere I looked.

We also talked about books. Michael's library is packed floor to ceiling with books, representing a staggering array of tastes, from Richard Russo to Stanislaw Lem to Vladimir Nabokov to José Saramago. (Fun fact: he read all of Dostoevsky in high school, while also teaching himself vector calculus.) We swapped quotes and stories we knew, among them Borges'

"The Garden of Forking Paths," which mirrored the bifurcation fractals on which he and Mandelbrot had collaborated in their first joint project. I introduced Michael to the poems of Louise Glück and Anne Sexton, whose "Riding the Elevator into the Sky" reminded him of staring into the night sky as a child, thinking about the immensity, the depth, of the universe. These things were all connected.

The following year, I wrote my thesis in poetry, a collection of poems in which traces of these conversations can be found. Borges is there, and Tolstoy, as is the night sky. Even the cats make an appearance (though somewhat more sinister in the poetic context). More than one poem is for Michael himself. It felt to me that we were constantly asking what lay behind the objects and events of the world. Often, there was sadness, loss, but there was always wonder as well. *What is this? What is it?* I wrote. The questions more important than the answers, if answers could be found. Meanwhile, we were writing this book. We were asking the same questions. What is this world? What logic can make it sing?

Here we are with a cat (Bopper) in a room full of books.

What you need to know to read these chapters

Chapters 1 through 4 are our general introduction to fractals: we discuss their several mathematical forms and provide a survey of their appearances in the world. For these chapters you will need just some basic algebra and geometry and, of course, a good dose of patience. If you are interested, more mathematical details can be found in Appendix A.

Sections using more than basic algebra and geometry are easy to recognize by the greater density of math. In these sections, we give a roadmap to what results are developed there, so even if you don't want to follow the details, you still can keep track of the flow of ideas.

In describing below which chapters need what math, we will mention terms introduced only later. Don't be frightened by terminology that may be unfamiliar now: everything you need to know will be explained. One way to use this guide is to read comments about a chapter only just before reading the chapter.

The Mandelbrot set and Julia sets live in the plane of complex numbers, so in Chapter 5 we make extensive use of complex numbers. In Section 5.6 we discuss Newton's method and use the fact that the slope of the tangent line to the graph of $y = f(x)$ is the derivative $f'(x)$. Section 5.7 is about the Mandelbrot set in 4 dimensions, but we don't expect familiarity with 4-dimensional geometry. Rather, we develop the idea a bit in that section, and a bit more in Sections A.72 and A.73 of Appendix A.

Chapter 6 is about dimensions. We'll see that dimensions are exponents, and that we calculate dimensions by using logarithms. We'll develop all the logarithm properties you'll need, but this chapter will go more smoothly if you've seen logarithms already, even if you've forgotten some of the details. In addition, we use the quadratic formula, and solve some cubic equations by first dividing out a linear factor and then applying the quadratic formula. We sum geometric series, though we review the derivation of the formula in Appendix A.2. In Section 6.2 we calculate some limits. In Section 6.4 we use computations of probabilities and of eigenvalues of matrices, which we review (or introduce) in Appendix A.83 and A.84. So you see, the mathematical requirements for the first few sections of Chapter 6 are relatively modest, but increase as we go through the chapter.

The first two sections of Chapter 7 require nothing beyond familiarity with iterated function systems (IFS). Section 7.3 is another story. Here we'll use more probabilities, and calculations involving logarithms and derivatives.

Appendix A is the most mathematically demanding part of the book. Here we derive topological properties of some of the classical fractals, show how to speed up the random IFS algorithm by using de Bruijn sequences, use integrals to derive the scaling property of Brownian motion and fractional Brownian motion, illustrate the relation between correlation functions and power spectra using the Fourier cosine series, and give some of the arguments explaining allometry. Our brief investigation of the distribution of epidemics uses differential equations.

In studying the Mandelbrot set, we use the triangle inequality to derive the escape criterion. A clever change of (complex) variables shows that the point at ∞ is a fixed point for the Mandelbrot function. We review the polar representation of complex numbers to clarify the dynamics on Julia sets. This takes us on a detour through some chaotic dynamics, where we establish the topological properties of chaos. Our investigation of stable cycles in the Mandelbrot set uses derivatives. Here we must take square roots of complex numbers, so the polar representation is used again. Newton's method for complex functions uses derivatives of complex variables – hardly a surprise.

In studying dimensions, we use calculus to prove the Moran equation has a unique solution, and use geometric series arguments to extend the Moran equation to fractals with infinitely many scaling factors. We show a sequential limit suffices in computing the box-counting dimension. Computing dimensions of random fractals is extended to the case where the scaling factors can take on values in an interval, not just a few values. We build up the bits of linear algebra needed to compute eigenvalues. The section on algebra of dimensions involves a fair amount of geometry.

Estimating how certain we are that an empty part of a driven IFS is a result of the dynamics of the system, and not of a too short data set, uses a Markov chain argument. Our study of synchronization makes use of Liapunov exponents to establish chaotic dynamics. Calculus is used freely in deriving some properties of multifractals. Appendix A, then, makes fairly heavy mathematical demands in some sections. Even there, we think the architecture of the ideas is presented in a straightforward way.

Appendix B is where you can find answers to the exercises included in the text.

In the References we provide a list of sources to material in the chapters, in the appendices, and in the credits list. The References are organized section by section to match the chapters. Where we think it will be useful, we introduce a section's sources with a description of why we cite them.

Inevitably, we will have omitted sources beloved by some. Blame for this crime belongs entirely to the older author, who has been studying fractal geometry longer than the younger author has been in the world, so some references he once treasured now are hiding.

Many worlds

While talking through the ideas we were sorting and organizing for this book, we noticed they fit nicely into three drawers: MathWorld, PhysicalWorld, and FractalWorld. Also, we've spent much of the last four years in WritingWorld.

MathWorld is the home of mathematical ideas: triangles, circles, surfaces, prime numbers, differential equations, groups, cohomology rings, and also (mathematical) fractals. MathWorld is potentially infinite, maybe actually infinite. It's a world of ideas: flights of intuition building our impressions, proofs established by crystalline logic. This is where the older author has spent most of his life, with occasional excursions into CatWorld, an important part of PhysicalWorld.

PhysicalWorld is where most people spend much of their lives. This is the world of rocks, trees, birds, cats, rain, gardens, the night sky, piles of books, and people. It's the world of gravity and of light, of molecules and crystals and microbes, of kitchens and frying pans and crepes with mushrooms and asparagus. The body and the brain live here. The mind can, but need not.

FractalWorld lies partly in MathWorld and partly in PhysicalWorld. Everything fractal belongs here, but some fractals are ideas and others are physical objects. FractalWorld groups together those ideas and objects made of pieces that look like shrunken copies of the whole. FractalWorld was discovered by Benoit. Now this world, once invisible, is seen clearly by so many.

WritingWorld holds the collections of words written by Jorge Luis Borges, Italo Calvino, Leo Tolstoy, Mary Oliver, Louise Glück, Anne Sexton, José Saramago, Yasunari Kawabata, James Joyce, Joyce Carol Oates, Flann O'Brien, Alice Munro, John Barth, and on and on and on. We don't mean the books as objects – these live in PhysicalWorld – but the strings of ideas and words. After all, words are more than the ideas they represent. If the older author was the younger author's guide to FractalWorld, the younger author was the older author's guide to WritingWorld.

Web supplement

We've written a web supplement, located at

http://yalebooks.com/fractalworlds

Organized by book chapter, here you'll find additional exercises, color versions of some of the figures, illustrative animations, links to Java software, and any new topics we've found since the book went to press. For us, the most difficult aspect of writing this book – or any book – is its finality. The last set of corrections is made, the book files go off to the printer, and we're done. From that moment on, all comments of the form "Look at this new application of fractals. Isn't this cool? We've got to include a paragraph or two about it" become unwelcome by the publisher. But we can change the web supplement whenever we want, keeping it more a process than an object – which is quite appropriate for something involving fractals, don't you think?

Chapter 1

What is the shape of the world around us?

This chapter builds visual skills needed for recognizing fractals, and establishes a vocabulary for studying them. Here you will find just descriptions, no formulas or equations. Those come aplenty in the next chapters; this one is where we learn how to read the pictures.

We will begin by investigating the familiar symmetries, those under translation, rotation, and reflection, illustrating each with examples from nature. However, many examples from nature exhibit, and often are dominated by, another symmetry, symmetry under magnification. These shapes are called self-similar. We will present geometrical examples, and then variations: self-similar with memory, self-affine, nonlinearly self-similar, and statistically self-similar. Each is illustrated with examples from geometry and from nature or manufacture.

1.1 Types of symmetries

The world is made up of shapes, some easy to describe, some more challenging. A pebble dropped into a still pond produces roughly circular ripples. A grain of salt is approximately a cube. To be sure, these are not exact Euclidean circles and cubes, but then Euclidean geometry describes the ideal types of physical objects battered and trodden upon by forces of the world. The real world contains no exactly perfect square, but it does contain many shapes very close to a perfect square. Rather than investigate each approximate square by itself, it's simpler first to understand the *squareness* of the shape – four right angles, four sides of the same length – and only then study that object's particular quirks.

Often understanding squareness, or circularity, or triangiosity, is helped by recognizing any symmetries the shape exhibits. Three kinds of symmetry are familiar:

- *Symmetry under rotation.* Rotating a circle around its center doesn't change the circle at all. Short of painting part of the circle some color, how can we tell if it has been rotated? In nature, snowflakes are (approximately) unaltered by 60° rotations.

- *Symmetry under translation.* A field of flowers looks more or less the same if we look left or right. What else? A honeycomb. Sand on a beach. A sky filled with low clouds. A brick wall.

- *Symmetry under reflection.* Reflecting a human body across a vertical plane through its middle produces a recognizably similar replica, at least on the outside of the body. (The inside is a bit more complicated.)

Figure 1.1: Symmetry under rotation (first), translation (second), and reflection (third). Fourth: OK, faces aren't really reflection symmetric.

Informed by our perception of appropriate symmetries, simple descriptions of many shapes can be written in the language of Euclidean geometry. But is all the world just roughed-up high school geometry?

1.2 Symmetry under magnification

More complicated shapes in nature seem to defy description by the familiar symmetries. Think about the silhouette of an old tree, or the twists and turns of a coastline, or the branches off branches off branches of your circulatory system, all shown in Fig. 1.2. Or this passage from Olive Schreiner's *The Story of an African Farm*.

> A gander drowns itself in our dam. We take it out, and open it on the bank, and kneel, looking at it. Above are the organs divided by delicate tissues; below are the intestines artistically curved in a spiral form, and each tier covered by a delicate network of blood-vessels standing out red against the faint blue background. Each branch of the blood-vessels is comprised of a trunk, bifurcating and rebifurcating into the most delicate, hair-like threads, symmetrically arranged. We are struck with

1.2. SYMMETRY UNDER MAGNIFICATION

its singular beauty. And, moreover – and here we drop from our kneeling into a sitting posture – this also we remark: of that same exact shape and outline is our thorn-tree seen against the sky in mid-winter; of that shape also is delicate metallic tracery between our rocks; in that exact path does our water flow when without a furrow we lead it from the dam; so shaped are the antlers of the horned beetle. How are these things related that such a deep union should exist between them all? Is it chance? Or, are they not all the fine branches of one trunk, whose sap flows through us all? That would explain it. We nod over the gander's inside.

  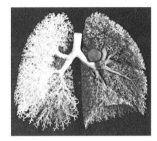

Figure 1.2: Tree, coastline, and lung and circulatory casts.

None of the three standard symmetries gives us any traction for finding a simple description of a tree in the winter, which cannot be rotated, translated, or flipped without serious alteration. But there is another symmetry, vaguely familiar for centuries to artists and to children, but only recently added to the toolbox of scientists and mathematicians:

- *Symmetry under magnification*. When magnified by an appropriate factor, a small piece of an object looks very much like the whole object.

Many plants exhibit similar structures over several levels. Trees, ferns, Queen Anne's lace, all do this, as does broccoli, which is made up of florets of florets of even smaller florets. Think about it. What other vegetable fractals can you find?

Figure 1.3: Broccoli, floret, floret of floret.

In the same way that an old floor tile illustrates the notion of squareness, but only roughly, variations on self-similarity in natural fractals can be a bit complicated. So we'll start with geometrical ideals of self-similarity before venturing out into nature. Fig. 1.4 illustrates the types of geometrical self-similarity we study.

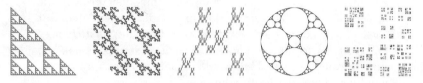

Figure 1.4: Fractals that are self-similar (first), with memory (second), self-affine (third), nonlinear (fourth), and statistically self-similar (fifth).

1.3 The Sierpinski gasket

One of the simplest, and also the most prevalent, of the self-similar fractals is the Sierpinski gasket. A kind of mascot for fractals in general, the Sierpinski gasket is a triangle with a sequence of ever-smaller triangular holes. Often we call it just *the gasket*.

In Fig. 1.5, in the second image we can see the gasket broken into three outlined pieces, each 1/2 as wide and 1/2 as tall as the whole, but otherwise identical. We say that these pieces are copies of the gasket, *scaled* by 1/2. Once we have split the gasket into three pieces, we can see that each of these pieces can be split into three pieces again and again. In the third and fourth images of Fig. 1.5, the gasket is broken into nine pieces scaled by 1/4 and twenty-seven pieces scaled by 1/8. Mathematically, this process of subdivision into smaller pieces can be carried on without end.

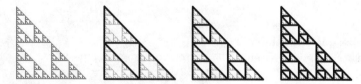

Figure 1.5: The Sierpinski gasket. Outlines indicate several decompositions into smaller gaskets.

Although we can find the gasket as part of decorative motifs in 900-year-old churches (and some younger ones), the geometrical properties of the Sierpinski gasket were studied first in 1915. Along with the Koch curve and the Cantor set, the gasket was developed to answer specific

1.4. FRACTALS WITH MEMORY

mathematical questions. Many early–twentieth–century mathematicians regarded these irregular shapes as monsters, anomalies that did not submit to standard analysis. Benoit was the first to observe that these share a common property, self-similarity, an organizing principle whose importance was not recognized until fractals made it visible.

The gasket gives a clear picture of the meaning of the term "self-similar." You may recall from geometry that triangles are *similar* if they have the same shape but perhaps different sizes. The gasket is made of three copies similar to itself, which are in turn made up of smaller similar pieces, and so on.

This notion of self-similarity admits many variations. We will explore some others in the next sections.

1.4 Fractals with memory

This next fractal is not as easily identified as being self-similar. In fact, it consists of four pieces which seem to replicate the whole, but with certain parts missing. In the first image of Fig. 1.6, we can see that each of the four quarters is missing one corner, but not the same ones. This same pattern of missing corners continues for all levels; each piece is made up of four pieces, each with corresponding corners left blank. That is the sense in which this is a fractal, an instance of Benoit's statement that often a fractal can be described just as well by what has been removed as by what remains.

Figure 1.6: A fractal with memory, the missing pieces indicated with an addressing scheme to exhibit the pattern of empty regions.

To decribe this pattern of missing corners, we can number the four quarters of the square as shown on the second image of Fig. 1.6. These are the *length*-1 *addresses* of this fractal. The third image of Fig. 1.6 shows that each length-1 address region can be subdivided into four length-2 address regions. For example, the lower left corner of the lower left quarter has address 11; it is the address 1 part of the address 1 quadrant. As we see in the third image, the empty length-2 addresses are 11, 23, 32, and 44. In fact, this fractal can be described just by specifying that the empty squares are all those with addresses *containing* 11, 23, 32, or 44, such as 3<u>11</u>, 1<u>23</u>, 4<u>32</u> and 343<u>44</u>. The fourth image of Fig. 1.6 shows each of the empty length-3 addresses, each containing an empty length-2 address.

1.5 Self-affine fractals

Self-affine fractals are characterized by different scaling factors in different directions. The gasket is scaled both horizontally and vertically by 1/2, but it is possible to have a fractal – no longer self-similar – whose proportions are not so balanced.

Every example of Fig. 1.7 consists of six pieces, each scaled by a factor of 1/4 horizontally and 1/3 vertically. Different scaling factors in different directions may seem to be a

Figure 1.7: Examples of self-affine fractals, with 4×3 grids to emphasize the scalings.

minor geometrical variation, but these differencs cause sobering, perhaps insurmountable, difficulties in computing fractal characteristics, including one kind of measurement called *fractal dimension*, which is roughly the amount of complication, or roughness, a fractal exhibits. Since, in most cases, fractal dimension can be calculated solely from the number and scalings of the pieces, self-affine fractals throw a wrench into the system. Each fractal of Fig. 1.7 consists of the same number of pieces, of the same size, but in Appendix A.89 we see they have different fractal dimensions.

1.6 Nonlinear fractals

In the examples we have seen so far, each little bit of the fractal looks pretty much like the whole fractal, or at least a large piece of it. Once perceived, scaling symmetry usually is very straightforward. But suppose the symmetry is distorted by a subtle but patterned logic? The fractal shown above is produced by many iterations of a process called *inversion in a circle*.

If straightforward symmetry is like reflection in a flat mirror, inversion in a circle is based on reflection in a curved mirror. Objects near the mirror experience only mild distortion, while objects farther away are distorted

1.7. RANDOM FRACTALS

more severely. Imagine looking into a polished metal bowl: your nose is relatively undistorted at the center of the bowl, while the rest of your face stretches away down the sides. Fig. 1.8 illustrates this distortion.

As an example, take the circles C_1 through C_5 in the first image of Fig. 1.9 as the curved mirrors. For each circle, the part of the fractal inside that circle is the inverse of everything outside that circle. Inside each of the circles C_1 through C_4 are four shapes we could call *curved* gaskets, and inside C_5, four much smaller curved gaskets. Look closely.

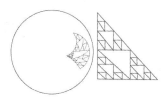

Figure 1.8: An early stage of a gasket, inverted in the circle.

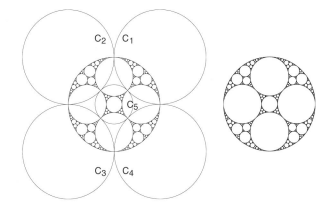

Figure 1.9: A circle inversion fractal, with the circles that generate it (first), and without (second).

The Sierpinski gasket becomes predictable as soon as we recognize that every little bit is an exactly scaled copy of the whole. Circle inversion fractals are more subtle, and so may hold our interest longer. The scaling varies, depending on the location. Each little piece shares some features with the whole, but also differs from the whole, and from other pieces, in intricate ways. Though we understand it, it still may surprise us. And we like to be surprised.

1.7 Random fractals

This strange patchwork quilt is an example of a *statistically self-similar* fractal. The fractal is divided into four copies, each of which is divided into four copies, and on and on and on, forever.

However, instead of each copy being scaled by a constant proportion, here the scaling factors at each level vary randomly between 1/4 and 1/2. Fig. 1.10 illustrates the first, second, third, and, skipping a few, sixth iterations of generating a statistically self-similar fractal.

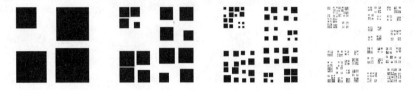

Figure 1.10: The first, second, third, and sixth stages in generating a statistically self-similar fractal.

Learning to recognize random fractals takes practice. Each piece is not an exact scaled copy of the whole; rather, an appropriate average of the many pieces presents a scaled copy of the whole. Most, if not all, natural fractals follow this general rule, as environmental factors influence the growth of the fractal in variable but consistent ways. Randomness is the key to bridging the gap between MathWorld and PhysicalWorld. Of course, it's all part of the landscape of FractalWorld.

1.8 Some non-fractals

Before continuing, we shall pause to point out some mistakes, still far too common, in deciding whether an object is fractal or not.

Figure 1.11: Three triangles do not a Sierpinski gasket make. A raspberry is not any sort of fractal, nor is a brick wall. A spiral shell is self-similar only about its center.

The first issue is whether an object exhibits similar structures over enough levels. No physical object can exhibit substructures on arbitrarily small scales, but we should see enough levels that self-similarity represents a significant symmetry. Three triangles do not constitute a gasket, as we see in the left image of Fig. 1.11. There is no hint of any process being carried out over several levels; therefore, no fractal in sight. Similarly, a raspberry, despite being a round fruit made up of round seeds, exhibits no further levels of self-similarity, and cannot be said to be truly fractal.

1.8. SOME NON-FRACTALS

Another common mistake is the claim that any repeating pattern is a fractal. This is nonsense. Fractals involve symmetry under magnification, not mere translation. A brick wall is not a fractal; neither is a field of flowers. Pretty, but not fractal.

More subtle is the spiral of a seashell, seen in the last image of Fig. 1.11. While it is true that magnifying the center of the spiral does give another spiral, magnifying any other point of the spiral gives a smooth curve, which gradually flattens out under successive magnification. Self-similarity that holds at only one point does not reveal a fractal structure.

Figure 1.12: Russian dolls and the cat with a tie are self-similar about only one point. These are *not* fractals. The pattern of cat earrings contains a fractal.

Related mistakes include nested Russian dolls (first image of Fig. 1.12) and a cat wearing a tie which has a picture of a cat wearing a tie, and so on (second image of Fig. 1.12). Though these do exhibit a kind of self-similarity, like a Nautilus shell, the self-similarity holds only for magnification about one point. Along these lines, the sequence of ever-smaller cats under hats in *The Cat in the Hat Comes Back* has been misinterpreted by some as fractal, but this is another example of Russian dolls.

We must admit we're a bit sad that Russian dolls are not fractal, because if they were, then in Flann O'Brien's *The Third Policeman*, Office MacCruiskeen's potentially infinite collection of nesting chests would be an example of a fractal in literature. It remains a fascinating passage – beautiful fading to strange and then to terrifying – but not, alas, fractal.

Jennifer Lanski's drawing of a cat wearing earrings (third image of Fig. 1.12) is different. The cat wears two earrings, each of which has a picture of a cat wearing two earrings, and so on. That each earring contains *two* smaller earrings is good enough for us to see a pattern that continues to produce a perfectly good fractal, of a kind called a *Cantor set*.

1.9 Self-similarity in the physical world

Although fractals are most cleanly seen in MathWorld, battered versions of self-similarity also live in PhysicalWorld. For example, visual instances of some fractals can be generated by video feedback with mirrors. The first image of Fig. 1.13 shows the experimental setup: a video monitor with mirrors lined up along two edges. A camera is aimed at the common point of the mirrors and monitor, and the video shown on the monitor is captured by the camera. The resulting cascade of images within images can produce fractals. Camera zoom, angle, and position are the experimental parameters, and the feedback loop is the iteration. The second image of Fig. 1.13 is an example of a fractal generated by this arrangement. An Internet search will turn up more dynamic video feedback effects, but for the moment we are interested in the intricacies of self-similarity captured in this stationary fractal image. All this, just from positioning a camera and some mirrors and playing around.

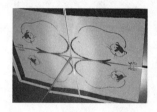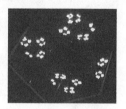

Figure 1.13: video feedback with mirrors: the experimental arrangement (first), a fractal image (second), and a fractal with memory (third).

1.9.1 Fractals with memory

In the third image of Fig. 1.13, the camera is positioned so it shows only a portion of one image. This exclusion is carried through all levels of the fractal. The whole image has four pieces, three of which have four sub-pieces, while one has only two sub-pieces. This video feedback arrangement has constructed a fractal in which some pieces go through the feedback loop, while others do not. Consequently, we can say that this is a video implementation of a fractal with memory.

1.9.2 Self-affine fractals

Often mountain ranges scale differently in vertical and horizontal directions. Geological forces, about the same along the length of the mountain range, push up the rocks with one scaling factor; meteorological forces weather down the rocks with a different scaling factor. Different horizontal and vertical scalings make these fractals self-affine, just as other objects

1.9. PHYSICAL WORLD EXAMPLES

grown or sculpted by different forces in different directions often exhibit self-affinity.

1.9.3 Nonlinear fractals

For an example of a nonlinearly scaled fractal, we can look at curved mirrors, specifically at four silvered balls, three arranged in a triangle and the fourth placed on top, as shown in the first image of Fig. 1.14. By illuminating the space in between with a camera flash or a laser pointer, we can see a pattern of reflections forming a curved gasket. Here the iteration is achieved through multiple reflections, the nonlinear version of sitting between parallel mirrors in a barber shop.

Figure 1.14: The setup and two examples of an optical gasket.

1.9.4 Random fractals

Finding examples of random fractals in the world is just too easy. What object in the wide world wasn't grown or sculpted with some environmental effects, the hand of forces unseen?

Some of these forces are so massive we can only get a look at them from a low Earth orbit. Fig. 1.15 shows a fractal distribution of clouds: a few very large, more medium, many small. How does this make the clouds fractal? Think of this: the gasket has 1 large hole, 3 smaller holes, 9 still smaller, 27 even smaller, and so on.

Figure 1.15: Clouds from space.

Another way to see the fractality of clouds is to take a look at the next massive thundercloud that rolls by: you'll see billows on billows on billows, Or compare the small puff of condensing steam from a teapot to a cumulus puff outside your window.

In the first image of Fig. 1.16 we see a coastline, quite similar to a random fractal, called a randomized Koch curve, that we'll look at in Chapter 2. In Chapter 6 we'll talk about the role coastlines played in the

 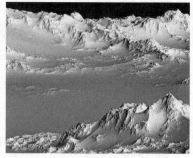

Figure 1.16: First: Coastline and river. Second:A fractal forgery of mountains .

early growth of fractal geometry. Also in this figure are rivers and other branching structures exhibiting the same kind of intense complication. Yes, the world is filled with beautiful random fractals.

Properly combined, these variations on self-similarity can produce artificial images of such startling reality. Alone, that may not be so surprising. Most of us have seen the fantastical computer-generated landscapes of films like *Avatar*. What is interesting about fractal forgeries in particular is the small amounts of data needed to produce the images. Without having to consider the detailed mechanisms of tectonics and weathering, realistic forgeries can be generated merely by understanding the fractal statistics of the world. Ken Musgrave, one of Benoit's students and long time collaborators, developed many of the techniques to create amazingly believable landscapes, one of which is shown in the second image of Fig. 1.16.

Figure 1.17: Three fractals in the kitchen.

Finding other examples of random fractals in the world is a good exercise. We leave you with three examples from the kitchen: lettuce leaves with crinkles on crinkles on crinkles, a crepe with a distribution of browned circles representing the range of sizes and numbers of bubbles (lots of small, a few big bubbles) produced when the batter cooks, and the branching pattern formed when cold butter is pressed between two surfaces and the surfaces are pulled apart. Not everything in your kitchen is fractal; for example, a bunch of grapes is well described as a collection of approximate spheres with about the same radii. But many things in the kitchen, the house, the world are well described by fractals. Take a look. What else can you find?

Chapter 2

Self-similarity in geometry

For the simplest geometrical fractals, we begin by puzzling out what rules will produce a given fractal. This is a fractal version of *Jeopardy*: the fractal is the answer, and the question is "What happens if I apply these rules over and over again?" Each rule can be described as a combination of four types of operations: scaling, reflection, rotation, and translation. After applying this technique, called Iterated Function Systems (IFS), to several examples, we see how to extend this technique to the variations on self-similarity presented in the previous chapter.

2.1 A simple way to grow fractals

Many fractals are quite beautiful: complex structures, built layer upon layer of intricate detail, drawing us ever deeper. Less easily recognized is the beautiful – and simple – mathematics that generates these shapes. Rather than just look at pictures, we'll learn how to grow some fractals by using easy manipulations of geometrical objects. When you can build these shapes, when you know how the pieces fit together and why, you will possess a key to decoding fractals in the world.

Building the simplest fractals involves two steps:

- Recognize the fractal A as made up of some number of smaller copies $A_1, A_2, \ldots, A_n$ of the fractal.

- For each smaller copy A_i, find a way to transform the whole fractal A into A_i.

The first step (usually) is easy. For example, each of the fractals in Fig. 2.1 is made up of three copies of itself, each scaled by 1/2 in both directions. Of course, this decomposition could be carried out forever: the fractal consists of three smaller copies of itself, each of which consists of three still smaller copies, each of which consists of three even tinier copies, and so on. Recall the Sierpinski gasket Fig. 1.5. This repeated splitting into smaller copies is a geometric version of the story that the world rests on the back of four elephants standing on the shell of the world turtle, which stands on the shell of another turtle, and so on. It's turtles all the way down. Every one of these decompositions into smaller pieces is correct, but for our purposes, the simplest description usually is enough: three pieces each scaled by 1/2.

To see the second step, we investigate the Sierpinski gasket, the first shape in Fig. 2.1. This, and all the fractals in this section, are drawn in the unit square

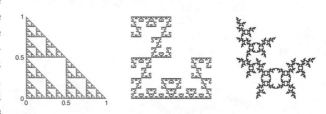

Figure 2.1: Fractals composed of three pieces, each scaled by 1/2.

in the xy-plane, with the origin at its lower left corner. The coordinates of the x- and y-axes are shown along the axes for the gasket; you'll have to imagine that the other shapes are similarly inscribed. To list the transformations that take the whole shape to each piece, recall that in place of the phrase *shrink vertically and horizontally by* 1/2, we say *scale by* 1/2, and that every scaling shrinks toward the origin. The locations of the scaled copies are determined by their translations. Then the rules that take the entire Sierpinski gasket to each of its three pieces are

- *lower left:* scale the whole shape by 1/2 and leave it at the origin,
- *lower right:* scale by 1/2 and move (translate) 1/2 to the right, and
- *upper left:* scale by 1/2 and translate 1/2 up.

Applied together, these three rules return the gasket unaltered, identical in every detail. This is no accident. When we find the rules for any fractal, applying all those rules to the fractal returns exactly that fractal, while applying the rules to every other shape alters that shape. This is the *invariance property* of fractals.

(Well, not absolutely *every* other shape. Applying any collection of fractal rules to the whole plane will give us back the whole plane. But the invariance property does hold for every shape you can draw on a computer screen.)

To illustrate this point, consider what happens if we apply these three rules to some other shape. Look at the first and second images of Fig.

2.1. A SIMPLE WAY TO GROW FRACTALS

2.2. In one iteration, the first cat turns into three *identical* cats. Hardly a surprise, at least not yet.

The magic comes next. If we apply the gasket rules to the three small cats they become nine still smaller cats, then twenty-seven even smaller cats. Again and again and again we can repeat the process, until the cats vanish and the gasket appears. For this reason, the gasket is called the *attractor* of these three rules. Starting from any shape on a computer screen and applying the gasket rules again and again will give a sequence of images that always produce the gasket. In that sense, the gasket is *defined by* the three gasket rules, which is a much easier description than the infinite collection of nested triangles itself. By characterizing complex shapes not as static objects, but as the processes by which they grow, fractal geometry can uncover the simplicity that is hidden in apparent complexity.

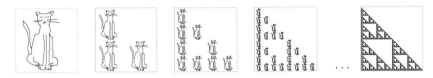

Figure 2.2: The gasket rules applied to a cat, then repeated.

We should mention that, for a given set of rules, in order to be sure that sequence of images produced during "again and again and again" converges to the same shape, each of the rules must be a *contraction map*. That is, each rule must make the shape smaller. This requirement can be relaxed a bit if the image is produced by the random IFS algorithm of Sect. 2.6.

When finding the rules for the gasket, we encountered scaling and translation, two of the four ways we'll use to manipulate shapes in the plane. To find the rules for the middle and right fractals of Fig. 2.1, we must understand the other two ways, reflection and rotation, to transform shapes.

Reflections first. See the first image of Fig. 2.3. Start with a shape, a square containing an L, in the first quadrant. Image A is the left-right reflection, that is, reflection across the y-axis. Image B is the up-down reflection, that is, reflection across the x-axis. Image C is the compo-

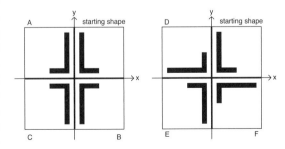

Figure 2.3: Reflections then rotations.

sition of these, reflection across the x-axis and reflection across the y-axis, in either order. Each reflection involves two pieces of information: how the orientation of the shape changes and where the reflected image lies. The location of the image is important for determining the subsequent translations to assemble the pieces into the whole.

Rotations next. Look at the second image of Fig. 2.3. By rotations we mean rotations about the origin, with counterclockwise rotations represented by positive angles. Quadrant sections D, E, and F show rotations by 90°, 180°, and 270°. Note, for example, that rotation by 270° is equivalent to rotation by −90°, that is, a clockwise rotation by 90°. Also, comparing images C and E illustrates the fact that the composition of reflections across the x- and y-axes is equivalent to rotation by 180° (whether clockwise or counterclockwise doesn't matter). As with reflection, both the orientation and the location of the image are important to note.

With these types of transformations in mind, we can characterize every plane transformation by six numbers.

- r denotes scaling in the horizontal direction. Think of multiplying the x-coordinates of each point by r. Then we see that negative r reflects the shape across the y-axis.

- s denotes scaling in the vertical direction. Think of multiplying the y-coordinates of each point by s. Negative s reflects the shape across the x-axis.

- θ denotes rotation of horizontal lines about the origin, with positive angles indicating counterclockwise rotations.

- φ denotes rotation of vertical lines. (For almost all of our examples, $\theta = \varphi$ and the image rotates about the origin without distortion.)

- e denotes horizontal translation (motion). Think of adding e to the x-coordinate of each point.

- f denotes vertical translation (motion). Think of adding f to the y-coordinate of each point.

r	s	θ	φ	e	f
1/2	1/2	0	0	0	0
1/2	1/2	0	0	1/2	0
1/2	1/2	0	0	0	1/2

Table 2.1: IFS rules for the gasket.

The rules to grow a fractal can be presented in a table of these values with the r, s, θ, φ, e, and f values of each row determining instructions

2.1. A SIMPLE WAY TO GROW FRACTALS

on how to change a shape. For example, IFS rules for the gasket are given in Table 2.1.

Of course, the gasket is a particularly simple fractal. What rules describe the fractal shown in Fig. 2.4, the second shape of Fig. 2.1? To help recognize the orientation of each piece relative to the whole, the three main pieces of the fractal are outlined in the second image of the figure.

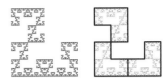

Figure 2.4: The second fractal of Fig. 2.1, decomposed.

- *lower left*: scale by 1/2.

- *lower right*: scale by 1/2, reflect across the y-axis, and translate 1 to the right. (The whole 1 is necessary for translating to coordinate 0.5 lower left corner of this piece, located at -0.5 after reflecting.)

- *upper left*: scale by 1/2, rotate 180°, and translate 1/2 to the right and 1 up.

The rules are collected in Table 2.2.

r	s	θ	φ	e	f
1/2	1/2	0	0	0	0
-1/2	1/2	0	0	1	0
1/2	1/2	180	180	1/2	1

Table 2.2: IFS rules for the fractal of Fig. 2.4.

Of course, the upper left piece also can be obtained by scaling by 1/2, reflecting across both the x- and y-axes, and translating 1/2 to the right and 1 up. With this choice for transformations, we see the table, Table 2.3, could look a little different, but still will produce the same fractal.

r	s	θ	φ	e	f
1/2	1/2	0	0	0	0
-1/2	1/2	0	0	1	0
-1/2	-1/2	0	0	1/2	1

Table 2.3: Different IFS rules for the fractal of Fig. 2.4.

What about the fractal of Fig. 2.5, the third shape of Fig. 2.1? That we leave as practice; the decomposition into three pieces in Fig. 2.5 may help. The solution can be found in A.1.

If that's not enough for you, try some of the IFS exercises of Fig. 2.6. (Answers to these, and all exercises, are in Appendix B.) You can test your ideas with the Deterministic IFS Java software, and also find more practice problems, on the book webpage

http://yalebooks.com/fractalworlds

Feel free to see what else you can build, just by tweaking the rules.

The mathematician John Hutchinson formalized the IFS method for analyzing and building fractals, and Michael Barnsley, founder of the company Iterated Systems Inc., coined the term IFS and popularized the technique. IFS is the only part of fractal geometry that can rival the Mandelbrot set in its power to mesmerize. All we need to master is how to recognize symmetry under magnification and how to manipulate shapes in the plane. This skill opens up a world of wild images which can be tamed by our understanding of scaling and geometry.

Figure 2.5: The third fractal of Fig. 2.1, decomposed.

Exercises

Find IFS rules to generate these fractals.

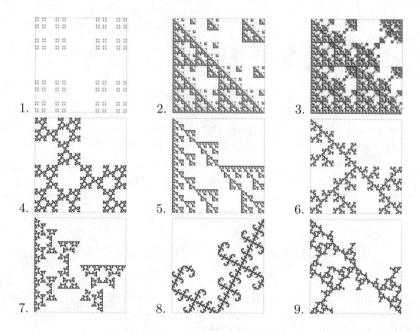

Figure 2.6: IFS exercises.

2.2 Some classical fractals

Recall that we mentioned (in Sect. 1.3) that some mathematical examples of fractals were known, not as fractals, of course, many years before Benoit claimed them for his new field. Using IFS, these examples have simple, but interesting, constructions.

Not all fractals live in the plane. Another point of these examples is to extend IFS to 3 dimensions. We give four examples in the plane, and a 3-dimensional relative of each example. In the top row of Fig. 2.7, we see the Cantor middle-thirds set, the Sierpinski gasket, the Sierpinski carpet, and the Koch snowflake. In the bottom row, we see versions of these in 3 dimensions, respectively: the Cantor cube, the Sierpinski tetrahedron, the Menger sponge, and the Koch tetrahedron. For the first three, we show a fairly high number of iterations; a high iteration of the Koch tetrahedron holds a surprise, to be revealed soon.

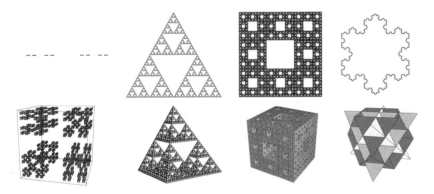

Figure 2.7: Eight classical fractals.

2.2.1 The Cantor middle-thirds set and the Cantor cube

In 1883 Georg Cantor published one of the earliest mathematical constructions of the class of shapes that would become fractals. This relatively unflashy fractal can be visualized easily as the limit of the process starting with the line segment $[0, 1]$ and removing the open middle third, $(1/3, 2/3)$, leaving two intervals, $[0, 1/3]$ and $[2/3, 1]$. Iterating, the next step is to remove the open middle thirds of both these intervals and continue. The first few iterates are shown in Fig. 2.8.

Figure 2.8: The first four iterates in constructing a Cantor set.

Because this Cantor set lives in the line, not in the plane, the two IFS transformations depend only on x and expressed easily by formulas

$$T_1(x) = x/3 \quad T_2(x) = x/3 + 2/3.$$

Denoting the Cantor set by $\mathcal{C}$, we see that $T_1(\mathcal{C})$ is the left piece of the Cantor set and $T_2(\mathcal{C})$ is the right piece, that is,

$$\mathcal{C} = T_1(\mathcal{C}) \cup T_2(\mathcal{C}),$$

which shows the Cantor set is the fractal generated by the IFS with these two transformations.

The late 19th and early 20th centuries saw significant efforts to clarify the logical foundations of mathematics, which are not as straightforward as you might expect. The Cantor set was an important early example of a geometry that challenged the fundamental concepts of infinity and length.

The Cantor set is interesting because it expresses three seemingly mutually contradictory properties:

- the length is 0,
- the set is uncountable, and
- every point is a limit point.

You can find proofs of each of these properties in Sect. A.2. For now we'll build intuition and give a hint of just why these three properties might seem mutually contradictory.

Showing that the length is 0 is easy. We'll see that the lengths of the intervals removed add up to 1, which was the length of the interval from which we started removing the middle thirds. The Cantor set consists of nothing but a sparse set of lengthless points. For this reason, Benoit called the Cantor set a *fractal dust*.

Uncountability is a bit trickier. First, we must recognize that infinities come in different sizes. Smallest (we agree, "smallest" is a peculiar descriptor for something infinite) is the infinity of the counting numbers, the positive integers. A larger infinity is the measuring numbers, the real numbers. Despite the fact that when forming the Cantor set, all the length of the interval $[0, 1]$ has been removed, the points that remain to make up the Cantor set are the same size infinity as that of the points making up the whole interval $[0, 1]$.

On the other hand, the endpoints (1/3, 2/3, 1/9, 2/9, 7/9, 8/9, and so on) of the open intervals removed when carving out the Cantor set

2.2. SOME CLASSICAL FRACTALS

constitute a countable collection of points in the Cantor set. Every Cantor set picture we draw shows only endpoints, leading to the inevitable, but unsettling, conclusion that most of the Cantor set is invisible.

For the third property, you need to know what a *limit point* is. This is most easily understood by an example. Let's say we build a set consisting of these numbers,

$$0, 1/2, 2/3, 3/4, 4/5, 5/6, 6/7, 7/8, 8/9, \ldots 1$$

This set has exactly one limit point, the number 1. This number is the limit of the sequence consisting of all the other numbers of the set. Another way to say this is that no matter how closely to the number 1 we look, we can find some number of the sequence, different from 1, lying closer to 1. For instance, suppose we pick the distance $1/100$. Then the number $100/101$ is closer than $1/100$ from 1. Why do we include "different from 1"? Because without that, every number in the set would be a limit point, and this makes no sense. The number 0 isn't a limit of anything in the set – it's completely isolated.

That every point of the Cantor set is a limit point means the Cantor set has no hermits, that each Cantor set point has infinitely many nearby neighbors. For instance, the Cantor set point 0 has neighbors $1/3$, $1/9$, $1/27$, $1/81, \ldots$ among many others.

Let's see which of these three Cantor set properties – length 0, uncountability, and every point is a limit point – are shared by other sets. Do any familiar sets have all three? The real numbers in $[0, 1]$ are uncountable and every point is a limit point, but their length is 1. The rational numbers (fractions) in $[0, 1]$ have length 0 and every point is a limit point, but they are countable. Can you think of an example that has length 0 and is uncountable, but not every point is a limit point? Here's a hint: start with a Cantor set. (The answer is in A.3.)

That the Cantor set has all three properties emphasizes what a strange set it is. Fairly directly, all three are consequences of the self-similarity of the Cantor set.

We can build the *Cantor cube* $\mathcal{C}^3$, a version of the Cantor set in 3 dimensions, by putting a Cantor set $\mathcal{C}_x$ along the x-axis, a Cantor set $\mathcal{C}_y$ along the y-axis, and a Cantor set $\mathcal{C}_z$ along the z-axis. Then a point (x, y, z) belongs to $\mathcal{C}^3$ if x belongs to $\mathcal{C}_x$, y belongs to $\mathcal{C}_y$, and z belongs to $\mathcal{C}_z$. This is called the *product* of three Cantor sets and is an example of a way to make fractals in higher dimensions from those in lower dimensions.

This fractal can be generated by an IFS with eight transformations. Each shrinks the shape to $1/3$ its size in each direction. Then, as shown in the first image of Fig. 2.9, the copies are moved so each shares a corner with the unit cube. This figure shows the first four iterates of constructing the Cantor cube.

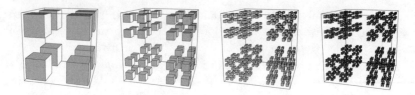

Figure 2.9: The first four iterates in constructing a Cantor cube.

2.2.2 The Sierpinski gasket and Sierpinski tetrahedron

  Although the first mathematical description of the gasket was given in 1915 by the Polish mathematician Wacław Sierpiński, this shape has been used as an architectural decoration since at least 1104 (in the cathedral of Anagni, Italy) and perhaps much earlier. The Sierpinski gasket in the top row of Fig. 2.7 is based on an equilateral triangle, instead of the right isosceles triangle used earlier. Like the Cantor set, the Sierpinski gasket was constructed to answer a mathematical question. A point q on a curve is called a *branch point* if the curve crosses every small circle, no matter how tiny, centered at q, in more than two points. In general, these curves may be angular and more complicated than the smooth curves of classical geometry. In Fig. 2.10, p is not a branch point, while q is. A century ago, intuition suggested that no curve can be made entirely of branch points. The Sierpinski gasket is a counterexample: every point of the gasket is a branch point.

The *Sierpinski tetrahedron*, the extension of the Sierpinski gasket into 3 dimensions, consists of four copies of itself, each scaled by 1/2 in all three directions. The copies are translated so each shares a corner with the original tetrahedron. The first four iterates are shown in Fig. 2.11. Notice that with each additional iterate, the shape becomes more ephemeral.

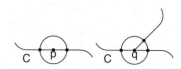

Figure 2.10: The point p is not a branch point; the point q is.

2.2.3 The Sierpinski carpet and the Menger sponge

  First described in 1916 by Sierpiński, the Sierpinski carpet, made of eight copies of itself, each scaled by 1/3, is an example of a *universal plane curve*: every

2.2. SOME CLASSICAL FRACTALS

Figure 2.11: The initial tetrahedron, and the first four steps of building the Sierpinski tetrahedron.

curve in the plane can be continuously transformed – stretched, shrunk, slid, and rotated, but not broken – to lie on the Sierpinski carpet. Fig. 2.12 illustrates an example of this.

Finding IFS rules to generate the carpet is straightforward, because the carpet consists of eight pieces, each with a scaling factor of $1/3$ in both the x- and y-directions. The translations are easily deduced from the placement of the pieces.

The most common way the Sierpinski carpet can be extended into 3 dimensions is to remove the middle $1/3 \times 1/3 \times 1/3$ cube, together with the six sharing a face with it. This can be generated by an IFS with twenty transformations, each of which scales the original by a factor of

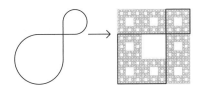

Figure 2.12: A curve in the carpet.

$1/3$ in all three directions. The translations of the IFS rules are shown by the placements of the cubes in the first image of Fig. 2.13.

Figure 2.13: The first three iterates in constructing a Menger sponge.

The first three iterates of the construction are shown in Fig. 2.13. The fractal generated is called the *Menger sponge*; it was described first in 1926 by the Austrian mathematician Karl Menger, as part of his study of ways to generalize the notion of dimension. Like the Sierpinski carpet, the Menger sponge is a universal curve: every curve that can be drawn in 3-dimensional space can be continuously transformed to fit in the Menger sponge.

2.2.4 The Koch snowflake and the Koch tetrahedron

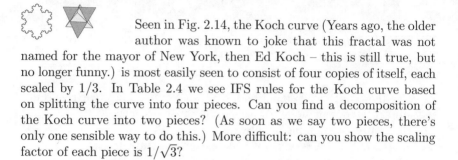

Seen in Fig. 2.14, the Koch curve (Years ago, the older author was known to joke that this fractal was not named for the mayor of New York, then Ed Koch – this is still true, but no longer funny.) is most easily seen to consist of four copies of itself, each scaled by 1/3. In Table 2.4 we see IFS rules for the Koch curve based on splitting the curve into four pieces. Can you find a decomposition of the Koch curve into two pieces? (As soon as we say two pieces, there's only one sensible way to do this.) More difficult: can you show the scaling factor of each piece is $1/\sqrt{3}$?

r	s	θ	φ	e	f
1/3	1/3	0	0	0	0
1/3	1/3	60°	60°	1/3	0
1/3	1/3	-60°	-60°	1/2	$\sqrt{3}/6$
1/3	1/3	0	0	2/3	0

Table 2.4: An IFS table for the Koch curve.

In 1904, the Swedish mathematician Helge von Koch gave this construction as an example of a curve that has tangents at no point. This seems intuitive when we view a Koch curve picture: it appears to be made up of little Vs in various orientations, and the tip of a V has no tangent. But the tips of the Vs form a countable set, so most of the uncountable Koch curve (One way to see it is uncountable is to note that the part of the Koch curve that lies on the x-axis is a Cantor set, which we've seen is uncountable.) must *not* be tips of little Vs, the vertex points.

A more subtle approach is necessary to show that the Koch curve has no tangent anywhere. Recall that a *secant* of a curve at a point p on the curve is a line between the point p and any other point q on the curve. Then one of the fundamental observations of calculus is that for every smooth curve, the tangent at any point p on the curve can be approximated ever more closely by secants. This is illustrated by the first image of Fig. 2.15.

Figure 2.14: The Koch curve.

At vertex points, secants swing through a 30° range. For example, in the second image of Fig. 2.15, we see a vertex point (tip of a V) p on the Koch curve, together with secant lines through a collection of points on the Koch curve. All the secants lie between the left-most and right-most secants. The angle between these is 30°, and no matter how closely to the

2.2. SOME CLASSICAL FRACTALS

Figure 2.15: First: Secants approximate a tangent for a smooth curve. Second: why the Koch curve has no tangent at vertex points.

point p we look, we find Koch curve points on the left-most secant and Koch curve points on the right-most secant. Certainly, the Koch curve has no tangent at this vertex. (Looking to the right of the point, we'd get a similar collection of secants, so again, no tangent.) In fact, every vertex point of the Koch curve behaves this way, with secants oscillating over a 30° range. No vertex point of the Koch curve has a tangent.

But we're not finished, because most of the Koch curve isn't vertices. What about everything that is not a vertex point? In Fig. 2.16, we illustrate the fact that every non-vertex point is a limit of a sequence of vertex points. In this figure, the second and third images are magnifications of the boxed part of the picture to their left. We see the secant lines between q and this sequence of vertex points spin around and around q without resolution, so certainly the Koch curve has no tangent line at q.

Figure 2.16: Zooming in to the point q, secants to q spin round and round, precluding a tangent for a reason different from that at vertex points.

We can see that, indeed, the Koch curve has no tangent at any point, but the reasons are different at vertex points and at limits of sequences of vertex points. In fact, the whole Koch curve is made only of vertex points and limits of sequences of vertex points, so these two cases suffice to show the Koch curve has no tangents anywhere.

Figure 2.17: The Koch snowflake.

If we glue together three copies of a Koch curve along the sides of an equilateral triangle, we get the Koch snowflake.

To extend the Koch snowflake into 3 dimensions, we replace the equilateral triangle with a regular tetrahedron, a shape having four faces, each an equilateral triangle. Think a pyramid with a triangular base. To each of these equilateral triangles we glue something analogous to a Koch curve, but in 3 dimensions instead of 2. The sides of a tetrahedron will do. The first four iterates of building a face of the Koch tetrahedron are shown in Fig. 2.18. Call this the *Koch pyramid* construction.

Figure 2.18: The first four iterates in constructing the Koch pyramid.

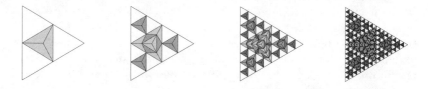

Figure 2.19: View from below of the first four iterates in constructing a face of the Koch pyramid.

Fig. 2.19 shows a different perspective on building the Koch pyramid, with each added chamber jutting away from the viewer. We omit the base of each protruding tetrahedron, so a complex series of rooms off rooms can be seen to fill up the whole pyramid as more and more tetrahedra are added to the exterior. In the meantime, this exterior eventually reduces to a smooth-sided pyramid that betrays no hint of the gaskets hidden within. To help see this, look at a paper model of an early iterate of the Koch pyramid construction in Fig. 2.20.

Last but not least, the *Koch tetrahedron* is obtained by applying the Koch pyramid construction to each face of a tetrahedron. Fig. 2.21 shows the first four iterates of building the Koch tetrahedron. Like the Koch pyramid, the outside of the Koch tetrahedron is deceptively simple: a cube, one of the most basic shapes in Euclidean geometry. How could we expect that the construction giving the beautiful Koch snowflake when pushed into 3 dimensions would produce only a humble cube? Geometry is full of surprises.

There are still more examples of peculiar curves, including the constructions of Bolzano, Weierstrass, Peano, and Hilbert. The few instances

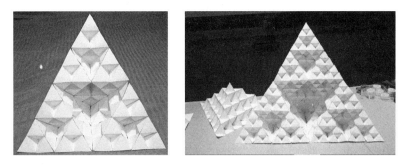

Figure 2.20: Inside and outside views of a model of the Koch pyramid.

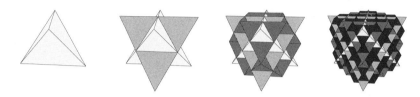

Figure 2.21: The first four iterates in constructing the Koch tetrahedron.

we have seen give a sketch of the kinds of mathematical examples built before Benoit. More importantly, we see that each was designed to solve a particular problem, and that these problems did not have any obvious relation to one another. People often miss the point when they say, "Yes, Mandelbrot came up with the name 'fractal,' but many examples were known by those who came before him." These examples demonstrate particular and unrelated mathematical properties – they were never seen as instances of a single process until Benoit united them. Recognizing this was the beginning of Benoit's reimaging of our understanding of much of the world. It is impossible for us now to think of these examples as being unrelated. *That's* how fundamental Benoit's discovery was.

2.3 Fractal trees and ferns

Natural shapes are less regular than mathematical examples, but they can be approximated using the same set of principles. That is, for each smaller copy of the whole, find the transformation which turns the whole into that piece.

The first image of Fig. 2.22 shows a fractal tree with two branches on the left and two branches right, each of these having two branches on each side, and so on. The second image, a fractal fern, displays a more subtle and complicated fractality, but can be decomposed into a (small, it turns out) number of scaled copies as well. As before, if you can find the rules, you can build the fractal.

Unlike the examples we've already seen, natural shapes do not often have scaling factors as regular as 1/2 or 1/3. Instead, we might find 0.95 or 0.34. Luckily, visual inspection and a ruler and protractor are enough to get you through these examples.

Shown in Table 2.5, the tree rules are relatively simple. The values are obtained by measuring features in the whole and the corresponding features in each piece.

Figure 2.22: Tree and fern.

We describe each rule in turn. For now, we'll let you think about what to measure and how to calculate the IFS table entries. A few moments spent puzzling about this is worth more than a page of explanations by us. Details are given in A.4. In Sect. 2.4 we'll run through an example step by step.

	r	s	θ	φ	e	f
1	0.60	0.50	40°	40°	0	0.60
2	0.50	0.45	20°	20°	0	1.10
3	0.50	0.55	-30°	-30°	0	1.00
4	0.50	0.40	-40°	-40°	0	0.70
5	0.05	0.60	0	0	0	0
6	0.05	-0.50	0	0	0	1.0

Table 2.5: IFS for the tree of Fig. 2.22.

The first four rules make the branches: lower left, upper left, upper right, and lower right, respectively. Each is a scaled copy of the tree, rotated to achieve the orientation of each of the four main branches and translated by sending the base of the trunk to the point where each branch meets the trunk. To make the tree appear more natural, the branches are scaled by different amounts horizontally and vertically, destroying the exact similarity between the branches and the tree.

Figure 2.23: Tree with half-trunk, tree with its main pieces pulled apart.

2.3. FRACTAL TREES AND FERNS

The only difficulty is the trunk itself. Though physical tree trunks are more classical Euclidean than fractal (except for the roughness of their bark), we can at least render the whole image using the same transformation process, albeit with some significant distortion. A first guess is to make the trunk by shrinking the tree a lot in the x-direction and by about half in the y-direction, as we do in the fifth IFS rule. The first image of Fig. 2.23 shows the resulting picture, which, as you can see, is not quite there yet. One repair, achieved with the sixth row of the table, is to shrink the tree again by amounts like those in the fifth row, reflect it vertically so the larger part is on the bottom, and then translate vertically. The second image of Fig. 2.23 shows the six copies of the tree generated by this IFS. Here the pieces have been pulled apart to more clearly reveal their individual natures.

The fern IFS, inspired by the famous 1986 image by Barnsley and his coworkers, is a bit more difficult. The first attempt to decompose the fern might be discouraging: the bottom left and right fronds are small copies of the fern, as are the fronds left and right above those, as are the fronds above those, and so on. Viewed this way, the fern is made of dozens, maybe hundreds

 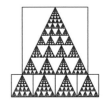

Figure 2.24: A relative of the gasket, split into five pieces.

(in fact, infinitely many, if we think of the mathematical fractal represented by this image) of smaller copies of itself. In this case, although this may be the most obvious way to decompose the fern, it leads to an IFS table of impractical or impossible length. Time to look for another approach.

A relative of the gasket will guide us in finding a more efficient solution. The shape in Fig. 2.24 can be decomposed into a row of four smaller copies along the bottom, a row of four slightly smaller copies above those, a row of four still smaller copies above those, and so on. Clearly, as with the fern, this is not a useful decomposition. The second image of Fig. 2.24 shows a much better approach: take the four smaller copies in the bottom row, each scaled by $r = s = 0.25$, and the rest of the shape can be described as a copy of the whole, scaled by $r = s = 0.75$. Find ap-

Figure 2.25: Fern without its stem; fern pulled apart.

propriate translations for each piece, and the description is complete.

If we use this same approach to render the fern, we can go from infinitely many copies to just three (well, four, as we'll see). The bottom left and right fronds are small copies of the fern. The right copy is reflected across the y-axis to achieve the proper orientation, and both are rotated (counterclockwise for the left copy, clockwise for the right), then translated vertically (the left copy higher than the right). Taking the bottom of the stem to be the origin of the coordinate system, the bottom left and right fronds are not translated horizontally. By making the shrinkings affinities instead of similarities, and by rotating through angles of slightly different absolute values, we can make the fern look less perfectly symmetrical and so more natural. Rules for the bottom two fronds are in Table 2.6. Now

r	s	θ	φ	e	f
0.30	0.34	49°	49°	0	1.60
-0.30	0.37	-50°	-50°	0	0.44

Table 2.6: Rules for the bottom left and right fronds.

imagine snipping off the bottom left and right fronds and all the stem below them. What remains is a copy of the fern, not much smaller than the original, rotated a little to give the fern its overall curl, and translated vertically, described in Table 2.7.

r	s	θ	φ	e	f
0.85	0.85	-2.5°	-2.5°	0	1.60

Table 2.7: Rule for everything above the bottom two fronds.

Combining these three rules gives the first image of Fig. 2.25. Pretty close, but not quite the whole fern. Obviously, we've not written a rule to draw the stem of the fern. But notice a consequence of fractality: if the bottom of the stem of the fern is missing, so is the bottom of the stem of each frond to attach it to the body of the fern. These missing stem pieces continue to the fronds of the fronds, the fronds of the fronds of the fronds, and so on forever. In fact, if we leave out the bottom of the stem, there is no stem anywhere So we add a fourth rule to make the bottom of the stem, and the rest is filled in by self-similarity. The second image of Fig. 2.25 shows the fern disassembled into these four pieces: bottom of the stem, lowest left and right fronds, and all the rest. Table 2.8 gives the complete fern IFS.

The tree image is a reasonable fractal forgery of a natural object, until it is examined closely. When the older author showed this image to a botanist, the response was, "Well, it doesn't exhibit heliotropism. The branches don't bend upward toward the sun." Still, except for the trunk,

2.4. FRACTAL SPIRALS

the tree is a straightforward fractal, not much more complicated than the Sierpinski gasket. The fern represents a different level of fractal complexity. Iterating the "everything above the bottom two fronds" rule fills in all the other fronds with successively smaller copies of the bottom two. Understanding this trick opens our eyes to simple fractal descriptions of many more natural objects.

r	s	θ	φ	e	f
0.30	0.34	49°	49°	0	1.60
-0.30	0.37	-50°	-50°	0	0.44
0.85	0.85	-2.5°	-2.5°	0	1.60
0	0.16	0	0	0	0

Table 2.8: An IFS for the fern.

2.4 Fractal spirals

While it is true that magnifying a spiral shell about its center does give another spiral, magnifying about any other point of that spiral looks more like its tangent line. This is why, as mentioned earlier, this simple spiral is not fractal.

A properly fractal spiral must be a spiral made up of smaller spirals that are themselves made up of still smaller spirals, and so on forever. Fig. 2.26 shows an example of one such spiral. While we might at first see this as made up of many smaller copies of the entire spiral, experience with the fern in the previous section suggests we may decompose the spiral into just two pieces: the right-most subspiral, and everything else. Easy enough – we just have to do some measuring to find the IFS rules.

Figure 2.26: A fractal spiral.

To determine the scaling, rotation, and translation of the pieces, Fig. 2.27 indicates some relevant points to measure. The segment ac in the whole spiral corresponds to the segment ab in the right-most subspiral. Measuring these lengths and dividing gives the scaling factor 0.3. The corresponding calculation with the segments ec and ac gives the scaling factor 0.85 for everything else. The translation

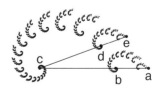

Figure 2.27: Fractal spiral with labels.

of the right-most subspiral is $bc/ac = 0.7$. (For the translations there are choices in addition to dividing the lengths of the reference lines. This is the scaling rule introduced in A.4.) The rotation of everything else is the angle $\angle eca = 20°$.

Since the spiral is symmetrical in a way that our unit square IFS rules have not been, it will be much easier to use the center of the spiral, marked c in the second top figure, as the origin of the coordinate system. From there, we may define the transformations accordingly. Thus the IFS for this spiral is given in Table 2.9.

r	s	θ	φ	e	f
0.30	0.30	0	0	0.7	0
0.85	0.85	20°	20°	0	0

Table 2.9: IFS for the spiral of Fig. 2.26.

Figure 2.28: Two more complicated fractal spirals.

Once understood, this idea is easy to generalize to spirals with more arms. Two examples are given at the bottom of Fig. 2.28. For the curious, the IFS rules for these spirals are given in A.6.

2.5 IFS with memory

In our approach to IFS so far, *all* transformations are applied with each successive iteration. That is, which transformations are applied at each step does not depend on any prior conditions. Next, we'll see how forbidding some *combinations* of transformations can give images more complicated than those we have seen so far.

2.5. IFS WITH MEMORY

To simplify the parsing of these images, we focus on these transformations, the *square IFS rules*:

$$T_1(x,y) = (x/2, y/2) + (0,0) \qquad T_2(x,y) = (x/2, y/2) + (1/2, 0)$$
$$T_3(x,y) = (x/2, y/2) + (0, 1/2) \qquad T_4(x,y) = (x/2, y/2) + (1/2, 1/2)$$

Allowing all compositions of transformations, this IFS produces the filled-in unit square S. That is, we have the *square decomposition*

$$S = T_1(S) \cup T_2(S) \cup T_3(S) \cup T_4(S)$$

Of course, each of these small squares can be divided into four still smaller squares, for example,

$$T_1(S) = T_1\Big(T_1(S) \cup T_2(S) \cup T_3(S) \cup T_4(S)\Big) \quad \text{(square decomposition)}$$
$$= T_1(T_1(S)) \cup T_1(T_2(S)) \cup T_1(T_3(S)) \cup T_1(T_4(S))$$

To each of these subsquares we assign *addresses*, which function as directions for locating the subsquare within the large square. The length-1 and -2 addresses are shown in Fig. 2.29. The pattern may be continued to whatever address length you'd care to see. For a length-3 example, $T_1(T_2(T_3(S)))$ has address 123.

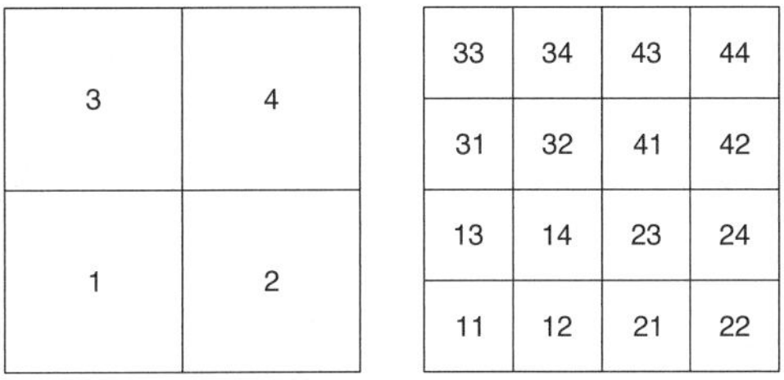

Note that when viewed as *sequential coordinates*, the sequence of transformations to arrive at the region, addresses are *read right to left*. Remember that the order of search must agree with the order of the transformations performed. In the formula, the innermost parentheses are applied first, then the next, working outward. From any point in the square S, to reach the region 123, first apply T_3, which takes us to region 3. Next apply T_2, which takes us to region 23. Finally, T_1 takes us to region 123.

Figure 2.29: Length-1 and length-2 addresses for the square IFS rules.

Addresses can be interpreted in another way, as *spatial coordinates*. With this interpretation, they are *read left to right*. First, address 123 lies in square 1. Within 1, the region lies in the 2 part of 1. Within 12, the region lies in the 3 part of 12. This should remind you of decimals: the left-most digit specifies a tenth of the interval. The next digit specifies a tenth of a tenth, and so on.

Both the sequential coordinate and the spatial coordinate interpretations are useful; which to use depends on what you are doing. In each situation, we'll be careful to specify which we are using.

Now, instead of focusing on what is built, as with standard IFS rules, we will focus on what is taken away. Any departure from a filled square signals some forbidden combinations of transformations. For example, in the first image of Fig. 2.30 we see the fractal generated by forbidding the composition $T_4 \circ T_1$. Certainly, this exclusion leaves out the inside of the square with address 41, but what about all the other empty squares in this picture?

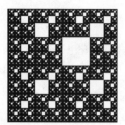 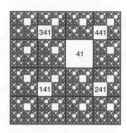 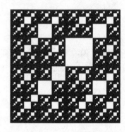

Figure 2.30: First: The fractal generated by forbidding the composition $T_4 \circ T_1$. Second: Labeling the empty length-2 and length-3 addresses of the first image. Third: The fractal generated by forbidding $T_4 \circ T_1$ and $T_1 \circ T_4 \circ T_4$.

In the second image of Fig. 2.30, we label the empty length-3 address squares, 141, 241, 341, and 441. Since points inside squares with addresses containing 41 must result from applying transformations to points inside square 41, and there are no points inside that square, we should not expect any of these length-3 squares to be filled either. Extending this observation in the obvious way, we see that for any forbidden composition of transformations, every address *containing* that forbidden composition is also empty.

The third image of Fig. 2.30 is generated by the forbidden compositions $T_4 \circ T_1$ and $T_1 \circ T_4 \circ T_4$. Note that forbidding $T_1 \circ T_4 \circ T_4$ is not a consequence of forbidding $T_4 \circ T_1$; therefore (for every address length greater than 2) the third picture has more empty squares than the first or second.

Because which transformations occurred in the past determines which transformations are allowed to occur next, we call this construction *IFS with memory*. With the standard IFS we spent some time finding the transformations to produce the given IFS. Here we have analogous problems: given a fractal generated by the square IFS rules, what are the forbidden compositions that produce the fractal?

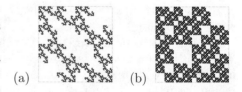

Figure 2.31: Empty address practice.

Before looking at some more subtle issues of IFS with memory, we'll

2.5. IFS WITH MEMORY

pause to give you a chance to test your grasp of the ideas presented so far. For both of the fractals pictured in Fig. 2.31, find the empty length-2 addresses, and test if all the empty addresses are determined by these. Try to do these before looking at the solutions, presented now.

Practice (a). First, identify the empty length-2 addresses by dividing the unit square into a 4 × 4 grid of smaller squares and then noting the length-2 addresses of all empty grid squares. In the second image of Fig. 2.32 we see these are 11, 23, 32, and 44. Because these squares are empty, we say their addresses are *forbidden pairs*.

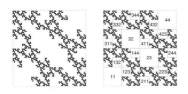

Figure 2.32: Practice (a).

Next, find all the empty length-3 addresses and see if every empty length-3 address contains a forbidden pair. We know that because address 23 is empty, then addresses 123, 223, 323, and 423 must also be empty, since the only way to get a point in these address is to apply one of the transformations to a point in square 23, which is empty.

Also in the second image of Fig. 2.32 we see that addresses 123, 223, and 423 are empty, but we don't have to bother labeling 323. It lies within square 32, which is already empty. In fact, each of the forbidden pairs, 11, 23, 32, and 44, is made up of four empty length-3 subsquares, all unlabeled in the figure.

In addition, in the figure we see the 12 labeled empty length-3 addresses each contain a forbidden pair. For example, 144 contains the forbidden pair 44, so 144 is empty. Since every empty square in this fractal can be explained by the four initial forbidden pairs, at least up through length-3 addresses, we can say that this fractal is generated by forbidden pairs.

If we had found some empty length-3 addresses not containing an empty length-2 address, then the fractal would not have been generated by forbidden pairs and we would need to list forbidden triples to specify the fractal.

Of course, to test if this fractal really is determined by forbidden pairs, we would have to check arbitrarily long addresses. For now we'll be content with checking that every empty length-3 address contains an empty length-2 address.

Figure 2.33: Practice (b).

Practice (b). From the second image of Fig. 2.33 we see the empty length-2 and length-3 addresses are

1$\underline{4}$, $\underline{44}$, 1$\underline{14}$, 2$\underline{14}$, 2$\underline{44}$, 3$\underline{14}$, 3$\underline{44}$, 4$\underline{14}$,

111, 411.

We have underlined the forbidden pairs, including where they appear in empty length-3 addresses. Note that 111 and 411 do not contain any of the empty length-2 addresses. Clearly, not all the empty length-3 addresses of this fractal are the result of forbidden pairs. As soon as we have even one empty address without a forbidden pair, we know that the fractal cannot be determined by forbidden pairs.

Since it would be tedious to write out all of the rules for building these fractals with memory, we instead use a simple visual representation. Draw a graph with four vertices, labeled 1, 2, 3, and 4, each corresponding to T_1, T_2, T_3, and T_4. Then draw an arrow from i to j if the composition $T_j \circ T_i$ is allowed. Note the direction of the arrow and the order of the composition. The arrow $i \to j$ means T_j may follow T_i; that is, T_j may be performed upon T_i, so $T_j \circ T_i$ is allowed. If a combination is not allowed, there will be no arrow shown moving from the first transformation to the second. This picture is called the *transition graph* of the IFS. Some examples are shown in Fig. 2.34.

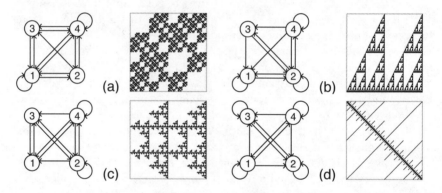

Figure 2.34: Four examples of forbidden pair IFS, with their transition graphs.

Here is an interesting chance to backtrack. Looking at the fractals of Fig. 2.34, we can see that example (a) can be drawn in another way, *without memory*. Look at the subdivision in Fig. 2.35. The parts labeled 1 and 2 are copies of the whole fractal, each scaled by 1/2. The parts labeled 3, 4, 5, and 6 are copies, each scaled by 1/4; those labeled 7 and 8 are scaled by 1/8. Because it can be decomposed entirely into smaller copies of itself, this fractal can be made using IFS with no memory. This is not exciting on its own: subdividing a fractal into pieces and finding the rules to take the whole shape to each piece – been there, done that. What is interesting is that we can tell this just by looking at the transition graph.

To see how, we need one bit of terminology. A vertex of the transition graph is called a *rome* if there are arrows to that vertex from every vertex, including itself. For instance, 1 and 4 are romes in (a) of Fig. 2.34. As the

2.5. IFS WITH MEMORY

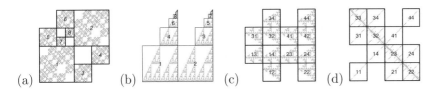

Figure 2.35: Decompositions of the fractals of Fig. 2.34.

term "rome" implies, all roads lead to Rome.[1] This means that the part of the fractal in the square of that vertex is a copy of the whole fractal, scaled by 1/2. Here's why. Suppose 1 is a rome, so the transition graph has arrows $1 \to 1$, $2 \to 1$, $3 \to 1$, and $4 \to 1$. The arrow $1 \to 1$ takes the part of the fractal in address 1 and scales it by 1/2, leaving it in address 11. The arrow $2 \to 1$ takes the part of the fractal in address 2 and scales it by 1/2, leaving it in address 12. Similarly for the parts of the fractal in addresses 3 and 4. Taken together, the parts of the fractal in addresses 1, 2, 3, and 4 constitute the whole fractal, so because 1 is a rome, address 1 contains a copy of the whole fractal scaled by 1/2.

Now we are ready to list the transition graph characteristics guaranteeing that a fractal produced by an IFS with memory also can be produced by a IFS of the kind we introduced first, an IFS without memory.

Expressed in terms of the transition graph, the memory reduction conditions are these.

1. The graph must have at least one rome.

2. For each non-rome vertex, there is a path in the transition graph from some rome to that non-rome.

These conditions guarantee that the fractal generated by an IFS with memory also can be generated by an IFS without memory. The first condition guarantees at least one part of the shape contains a scaled copy of the whole shape; the second condition shows that every part of the shape is a scaled copy of one of these scaled copies of the whole shape.

To avoid potential problems involving practicality, we impose an additional condition.

3. There is no loop passing through only non-rome vertices.

If conditions 1, 2, and 3 are satisfied, the fractal can be generated by an IFS without memory and with a finite set of transformations.

Now back to example (a). We saw that vertices 1 and 4 have arrows to them from all vertices, including themselves, so 1 and 4 are romes.

---
[1]That's really the reason, though when Doug Lind, a colleague from the University of Washington, mentioned this name, the word was mistaken for "roam," resulting in some Google searches that, while annoying at the time, now seem mildly entertaining.

r	s	θ	φ	e	f
1/2	1/2	0	0	0	0
1/2	1/2	0	0	1/2	1/2
1/4	1/4	0	0	1/2	0
1/4	1/4	0	0	3/4	1/4
1/4	1/4	0	0	0	1/2
1/4	1/4	0	0	3/4	1/4
1/8	1/8	0	0	1/4	1/2
1/8	1/8	0	0	3/8	5/8

Table 2.10: IFS rules to generate the left fractal of Fig. 2.34

In addition to the two romes, we have arrows $1 \to 2$ and $4 \to 2$, giving the copies labeled 3 and 4; arrows $1 \to 3$ and $4 \to 3$, giving the copies 5 and 6; and arrows $1 \to 2 \to 3$ and $4 \to 2 \to 3$, giving the copies 7 and 8. These are all the paths in the graph, and all the small copies in the fractal, illustrated in Fig. 2.36. Consequently, example (a) is generated by the IFS of Table 2.10.

Figure 2.36: Transition graph spatial paths.

To relate this IFS table to the addresses of the parts of the fractal, note, for example, that the part labeled 4 (fourth row of the IFS table) occupies address 24 and so is given by the composition $T_2 \circ T_4$. From there, we can calculate the transformation parameters with some easy algebra.

$$T_2(T_4(x,y)) = T_2(x/2 + 1/2, y/2 + 1/2) = (x/4 + 1/4 + 1/2, y/4 + 1/4)$$
$$= (x/4 + 3/4, y/4 + 1/4)$$

that is, $r = s = 1/4$, $\varphi = \theta = 0$, $e = 3/4$, and $f = 1/4$.

What about fractal (b) of Fig. 2.34? From the transition graph we see that states 1 and 2 are romes. Can we find an IFS without memory to grow this fractal? This can be a bit tricky; our first thought is that this is just two gaskets. That's true, but not so helpful. We're trying to decompose the fractal into small copies of the whole shape, and the whole shape is two gaskets, not one gasket.

As we see in Fig. 2.35, the parts of (b) labeled 1 and 2 are copies of the whole fractal scaled by $1/2$, the parts labeled 3 and 4 are scaled by $1/4$, the parts labeled 5 and 6 are scaled by $1/8$, the parts labeled 7 and 8 are scaled by $1/16$, and so on. Unlike the first example, this one does not seem as though it will be easily resolved without memory. What's the difference between the transition graphs of (a) and (b)? In (a) we find

2.5. IFS WITH MEMORY

only a few short paths beginning with romes and continuing only through the non-romes. These are all the paths that begin in romes and continue through only non-romes 2 and 3.

$$1 \to 2, 1 \to 3, 4 \to 2, 4 \to 3, 1 \to 2 \to 3, 4 \to 2 \to 3$$

For the transition graph of (b), the loop $4 \to 4$ gives rise to arbitrarily long paths through non-romes 3 and 4. For example,

$$\begin{array}{llll} 2 \to 4, & 2 \to 4 \to 4, & 2 \to 4 \to 4 \to 4, & \ldots \\ 2 \to 3, & 2 \to 4 \to 3, & 2 \to 4 \to 4 \to 3, & 2 \to 4 \to 4 \to 4 \to 3, & \ldots \end{array}$$

Each of these is a small copy, requiring its own rule if we try to build this fractal without memory. Even though this fractal has two romes, the presence of a loop (here, $4 \to 4$) through non-romes gives rise to an infinite cascade (two infinite cascades in this example) of smaller copies, and the corresponding IFS would have to have infinitely many rules. This is not particularly useful.

However, fractal (c) of Fig. 2.34 looks simpler. The transition graph has no romes, so we should not expect to be able to grow this fractal without memory. Note the fractal has two horizontal lines, one at $y = 1/3$, the other at $y = 2/3$. No piece of the fractal has two horizontal lines from left side to right side, as we can see with some squinting, so this fractal cannot be generated by IFS without memory.

Finally, consider (d) of Fig. 2.34. We see that 2 and 3 are romes, and the loops at 1, at 4, and between 1 and 4, give us arbitrarily long paths between non-romes. We saw this is a problem in example (b), that producing the IFS without memory took an infinite collection of transformations. But here the situation is even worse. The part of the IFS attractor in squares 1 and 4 is a single straight line, not at all a copy or copies of the whole shape, no matter how reduced. The problem here is that neither 2 nor 3 feeds into 1 and 4. In the transition graph, the only arrows going into 1 and 4 are from 1 and 4. This generates the line between corners 1 and 4, and nothing else. Avoiding this problem is the reason for condition 2.

Of course, we can build IFS with longer forbidden combinations, for example, forbidden triples that do not have to contain forbidden pairs. Fig. 2.33 showed one example, Fig. 2.37 shows others, selected mostly because they are nice to look at.

However, these pictures and the methods producing them have applications beyond aesthetics. In Sect. 7.1 we will use a variation of IFS to seek patterns in data strings. These investigations of IFS with memory will inform our attempts at finding how much of the past we need to know in order to make predictions, perhaps only probabilisitc, about the future.

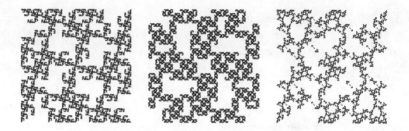

Figure 2.37: Some IFS generated by forbidden triples.

Exercises

1. Considering addresses only up to length 3, which of the fractals of Fig. 2.38 are determined by forbidden pairs?

2. Draw the transition graphs for the forbidden pair IFS of Fig. 2.39. Determine if each can be generated by a memoryless IFS with a finite collection of transformations, and if it can, find the IFS.

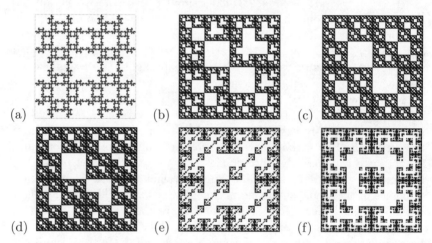

Figure 2.38: Fractals for Exercise 1.

While the range of images in Fig. 2.39 is substantial, you may notice that we can begin to sort the fractals into groups based on the answers to a few simple questions: Is it connected (all one piece)? Does it have loops? We explore these matters a bit in A.7.

2.6 Random rendering of fractal images

Every example so far has been created with a fractal rendering method called the *deterministic algorithm*. Like a recipe, an algorithm is a set of instructions, here for making pictures, not soufflés. Think of how we made

2.6. RANDOM RENDERING OF FRACTAL IMAGES

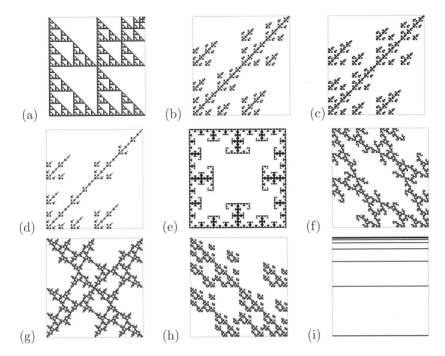

Figure 2.39: Fractals for exercise 2.

the gasket. At every step, the shape is turned into three smaller copies of itself. We keep applying all the gasket rules again and again, each time to the result of the previous application. The label "deterministic" indicates that no choices are made once the process is started. We know exactly what will happen every step of the way.

Another way to render the gasket is to start with a point (x_0, y_0) and generate a sequence of points $(x_1, y_1), (x_2, y_2), (x_3, y_3), \ldots$, where each of these points is obtained from the one before it by applying one of the gasket transformations T_1, T_2, and T_3, selected randomly each time. Depending on which transformation is applied, the next point will be drawn in a different section of the gasket. For a while, this will look like a random assortment of points, but if you wait long enough, the gasket will appear: a disappearing Cheshire cat, but in reverse. This is called the *random algorithm*, illustrated in Fig. 2.40, where the first few random choices are

$$(x_1, y_1) = T_3(x_0, y_0), (x_2, y_2) = T_1(x_1, y_1), (x_3, y_3) = T_2(x_2, y_2)$$

The first and second images of Fig. 2.41 illustrate the random algorithm where each T_i is applied with probability 1/3, giving an even distribution of points over the shape.

But strange things can start to happen when you tweak these probabilities. The third image of Fig. 2.41 shows 5,000 points with p_3, the

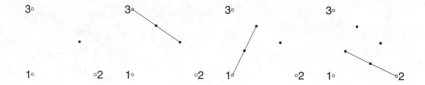

Figure 2.40: Four steps of the random algorithm. Circles mark the vertices of the gasket. In the first picture, the point (dot) is (x_0, y_0). Successive images add $(x_1, y_1) = T_3(x_0, y_0)$, $(x_2, y_2) = T_1(x_1, y_1)$, and $(x_3, y_3) = T_2(x_2, y_2)$.

probability of applying T_3, much greater than p_1 and p_2, the probabilities of filling in the other pieces of the gasket. Changing the probabilities alters the rate at which the picture fills. Here, the top piece of the gasket holds more points than the other two combined. So long as all the $p_i > 0$, the dance of points across the screen eventually will fill in the whole fractal, as we'll see in a moment.

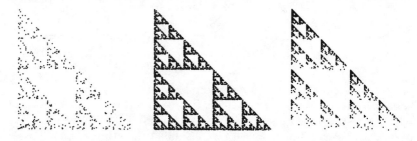

Figure 2.41: Three renderings of the gasket by the random algorithm. First: 500 points. Second: 5,000 points. Third: 5,000 points with different probabilities for each transformation.

Now we can use the notion of addresses to show that the deterministic and random algorithms for rendering IFS images produce the same shape. Though any IFS shape would work, we'll use the eminently simple example of the square IFS rules.

With these transformations, the deterministic IFS produces the filled-in unit square, decomposed into four length-1 address squares, 16 length-2 address squares (see Fig. 2.29), and in general, 4^n length-n address squares. Note that the length-1 address squares have side length $1/2$, the length-2 address sqaures have side length $1/4 = 1/2^2$, and in general, the length-n address squares have side length $1/2^n$.

In order to be sure that the random algorithm will visit every pixel of the square, first find a value of n large enough that $1/2^n$ is less than the side length of a pixel. As long as every length-n address square is visited by points produced by the random algorithm, every pixel of the shape will

2.6. RANDOM RENDERING OF FRACTAL IMAGES

be filled in. To do this, suppose the point (x_0, y_0) lies somewhere in the square S. In this random running of the algorithm, it just so happens that the first few transformations applied are T_a, then T_b, then T_c. The results would look like this:

$(x_1, y_1) = T_a(x_0, y_0)$ lies in the square with address a
$(x_2, y_2) = T_b(x_1, y_1)$ lies in the square with address ba
$(x_3, y_3) = T_c(x_2, y_2)$ lies in the square with address cba
and so on

That is, the effect of each additional transformation is to shift the old address of the point one place to the right and insert the new address on the left.

For example, suppose $(x_0, y_0) = (1/2, 1/2)$, the center point of the square. Then $(x_1, y_1) = T_1(x_0, y_0) = (1/4, 1/4)$ would be the center of the square with address 1. Then the point $(x_2, y_2) = T_3(x_1, y_1) = (1/8, 5/8)$ would be the center of the square with address 31, and so on. Fig. 2.42 illustrates this.

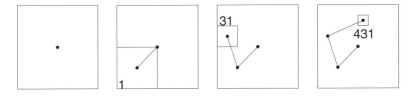

Figure 2.42: The movement of points under the random IFS algorithm.

One more ingredient is needed: the role of randomness. Every adequately long random sequence of 1s, 2s, 3s, and 4s contains every length-n sequence of 1s, 2s, 3s, and 4s. Why is this? When we read a random sequence, at every point the next number can be anything. We can call this the *anything goes* property of random sequences. However, if we knew that a number, 121431 for example, never occurred in the sequence, then when we got to 21431, we would know that the next number could not be 1 (remember, the next number is put on the left end of the sequence), violating the anything-goes property. So if we wait long enough, the random IFS algorithm will visit every address length n square, and so every pixel of the image. In A.8 we present a way to estimate how many iterates are needed to visit a given address.

2.7 Circle inversion fractals

So far, we've built fractals with pieces made from simple rules: shrink, reflect, rotate, and translate. Adding memory gave us a larger collection of fractals, but still each piece was roughly similar to the whole. Now suppose we use rules that allow stretching or shrinking by different amounts in different locations. Will this give us new fam-

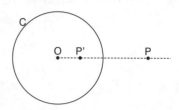

Figure 2.43: Points P and P' are inverses in the circle C.

ilies of fractals, complex shapes we have not seen before? Or put differently, will this allow us to recognize fractality in a new family of shapes? Of course, the answer is yes.

Circle inversion gives us one such family. Here, we can think of the small copies as distorted versions of the whole, recognizable but strange, like your reflection in a funhouse mirror. These fractals are more visually interesting than, say, the humble gasket, but to understand them, we must learn how to invert in circles.

We can think of inversion in a circle as what happens to a reflection in a curved mirror, taking the circle itself as our mirror. This is close to the original formulation of inversion, by Apollonius of Perga, about 2,300 years ago. This process is more easily understood by showing first what it does to a point. Suppose C is the circle with center O and radius r, as illustrated in Fig. 2.43. The point P is inverted to the point P' across the circle C if the points P, P', and O satisfy these two properties.

- *ray property*: P and P' lie on the same ray from O
- *radius property*: $OP \cdot OP' = r^2$.

From this definition, we can show that inversion in the circle C, denoted by I_C, acts in these four basic ways.

- I_C interchanges the inside and the outside of C.
- I_C leaves fixed every point of C itself.
- Inverting twice in C leaves every point of the plane unchanged.
- When applied to two points outside of C, I_C reduces the distance between the points. When applied to two points inside C, I_C increases the distance between the points.

2.7. CIRCLE INVERSION FRACTALS

These properties are illustrated in Fig. 2.44, where we can see that the part of the cat outside the circle inverts to inside the circle and shrinks in the process. By the same rules, the part of the cat (the tip of the tail) inside the circle is inverted outside the circle and grows in the process. Note how the bend of the cat's tail changes with inversion. Like reflection, inversion

Figure 2.44: Inversion of a cat across a circle.

reverses left and right. The only points left unaltered by inversion are the exact points where the cat's tail crosses the circle. Finally, inverting again returns the picture to the original.

Our principal business with inversion is what it does to other circles. Simply enough, inversion in the circle C transforms most circles into other circles. In Fig. 2.45 the circle C is solid, points evenly spaced around the circle S are shown as tiny circles, and points along the circle $I_C(S)$ are shown as tinier discs. Note that the little discs are not evenly spaced around $I_C(S)$. Do you see why?

The third picture of Fig. 2.45 shows an important case: the circle S (little circles) intersects C perpendicularly. That is, at both points where S crosses C, the tangent line of S is perpendicular to the tangent line of C. In this case, S and $I_C(S)$ (little discs) are the same circle, though, to be sure, the part of S outside of C inverts to the part of $I_C(S)$ inside C.

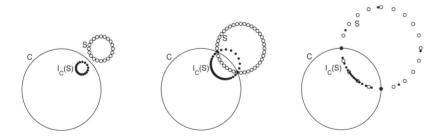

Figure 2.45: Inversions of several circles, illustrating that usually circles remain circles when inverted.

The "circles invert to circles" rule has one exception, or maybe "extension" is a more apt word. If the circle S to be inverted passes through the center O of the inverting circle C, then according to the radius property, the inverse of O must lie infinitely far from O. We explain this by saying that O inverts to the *point at infinity*, a single point that is infinitely distant from every point of the plane, in every direction. Then the inverse of

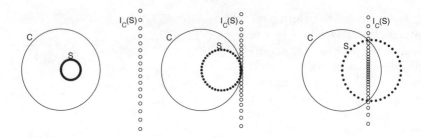

Figure 2.46: Inversion of a circle passing through the center of the inverting circle gives a straight line.

the circle S must include the point at infinity, so it is a circle of infinite radius. Since the curature of a circle, higher for small circles than for large ones, is defined as the reciprocal of its radius, a circle with infinite radius must have 0 curvature, a straight line. The line produced by these kinds of inversions depends on where the circle S lies in relation to the inverting circle C. Three examples are shown in Fig. 2.46.

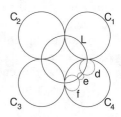

Figure 2.47: The attractor of inversion in four circles.

All of this can be applied to find attractors of circle inversions as with simple IFS. One way to do this is to remember that the attractor of an IFS is the only shape left unchanged by the simultaneous application of all the IFS transformations. We called this the invariance property of fractals: apply the gasket rules to the gasket and you get the lower left corner of the gasket, the lower right corner of the gasket, and the upper left corner of the gasket. Taken together, these form the gasket. Let's try to find something left unchanged by inversion in the four circles C_1, C_2, C_3, and C_4 of Fig. 2.47. We know that a circle L intersecting an inverting circle C at right angles is preserved, though inversion interchanges the part of L inside C with the part of L outside C. In Fig. 2.47 the circle L intersects all four inverting circles, C_1, C_2, C_3, and C_4, at right angles. Consequently, inversion in all four circles takes L to itself. As for why the attractor is exactly L, the key idea is this. Look at the inverses in, say, C_4 of C_1, C_2, and C_3. These are labeled d, e, and f in the figure. We can argue that the attractor lies inside the circles C_1, C_2, C_3, and C_4. Then the part in C_4 lies inside d, e, and f. Continuing this inverting process, we see the attractor is contained inside a family of shrinking circles that go all around L, as necklaces made of smaller and smaller beads.

None of our previous IFS could generate a circle, so producing a circle this way is nice. But a slight change in the inverting cricles can turn the attractor into a fractal: suppose we keep the centers of C_1, C_2, C_3, and C_4

2.7. CIRCLE INVERSION FRACTALS

where they are and reduce their radii. These circles are no longer tangent to each other. There still is a circle that intersects all four inverting circles at right angles, but this whole circle is not the attractor. As we see in Fig. 2.48, the attractor appears to be a Cantor set wrapped around this circle.

While inversion reduces the distances between points when we invert points outside the circle into the circle, it increases distances when we invert points inside the circle to the outside. Because of this, it isn't so clear how we can adapt the deterministic IFS algorithm to circle inversions. Benoit developed a method, which we'll describe in A.9, but it works only for

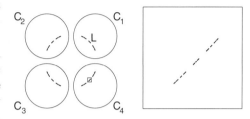

Figure 2.48: The attractor of inversion in four disjoint circles. The large box is a magnification of the small box.

some collections of inverting circles. Instead, we can almost always use the random IFS algorithm with circle inversions.

In adapting the random IFS algorithm to circle inversions, we must be careful about two issues. First, as we've just mentioned, to guarantee that an IFS does not produce points wandering all over the plane, all the transformations must reduce distances. To make sure the inversions we use always contract and never expand the shape, when applying the random IFS algorithm to circle inversion, we never invert in C_i a point that is already inside C_i, though we may of course invert it in some other circle.

Another issue comes from the fact that inverting twice in a circle returns to the original point. This is incorporated easily: in adapting the random IFS algorithm to circle inversions, we add the condition that we may never invert twice consecutively in the same circle.

For example, we have applied the random IFS algorithm to inversion in the five circles shown in the first image of Fig. 2.49. Four of these circles, C_1, C_2, C_3 and C_4 have radius 1 and centers $(\pm 1, \pm 1)$. The fifth, C_5, has center $(0, 0)$ and radius $\sqrt{2} - 1$, so it is tangent to the other four. The attractor of this ar-

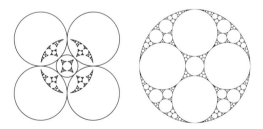

Figure 2.49: First: The random IFS algorithm with circle inversions. Note the uneven fill. Second: The complete fractal, gaps filled.

rangement is an intricate set of circles reflecting circles reflecting circles, and so on.

Something is still unsatisfactory about this rendering. Though the program ran a long time, the shape is not sufficiently filled. The problem lies in the fact that points near the inverting circle invert to other points only very slightly closer to the inverting circle, creeping toward the edges at a glacial pace. Many, many inversions are needed to fill in the small gaps. One solution is to try a different approach, similar to the idea that we can make the gasket by removing middle triangles. Starting with a solid shape and hollowing it out is much faster and, as you'll see in the second image of Fig. 2.49, much more attractive. This is an instance of the approach Benoit developed, described in more detail in A.9.

There are many variations on this standard circle inversion. In Fig. 2.50 we see two examples of circle inversion fractals where the inverting circles are made to overlap, intersecting tangentially or at 90° on the left, and tangentially or at 60° on the right. Some other angles will give messier, less symmetrical pictures. Long ago, circle inversion was part of every high school geometry course, but it has been abandoned. We hope that the beauty of circle inversion fractals will lead more people to learn about this ancient, but now oddly modern, branch of geometry.

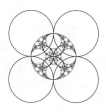 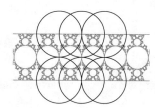

Figure 2.50: Circle inversion fractals for overlapping circles.

Also, circle inversion can be replaced by other nonlinear transformations, giving even more beautiful results, some of which are of central importance in modern topology. In Sect. 1.9 we saw a physical example of nonlinear fractals: the optical gasket, formed by shining a light into an arrangement of four reflecting spheres, a model of a tiny portion of Indra's pearls. In Buddhist mythology, the god Indra created the world as a net and placed a pearl at each knot of the net. Each pearl is connected physically (by the lines of the net) to only a few other pearls. But on each pearl is reflected the image of every other pearl, illustrating the interconnectedness of all things. Less mystically, it's a beautiful fractal picture of circles within circles within circles. Geometry can possess the wonder than many people may feel is the purview of mysticism

These few examples do not begin to exhaust all the generalizations of IFS. Every instance we find gives rise to others more subtle still, and those lead us to still more, on and on, level upon level. Yes, it shouldn't surprise us that the study of fractals is itself fractal.

2.8 Random fractals

Benoit understood fractals as a way to study the roughness of messy nature, that (as discussed in Chapter 6) fractal dimension was the first reliable, reproducible measure of roughness. Though the examples we have seen so far are generally too regular to be taken as models of any natural phenomenon, with the introduction of a certain amount of randomness, these rigid geometrical shapes give way to complicated models of natural self-similarity. Three examples are shown in Fig. 2.51.

Figure 2.51: Three randomized fractals: a Sierpinski gasket, a product of Cantor sets, and a Koch curve.

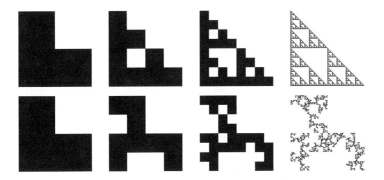

Figure 2.52: First, second, third, and eighth stages of constructing a Sierpinski gasket: nonrandom (top) and randomized (bottom).

How are these grown? Consider first the random gasket. One way to build the regular Sierpinski gasket is to start with a filled-in square and remove the upper right corner, leaving three smaller squares arranged in an L shape. Remove the upper right corner of each of these squares, and repeat, as illustrated by the top images of Fig. 2.52. This process generates a sequence of shapes that converges to the Sierpinski gasket. One way to randomize this process is to vary which square is removed. Instead of always removing the upper right square, try randomly selecting an upper right, an upper left, a lower left, or a lower right square to remove. The lower images of Fig. 2.52 show the result of one sequence of random choices.

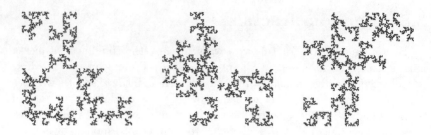

Figure 2.53: Three random Sierpinski gaskets.

Different random choices of which squares to remove result in random fractals different in detail but similar in general texture and pattern, as seen in Fig. 2.53. We say all randomized gaskets constructed in this way are *statistically self-similar*. If you can recognize statistical self-similarity, you are on your way to recognizing fractals in the world around us.

r	s	θ	φ	e	f
1/3	1/3	0	0	0	0
1/3	1/3	0	0	2/3	0
1/3	1/3	0	0	0	2/3
1/3	1/3	0	0	2/3	2/3

Table 2.11: IFS rules for the top images of Fig. 2.54.

For the second example in Fig. 2.51, we will again start with a non-random construction turning the filled-in unit square into a fractal made of four pieces, each scaled by 1/3, and then we will randomize it. The nonrandom construction, is illustrated in the top images of Fig. 2.54. In IFS speak, it is expressed in Table 2.11. We call this the *product* of two Cantor sets because looking along the x-axis, we find a Cantor set, and looking along the y-axis, we find another Cantor set.

How can we randomize this? One approach is to scale each replacement square by a factor drawn randomly from the interval $[1/4, 1/2]$. As illustrated in the bottom row of Fig. 2.54, we will continue to place the four subsquares so each shares a corner of the larger square containing it, but now the subsquares will cover a different proportion of the space.

We can examine a similar process based on the Koch curve. Starting from the horizontal line between $(0,0)$ and $(1,0)$, the standard Koch curve adds a Λ-shaped projection to each middle section. We can achieve randomization by reversing or preserving the orientation – up or down – of this middle Λ for each segment, as shown in the second row of Fig. 2.55.

These are a tiny sample of the variety of random fractals. A more complete catalog includes Brownian motion, fractional Brownian motion, Lévy flights, diffusion-limited aggregation, viscous fingering, dielectric break-

2.9. AND FLAVORS STRANGER STILL

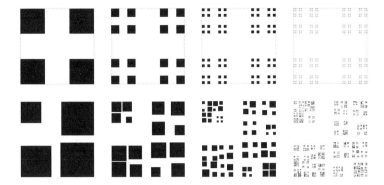

Figure 2.54: First, second, third, and sixth stages of constructing a product of Cantor sets, nonrandom (top) and randomized (bottom).

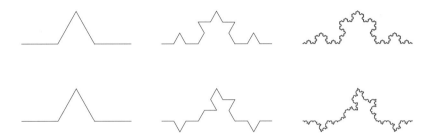

Figure 2.55: First, second, and sixth stages of constructing a Koch curve, nonrandom (top) and randomized (bottom).

down, percolation, and on and on and on. Some of these are described in A.10 and A.19. Every fractal we see in the physical world was grown or sculpted by forces similar across many scales but also perturbed by other interactions, environmental or internal. Learning to recognize randomized fractals is an essential step to seeing fractals in nature.

2.9 And flavors stranger still

Our list of types of self-similarity is not nearly complete. New examples have been discovered; newer will be discovered, we expect. Here we'll mention two examples: limit sets of Kleinian group actions and BPI (Big Piece of Itself) spaces.

The fractal of Fig. 2.56 looks a bit like a circle inversion limit set, but it isn't quite that. In A.11 we'll describe the transformations used to make it. For now we note that more than just a new type of beautiful fractals, these are an important ingredient in the current understanding of 3-manifolds, spaces assembled from pieces of 3-dimensional space. This stage of modern mathematics was launched by William Thurston in his brilliant geometrization program.

This point is worth repeating: fractals are far more than pretty pictures. They are deep and subtle mathematics.

The next example, a BPI space, is unlike anything we've seen. Introduced by Guy David and Stephen Semmes, the notion is a generalization of our familiar self-similarity. BPI spaces are characterized this way: a substantial portion of any piece of the space looks like a substantial portion of any other piece of the space. It is made of "big pieces of itself." Drawing even a simple example of a BPI space is tricky, so we save it for A.11. But we'll mention that the motivation for developing BPI spaces was to create a class of fractals on which something like calculus can be done. If you're asking why we can't do calculus on familiar fractals, just remember the effort we expended to show the Koch curve has no tangent lines, and that calculus is all about tangent lines.

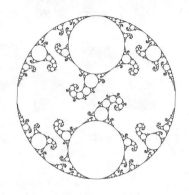

Figure 2.56: A Kleinian group limit set.

Fractals began as a way to measure an infrequently noticed symmetry of nature. Since then they have grown into almost every branch of science and culture. Now they seem to be self-generating: new kinds of fractals enable new categories of questions, which lead people to find still newer kinds of fractals. Fractals leading to more fractals. Did you expect anything else?

Chapter 3

Self-similarity in the wild

Our insides are a good place to start. Respiratory, circulatory, and nervous systems are all modeled, at least in part, by fractals. Some plants have similar structures: root systems and clusters of blossoms like goldenrod, Queen Anne's lace and its impressive, if hazardous, cousin the giant hogweed. We'll explore possible reasons for this, and more examples. We will also spend some time puzzling over the fact that across many species the metabolic rate per unit mass exhibits scaling, but not the scaling we might expect based on an animal's skin area and volume. The explanation may lie with fractals inside animals, plants, and even single cells.

Trade a microscope for a telescope to ask for the distribution of galaxies in the universe, and you will find a fractal structure spread over a large range of scales. Closer to home, fractal dust clumps, similar to the dust bunnies that grow under the beds of less thorough housekeepers, may catalyze the formation of planets. It seems that fractals make our bodies work, give us the ground under our feet, and may even let us recognize the scatter of galaxy clusters as a cosmic fractal drip painting.

Across this whole range of scales, a central issue emerges: how can instances of complicated shapes, branches off branches off branches or crinkles and whorls made of smaller crinkles and whorls, form by manifestly different processes? That raindrops and cells are roughly spherical is no surprise, a predictable result of simple physical constraints. But complicated shapes? Metal veins in a rock, a tree growing in the air, water flowing in the dirt, arteries growing on the intestine – what common process could these have? Nothing obvious, but we offer this possibility: that evolution discovers and uses the laws of geometry, both Euclidean and fractal. We mention this here to underscore the appeal of this notion, its potential for organizing part of our thinking about the world.

3.1 MathWorld vs. PhysicalWorld

To be able to recognize PhysicalWorld examples, we must clarify the difference between abstract mathematical fractals, now familiar from these earlier chapters, and the manifestations of fractals in PhysicalWorld. MathWorld fractals can be magnified forever and will always exhibit some exact form of self-similarity. PhysicalWorld fractals cannot be magnified without end. At some scale, the forces that grew or sculpted the object's fractal properties no longer are the dominant forces. Put another way: the atoms making up a fern do not look like little ferns. Also, nature is messy and complicated. Nothing grows in complete isolation from everything else. Environmental forces will distort the small pieces, keeping them from being exact copies of the whole. Rather than seeking exact copies within the larger shape, in our search for physical fractals, we benefit from squinting, looking for approximate patterns over a limited range of scales.

Nevertheless, PhysicalWorld fractals abound: the edges of clouds, the pores of sea sponges, branched lightning, river networks, some sand dunes, and on and on and on. Many objects that are rough, or branched, or have holes of many sizes, are fractal over some range of scales, and any claim of fractality must be accompanied by an estimate of this range. Whether or not fractality is a useful description of an object depends on this range. The configuration consisting of a brick, a space, and another brick is not described sensibly as a Cantor set. Without repetition across at least a few scales, any claim of fractality is difficult to support. Benoit was adamant about this, merciless in his criticism of early papers that asserted fractality from less than one order of magnitude of data. In PhysicalWorld, we are looking for objects made of approximate smaller copies of themselves, with enough complication to reveal the imprint of a force working over a considerable range of scales. The minimum range varies with the category of object, but a good rule of thumb is to look for a 100-fold magnification.

3.2 A foreshadowing of fractal dimension

Fractals model broken, jagged, complex, wiggly, rough shapes. But some shapes are rougher than others. The tip of your tongue is quite smooth, but have you ever felt the tip of a cat's tongue? It's much rougher, covered with tiny spines. In Chapter 6 we will learn how to calculate a number, called the *fractal dimension*, that can quantify this roughness. Even without the full development of that chapter, three examples will give an intuition sufficient for the mentions of fractal dimension presented in the next two chapters.

For a shape that is broken or rough, as we look more closely we see that its jaggedy edges fill up some, but not all, of the space around the shape. We'd like to measure how the fraction of the space occupied by the shape scales with how closely we look. If this fraction increases, the

3.2. A FORESHADOWING OF FRACTAL DIMENSION

roughness or complexity of the shape increases. One way to measure this complexity is to imagine a grid of boxes surrounding the shape. Count the number of occupied boxes, and look for a pattern in this number as the shape is covered with smaller and smaller boxes. In Fig. 3.1 we do this for a line segment, a filled-in square, and a gasket.

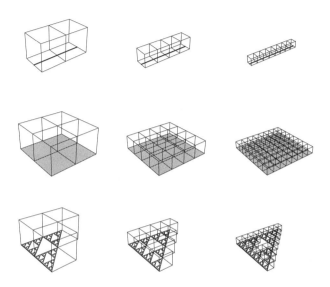

Figure 3.1: Boxes covering a line segment, a filled-in square, and a gasket.

Suppose the line segment has length 1, and the square and gasket have sides of length 1. For the line segment we see

$$N = 2 \text{ boxes of side length } r = 1/2$$
$$N = 4 \text{ boxes of side length } r = 1/4$$
$$N = 8 \text{ boxes of side length } r = 1/8$$

That is, $N = 1/r$.

Counting boxes for the filled-in square gives

$$N = 4 \text{ boxes of side length } r = 1/2$$
$$N = 16 \text{ boxes of side length } r = 1/4$$
$$N = 64 \text{ boxes of side length } r = 1/8$$

So for the square we see $N = (1/r)^2$.

Since a line segment is 1-dimensional and a filled-in square is 2-dimensional, both shapes satisfy this rule, with d denoting the dimension of the shape:

$$N = (1/r)^d$$

For the moment, we'll take this relationship to be the definition of the dimension, even though we know physical dimension and our exponent are about different aspects of the object. What do we get for the gasket?

$N = 3$ boxes of side length $r = 1/2$
$N = 9$ boxes of side length $r = 1/4$
$N = 27$ boxes of side length $r = 1/8$

That is, when $r = 1/2^k$, $N = 3^k$. For the gasket, the dimension formula becomes

$$N = 3^k = ((1/2)^k)^d = (2^k)^d$$

In Chapter 6 we'll study this exponent d and ways to solve for it, but for now we'll just note this relationship is called a power law (we'll say more about power laws in Sect. 3.5) and the solution is $d = \log(3)/\log(20 \approx 1.58$, which you can find if you remember how to use logarithms. But even without solving anything, we do notice this: N for the gasket grows more quickly than it does for a line segment (1-dimensional) and more slowly than it does for a filled-in square (2-dimensional). Therefore we can say that the fractal dimension of the gasket lies somewhere between 1 and 2.

We will have much more to say about this later. But for now, understand dimension this way. The higher the dimension, the more completely a shape's folds, branches, and wiggles fill the space that surrounds it. Or put another way, the higher the dimension, the higher the roughness, the higher the complexity.

3.3 Coastlines, mountains, and rivers

One of Benoit's first papers to gain a wide audience was titled "How long is the coast of Britain? Statistical self-similarity and fractional dimension." Published in *Science* in 1967, the paper showed that fractal geometry is the natural language for quantifying, and organizing our thinking about the causes of, the roughness of landforms.

A.12 gives a bit of technical detail on these three kinds of natural fractals.

3.3.1 Coastlines

The standard approach for measuring the length L of a curve C is to approximate the curve by line segments of length ℓ. If there are $N(\ell)$ of these segments at this scale, the length of the curve is

$$L(\ell) = N(\ell) \cdot \ell$$

For example, the first and second images of Fig. 3.2 show approximations of a circle by 5 and 10 segments of equal length; the third is the graph of

3.3. COASTLINES, MOUNTAINS, AND RIVERS

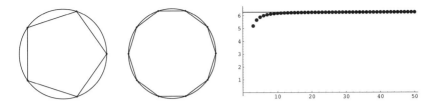

Figure 3.2: First and second: Approximating a circle with 5 and 10 segments. Third: Lengths of circle approximations using 3 to 50 sides.

length approximated through $N = 50$ segments. We can see that as the number of segments increases, the lengths approximate, ever more closely, 2π, the true circumference of the circle.

This method works for smooth curves, because any part of a smooth curve sufficiently magnified looks like its tangent line. The length of each of these straight line segments is pretty close to the length of the corresponding part of a tangent, and so is close to the length of a small portion of the curve.

Figure 3.3: Approximating the coastline of Britain by line segments. Shorter segments better approximate the coastline. (NASA photograph, our lines.)

The problem is that coastlines are not made up of smooth Euclidean curves that can be approximated in this way. Coastlines, like many other geological features, exhibit similar kinds of structure over a range of scales. That is, they are scale invariant, at least for some range. This is expected because the forces that sculpt coastlines – wind, tides, erosion – operate in approximately the same way over a wide range of scales. In the case of certain complicated curves, finer details can be picked up only by smaller measuring scales, and so the measured length of something like a coastline varies based on the scale of the unit used to measure it. Fig. 3.3 illustrates this.

An early quantitative assessment of this effect was made by British scientist Lewis Fry Richardson in 1961. For the coasts of Australia and South Africa (first image of Fig. 3.5), as well as the west coast of Britain (second image of Fig. 3.5) and the land frontiers of Germany and Portugal, Richardson measured the lengths $L(\ell)$ at different scales ℓ. The first graph of Fig. 3.4 shows his data for the west coast of Britain. Richardson observed that the length increases rapidly as the scale is reduced. This and the other graphs looked a lot like power laws. To examine them more precisely, Richardson plotted log(length) vs. log(scale), as shown in the second graph of Fig. 3.4.

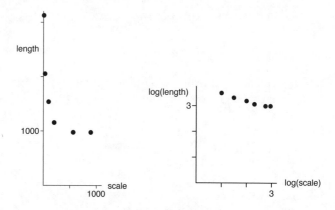

Figure 3.4: First: Richardson's calculation of the length of the coastline of Britain using different measuring scales. Second: a log-log plot of the same data.

In all his examples, Richardson found the points on the log-log graphs were well fit by straight lines, with slopes

- $s = -0.25$ for the west coast of Britain, one of the roughest in the atlas,
- $s = -0.15$ for the land frontier of Germany,
- $s = -0.14$ for the land frontier of Portugal, and
- $s = -0.02$ for the South African coast, one of the smoothest in the atlas.

For circles and other smooth curves, at a small-enough scale the points lie on a line of slope 0. Of those examined, only South Africa gets close, but even so it is clear that something more complicated is going on with coastlines.

In his 1967 *Science* paper, Benoit interpreted Richardson's results as showing that coastlines are so rough across so many scales that they are not smooth curves at all. They are, as you've guessed, fractal. In Chapter 6 we'll see how an exponent related to Richardson's s can quantify the degree of roughness of each coastline curve.

French physicist Bernard Sapoval, after studying the ability of fractal drums (drums

Figure 3.5: Smooth South Africa and rough Great Britain, at the same scale.

3.3. COASTLINES, MOUNTAINS, AND RIVERS

with fractal perimenters) to damp vibrations, suggested a similar mechanism that may be responsible for the fractal form of coastlines. Sapoval writes

> the reason why fractal coastlines exist is precisely because they are best at damping ... waves. As the coast damps down the waves, then the erosion to which it is subjected is reduced. They are thus stabilized by their fractal structure.

3.3.2 Mountains

Built by tectonic forces, sculpted by water and wind acting over many scales, the fractality of mountains should not be a surprise. Even a quick glance at a picture of barren mountains – after all, tree cover softens the geological details – shows peaks and valleys consisting of smaller peaks and valleys, and so on, over several levels. Once the roughness of these features is understood, although randomness prevents us from predicting the location of valleys and peaks, we can estimate at least their sizes, and perhaps the magnitude of countermeasures against erosion.

Figure 3.6: Tectonics, erosion.

Many years before Benoit gave fractals their name, geologists knew one aspect of fractality: that photos taken of stream beds and other rough features should always contain some manufactured artifact (common choices were a hammer or a camera's lens cap). Otherwise, it can be very difficult, sometimes impossible, to tell if we are viewing a nearby stream bed or a distant cliff face. Two examples of this *scale ambiguity* are seen in Fig. 3.7.

Figure 3.7: Two road cuts, with lens caps for scale. From Benoit's old slides.

The images in Fig. 3.8 show the same pictures, but this time with human scaling references. In the second, what looked like a lens cap turns out to be a garbage bag stretched over a hula hoop, with the word "Olympus" written in masking tape. Here we have a glimpse of Benoit's sense of humor, and a compelling example of the way fractals compose our world.

The roughness of mountains can be used to estimate their age and state of weathering. In addition, once the roughness has been measured, realistic fractal forgeries of mountains become feasible. Benoit's

Figure 3.8: Another view.

colleagues Sig Handelman, Richard Voss, and Ken Musgrave have incorporated measured values of roughness into programs for generating fractal mountains. With very small data sets, they are able to make convincing forgeries, Fig. 3.9 for example. The success of these compact descriptions of realistic, complex images suggests we have understood something important about the processes building the real-world objects.

At Benoit's suggestion, Musgrave showed some of his landscape forgeries in the first fractal geometry course the older author taught at Yale, where he discussed how the randomization was fine-tuned for the desired roughness. After Musgrave left, one of the students asserted that one of the pictures was a fake. Of course, all were fakes. No, the student said, I mean a fake fake, that is, one of the pictures was real, put in to fool us.

Figure 3.9: Artificial mountains.

Some others nodded, agreeing. When asked which they thought was real, the first student offered his suggestion; others disagreed. No, that one was a clear fake. This one is real. Why? Reasons and counter-reasons flew for a while. At the end of the energetic discussion, the class understood that real mountains and fractal forgeries of mountains follow the same rules. This is why the forgeries are convincing. The construction of the fractal mimics the process, not merely the appearance.

3.3.3 Rivers

When Benoit began refining the ideas that would grow into fractal geometry, satellite photos of large landforms – mountain ranges and river networks – were not readily available. Now mapping apps like Google Maps give anyone with a reasonable Internet connection the ability to go investigating. Fig. 3.10 shows three examples of some intriguing images of rivers seen from space.

  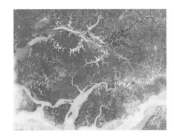

Figure 3.10: River networks from Norway (first), Mongolia (second), and North Africa (third).

Fractals enter the study of rivers in at least two ways. The complexity of the patterns of meandering streams off of meandering streams is often fractal, and this complexity gives a measure of the roughness of the landforms through which the river flows. Second, the watersheds of many rivers have fractal boundaries, reflecting twists and turns of ridgelines separating the basins of different rivers. Because the actual intricate boundaries between watersheds are better modeled by fractal geometry than by Euclidean geometry, fractals are an important tool for ecologists and resource managers.

Figure 3.11: Waterfall with self-similar cascades.

On a smaller scale, some rivers display visual fractals in the beautiful branching patterns of the waterfall shown in Fig. 3.11. Perhaps the rocks must be fractal to produce fractal waterfall branching, or maybe the chaos of falling water suffices.

In Fig. 3.12 we see some Google Maps views of North African sand dunes. The similarity of these branching patterns to those of river networks reinforces the impression that deserts are rivers of sand. Of course, sand isn't a liquid, but neither is it a solid. It's a collection of particles, driven by gravity and

Figure 3.12: First: North African sand dunes. Second: A magnification of the first.

wind, slowed by friction. How can such a system produce these amazing patterns? A good question, whose answer we await still.

Fractals abound in the landforms of the world. They represent, of course, the action of certain forces over distance scales spanning several orders of magnitude. Our ability to zoom in effortlessly from the scale of a continent down to the scale of a person makes clear – literally clear – that the roughness of a coastline is about the same over a considerable range of scales. Discovering this without high-resolution satellite photos was challenging. But with these images readily available on the web, we all can discover coastline scaling and its variations. Every kid can be an explorer. Really.

3.4 How (and why) the lungs are fractal

Our lungs are remarkable structures. With each breath we inhale and exhale about half a liter of air (our *tidal volume*), exchanging oxygen from the air for carbon dioxide from the blood in under a second during heavy exertion. To do this, the lungs must have a very large surface area contained in a very small volume. Typically, this area is about 130 square meters, three-quarters the surface area of a tennis court, contained in a volume of five or six liters. To give a sense of the complexity of this structural contortion, University of Bern physiologist Ewald Weibel pointed out this is equivalent to folding an envelope into a thimble. And of course the physiological problem is not just how to fold the area of the lungs into their volume; the problem is how to do this and still have efficient gas exchange. This, as you can imagine, is quite a puzzle.

Here is the solution that has evolved: starting with the trachea, the airway splits into two branches, each of which splits into two, and each of those split again and again and again, between 18 and 30 times, with 23 splittings on average. The first 15 generations of branching are designated the *conducting airways*. The branches beyond this level constitute the *air exchange zone*, and each is decorated with *alveoli*, little sacks about 1/4 mm in diameter, about half a billion in total for the lungs. The alveoli constitute the elaborate, foam-like surface across which gas exchange occurs.

When considering the surface area of hundreds of millions of little bubbles, computation starts to look tricky. It is much easier to start focusing on the self-similar structure of the branches. At each branching, a mother branch of diameter d_0 splits into two daughter branches of diameters d_1 and d_2. Early in the 20th century, biologists W. R. Hess and C. D. Murray showed that air will flow most easily between mother and daughters if their diameters are related by

$$d_0^3 = d_1^3 + d_2^3.$$

3.4. HOW (AND WHY) THE LUNGS ARE FRACTAL

If the daughter branches have about the same diameters at each splitting, then the Hess-Murray equation gives

$$d_1 = d_2 = 2^{-1/3} \cdot d_0 \approx 0.79 \cdot d_0$$

Indeed, this relationship is observed for the conducting airways. Evolution has discovered how to maximize the ease of airflow.

Or almost. Careful measurements by Weibel and others show that the actual mother-daughter diameter ratio is closer to 0.85 than it is to the theoretical optimum of 0.79. This slight difference is thought to be a safety factor against the effects of airway constriction due to asthma, which would suggest that even departures from optimality serve a purpose. Evolution, it seems, has discovered nuance as well.

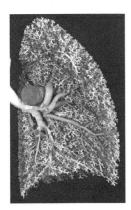
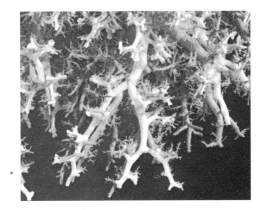

Figure 3.13: First: A cast of a lung and of the arterial and venous circulatory systems. Second: A magnification of a lung.

In the air exchange zone, the diameters of mother and daughter branches change very little, slowing the air flow to favor the diffusion of oxygen and carbon dioxide across the alveolar membranes. In the second image of Fig. 3.13 we see airway branching is accompanied by the branching of pulmonary arteries and veins. The bronchial and arterial trees branch in parallel throughout the levels of the conducting airways. After that, the arteries branch more frequently, on average 28 times, putting the number of capillaries at about 30 times the number of terminal airway branches. The capillaries fill the spaces between alveoli, so that each capillary contacts alveoli on two sides. The interlocking trees of the respiratory and circulatory systems have evolved beautifully to assure that air and blood are delivered to every part of the lungs. Moreover, this branching architecture guarantees that every part of the air exchange membrane lies within about 20 cm of the trachea, a short distance to have to move air in and out of the lungs.

We should mention one more design feature. The alveoli are spread along the last eight generations of branching, each of which is surrounded by its own network of arteries. As we progress down the branching levels of the lungs, the oxygen concentration of the air drops because some oxygen and some carbon dioxide have already been exchanged in alveoli higher up. If the oxygen concentration drops too much, gas exchange at the later alveoli is inefficient and the circulatory system will grow shunts, direct connections between arteries and veins bypassing some of the alveoli. Weibel showed that this damaging drop in oxygen concentration need not occur if the ratio of physical factors (the diffusion rate of O_2 and the rate at which it permeates the alveoli-capillary membrane) equals the ratio of morphological factors (the gas exchange area and the size of the group of alveoli descending from a single branch). For mammal lungs, these ratios are quite close. It looks as though evolution has discovered this guideline, too.

All this leads to a natural question: how is the lung surface area measured? Certainly not by slicing open and flattening all half-billion alveoli. The answer goes back to something called the *Buffon needle problem*, a question posed in 1777 by French mathematician Georges-Louis Leclerc, Comte de Buffon. Buffon's question was What is the probability that a needle of length L, when dropped on a floor of parallel boards of width d, will cross one of the seams between the boards? Buffon answered his own question: For a short needle, $L < d$, the probability is $(2/\pi) \cdot (L/d)$. To find the surface area of the lungs, the needle problem only needs to be adapted to a different "floor." Weibel sliced thin cross-sections of lung tissue, producing curves that he overlaid with a grid of test lines of total length L. This process is illustrated in Fig. 3.14. As Weibel showed, if I lines cross the curves, the lung surface area A and the volume V the lungs enclose satisfy the relation

$$\frac{A}{V} = 2 \cdot \frac{I}{L}.$$

In A.13 we'll derive this in a special case, illustrating the general principle.

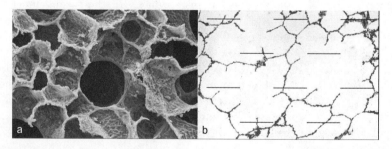

Figure 3.14: Measuring the alveolar area. First: A photomicrograph of a lung cross-section. Second: The edges of another portion, with lines overlaid for the area measurement.

3.5. POWER LAWS

In 1963, Weibel used this approach to measure the alveolar surface area using a light microscope, obtaining a value of 80 square meters, which was certainly impressive for an area folded into a volume of five or six liters. But then, in the 1970s, Weibel repeated these measurements with an electron microscope. This time he obtained an area of 130 square meters. After hearing Benoit lecture about measuring the length of the coastline of Britain, Weibel realized that the different area measurements had to be a consequence of the fractality of the lungs.

Additionally, Weibel speculates that the genetic instructions to build the lung likely are similar to IFS rules. Fractals may be more than a convenient language for us to discover patterns: they may explain the logic our DNA uses to build parts of our bodies

3.5 Power laws

One more tool is needed to measure the fractal relation of a part to the whole, especially when the whole is not visible in its entirety. Geometrical self-similarity has gotten us so far. Now we need power laws.

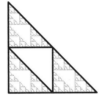 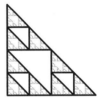

Figure 3.15: The gasket, divided into three and nine pieces.

Fractals that occur in time are one example of data for which we may not be able to get a big-picture view. For this, for any observations that occur in some ordered sequence, we need power laws to tell us if the bit of sequence we can see has fractal scaling.

We say values y and x are related by a *power law* if

$$y = a \cdot x^d$$

for some constants a and d. The exponent d is why this relation is called a power law: y depends on some power of x.

At the moment, though our main goal is to understand non-geometric power laws, we will start with a familiar geometric example. As we know, the gasket can be divided into three pieces, indicated by dark outlines in Fig. 3.15, where each piece is a copy of the whole gasket scaled by $1/2$. On the right we see $9 = 3^2$ pieces, each scaled by $1/4 = 1/2^2$. Continuing in this fashion, we find the gasket can be divided into 3^n pieces, each scaled by $1/2^n$. Let y denote the number of pieces and x the scaling factor. The gasket power law, which we saw in Sect. 3.2, is

$$3^n = (1/2^n)^d$$

The same exponent d holds true for all scales $1/2^n$. This is a signature of self-similarity: no matter what size piece we see, its subpieces are always related to it in the same way.

In Chapter 6 we will see how to determine and interpret this exponent d. For now, we are content to say that the presence of such an exponent strongly suggests fractality.

To apply this rule to non-geometric measurements, we will start with a sequence of measurements and look for patterns in time. For geometrical fractals, power laws quantify how the number of pieces of a fractal scales with the size of those pieces. For sequential data, power laws quantify how the correlation between sequence entries scales with the time between the entries, and we say a sequence of data is fractal if the correlations reveal a scaling in time analogous to the spatial scaling of geometrical fractals.

Suppose we have a sequence $z_1, z_2, z_3, \ldots, z_N$. These could be intervals between successive heartbeats, or nucleotides in DNA sequences, or any collection of measurements subject to some natural ordering. If the ordering is by time, this is called a *time series*. Basically, we need to quantify how z_i is related to z_{i+w} and how this relationship varies with the interval w. If the z_i are numbers, we can use the *correlation function*

$$C(w) = \langle z_i z_{i+w} \rangle,$$

where the pointy brackets indicate the average of the products over the whole data set. That is,

$$\langle z_i z_{i+w} \rangle = \frac{1}{N-w} \left(z_1 z_{1+w} + z_2 z_{2+w} + \cdots + z_{N-w} z_N \right)$$

This is a long formula. Fig. 3.16 should help unpack it. The first upper image is a graph of (the beginning of) a time series, with z_i values plotted vertically, i values horizontally. To compute $\langle z_i z_{i+1} \rangle$, $w = 1$ in the formula above, we multiply $z_1 z_2, z_2 z_3, z_3 z_4, \ldots$, add

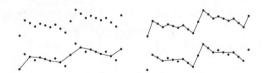

Figure 3.16: Top first: The time series. Second: lines connect points separated by one time step. Bottom: connecting points separated by two time steps.

them up, and divide by the number of products we're adding. In the second upper image, lines connect z_1 to z_2, z_2 to z_3, and so on. That is, lines connnect the numbers we multiply to compute $\langle z_i z_{i+1} \rangle$. In the second row of graphs, lines connect the numbers we multiply to compute $\langle z_i z_{i+2} \rangle$. In the first lower graph, lines connect z_1 and z_3, z_3 and z_5, and so on. In the second lower graph, lines connect z_2 and z_4, z_4 and z_6, and so on. Then we can add up these products and divide by their number to compute $\langle z_i z_{i+2} \rangle$.

3.5. POWER LAWS

Roughly, the correlation function $C(w)$ measures the strength of the relation between the present and the past that is w time steps away. By itself, a strong correlation is not a sign of fractality. If a power law relation can be found between $C(w)$ and w, we should look for some fractal dynamic at work.

To see the next step of this relation between power laws and fractals, we must understand a bit more about $C(w)$. We'll use a simple example. Suppose the z_i are the amplitudes of the wave that repeats exactly w time measurements later. That is, for every i, $z_i z_{i+w} = z_i^2$. The square of the amplitude of a wave is called the *power* of that wave. In terms of the wave, what, exactly, is w? Typically, time series measurements are made at constant intervals. Between z_i and z_{i+1} a time Δt has elapsed. Then $w \Delta t$ is the time the wave takes to return to the same amplitude. This is the *period* of the wave; the *frequency* f is the reciprocal of the period. This agrees with our intuition: the less time the wave needs to repeat, the more frequently it repeats. If w corresponds to the period of a wave, $C(w)$ is high. If not, some of the products $z_i z_{i+w}$ will be positive and some will be negative. Cancellations will occur in the computation of $\langle z_i z_{i+w} \rangle$, giving a lower value of $C(w)$.

Most signals don't consist of a single wave, but are a combination of waves of many frequencies and amplitudes. If you are near a piano, you can test this. Open the piano lid so the strings are visible and make some loud noise – shout, clap your hands, drop a large book – and look at the strings. Each vibrating string represents the frequency of one component of the sound, and the amplitude of each string's vibration is proportional to the amplitude of that frequency within the original sound.

Rather than carry around a piano to look at vibrating strings, we can graph the amplitude – or, more often, the power – as a function of the frequency f. The resulting picture is called the *power spectrum* $P(f)$ of the sound. This is a graph of the power of each frequency present in the sound. Of course, we can do this plot for any signal, not just for sounds. In A.14 we derive a relation between correlation functions and power laws.

A *scaling signal* has a power spectrum $P(f)$ proportional to $1/f^\beta$ for some number β. We write

$$P(f) \propto \frac{1}{f^\beta}$$

Scaling signals are divided into three broad classes of behavior, depending on the value of β. In the early history of electronics, this classification was applied to different kinds of noise in circuitry, and the noise terminology persists.

- *White noise.* $\beta = 0$. Each measurement is uncorrelated to that coming before it. This is "too random." The past has no influence on the present.

- *Brownian noise.* $\beta = 2$. Each measurement is the previous measurement plus a bit of white noise. This is too boring: gradual movements go up or down, and are not random enough. The past, but not much of the past, influences the present.

- $1/f$ *noise.* $\beta = 1$ lies midway between white noise and Brownian noise. This exhibits a balance of familiar patterns and randomness, but just enough randomness. Goldilocks randomness. Much more of the past can influence the present.

Of course, the exponent β can take on values other than just 0, 1, and 2. In practice, a signal is called white, Brownian, or $1/f$ noise if β is pretty close to 0, 2, or 1, though how close depends on the application. Fig. 3.17 shows instances of each of these three: first, $P(f)$ vs. f; second, $\log(P(f))$ vs. $\log(f)$; and third, the time series.

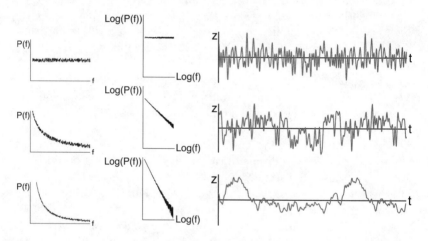

Figure 3.17: First: Power spectrum. Second: Log power spectrum. Third: Time series. Top to bottom: Plots for white noise, $1/f$ noise, and Brownian noise.

Together with the geometry of previous chapters, power laws round out the collection of tools we need to find fractals in the wild. We will start with some power law examples and then move on to more geometrical fractals.

3.6 Forests and trees

Suppose you want to estimate how much carbon dioxide a tropical rain-forest can remove from the atmosphere. Finding how much a single leaf can remove is straightforward, once you know how. But then what? Count the number of leaves in a typical tree and count the number of trees

3.7. OUR COMPLEX HEARTS

in the forest? But really, no one wants to count the number of trees in the forest. Rather, measure the area occupied by a single tree in a satellite view of the forest, and divide this into the area occupied by the forest. This gives about the number of trees in the forest. To find the number of leaves in a tree, count the number of leaves on a typical branch and multiply by the number of branches in a tree. Then the number of leaves in the forest is

leaves per branch × branches per tree × trees per forest

But this calculation has at least two problems: what's a typical branch of a tree, and what's a typical tree in a forest? Theoretical physicist Geoffrey West, evolutionary biologist Brian Enquist, and ecologist James Brown approached this problem in a different way. For trees in a Costa Rican rain forest they found a power law relation between the diameters of the branches of a tree and the number of branches having that diameter. Then, tree census data revealed a power law relation between the trunk diameter of a tree and the number of trees having that trunk diameter. Remarkably, both power laws have the same exponent. In this sense, branch is to tree as tree is to forest. The resource distribution pathways of a tree, discovered and fine-tuned by evolution, match those of the whole forest.

Armed thus, ecological and environmental scientists can make more reliable calculations. Similar analyses have given better estimates of the number of herbivores, and even the number of tiny arthropods, a forest can support. Since smaller creatures sample the forest on smaller scales, the fractality of plants revises bug population estimates upward.

3.7 Our complex hearts

We like to imagine that our hearts, at least when we are at rest, beat at pretty constant rates. Unless startled, say by a loud noise or a cat leaping to our chest in the middle of the night, a healthy heart should be a decent clock, leaving more irregular heartbeats to the old and the infirm. But contrary to this common expectation, a graph of the intervals between successive beats does *not* consist of points a nice straight line. Fig. 3.18 shows two real examples, plotted with the same vertical scale. The second graph has several very low values, each signaling an interval of about 0.19 seconds between successive heartbeats, while the mean interval between successive heartbeats is 0.41 seconds. The first graph has a mean interbeat interval of 0.43 seconds.

In the 1980s the complex appearance of interbeat interval records suggested to many that heartbeats are chaotic. To respond to environmental and activity-induced variations – the cardiovascular spikes as you sprint away from a peckish saber cat – the heart muscles must respond rapidly, and as gracefully as a flock of birds turning in the air, and not rigidly, like

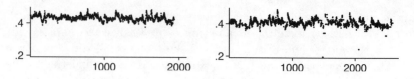

Figure 3.18: Intervals between heartbeats. The horizontal axis shows the beat number.

troops marching in step. Standard data analysis tools were adapted from physical sciences, papers were published, explanations advanced. Being more robust than their periodic cousins, chaotic systems were thought to be better able to adapt to environmental demands. A healthy heart is a chaotic heart, a healthy brain a chaotic brain, or so suggested one 1989 essay in *Science* and a 1990 essay in *Scientific American*.

But the heart is not a simple chaotic system. At the very least, it is chaos plus noise, with outside influences shouting over the top of our heartbeats. A collection of papers in the journal *Chaos* in 2009 demonstrated problems with these analyses, some techniques of which are unable to discriminate between chaos and periodicity with a small random perturbation. Rather than establishing pure cardiac chaos, careful analyses suggest the presence of several power laws over different time scales. The heartbeat is not a simple fractal but rather a more complicated combination of processes called a multifractal. (See Sect. 7.4 for some details.)

Fig. 3.19 shows the power spectra of the heartbeats of Fig. 3.18. We see a very noisy decay of power with increasing frequency, plausibly a roughened $1/f$, but some subtle work is needed to study these signals.

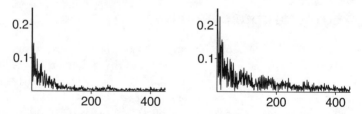

Figure 3.19: The power spectra of the time series of Fig. 3.18.

And one final relation between fractals and the heart: heartbreak, too, inhabits many scales. A soufflé falling because of mistakenly spraying oil into the ramekins before pouring in the custard. A mind being chewed up by the start of Alzheimer's, mental skills developed over half a century drifting out of the top of the head like heat through a poorly insulated roof in the winter. Friendships and loves hobbled by time. A desperate wish for more chances, more years, which is left unheeded. A sleepless night. A walk through a garden. These make up the fractal textures of heartbreak.

Of course, love has fractal characteristics as well, is felt on many levels. You've heard enough from us; time to think of your own examples.

3.8 Metabolic rates

The way animals, ourselves included, disperse heat has a fractal aspect. This and other biological quantities that exhibit power law scalings with size fall under the general heading of *allometry*. The reason for this metabolic scaling is not obvious – in fact the details remain contentious. Let's begin with the basic calculation.

The surface area, A, of an animal scales approximately as its linear size, L, squared: $A \propto L^2$. Similarly, the mass, M, of an animal scales as its volume, so $M \propto L^3$. Heat dissipation occurs across the surface, so the total metabolic rate of an animal is proportional to L^2, hence to $M^{2/3}$. Then the metabolic rate per unit mass must be proportional to $M^{2/3}/M = M^{-1/3}$, or so argued Max Rubner in 1883. Since pulse rate is related to metabolic rate per unit mass, larger animals should have slower pulse rates than smaller animals. Indeed, this is the case: a whale's heartbeat mimics the slow pulse of ocean waves, while the heartbeat of a baby field mouse asleep cupped in your palm is too fast to count. To this, add the observation that most mammal hearts beat one to two billion times during an animal's life. In the absence of external perturbation (predation, disease, random violence), the mouse nestled in your hand will have a shorter life than will you, and you will have a shorter life than the whale drifting far out at sea.

Logically, this makes perfect sense. But careful measurements published in 1932 by Max Kleiber (another Max – what are the odds?) showed something different: for most animals, over a range of sizes spanning an amazing 21 orders of magnitude, microbes to whales, the metabolic rate per unit mass varies as $M^{-1/4}$ rather than as $M^{-1/3}$. Other biological variables exhibit similar scalings with mass, and in all cases the exponent is a simple multiple of 1/4. Rate of heartbeat scales as $M^{-1/4}$, blood circulation time and lifespan as $M^{1/4}$, age of first reproduction as $M^{3/4}$, the time of embryonic development as $M^{-1/4}$, and the diameters of tree trunks and of aortas as $M^{3/8}$.

Rubner's argument appeared so clear; how could it be wrong? A practical reason for getting the metabolic power law right is to be able to properly scale doses of experimental drugs from small test animals, usually mice, to human test subjects.

It turns out that heat isn't dissipated just across the skin surface, but also through our breath, and therefore across the entire convoluted surface of our respiratory system. The complex, fractal branching of the lungs must figure in the metabolic rate calculation. But the question is more general than this, for plants also exhibit an $M^{-1/4}$ metabolic rate per unit mass scaling, so any attempt at a universal explanation must include plants as well as animals and, it turns out, the metabolic pathways of microbes.

We may be able to understand this scaling through a model in which metabolic resources are distributed by fractal networks, including vascular networks in plants and animals, the interior structures of cells, and even virtual hierarchical networks of resource flow from amoebas to forests. In addition, calculations should be based, not on exterior surface area and volume, but on biological area and biological volume. *Biological area* means total effective surface area, the area across which nutrients and energy are exchanged between inside and outside. Examples include gastrointestinal area, total leaf area, the area of mitochondrial membranes within cells. *Biological volume* is the volume of biologically active material, usually less than the volume enclosed by the surface because of the presence of empty regions within the organism. Also, it is important to note that the smallest scale of the resource distribution network, for example, capillaries in animals, is about the same across all species. Capillaries of a whale are not whale-size versions of capillaries of a mouse; both have capillaries of about the same diameters, so a whale's circulatory system must have more levels of branching than a mouse's circulatory system. This straightforward observation – obvious, really, when you think of comparing a mouse's capillaries and aorta to a whale's capillaries and aorta – shows that the mouse's circulatory system is not just a scaled version of the whale's circulatory system. This difference is central to the derivation of the $M^{-1/4}$ scaling by West, Brown, and Enquist. They conclude

> Unlike the genetic code, which has evolved only once in the history of life, fractal-like distribution networks have originated many times. Examples include extensive surface area of leaves, gills, lungs, guts, kidneys, chloroplasts, and mitochondria, the whole-organism branching architecture of trees, sponges, hydrozoans, and crinoids, and the tree-like networks of diverse respiratory and circulatory systems. ... Although living things occupy a 3-dimensional space, their internal physiology and anatomy operate as if they were 4-dimensional.

Four-dimensional? Really? Let's pretend we're 4-dimensional, in which case our boundary is 3-dimensional. A simple 4-dimensional object – a hypersphere or hypercube, for instance – with uniform density and diameter L has mass M scaling as L^4. The boundary, across which heat is dissipated, is 3-dimensional, so the metabolic rate scales as

$$\text{metabolic rate} \propto L^3 = (M^{1/4})^3 = M^{3/4}$$

Then the metabolic rate per unit mass scales as $M^{3/4}/M = M^{-1/4}$. So we see the hierarchical energy dissipation network of organisms simulates the metabolic rate scaling we'd see if the organism had a simple shape but lived in 4 dimensions. This may be just a numerical coincidence, but if not, we may find a new direction for understanding relations between biological function and form.

A different explanation of the $M^{-1/4}$ scaling is based on the assumption that natural selection tends to minimize the total blood volume of an animal. Jayanth Banavar, Amos Maritan, and Andrea Rinaldo constructed a model based on the geometry of spanning networks without any assumption of hierarchical structure.

Both these approaches, as sketched in A.15, give the same scaling law, but they use different methods and proceed under different assumptions. They do show that it is possible to find a universal explanation for this metabolic scaling law, though of course it may be that neither explanation is correct, and there may in fact be no single explanation at all. There may be some very deep science here, waiting to be discovered – or maybe not. In a 1983 *NOVA* episode, Richard Feynman remarked

> If it turns out there is a simple ultimate answer which explains everything, so be it, that would be very nice to discover.
> If it turns out it's like an onion with millions of layers and we're just sick and tired of looking at the layers, then that's the way it is.

We, at least, are not tired of the layers we have found so far.

3.9 Fractals and DNA

Biology has changed dramaticaly since the older author studied it, about fifteen years after James Watson and Francis Crick discovered the molecular structure of DNA, and since the younger author studied it, five years after the human genome was mapped. With the explosive growth of molecular biology and the sequencing of genomes, biology and geometry are talking with one another more deeply than ever before. At some point, fractals and biology began noticing one another. In this section we survey some examples of fractals in DNA, though it is only a small sampling of the many, many examples waiting to be explored.

3.9.1 Long-range correlations

The genetic code is written in an alphabet of four letters, C, A, T, and G, standing for the nucleotides cytosine, adenine, thymine, and guanine. Voss and others have studied, and in some cases graphed, the correlations of these sequences. Voss argued that any plot in a space of fewer than 4 dimensions will of necessity introduce spurious numerical correlations and therefore lead to false conclusions. Better, and certainly less taxing, than plotting in 4 dimensions is to compute the correlation function in order to find the strength of the relationship between nucleotides positioned a distance w apart in the sequence:

$$C(w) = \langle z_i z_{i+w} \rangle$$

To carry out this computation, we will need to multiply nucleotides, so we will need to find a way to assign some numerical value to each nucleotide. Voss approached this by defining the product this way:

$$z_i z_{i+w} = \begin{cases} 1 & \text{if } z_i \text{ and } z_{i+w} \text{ are the same nucleotide} \\ 0 & \text{if } z_i \text{ and } z_{i+w} \text{ are different nucleotides} \end{cases}$$

As an example, consider the DNA sequence for amylase, an enzyme found in saliva which serves to catalyze the breakdown of starches into sugars. It is represented by a string of 3,957 letters, beginning T,G,A,A,T,T,C, ..., so, for example,

$$z_1 z_2 = 0 \quad \text{because } z_1 = T \text{ and } z_2 = G$$
$$z_2 z_3 = 0 \quad \text{because } z_2 = G \text{ and } z_3 = A$$
$$z_3 z_4 = 1 \quad \text{because } z_3 = A \text{ and } z_4 = A$$

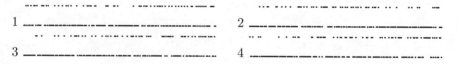

Figure 3.20: Graphical representations of the first 200 products $z_i z_{i+w}$, $w = 1, 2, 3, 4$, for amylase.

In Fig. 3.20 we see plots of the first 200 $z_i z_{i+w}$ for $w = 1, 2, 3,$ and 4. The correlation functions for each of these, respectively, are

$$C(1) = \frac{1}{3956}(z_1 z_2 + z_2 z_3 + \cdots + z_{3956} z_{3957}) \approx 0.3079$$

$$C(2) = \frac{1}{3955}(z_1 z_3 + z_2 z_4 + \cdots + z_{3955} z_{3957}) \approx 0.3016$$

$$C(3) = \frac{1}{3954}(z_1 z_4 + z_2 z_5 + \cdots + z_{3954} z_{3957}) \approx 0.3002$$

$$C(4) = \frac{1}{3953}(z_1 z_5 + z_2 z_6 + \cdots + z_{3953} z_{3957}) \approx 0.3018$$

and so on. From these, we can compute the power spectra and make some assessments about long-range correlations.

3.9. FRACTALS AND DNA

Voss applied this method to the complete genome of human Cytomegalovirus strain AD 169 (229,354 bases) and found that for high f (short w) the power spectrum looks like white noise, but for lower f (longer w) the power spectrum is close to $1/f$, more precisely $1/f^\beta$ with $\beta = 0.84$, establishing a fractal correlation in this nucleotide sequence.

Individual DNA sequences have relatively noisy power spectra, but by averaging over GenBank classifications, Voss obtained smoother long-range power spectra with the values of β, the power law scaling exponent, shown in Table 3.1.

bacteria	1.16	organelle	0.71	virus	0.82
plant	0.86	invertebrate	1.00	vertebrate	0.87
mammal	0.84	rodent	0.81	primate	0.77

Table 3.1: Some power spectrum exponents.

Our DNA is rife with fractals, but at the moment, we don't know why.

3.9.2 DNA fractal globules

Unwound into a linear strand, our DNA would stretch to nearly two meters in length, longer than some of us, older and younger author included, are tall. Placed end-to-end, the unwound DNA of each of our ten trillion cells would span the diameter of Pluto's orbit. The difficult task of folding this DNA into the cell nucleus, which has a diameter of about five mil-

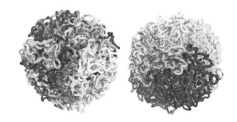

Figure 3.21: First: A tangled equilibrium globule. Second: A fractal globule. Similar grayscales represent nearby regions in the DNA strand. (Simulations by Leonid Mirny and Maxim Imakaev.)

lionths of a meter, is even more challenging than folding the 130-square-meter lungs into the 5-liter chest cavity. For DNA, this task is accomplished by *histones*, protein components of chromatin, the combination of proteins and DNA making up cell nuclei. Among the many crucial tasks it performs for us, chromatin compacts our DNA into a tiny volume, helps prevent damage to DNA, allows DNA replication during cell division, and controls gene expression. Chromatin occurs in two forms: more tightly folded heterochromatin and more loosely folded euchromatin. The looser folding of euchromatin allows gene regulatory proteins to bind with the DNA, to initiate gene transcription. More tightly folded heterochromatins also function in gene regulation, as well as the promotion of chromosome integrity, by both protecting the ends of chromosomes from deteriorating and preventing chromsomes from combining.

Until recently, the most widely accepted organizational scheme for folded DNA was the *equilibrium globule*, a model derived from the statistical mechanics approach to polymer folding. This model can achieve the needed size reduction, but at the expense of introducing many knots and tangles. These could inhibit the unfolding of the globule to make accessible parts of the DNA during gene activation, gene inactivation, and cell replication.

In 1988 Alexander Grosberg and coworkers proposed another folding conformation, the *fractal globule*, and in 1993 they suggested that this geometry was evolution's solution to the DNA folding problem. Grosberg reasoned that in order to fold such a long structure into such a small volume, DNA must be collapsed in an iterative process of folds on folds on folds on folds, allowing for extensive folding without risk of knotting. Fig. 3.21 shows simulations of an equilibrium globule (first) and a fractal globule (second). Knots and tangles in the equilibrium globule might prevent a cell from being able to read its own DNA. In the fractal globule, portions of the DNA strand can unfold and refold during gene activation and repression without unfolding the whole globule. Experiments reported in a 2009 *Science* paper by Erez Lieberman-Aiden et al. provided strong evidence to support the fractal globule hypothesis. Job Dekker, one of the coauthors, described this shape as a globule made of globules made of globules made of globules.

Another experiment, by Aurélien Bancaud et al., reported in the *EMBO Journal* in 2009, offers strong evidence for how the nucleus maintains two distinct regions with different levels of gene activity. Genes tend to be inactive in regions containing heterochromatin and active in regions containing euchromatin. The hypothesis was that chromatin has a structure like a sponge, blocking larger molecules and allowing smaller molecules to enter a region. The experiment involved tracking the movements of fluorescent molecules of different sizes. It turned out that, contrary to expectations, all molecules experienced the same degree of obstruction, suggesting that the nuclear environment has a similar structure across a range of scales: that is, the cell nucleus is fractal.

Moreover, euchromatin has a more complex structure, which means that it occupies more 3-dimensional space with its large, crinkled surface than a less-complex surface would. We can quantify this by saying that euchromatin has a higher fractal dimension. By contrast, heterochromatin has a lower fractal dimension and presents a smaller, smoother surface. Since proteins activate a gene by binding to specific, sparsely distributed DNA sequences, the rougher surface of euchromatins could encourage proteins to explore large portions of the DNA sequence, expediting the locating of the target. Since proteins that inactivate genes bind to histones, the smoother fractal surface of heterochromatin would encourage shorter movement. It is possible that just altering the fractal structure of the chromatin could change the behavior of different areas of the DNA.

In 2012, Bancaud and another team speculate about why chromatin might have evolved a fractal structure, and suggested additional experiments to test the fractal globule hypothesis.

That in some fundamental way our DNA may have fractal properties, not just in its sequencing but in its 3-dimensional structure, gives still more evidence that evolution discovers and uses the principles of fractal geometry. So far the evidence suggests that the more carefully we look, the more places we will find fractals in our cells.

3.9.3 DNA Sierpinski gaskets

Currently, computer circuit boards are produced by photolithography, a way to transfer tiny geometric patterns to substrates by light. To achieve smaller, and consequently faster, computer components, another approach is needed. One alternative may be provided by *algorithmic self-assembly*, a method for programming bits of DNA to grow into specified shapes or functional forms. A promising approach, DNA programming occupies the middle of the spectrum between natural chemical reactions too simple to program and living systems too complicated to control (at least for now).

We should be familiar with the dual role of DNA: as active instructions for synthesizing proteins and as passive data to be copied during cell replication. Consequently, a natural – though by no means easy – approach is to design DNA configurations whose pieces are instructions for assembling the whole configuration. That the pieces contain instructions for the whole makes fractals a good candidate for molecular self-assembly experiments.

Erik Winfree, a Caltech computer scientist who made significant contributions to the implementation of DNA computing, constructed short strands of artificial DNA that can assemble themselves into a Sierpinski gasket with a side length of about a micron. The molecules of the gasket directed their assembly into the gasket. The implications are remarkable: in principle, DNA can be programmed to self-assemble into any configuration, whether to carry out computations or construct molecular machines. The humble gasket gives us our first taste of an exciting new field of possibility.

3.9.4 DNA fractal antennas

That DNA may be sensitive to electromagnetic (EM) radiation is much on the minds of people who spend a noticeable fraction of each day with a cellphone surgically attached to an ear. One explanation of the EM sensitivity of DNA is the notion that DNA may function as a fractal antenna.

Physiologist Martin Blank and pathologist Reba Goodman, both at Columbia University, argue that the sensitivity of DNA to a wide range of frequencies – extremely low frequency (ELF) to radio frequency (RF) – is a characteristic of fractal antennas (which we will discuss in Sect. 4.3). This reinforces the theory of the fractal globule structure of folded nuclear DNA.

3.9.5 Extraterrestrial DNA

Some assert that some fractal patterns in DNA point to recent visitations by extraterrestrials... OK, this is a test of your critical thinking. You can, indeed, find this kind of assertion on the web. But come on, are you that gullible?

3.10 How planets grow

In our own experience, the Earth seems like a fairly ordinary lump. Much of the surface is covered with a layer of liquid water, but still, mostly we think of the Earth as solid. How did this pretty blue marble of matter come to be in mostly empty space?

We know that the sun condensed from a nebula of gas and dust, with gravity pulling everything toward the center until it became hot enough to ignite the thermonuclear alchemy that turns hydrogen to helium, plus some extra light and heat. Of the rest of the gas and dust still orbiting the sun, some turned into meteoroids, some into comets, and some into planets. The gas giants – Jupiter, Saturn, Uranus, and Neptune – grew more or less the way the sun grew, but the story must be different for the terrestrial planets – Mercury, Venus, Earth, and Mars. How do you get from gas and dust to large, and largely solid, lumps?

One approach to understanding planet formation is called *ballistic aggregation* (BA). Here, particles travel in straight lines (ballistic trajectories) until they collide and then stick togther. BA forms dense, compact clusters, the first picture of Fig. 3.22 is an example. This growth suffers

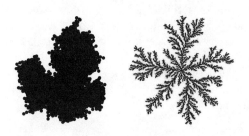

Figure 3.22: Aggregation simulations: BA (first) and DLA (second).

from a problem. To move in about straight lines through the nebula, particles must be fairly large and moving relatively quickly. Under these circumstances, collisions tend to fracture clusters, not add to them.

3.10. HOW PLANETS GROW

At least for the growth of small clusters, a better model is *diffusion-limited aggregation* (DLA), illustrated on the second image of in Fig. 3.22 and discussed in Sect. 6.6. This model begins with micrometer-size particles performing a random walk (Brownian motion, Sect. A.10) caused by thermal impacts of other particles. These random walkers collide and stick, growing branches, and side branches, and side branches off side branches, and so on. Most growth is around the periphery of the cluster because the outer branches screen the inner from particles which randomly wander in and are more likely to get caught near the edges than they are to miss the outer edges and meander randomly toward the inner regions of the growing clump. Consequently, DLA forms wispy fractal clusters described as "fluffy" by some astrophysicists. Basically, they're dust bunnies. (The fractal clusters, not the astrophysicists.)

As these clusters grow, they are less likely to stagger randomly into one another. The impacts of small dust particles have less effect, and the clusters begin to zip along in straight lines until they collide with each other and stick. This is *ballistic cluster-cluster aggregation* (BCCA). Because wispy clusters collide and stick, screening effects are even more pronounced than they are for single particles in DLA. Consequently, BCCA clusters are even wispier than DLA culsters, which are in turn wispier than BA clusters.

Clusters continue to collide, stick, and grow more wispy, up to some critical aggregate radius (maybe about 3 cm, but perhaps quite a bit larger). Above this critical radius, when clusters collide they absorb collisional energy by partially collapsing and melting, becoming denser with each collision. Experiments show that, unlike in a collision between two dense BA clusters, an object colliding with a dust clump does not scatter debris in every direction, but instead compresses the clump in the direction of the collision, much the way a snowplow compresses the snow forward. In addition, dust clumps are good at holding in the heat of collision, enhancing the melting and restructuring.

Dutch astrophysicists Carsten Dominik and coauthors sum this up:

> small solid particles grow rapidly into aggregates of quite substantial size, while retaining their fractal nature (in the early growth stage) or a moderate to high porosity (for later growth stages).

All this depends on the physical properties of dust clumps and how they restructure under collision. Sophisticated experiments have been, and are being, performed, but one of the earliest, designed and performed by physicist Bertram Donn, involved throwing snowballs into a snowbank to test how snow compresses under impact by objects of similar density.

Donn and physicists Joseph Nuth of NASA Goddard and Peak, of Utah State University, hypothesized that fractal dust aggregates grow quite large before restructuring to planetesimals. We will sketch some calculations in Sect. 6.6, suggesting that the density of DLA and BCCA clusters decreases with size. The lower density works to slow the movement of these clusters and allows them to absorb more energy by restructuring.

Behavior of these clusters under collision are investigated by experiments in vacuum and microgravity carried out in drop towers, in parabolic flights, and on the space shuttle. Detailed descriptions of this work are given by German astrophysicist Gerhard Wurm and his colleagues.

Finally, three points of interest. First, fly-by observations show comets also have low density. Essentiallly big dirty snowballs, many comets formed about the same time as the solar system, so we may expect them to be similar in structure to the early planets: low-density fractal clumps.

Second, observations of stellar discs – pancake-shaped collections of dust, gas, and clumps of matter with a star at the center – show infrared emission and scattering by particles quite a bit larger than interstellar dust and more like fractal dust clumps. These may be early stages of a solar system not unlike our own, though of course much more observation is needed to understand this.

Figure 3.23: Peggy.

Third, in April 2014, planetary scientist Carl Murray and coauthors reported an extended bright region on the outer edge of Saturn's A ring, the outermost of Saturn's large rings. The region is thought to be the track of a nascent moon, nicknamed "Peggy." It is a small object about 1/1000 the size of the bright region circled in Fig. 3.23, a photo taken from the Casini spacecraft.

Similarities between Saturn's ring system and the early solar nebula suggest that the mechanisms responsible for planet formation may be involved in the production of Saturn's moons. Watching Peggy's birth may give us data about the fractal aggregation model of planet formation.

However it unfolds, it seems likely that fractals were an essential step in giving us a place to stand in the universe.

3.11 Reversals of the Earth's magnetic field

Earth's magnetic field is generated by currents of molten iron in its outer core. This field helps keep us safe from cosmic rays; of course, our atmosphere helps a lot with this, too. Geological records show that the magnetic field reverses direction from time to time. The reversals do not occur at uniform intervals, but in a pattern that resembles a Lévy flight (a path whose jumps scale according to a power law, more in A.10). The times of reversals are shown in Fig. 3.24, with a time scale of about 169 million years. Note the long intervals of constant orientation, interspersed with many rapid reversals. This does resemble a Cantor set. But why?

3.12. THE DISTRIBUTION OF GALAXIES

For that matter, why do the reversals occur at all? There are theories, but no agreement. Maybe the fractal structure of the reversal pattern will shed some light on the mechanism of the reversals.

Figure 3.24: Reversals of the Earth's magnetic field, over the last 169 million years, from now back to the middle Jurassic. The lower picture is an expansion of the boxed region of the upper picture.

3.12 The distribution of galaxies

Galaxies are aggregates of, among other things, dust, gas, and stars: millions and millions of stars. There are a few million stars in dwarf galaxies and up to 100 trillion in the giant galaxy IC1101 in the Abell 2029 cluster. In diameter, galaxies range between about $1,000$ pc (pc = parsec, about 3.26 light years, or 30.9 trillion km, so yes, even little galaxies are big) and $2,000,000$ pc. Typically, galaxies are separated from nearest neighbors by several Mpc (million parsecs). At their centers, most – maybe all – galaxies contain supermassive black holes. This isn't all: relatively recent discoveries suggest that 90 percent of the mass of each galaxy consists of dark matter. Galaxies are complicated, beautiful structures – but for now we're going to ignore all this and treat them as points.

A.16 contains some additional information, including brief descriptions of some of the tests of power law scaling in galaxy distributions.

The visible universe may contain 200 billion galaxies. The cosmologies of Newton and Einstein were homogeneous, galaxies spread uniformly throughout the universe, like blueberries in a blueberry muffin. Instead, it seems that galaxies are actually clumped together in clusters, which are themselves organized into superclusters. If homogeneity occurs, it must hold on a distance scale larger than superclusters. On the largest observable scales, these clusters are arranged in sheets and filaments, winding through immense voids. The space between these sheets and filaments is profoundly darker, emptier, than the emptiness of the dark space between stars. Most of the universe is filled with lonely emptiness.

Another possibility is *hierarchical clustering*, clumps of galaxies grouped together into larger clumps, themselves organized into still larger clumps. The Swedish scientist Emanuel Swedenborg was an early proponent of hierarchical clustering; his work anticipated much of self-similarity.

Because hierarchical clustering gives rise to a power law distribution of mass, if the power law exponent is small enough, hierarchical clustering could be a resolution of *Olbers' paradox*: why is the night sky dark instead of fillled with light? Theoretically, if space is infinite and doesn't run out of stars, then in a static universe we should be able to see a star in every direction we look, at some distance or other. Perhaps surprisingly – at least, this surprised us – one of the clearest statements of Olbers' paradox was given by Edgar Allan Poe, in his essay *Eureka*:

> Were the succession of stars endless, then the background of the sky would present us an uniform luminosity, like that displayed by the Galaxy – since there could be absolutely no point, in all that background, at which would not exist a star. The only mode, therefore, in which, under such a state of affairs, we could comprehend the voids which our telescopes find in innumerable directions, would be by supposing the distance of the invisible background so immense that no ray from it has yet been able to reach us at all.

Indeed, as Poe supposed, there is such a horizon. We can see stars no further away than 14 billion light years, the distance equivalent to the approximate age of the universe. However, Poe may have been wrong about the "uniform luminosity" of the night sky. Not only is there a horizon, but there may not be a star in every direction, if hierarchical clustering holds and the power law exponent is low enough. To see why, we'll need to know a bit more about measuring the complexity of fractals, so we'll do this calculation in Sect. 6.6. Instead, stars are clumped together, and emptiness is clumped together, over a huge range of scales, ensuring that wherever we look we will find patches of light and patches of dark.

The question of how far out this fractal clustering may extend has not yet been answered to universal satisfaction. Canadian-American physicist P. J. E. Peebles believes the data support fractality out to several tens of megaparsecs (Mpc). Italian physicist Luciano Pietronero thinks fractality extends at least to 100 Mpc, and maybe quite a bit further. Part of the controversy arises from their different methods to calculate distances to remote galaxies. An excellent summary of these arguments, and potential problems with the calculations, is given by Russian physicist Yurij Baryshev and Finnish physicist Pekka Teerikorpi in their 2002 book *The Discovery of Cosmic Fractals*.

Before leaving this section, we will mention theoretical fractal constructs on two vastly different scales: larger than the (observable) universe and far smaller than subatomic particles.

In the late 1980s, Russian-American theoretical physicist Andrei Linde proposed the self-replicating, chaotically inflating universe. In this model, inflating patches of space pinch off to form bubble universes, which give rise to other bubble universes, and on and on, forever. In 2007, American theoretical physicist and cosmologist Alan Guth, architect of the original inflationary model of cosmology, gave a clear description of how chaotic inflation produces a collection of universes arranged according to a branching fractal tree. At any point, the universes branching off a given universe resemble larger bits of the branching pattern.

At the other end of the scale spectrum, at the *Planck length*, 10^{-35} m, the realm of string theory, some approaches to quantum gravity involve a fractal structure of spacetime itself. One fascinating consequence of the work of Italian theoretical physicist Carlo Rovelli and French mathematician and Fields Medalist Alain Connes is that time may be an emergent property, not part of the underlying structure of the universe. Others have suggested that on the smallest scales space is 2-dimensional, and with increasing scale the dimension gradually increases, necessarily progressing through fractal realms, to what we observe as multiple distinct dimensions.

A fractal structure for spacetime was proposed by French physicist Laurent Nottale in his *scale relativity theory*, an intriguing approach imposing the requirement of scale invariance on physical laws, extending the principle of relativity from states of motion to also include scale. Nottale showed, for example, that Schrödinger's equation, one of the central tools of quantum mechanics, can be derived from the assumption that spacetime is fractal. Nottale's ideas are quite interesting, but have not been widely embraced by the physics community.

Fractal structures may occur from the Planck length to the size of the entire universe, and maybe to the bubbling, branching growth of all universes. As far as we know now, a larger range of scales is, literally, impossible. "Fractals everywhere" may mean fractals *everywhere*.

3.13 Is this all?

The answer, of course, is "No." The list of fractals found in the world is immense, and growing. Any attempt at encyclopedic listing being doomed, we just picked a few of the topics most interesting to us and discussed these in enough detail to give some sense of how fractals are relevant. To give you a chance to explore at your own speed, we'll close with a fragment from our larger topics list, with an encouragement for you to exercise your Internet search muscles.

By attaching electrodes to the scalp and recording voltage fluctuations, electroencephalography is a method for detecting patterns of electrical activity in the brain. Electroencephalogram (EEG) traces are the results of the synchronized movement of ions across neuron membranes and can be read to reveal some of what goes on inside our heads. Based on a study of how rabbits store memories of familiar scents, fractals may play a role in how familiar memories are encoded. This is Walter Freeman's explanation for pre-attentive perception. If successive exposure scuplts the EEG-revealed memory into a fractal, then sparse perception (a part) can trigger the full memory (the whole). From a block away, we can recognize the face of an acquaintence, with far less detail than we'd need to describe a single feature of the face. Rabbits learn to recognize the scents of food, fleet foxes, and friends from afar, using fractals (A.17).

The shells of ammonites, extinct mollusks, have fractal sutures (Fig. 3.25, first image) between chambers. This has been known for many years, Raymond Moore's gorgeous *Treatise on Invertebrate Paleontology* (1957) shows hundreds and hundreds of detailed drawings.

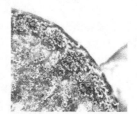

Figure 3.25: Ammonite and fingerpaint.

In 1992 G. Boyajian and T. Lutz published a quantitative analysis (using the box-counting dimension, described in Chapter 6) of the complexity of the sutures of all 615 genera of ammonites. They found an evolutionary increase in complexity, which Harvard paleontologist Stephen Jay Gould suggested was not the result of a hypothetical survival advantage conferred by greater complexity, but rather is the result of random genetic drift and what he called the *left wall* of simplicity. Because sutures separate chambers of the ammonite shell, they cannot be less complex than smooth curves. Random genetic drift can only increase suture complexity. This unidirectionality is Gould's left wall. Indeed, some descendants of genera with complex sutures have even more complex sutures, but some have less complex sutures (A.18).

Lightning and other forms of dielectric breakdown, electrodeposition, and viscous fingering all exhibit fractal patterns. For example, as seen in the second image of Fig. 3.25, models of viscous fingering can be easily produced with fingerpaints (A.19).

Swiss astronomer Fritz Zwicky thought outside many boxes. For example, in 1937 he deduced that the gravitational deflection of starlight proposed by Albert Einstein in 1917 and verified by Arthur Eddington in 1919 could cause clusters of galaxies to act as gravitational lenses, an effect observed in 1979. In 1933 Zwicky analyzed the motion of galaxies in the Coma galaxy cluster and deduced that much more matter must be present than is seen in the galaxies. He called this dark matter. Well, he called it *dunkle Materie*, dark matter is the English translation. Dark matter in our galaxy should cause its satellite galaxies to orbit in a roughly spherical distribution. Instead, they lie mostly in a disk. One explanation, proposed by Noam Libeskind and his coworkers, is that dark matter

3.13. IS THIS ALL?

is not distributed uniformly, but rather along a *dark web*, an immense network of filaments and nodes spread throughout the universe. In this model, dark matter on the web gravitationally attracts dust and gas to the nodes, forming large galaxies. Small galaxies form along the filaments, then travel along them to become satellites of large galaxies. Not everyone is satisfied with this explanation, but we find it fascinating: a fractal web of dark matter condensing galaxies and directing their slow waltz across the cosmos.

In nutrient-poor environments, some bacteria (for example, *Bacillus subtilis*) grow DLA-like colonies with fractal branches (A.20). Earthquake distributions plotted by position as well as frequency of earthquake magnitudes are fractal (A.21). EEG and electrocardiogram (EKG) patterns exhibit fractal scaling (A.22). So do sunspots (A.23), solar flares (A.24), solar prominences (A.25), and patterns of filaments in supernova remnants (Cassiopeia A is a good example) (A.26). The cross-section of Saturn's rings is a (fat) Cantor set, according to some analyses (A.27). Also, cracks in rocks (A.28), pores in soil (A.29), the distribution of resources across the Earth (A.30). Clouds meteorological and volcanic (A.31); all manner of atmospheric, oceanic, and extrasolar turbulence (A.32). Snowflakes, frost on a windowpane (A.33). Foraging paths of albatrosses and of mountain goats; search patterns of suburban kids playing hide-and-seek (A.34). The distributions of some epidemics in space and in time (A.35). Time records of ion channel kinetics (A.36). Fitness landscapes (A.37), Sewall Wright's pictorial representation of the fitnesses of related species, of nearby genotypes, sometimes are fractal over a significant range of scales. There are more fractals – many more – doubtless more when you read this than there were when we wrote it. But by now the point should be clear. Fractals are all over the place, under every rock and leaf, out into the dark space between galaxies.

Chapter 4

Manufactured fractals

Humans have been using fractals since long before we knew what they were, and now we are finding more uses than ever before. We build fractals for compact multiband antennas, microwave invisibility cloaks, more efficient capacitors and fuel cells, fiber optics with less distortion, image compression algorithms, camouflage patterns, even the infrastructure of the Internet itself. We'll sample some of these and a few others, but an encyclopedic treatment of applications of fractals would be ... an encyclopedia.

Some argue that manufacture, at least simple manufacture, is necessarily Euclidean. Straight lines and circles are easy to produce. But "easy" depends on perception and available skills. Making a knife by chipping bits from the edge of a piece of flint involves repetition, but performed along a line. If repetition had been directed to be across distance scales, whisks might look more like branches of trees, all shoes would have rough soles, in mechanical processors levers would have a distribution of lengths and connections. Some of what looks complicated to us now would have been simple. Expectation and perception inform one another in fascinating ways.

A billion years of evolution discovered architectures that are efficient, robust and, when viewed with appropriate geometrical tools, simple. How very different would our physical constructs be if ancient geometry had noticed the scaling of nature before Euclid hammered all our perceptions into straight lines and circles?

Industry will not have the last word in this chapter, as we will also survey some examples of fractals in music, literature, and art. Because they so clearly illustrate the range and power of fractals, these examples outside of science were especially important to Benoit, and are to us. Artists, writers, and composers use the logic of small structures recapitulating larger structures to image deep patterns of the world around us.

4.1 Chemical mixers

Chemical engineer Marc-Olivier Coppens of University College London has developed a project called Nature Inspired Chemical Engineering (NICE) to incorporate some of nature's solutions to mixing, transport, and reaction problems into chemical engineering design. NICE applies biological paradigms, but ensures they are more tunable than are most biological systems. Major components of this approach include

- hierarchical networks to span scales,
- mechanisms to balance forces at different scales, and
- the use of dynamical processes to organize behavior at different scales.

The first of these is manifestly fractal, so our example comes from this category.

For example, the *fractal fluid injector* is modeled on the lungs. Fluid flows through a central conduit, corresponding to the trachea, then into even smaller tubes that branch and branch and branch again. The fluid finally enters the reaction vessel at the ends of the smallest tubes, sometimes as small as one-tenth of the diameter of the central conduit.

Unlike standard injectors, where chemicals enter at a high volume from one central point, fractal fluid injectors introduce the new chemical from many outlets spread evenly around the desired space. The injected fluid leaves the outlets simultaneously because all outlets are the same distance from the inlet. This helps assure that the chemical reactions occur uniformly throughout the vessel.

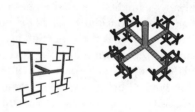

Figure 4.1: First: A planar injector at the bottom of a reaction column. Second: An injector designed to uniformly infuse a 3-dimensional reaction vessel.

The dimension of the injector in the second image of Fig. 4.1 is 2.6, less than 3 because the goal is to fill the 3-dimensional reaction vessel with the injected fluid, not with the fluid injector. One particularly interesting application of this technology helps increase efficiency of proton exchange membrane (PEM) fuel cells. In these, fractal injectors distribute air uniformly over the cell catalyst and drain water uniformly from the cell, streamlining the reaction process. Fuel cells likely will be an integral part

4.2 Capacitors

Capacitors, small electronic devices for storing energy, are essential components of our daily electronic lives, from cellphones, to computers, to food processors. Shown in cross-section on the top of Fig. 4.2, the basic capacitor design consists of two metal plates, usually discs or rectangles, separated by an insulator. An excess of electrons on one plate repels electrons on the other, resulting in a negative charge on one plate and a positive charge on the other. Electrical field lines between the plates indicate the density of the charges. By connecting the plates, a flow of electrons can be triggered from the negatively charged plate to the positively charged plate, producing a brief burst of energy.

Parallel plate Euclidean capacitors do not scale well for microelectronics: a narrower insulating layer can withstand only a weaker electrical field before the insulator breaks down. Lateral capacitors have a similar premise, but with a field penetrating the lateral space between side by side plates. Better still is the model of laterally paired parallel plate capacitors with an electrical field permeating the lateral and horizontal space between plates. More field lines mean more energy storage.

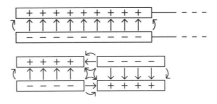

Figure 4.2: Sketch of the electrical field between capacitor plates.

Fractals enter the picture here. Because lateral flux depends on the perimeter length of the plate, fractal plate design which increases the perimeter for a given area can greatly increase the amount of energy stored in the system without increasing the size. In fact, the more wiggly the perimeter, the higher the perimeter to area ratio. In Chapter 6 we quantify wiggliness though the fractal dimension. For now, this intuition suffices: if its dimension is closer to 1, the perimeter is more like a smooth curve and has shorter length; if its dimension is closer to 2, the perimeter has wiggles off of wiggles off of wiggles

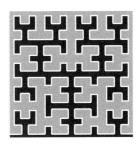

Figure 4.3: A schematic of a fractal capacitor. Fractal plates are black and gray.

and so has greater length.

Hirad Samavati and his coworkers show that with a horizontal spacing of 0.6μm, a physical (not simulated) fabricated fractal plate with boundary dimension of about 1.6 has over twice the capacitance of a Euclidean capacitor of the same area. Simulations show that an increase of the boundary dimension to 1.80 would give a capacitance five times that of the Euclidean capacitor. Illustrated schematically in Fig. 4.3, lateral capacitors consisting of fractal plates with interlocking boundaries are even better, achieving a 20-fold increase in capacitance for high-dimensional boundaries. See A.39 for more details.

This is a fascinating application: electronic component design improved by an interaction of electrical fields with fractal geometry. This is not the only such application. Wait till you see what comes next, and next after that.

4.3 Wide-band antennas

Antennas work when the varying electromagnetic field of radio waves causes the electrons of the antenna to oscillate in step with the waves, inducing a current along the length of the antenna. If the antenna length is some integer multiple of a half-wavelength of the ra-

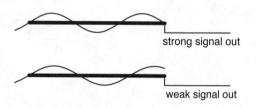

Figure 4.4: An antenna absorbing different frequencies.

dio signal, the antenna produces a strong signal, as seen in Fig. 4.4. Amplify this signal and all of a sudden you get the Fifth Brandenberg Concerto on *Sunday Baroque* in full-chorded glory. Adjust the antenna length and you'll find yourself listening to *What Do You Know* a few MHz away.

Of course, we don't usually change the physical length of the antenna. The effective number of wavelengths along the antenna, its *electrical length*, can be changed by adding electronic components to the antenna, an approach offering some success.

On the other hand, fractal antennas can be made to incorporate a range of length scales, making them more sensitive to a larger range of frequencies. Here the electrical length of the antenna is increased through the antenna geometry, without the addition of electronic components. The precise self-similar nature of these structures allows them to circumvent the classical design problems associated with complicated, compact antennas.

4.3. WIDE-BAND ANTENNAS

Several research groups, including those of Nathan Cohen and of Carlos Puente, have designed fractal antennas which are smaller than conventional (Euclidean) antennas. Fractal Antenna Systems, Cohen's company, holds patents on many fractal antenna designs, including cellphone and maritime antennas. A subtle symmetry of Maxwell's equations was applied by Cohen and Robert Hohlfeld to show that all antennas sensitive to a wide range of frequencies must be self-similar and symmetric about their center. Examples are shown in Fig. 4.5. An interesting recent development is the design of wearable broadband antennas, satellite and terrestrial communication fractal antennas that can be sewn into clothes.

Figure 4.5: Two self-similar and origin-symmetric fractal antennas.

Figure 4.6: Fractal antenna designs.

In Fig. 4.6 we see two designs by Fractal Antenna Systems. The left is a 3D-printed fractal space-filling curve. In addition to antennas, this design has been used for resonators and power and sensor shields. The right consists of 3D-printed Sierpinski carpet modules assembled into a paraboloid. The gain exceeds that of a Euclidean paraboloid twice as large.

Fractal antennas are a clever approach to a physical problem, an application in which self-similarity is the central design feature, not an instance of "Oh, wouldn't a fractal widget look cool!" Here fractality is central to the physics of how the antenna works. These are examples of strong applications of fractal geometry to manufacturing design. Some more details are in A.40.

Here are three applications of fractal antennas, recent when we finished writing this book:

- safety backup controls in the Boeing 787,
- Nextivity's active cellphone booster to fix dropped calls, and

- pickup antennas in subways, airports, universities, municipal buildings, to extend cell and mobile coverage.

And of course this story is just beginning.

4.4 Invisibility cloaks

On Tuesday, August 28, 2012, U.S. patent number 8,253,639 was issued to Nathan Cohen's group at Fractal Antenna Systems for a wide-band microwave invisibility cloak, based on fractal antenna geometry.

The cloak (Fig. 4.8) consists of an inner ring, the boundary layer, which prevents microwaves from being transmitted across it. This is the region that will be invisible to observers. Surrounding the boundary layer are six concentric rings that guide microwaves around the cloaked region to reconverge at the point antipodal to where they entered the cloak. Close-ups of both regions are shown in Fig. 4.8. Microwaves emerge from the other side as though they had traveled in a straight line, rendering invisible the cloak and any object inside it.

Figure 4.7: Cohen's microwave invisibility cloak and a closeup of the outer layer.

In late August 2012, Cohen's group successfully microwave-cloaked a person: Nathan's son Peter (Fig. 4.9). In these graphs, a gain of 0 indicates almost no microwaves are scattered. The radar beam splits around Peter, like a stream around a stone, seeing what lies behind him without some of the microwave signal being reflected back to the radar antenna. Effectively, the radar beam does not notice Peter.

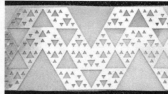

Figure 4.8: First: detail of the fractal pattern guiding microwaves around the target. Second: the boundary layer encompassing the target.

If fabricated at the sub-micron scale, instead of the current mm scale, this technology should act as an optical invisibility cloak. Combined with the flexible fractal antenna construction mentioned in the last section, we would have a wearable optical camouflage. Another application is mentioned in A.41.

4.5. FRACTAL METAMATERIALS

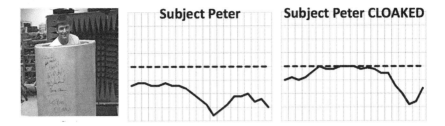

Figure 4.9: First: Peter in the microwave cloak. Microwave scattering of Peter without his cloak (second) and with his cloak (third). The horizontal scale is frequency, 500 to 1500 MHz; the vertical scale is gain, -10 dB to $+20$ dB.

Interesting times ahead. Is Harry Potter's invisibility cloak possible according to the laws of physics? Yes, using fractal design principles. Is optical invisibility an engineering reality? Not yet. But just wait a bit.

4.5 Fractal metamaterials

Substances can be fractal in space by exhibiting a hierarchical arrangement of pieces, or in time through dynamics occurring on no particular time scale. We'll give an example of each: some high-temperature superconductors are space fractals; lexan, the material of space helmets, is a time fractal. A final example, an extension of the fractal antenna and cloaking technologies of the last two sections, is a fractal metamaterial that moves heat.

Superconductivity, discovered in 1911 by the Dutch physicist Heike Onnes, is a quantum mechanical effect where electricity is conducted with zero resistance. So why aren't electrical transmission lines made of superconductors, zeroing the power loss between your house and the substation? Well, so far, superconductors only work if they are kept very, very cold. Until fairly recently, they had to be close to absolute zero (or about -460° F). The electromagnets at the Large Hadron Collider, instrument of the discovery of the Higgs boson, use superconductors cooled with liquid helium. This is all well and good for huge laboratories, not so useful around the house.

This situation improved a bit in 1986, when "high-temperature" superconductors, made of copper oxides, were discovered. These still required a brisk $-395°$ F to function, but recent experimental work has found superconductors that work at higher temperatures. This is an active field. We expect the situation will have changed between the time we write this book and you read it.

We're talking about superconductors in a book on fractals because in 2010 physicist Michela Fratini and her coworkers used a very sensitive scanning technique to study the structure of several copper oxide superconductors. They found that the crystal defects, the locations where the repeating pattern of the crystal is disrupted, exhibit fractal structure. These fractal patterns are the result of how rapidly the sample is cooled from high temperatures, so can be controlled to some extent. Especially interesting is the observation that the larger the scaling range of the fractal structure, the higher the temperature at which the material superconducts. We do not yet understand the mechanism relating fractals to superconductivity, but the importance of high-temperature superconductors makes this an interesting area to investigate.

Another area where fractal materials prove beneficial is space helmets. An astronaut wants a space helmet with two properties: it must be transparent, because the view of Earth from low orbit is wonder enough to last a lifetime, and it must be very strong, because a crack in a space helmet in low Earth orbit is very bad news indeed. To understand how strong space helmet plastics work, we look back to 1835, when physicist William Weber attached a weight to a long thread. The thread stretched almost immediately, then slowly stretched some more. This is called *stretched exponential relaxation*. In 1984, physicists Michael Shlesinger and Elliott Montroll proposed that stretched exponential relaxation can be explained by the migration of defects, irregularities in the material. A range of energy barriers to the movement of the defects gives rise to a range of relaxation times, from picoseconds to years for some polymers. Or put another way, relaxation times have no characteristic scale; relaxation occurs in fractal time. Lexan is made by cooling a liquid plastic, freezing an array of kinks in the molecular chain of the plastic. These kinks move along the chains in fractal time, allowing the lexan to absorb the mechanical energy of impacts.

Understanding diffusion in fractal time has applications back on the ground. When these energy-absorbing kinks reach the end of a molecular chain, they unkink and the material becomes more brittle. Modifying the chain ends could slow the aging process of polymers, essential components of many important constructions, in addition to space helmets.

Metamaterials are artificial constructs, assemblies of many tiny components arranged in particular patterns. Their operating properties are a consequence of the interaction of the components and the geometry of their arrangement. Applications include enhanced optical and acoustical resolution. Also, in 2013, Fractal Antenna Systems, which brought us the fractal cellphone antennas and invisibility cloaks of the last two sections, developed the ability to etch patterns of tiny fractal resonators on a surface, enabling the passive transfer of heat from one region to another. The possibility of an air conditioner with no moving parts, the ultimate in quiet technology, is awfully attractive on the absurdly hot July evening in New Haven when these lines were written.

4.6 Internet infrastructure

The Internet is the largest manufactured structure on the planet. Developed initially as a tool for sharing scientific information, the Internet has grown, in only a few decades, to encompass every aspect of human culture. All our science, almost all our literature (when we're writing this; when you read this, it may be all), art, music, history, personal stories, commerce – it provides an overwhelming snapshot of human life. But how big is the Internet?

In the November 2011 issue of *Scientific American*, the relative memory was given for some natural and manufactured information networks. Comparing the entries of Table 4.1 gives a sense of the astronomical scope of the Internet.

Human genome	750 MB, less than a laptop OS
iPad 2	64 GB (GB = gigabyte = 1000 MB)
Cat brain	98 TB (TB = terabyte = 1000 GB)
Human brain	3500 TB
Fujitsu K supercomputer	30,000 TB
Internet	1,000,000 TB

Table 4.1: Memory of the human genome, the internet, and some intermediate systems.

So the Internet is big, but how is it fractal? One way is through the distribution of Internet traffic. The statistics of voice traffic have been studied for years to inform construction of (landline) telephone infrastructure. These studies gave rise to an observation that the durations of most phone calls cluster about the average duration. The number of calls much shorter or much longer falls off very rapidly, decreasing according to the Poisson distribution, which is similar in shape to the familiar bell curve. Voice traffic is routed by *circuit switching*: when a phone call is placed, a circuit is opened between the phones and stays open for the duration of the call. Because of the low information rate of each call, many calls can share a common circuit.

A consequence of the Poisson distribution of voice traffic is that as we look over longer and longer time scales, the relative magnitude of the variations in the traffic drops. In Fig. 4.10 we illustrate this effect. Points on the first graph show the volume of traffic received per second, over a 60-second time span. Points on the second graph show the traffic received per 10-seconds, over a 600-second time span. The interpretation of the third graph is similar. Note that in each consecutive graph we have scaled both the time axis and the traffic axis by a factor of 10. The graphs flatten right out, so that the likely maximum range of variations is easily predicted. A network designed for three standard deviations above the mean never will fail from too much traffic.

The statistics for data traffic on the Internet are different, because they are transmitted by *packet switching*. The contents of a webpage, for example, are broken into autonomous packets. For each packet, routers find the best path from source to target, and the packets are reassembled when they arrive at the target computer that requested the webpage. Voice traffic travels an open pipeline for two-way communication; data traffic more closely resembles a swarm of bees, with each packet finding its own path between computers.

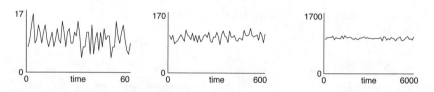

Figure 4.10: Poisson-distributed traffic, viewed over longer scales. In successive graphs, the scales of both the time (horizontal axis) and the traffic (vertical axis) are multiplied by 10.

As shown in Fig. 4.11, measurements of data traffic reveal that the longer the time scale, the larger the fluctuations. In this way, Internet traffic is self-similar, as revealed by its power law distribution. Most of this Internet traffic consists of very short emails and tweets (or whatever these have evolved into by the time you read this); some are larger files, webpages with complex embedded graphics; a tiny fraction are huge data sets or long YouTube clips. The same pattern can be seen in the distribution of webpage sizes, numbers of links pointing to or from webpages, and even the depth to which web users surf, demonstrating the deep fractal structure of the Internet. More examples are in A.42.

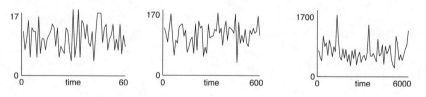

Figure 4.11: Fractally distributed data traffic, viewed over longer scales.

One practical application of the power law structure of the Internet has to do with Google's PageRank algorithm, which can sort billions of webpages very rapidly. Since running the search algorithms until they converge could take some time and computing power, Google instead exploits this self-similarity to find an approximate solution to ranking the pages it

4.7. MUSIC

discovers in your web search. Given the number of Google searches that are done every second, this is a practical application if ever there was one.

Visualization of the Internet's intersecting, complex networks can be beautiful and powerful. Walrus (not the animal), a tool developed in 1988 by Young Hong at CAIDA (the Center for Applied Internet Data Analysis), projects 3-dimensional graphs as a way to visualize this intricate macroscopic topology. A number of variables may be mapped in this way to produce different graphs. One approach, illustrated in the first image of Fig. 4.12, is to measure round trip time (RTT) the time between when a query is sent and a reply received. The graph is islands surrounding islands surrounding islands. These images are from
http://www.caida.org/tools/visualization/walrus/gallery1

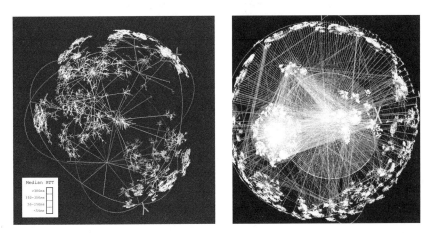

Figure 4.12: A round trip time (RTT) map and a connectivity map.

This spherical representation is a very effective, very clear way to accurately visualize such complicated networks. But there is another reason for studying these graphs: they are simply gorgeous. Making such breath taking geometry was not one of the goals of the architects of the Internet, because the Internet has no architects. This beauty is self-organizing, self-generating. What a wonderful world.

4.7 Music

Music is necessarily experienced in time, as a sequence of notes unfolding in the order in which they are played. So it may be difficult to perceive musical fractality without reading the score or listening to some piece again and again, until some aspects of the whole composition fit into our heads as a kind of graph or painting which may be explored outside of time.

Recognizing simple patterns in music can be satisfying and not too challenging. Most kids enjoy singing "Happy Birthday to You." Recognizing fractal patterns is much more difficult, since they enter through repetition across several time scales. Think of how you see the self-similarity of a Cantor set, all at once. Now imagine how you might do something analogous if you were listening to a Cantor set, trying to hold the whole pattern in your head as it trips by, and you'll see it's not so easy.

Thoughtful work by people on the border between math and music have found ways to understand, and sometimes even produce, fractal patterns in music. No single approach includes all the ways music can be fractal. This is no surprise: think of Bach, of Musorgsky, of Brubek, of Glass, of Death Cab for Cutie. Such a variety of styles begs for a variety of tools.

4.7.1 Harlan Brothers

Following a suggestion Benoit made to him, mathematician and musician Harlan Brothers has made a careful study of some ways in which music can be called fractal. In this study, "The Nature of Fractal Music," Brothers identified several categories of scaling in music. The meanings of some are obvious; we'll explain the others.

- *Duration scaling.*

- *Pitch scaling.*

- *Melodic interval scaling.* Melodies are sequences of pitches; melodic intervals are differences between successive pitches.

- *Melodic moment scaling.* Melodic moments are differences between successive melodic intervals.

- *Harmonic interval scaling.* Harmonic intervals are differences between simultaneous pitches.

- *Structural scaling.* The whole composition exhibits some form, melodic or harmonic for example, that is repeated on several structural levels of the piece.

- *Motivic scaling.* A melodic or rhythmic motif is simultaneously repeated on several time scales.

By plotting some value, such as duration or harmonic interval, against time, Brothers was able to find various fractal scalings in many different periods and genres of music. In the first bourrée of Bach's Cello Suite no. 3, structural scaling is at work in the repetition of musical phrase AAB (one passage, repeat, a different passage) over three levels. In addition, Brothers examined melodic interval and melodic moment scaling for the 36 Bach cello suites. In 14 of these he found evidence for the presence of these

4.7. MUSIC

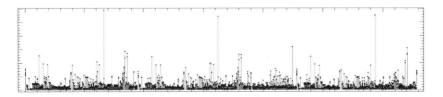

Figure 4.13: Duration plot of the Mozart Sonata in D Major, with time along the horizontal axis, duration along the vertical.

scalings. Also, Brothers studied duration scaling in Ellington's "Don't Get Around Much Anymore," Mozart's Sonata in D Major (K.284) (plotted in Fig. 4.13), and Ravel's *Le tombeau de Couperin*, obtaining reasonable power law scalings for all three composers. Such analyses are limited by the length of the data set – after all, we can't play more notes of the sonata than the composer wrote. Even so, this is an interesting way to look for fractals in music. In many places we look, we can find some sense of scaling in the layers of complex compositions.

4.7.2 György Ligeti

György Ligeti was a contemporary classical composer of serious pieces, some of which made the transition into popular culture. For example, portions of four of his compositions are included in the soundtrack of *2001: A Space Odyssey*. Ligeti won both the Wolf Prize and the Kyoto Prize.

About the mathematical aspects of Ligeti's work, John Rockwell quotes Ligeti:

> I do not use direct mathematical translation into my music. The influence is poetic: fractals are the most complex ornaments ever, in all the arts, like the *Book of Kells* or the *Alhambra*. They provide exactly what I want to discover in my own music, a kind of organic development.

We see Ligeti states, though does not describe how, he is guided by some aspect of fractals more general than those we have seen so far.

The citation for the Ligeti's 2001 Kyoto Prize mentions the influence of fractals on his work. In some form, every fractal is inherently complex. The fractality of Ligeti's music is revealed through its complex structure over many levels. Though recently some have attempted to quantify the degree of fractality of Ligeti's music by computing fractal dimensions of his compositions, we prefer to listen to the music and enjoy its levels.

4.7.3 Sergei Prokofiev

Martin Gardner, the journalist and author of *Scientific American*'s Mathematica Games column, recounted a surprisingly literal way that fractals in the environment gave rise to some of the earliest modern fractal music. Frank Greenberg described to Gardner how Sergei Prokofiev composed music for Sergei Eisenstein's 1938 film *Alexander Nevsky* using photographs from the film.

> Prokofiev then took these scenes and used the silhouette of the landscape and human figures as a pattern for the position of the notes on the staff. He then orchestrated around these notes.

Prokofiev converted visual to musical. Because the geological and weathering forces that build mountains work on a wide range of length scales, mountain profiles, and any musical composition directed by these profiles, necessarily will exhibit some scaling, some fractality, in time.

4.7.4 Heinrich Schenker

Heinrich Schenker was a German pianist, composer, and music theorist whose major work, *New Musical Theories and Fantasies*, presents a subtle and challenging method of musical analysis. Schenker's approach to musical analysis involved a dissection of the work into hierarchically ordered structural levels, these levels being ever-more-complex elaborations of some basis, from bar to theme to movement to concerto, for example. Schenker posited that great music exhibits the same form over several levels, though often these repetitions are concealed. Basically, though the language to express it was not yet invented, Schenker believed that great music has a hidden fractal form.

Central to Schenker's analysis is *motivic parallelism*. A *motive* is a melodic figure defined by pitch. Through expansion or contraction of this unit, a motive can appear at different structural levels of a piece. For example, in *Aspects of Schenkerian Theory*, David Beach describes a large-scale motivic expression:

> much of Beethoven's Sonata in A^b, Op. 110, is derived from relations stated in the opening four measures of the piece. (One might also note that the fourth measure contains a contraction of the middleground melodic structure of the entire phrase.)

In other words, the piece as a whole can be seen as an expansion of the opening notes.

In the same book, Charles Burkhart writes:

> A Schenkerian analysis of a musical work primarily reveals how that work is "composed" – that is, how its components may

4.7. MUSIC

> be viewed in terms of hierarchically ordered structural levels. ... [T]he theory sees the pitch organization of a work (or passage) as a series of progressively more complex elaborations of a simple foundation.

Though Schenker, who died in 1935, must have been unfamiliar with Cantor sets and Koch curves, he was sensitive to scaling, the central concept of fractality.

Schenker does not provide clear methodologies for locating these repetitions. Rather, this type of analysis, based on the recognition of subtle modulations, is mastered only with arduous study and musical intuition.

Perhaps the fractal aspect of Schenkerian analysis is most clearly stated by Schenker in *Free Composition*:

> Music was destined to reach its culmination in the likeness of itself.

4.7.5 Richard Voss and John Clarke

Another early example is the analysis conducted by physicists Richard Voss and John Clarke. Over a wide range of types of music, from Beethoven to jazz, Voss and Clarke observed that both volume and pitch fluctuations exhibit $1/f$ distributions: the size of the fluctuations has a power law dependence on the frequency of the fluctuations. Small fluctuations have high frequency and so occur often; large fluctuations have low frequency and occur much less often. As we mentioned in Sect. 3.5, distributions of this type often are associated with the hierarchical structure of fractals.

Voss and Clarke found $1/f$ distributions in loudness and pitch fluctuations for jazz and blues, classical and rock music, and talk radio. In addition, Voss found these correlations when analyzing the pitch of Ba-Benzélé music, Japanese traditional music, Indian ragas, Russian folk songs, and American blues; also in Medieval music, Beethoven's Third Symphony, Debussy piano pieces, a composition by Richard Strauss, and a composition by the Beatles.

Finding similar trends in so many different types of music is surprising, but Voss proposed an explanation. The Greeks believed art imitates nature. How this happens is relatively clear for painting, sculpture, and drama. Music, though, was a puzzle, until Voss speculated that music models the way nature changes in time: the variations of intensity of rain showers, gusts of wind blowing a field of flowers, the sound of waves lapping on a beach.

As described in A.43, Martin Gardner suggested another interpretation, based on the aesthetic balance of novelty and familiarity.

4.7.6 Charles Wuorinen

Charles Wuorinen, a highly regarded contemporary American composer, winner of a Pulitzer prize and a MacArthur Foundation grant, used similar ideas about the complex scaling of music to make highly formalized fractal compositions. In a time when so many composers turned to minimalism, the critic Michael Steinberg described Wuorinen as "maximalist through and through." Over the years Wuorinen developed a compositional language made up of scaled units whose logic formed a superstructure on which he built his complex compositions.

From early on, Mandelbrot had an interest in Wuorinen's music. When Wuorinen became aware of Mandelbrot's work, he recognized that self-similarity was a component of his composition process. He telephoned Benoit. "Hello, you don't know me, but my name is Charles Wuorinen." "Sir, I am well acquainted with your work, and am delighted to talk with you." This began an acquaintance that lasted for years.

In his essay "Music and Fractals", Wuorinen wrote that fractals in music arise in three ways:

- *Acoustically*: the signal has a $1/f$ power spectrum, as observed by Voss and Clarke.

- *Rhythmically*: the composition has a self-similar, hierarchical division of time; pitch generation is similar across several levels.

- *Structurally*: a single harmonic progression can determine a whole section, a portion of a section, or a single phrase.

Almost all of Wuorinen's work in the last four decades exhibits this last property, because he developed a compositional method based on nesting similar time and pitch structures within larger replicas of themselves. Of course, Wuorinen's composing is not the automatic generation of equations. Rather, this method is the preparation for composing, a kind of grid on which the final composition will be sketched. Then, Wuorinen says, he is free to compose creatively, with the large-scale coherence guaranteed.

And we must mention that *New York Notes: Music and Fractals* by Charles Wuorinen and Benoit Mandelbrot premiered on April 19, 1990, at the Solomon R. Guggenheim Museum in New York.

Music exhibits fractal patterns in other ways and with other composers. Of course, a great deal of nonsense has been written about fractals in music as well. And some people have even generated automatic fractal music by translating into music some of the patterns around the boundary of the Mandelbrot set or other fractals. While an interesting acoustical portrait of some aspect of these fractals, the examples we have heard are not musically interesting, nor are they fractal.

Some composers, including Wuorinen and Ligeti, deliberately build fractals into their compositions. Others, including Bach, Mozart, and Beethoven, intuitively noted, though maybe only indirectly, and without label or context, the fractal patterns that surrounded them, and guided some of their compositions to have fractal form. A theme repeated through this book is finding hidden patterns. Fractal aspects of music are patterns hidden by our sequential perception of music. What other fractal patterns will music reveal?

4.8 Literature

Like music, literature unfolds sequentially; unlike music, literature can be unfolded on the reader's schedule. We can stop, reread, skip ahead, refer to other books, muse on what we've read, discuss passages or chapters with friends. Thus freed from the rigid flow of time, literature can be more easily seen as a collection of patterns across scales. As far as we know, literary fractals may occur in at least three ways: as the subject, as the structure, and as a metastructure.

Here we survey some examples of *Fractals as subject*: fractals occurring in pieces by writers who are fascinated by geometry, by its power and elegance. Fractals serve as the subject of the writing, though often they stand in as metaphors for complexity or chaos in some other part of the narrative. For each example, we will give a quotation or two.

4.8.1 Stéphane Audeguy

In his novel *The Theory of Clouds*, Stéphane Audeguy explores many complicated themes, including, of course, clouds. Published in 2005, the English translation in 2007, the book contains this wonderfully clear description of scaling and the fractality of clouds, presented without using the word "fractal".

> If you zoomed in on one part of a cloud and took a photograph, then enlarged the image, you would find that a cloud's edges seemed like another cloud, and those edges yet another, and so on. Every part of a cloud, in other words, reiterates the whole. Therefore each cloud might be called infinite, because its very surface is composed of other clouds, and those clouds of still other clouds, and so forth.

Also of interest to us is the appearance of Lewis Fry Richardson, the British scientist whose study of how coastline length depends on the measuring scale was a key ingredient in Benoit's work on coastline dimension. In Audeguy's novel, Richardson appears primarily for his efforts to build a mathematical model of weather patterns with predictive power. Audeguy concludes his story about Richardson by saying Richardson's final project was an effort to model the causes of war. Benoit discovered Richardson's work on coastlines in precisely this project, his final publication, entitled "The problem of contiguity: an appendix of statistics of deadly quarrels."

The Abercrombie Protocol, Audeguy's book within a book, that began as a photographic atlas of clouds but then moved off in an entirely unpredictable direction, concludes:

> The enormous number of natural forms seemed irregular. ... A geometrician asked to measure the length of the coastline of England would calculate its irregularities as so many tiny segments of straight lines connected end to end. ... The tiniest irregularity itself consists of even tinier irregularities, and so on, such that we would have to conclude that the coastline was infinite.

As we'll see also in Stoppard's *Arcadia*, Audeguy gives a fictionalized account of how a careful observer could have discovered some aspects of fractals years before Benoit actually did. This emphasizes the important role that good questions play in the course of history.

4.8.2 Jorge Luis Borges

Jorge Luis Borges wrote compact short stories fixated on geometric paradoxes of infinities and of logic. In "The Garden of Forking Paths," Borges folds a story about two labyrinths into a branching story that is itself a labyrinth, manifested in the branching choices that define our futures. Describing the works of fictional author Ts'ui Pen, Borges' protagonist muses,

> In all fictional works, each time a man is confronted with several alternatives, he chooses one and eliminates the others; in the fiction of Ts'ui Pen, he chooses – simultaneously – all of them.

The labyrinth the men discuss, a labyrinth in which all men would become lost, is both the real lost maze built by the ancestor and the story itself, which meanders down one of the many branching paths. In fact, Borges' model of the branching universe resembles a bifurcating fractal whose largest scale is limited only by the imagination.

We won't say more. Read this story. Read it half a dozen times. Think.

We'll give another example, one more among so very many we might have chosen. In his story "The Aleph," Borges writes a gorgeous description of the Aleph,

> the only place on earth where all places are seen from every angle, each standing clear, without any confusion or blending.

4.8. LITERATURE

But of course the Aleph is part of the earth, so in the Aleph we see the Aleph, in which we see the Aleph, and on and on. Borges' Aleph echoes the monads of Leibniz (Sect. 4.8.6). For a less imaginative writer, the Aleph might seem more like a Russian doll than a Cantor set. But Borges' sleeves contain more tricks, for in his Aleph each thing appears from infinitely many points of view. Read his list of what he saw in the Aleph, but only after taking dramamine if you are prone to dizziness.

Borges, geometer of ideas, mapmaker of labyrinths mundane and subtle. The term "fractal" was still in the future when he wrote, but the ideas were as familiar to him as the branching veins on the back of his hand.

4.8.3 Arthur C. Clarke

In Arthur C. Clarke's *The Ghost from the Grand Banks* the Mandelbrot set makes several appearances, including as the shape of a lake outside a castle. Clarke imagined some peculiar psychological consequences of staring too long at zooms into the Mandelbrot set. He coined the term "Mandelmania" to describe the phenomenon wherein repeated viewings of deep zooms into the Mandelbrot set trap people in an internal world. This, of course, is fiction.

Clarke's interest in fractals was genuine, so much so that he agreed to narrate the film *The Colors of Infinity*, a guide to some visual aspects of fractals. Tangentially, the older author must mention that much of his interest in science was cemented by a copy of Clarke's *The Exploration of Space*, acquired during a wonderful family trip to the National Radio Astronomy Observatory at Greenbank, West Virginia, in the summer of 1961 – another example of the way cause and effect are nested within a web of connections and resonances.

4.8.4 Mark Cohen

There's no surprise that fractals appear in Mark Cohen's *The Fractal Murders*, with correct mathematics, including a description of Benoit's Trading Time Theorem, an important part of his study of the fractal properties of finance. In Cohen's novel, the Trading Time Theorem is extrapolated to have some predictive power, which becomes the basis of the murders in the title. Benoit liked this novel, although he asserted that his multifractal finance cartoons have only statistical accuracy, not real clairvoyant insight. Rather, he viewed his cartoons as a warning that standard (non-fractal) finance models underestimate risk.

Cohen's book contains accurate descriptions of fractality, perhaps to introduce a wider audience to technical notions central to the story, or perhaps to present a more complete picture of the characters. Regardless of the reason, he does get the details right. For example, one character summarizes a longish description of fractals this way:

"Yes, that's the interesting thing about fractal objects: their pattern remains more or less the same no matter how closely you examine them."

4.8.5 Peter Høeg

Peter Høeg's novel *Smilla's Sense of Snow* has an uncommon appreciation for matters geometrical, and late in the story the Mandelbrot set makes a surprising appearance:

> At the top, up against the bulkhead, lie three photographs. Two of them are black-and-white aerial shots. The third looks like a fractal detail of the Mandelbrot set.

Earlier in the novel Høeg gives a wonderful description the inner space where all geometries, including fractal, reside.

> Deep inside us is geometry. My teachers at the university asked us over and over what the reality of geometric concepts was. They asked: Where can you find a perfect circle, true symmetry, an absolute parallel when they can't be constructed in this imperfect, external world?
>
> I never answered them, because they wouldn't have understood how self-evident my reply was, or the enormity of its consequences. Geometry exists as an innate phenomenon in our consciousness. In the external world a perfectly formed snow crystal would never exist. But in our consciousness lies the glittering and flawless knowledge of perfect ice.

Indeed, all of mathematics, all of science, lives in our heads. The world around us is complex far beyond our grasp. Since our consciousness first woke up, blinked, and looked around, so much of our observations and thoughts have tried to sort these impressions into categories we can understand. Science and mathematics may seem inevitable, unique, but they are not. What tiny changes in the circumstances of our ancient ancestors would have given us an entirely different view of the world, a shadow projected along different axes?

4.8.6 Gottfried Wilhelm Leibniz

About a decade before your authors met, the older author and Benoit often discussed historical precursors of fractals, early efforts that revealed some aspect of fractality. Shortly before permanently relocating to New Haven, the older author recalled the monads of German philosopher Gottfried Wilhelm Leibniz. The Leibniz volume already was packed away in one of several hundred boxes of books, so Leibniz was postponed.

Benoit was away the summer of the move. When I found the Leibniz volume, I brought it to school and read *The Monadology*, only 22 pages containing 90 numbered paragraphs. The relevant paragraphs, capturing the essence of Leibniz's self-similar world, are these.

> 67. Every portion of matter may be conceived as like a garden full of plants and like a pond full of fish. But each branch of each plant, every member of an animal, and every drop of fluid within it, is also such a garden or pond.
>
> 68. And although the ground and air which lies between the plants of the garden, and the water which is between the fish of the pond, are not themselves plants or fish, yet they nevertheless contain these, usually so small however as to be imperceptible to us.

I put Leibniz back in my book bag. Later that morning, having cut his trip short, Benoit appeared in my office, and asked me to proofread an essay he'd written. In the essay, Benoit mentioned Leibniz, but referenced the wrong paragraphs.

When Benoit returned about an hour later, I mentioned the paragraphs. Leibniz came out of my book bag and I pointed out the relevant paragraphs. Benoit looked around the office – desk, laptop, two chairs, empty bookcase – and said, "You didn't expect me back for a couple of weeks. How in the world did you know to bring Leibniz today?"

It was just a coincidence, nothing more subtle, the kind of thing that happens every day. But sometimes my sense of humor misbehaves, "You know, Benoit, time is more complicated than even you can imagine." He looked puzzled, maybe slightly worried, and muttered something I didn't catch. But he said nothing more about it that day, or, as it turned out, any other day.

What did Benoit really think about this exchange? His sense of humor was subtle; maybe he was having a joke. Perhaps my joke on him contains his joke on me, maybe many others. Reminds us of monads. Or fractals.

4.8.7 Flann O'Brien

Flann O'Brien's brilliant novel, *The Third Policeman*, is filled with remarkable surreal images, presented in the most ordinary, matter-of-fact fashion. The part relevant to fractals is O'Brien's interpretation of the atomic theory of matter.

> "Now take a sheep," the Sergeant said. "What is a sheep only millions of little bits of sheepness whirling around and doing intricate convolutions inside the sheep? What else is it but that?"
>
> "That would be bound to make the beast dizzy," I observed, "especially if the whirling was going on inside the head as well."

The atoms of sheep are little sheep, a view echoed unintentionally by a student in the first iteration of the older author's fractal geometry course at Yale. At the end of the semester, the student admitted that he had expected the lesson of fractals would be that the atoms of a sheep were tiny sheep. When asked, the student said he was unfamiliar with O'Brien's writing; sheep had just popped into his head.

O'Brien's comic view of the world, or more precisely, of Ireland, is off-center, to say the least, and often takes a direction unfamiliar to most readers. But even so, he finds fractal aspects in his world. Don't think sheep and sheepness is an isolated example. Later in *The Third Policeman* we find another instance: the exchange of atoms of bicycles with those of their riders, with the consequence that riders show some traits of bicycles, and bicycles some traits of riders. The atoms of bicycles don't look like bicycles, but they act like bicycles – hint of a fractal geometry of behavior.

4.8.8 Richard Powers

A novelist with unusual dexterity in matters of science, mathematics, and technology, Richard Powers writes inventive stories of dazzling complexity. In *The Gold Bug Variations*, Powers investigates fractals in the music of J. S. Bach as well as in the structure of DNA. In Powers' imagination, Bach's music inspires the unraveling of the structure of DNA.

> I began to hear, too late, how the Base's symmetry ripples through the piece, unfolding ever-higher structures, levels of patterns, fractal self-resemblances.

And later:

> The double helix is a fractal curve. Ecology's every part – regardless of the magnification, however large the assembled spin-off or small the enzymic trigger – carries in it some terraced, infinitely dense ecosystem, an inherited hint of the whole.

One thread of Powers' later novel *Plowing the Dark* concerns the collaboration of an artist and a computer scientist to construct a convincing virtual reality environment, made up of the shapes of nature brought to a manageable level by fractal coding.

> Over the course of more makeshift sessions, he showed them how [to grow the image of a leaf]. He drew up genetic algorithms: fractal, recursive code that crept forward from out of its own embryo.

Fractal forgeries of natural objects achieve their economy by replicating – formally and geometrically, if not biologically and physically – how these objects grow, rather than how they appear.

Powers' fiction shows some ways that geometry – and explicitly fractal geometry – reveals the hidden dimensions of nature and of our own lives.

4.8.9 Olive Schreiner

Olive Schreiner gives beautifully clear descriptions of natural fractals in her memoir, *The Story of an African Farm*, written in 1883, but reading more like 1983. With elegant intuition, she addresses hierarchical structures and the universality of fractal processes. Although we have given this quote earlier, it is so remarkable that it's worth repeating here.

> A gander drowns itself in our dam. We take it out, and open it on the bank, and kneel, looking at it. Above are the organs divided by delicate tissues; below are the intestines artistically curved in a spiral form, and each tier covered by a delicate network of blood-vessels standing out red against the faint blue background. Each branch of the blood-vessels is comprised of a trunk, bifurcating and rebifurcating into the most delicate, hair-like threads, symmetrically arranged. We are struck with its singular beauty. And, moreover – and here we drop from our kneeling into a sitting posture – this also we remark: of that same exact shape and outline is our thorn-tree seen against the sky in mid-winter; of that shape also is delicate metallic tracery between our rocks; in that exact path does our water flow when without a furrow we lead it from the dam; so shaped are the antlers of the horned beetle. How are these things related that such a deep union should exist between them all? Is it chance? Or, are they not all the fine branches of one trunk, whose sap flows through us all? That would explain it. We nod over the gander's inside.

4.8.10 Tom Stoppard

Taking place on the fictional Coverly family estate in England, Tom Stoppard's play *Arcadia* includes some scenes in the 19th century, centered on the mathematical prodigy Thomasina Coverly and her tutor Septimus Hodge; other scenes are set in modern times, when the the Coverly heir, Valentine, is using the estate hunting books to search for chaotic patterns in the grouse population. Central to all the contemporary investigations is the uncovering of Thomasina's lesson book, complete with notes on her mathematical discoveries, including elements of current chaos theory and fractal geometry, or, as she writes, "the New Geometry of Irregular Forms discovered by Thomasina Coverly." This mathematical topic was a brilliant choice by Stoppard, because much of these fields can be developed with elementary mathematics, if only appropriate questions are asked. We think this is Stoppard's salute to Mandelbrot: compare Thomasina's statement

> Mountains are not pyramids and trees are not cones. God must love gunnery and architecture if Euclid is his only geometry.

with Benoit's iconic

> Clouds are not spheres, mountains are not cones, coastlines are not circles, and bark is not smooth, nor does lightning travel in a straight line.

Stoppard's grasp of science is solid; his descriptions of fractals, chaos, and thermodynamics are correct, elegant, and appropriate for both time periods. Another passage, especially lovely, describes a sequence of magnifications of the Coverly set, evidently the fictional equivalent of the Mandelbrot set. Describing a sequence of magnifications, Valentine says

> In an ocean of ashes, islands of order. Patterns making themselves out of nothing. I can't show you how deep it goes. Each picture is a detail of the previous one, blown up. And so on. For ever.

In Sect. 4.8.17 we'll speculate on how Valentine, so familiar with fractals, might not have known about the Mandelbrot set before reading Thomasina's lesson book.

One final note. We'll see the largest feature in the Mandelbrot set is the Main cardioid, a heart-shaped curve. The character who reads Thomasina's lesson book and writes the program to plot and zoom in on the Coverly set is named Valentine. We suspect this is not a coincidence.

We have not exhausted the kinds of ways in which fractals occur in *Arcadia*, and will continue our visit there when we study fractals as structure in Sect. 4.8.17.

4.8.11 John Updike

In John Updike's *Roger's Version*, a computer science graduate student lectures on cellular automata, fractal trees, the Koch curve, and the Mandelbrot set. Of course, Updike gets the math right. For example,

> A tree, like a craggy mountain or a Gothic cathedral, exhibits the quality of "scaling" – its parts tend to repeat in their various scales the same forms.

In fact, Updike's graduate student is searching for the face of god, which he believes he will find hidden deeply inside the Mandelbrot set.

Whether or not you believe in a god or gods (your authors do not), perhaps it is not so hard to believe that the infinite variations of the Mandelbrot set are rich enough to encompass everything, whatever that means to you.

4.8. LITERATURE

4.8.12 Kate Wilhelm

Kate Wilhelm's *Death Qualified* involves a psychological experiment through which viewing fractal images alters the archetypes of perception, opening views of amazing new worlds. This notion is similar to Clarke's *Mandelmania*, but Wilhelm pushes the idea in a different, more frightening direction. The story includes computers, chaos and fractals, and the Mandelbrot set. Because the images are so important, Wilhelm gives a description more complete than others written at this time.

> The screen cleared and a magnificent Mandelbrot set appeared with golden seahorse figures, and pale green dragons, flaring bands of fluorescent blue, and silver filigree. He zoomed in on a tiny section of the border to reveal the self-similarity again, and again. "It's infinite," he said softly ... Science has done a complete flip-flop. Reductionism is dead, holism lives. It's a brand-new game we're into."

Later in the story, we find that perception, even physical interaction, can be modified by staring at some particular sequences of images of the Mandelbrot set. The perceptual side of this did seem to occur in the late 1980s, as sequences of ever-higher magnifications into the Mandelbrot set mesmerized students and their parents, producing not so much oohs and ahhs, but silent staring without blinking. Many people unfamiliar with this sort of computer graphics experienced a minor tectonic shift in their thinking. Wilhelm's story dramatizes this behavior far beyond anything we have observed, of course. Or perhaps we just haven't looked at the right sequence of pictures.

Yet.

4.8.13 xkcd

Despite careful study of the entire coupus, we could find no reference to fractals in *Calvin and Hobbes*, and only one in *The Simpsons*, though we did find two references to chaos in *The Simpsons*. (We admit that we are not *Simpsons* completists, so may have missed something.) One is a statement of sensitivity to initial conditions in the episode "Time and Punishment". We'll discuss this in Sect. 7.2. The other occurred in the "Itchy & Scratchy Land" episode. Here Prof. Frink warns the park engineers that according to chaos theory the animatronic robots will turn on their masters. This may be a reference to the comment by Jeff Goldblum's character in *Jurassic Park* about the implications of chaos theory for the stability of the park. Chaotic, but not fractal. The fractals reference was peripheral, in the episode "Little Girl in the Big Ten," a project by two college students who befriend Lisa. Nothing more than a mention, really.

However, we hit gold with *xkcd*. In retrospect, this is not a surprise at all. Randall Munroe's brilliant, imaginative cartoons live deeply in the worlds of computer science, math, physics, astronomy, and many other interesting places. He is so scientifically literate, how could his view of these worlds not include fractals? Here are some examples. We're giving the urls to encourage you to visit the *xkcd* site if you haven't already.

http://xkcd.com/17	what if
http://xkcd.com/124	the blogofractal
http://xkcd.com/195	map of the internet
http://xkcd.com/543	Sierpinski valentine
http://xkcd.com/832	tic-tac-toe
http://xkcd.com/1095	crazy straws

Sierpinski gaskets, circle inversion fractals, and other patterns repeating on many scales. Munroe gets it right, really right. Not just a word or a passing reference, his examples represent self-similarity correctly. Usually funny, sometimes a bit sad, always engaging, he has opened up a new dimension of fractals in literature.

Now we turn from fractals as subject to fractals as an organizing principle, *Structural fractals*. Recognizing these is more difficult than noticing the term "Mandelbrot set" in a text. On the other hand, when we do find something, it is more interesting, and perhaps more important, than a simple mention. While no one can write "Mandelbrot set" unintentionally, writers can embed fractal structures in their texts without being aware they are doing so. Adjust the rhythms and emphases so the text sounds right, and you may have placed your words on top of a fractal.

4.8.14 James Cutting

James Cutting and his students Jordan DeLong and Christine Nothelfer studied the distribution of shot durations – long for soliloquies, middle for conversations, short for rapid action – for 150 films produced between 1935 and 2005. Every genre has examples in which the shot distribution is close to $1/f$, a power law scaling characteristic of fractal relationships. For example, *The 39 Steps*, *Rebel without a Cause*, and *Die Hard 2* have shot distributions about $1/f^{.93}$, $1/f^{.88}$, and $1/f^{1.06}$, respectively. All the films near $1/f$ contained strong narrative elements, so one possible explanation for the prevelence of $1/f$ distributions is that film makers discovered, perhaps unconsciously, that the rhythms of nature are a familiar superstructure on which to assemble narrative elements. Voss proposed a similar explanation for the ubiquity of $1/f$ distributions in music.

Not all films studied exhibited $1/f$ distributions: *Airplane!* has $1/f^{.20}$ and *Mr. Roberts* has $1/f^{.002}$. At least as far as shot duration distributions are concerned, fractality does not appear to correlate to how much we enjoy a film.

4.8. LITERATURE

4.8.15 Demetri Martin

When he was an undergraduate at Yale, Demetri Martin was a student in the older author's fractal geometry class. Back then, each student did a project in the class. For his project, Demetri wrote a fractal palindrome. Called "Dammit, I'm mad," the entire poem is a palindrome, that is, it reads the same forwards and backwards. For example, the title of the poem, which is also the first line, is a palindrome. The fractality resides in the levels where we find palindromes: the whole poem, the first line, the second line, the middle line (minus the last letter), the next-to-the last line, and the last line are palindromes. Palindromes at a few more levels would support a stronger claim of fractality, but given the effort involved in writing this, we were happy to view it as a proof of concept. Benoit was delighted by Demetri's poem.

"Dammit, I'm mad," appears in the memorial volume for Benoit edited by the older author and Nathan Cohen. This is Demetri's tip of the hat to fractals and comedy.

4.8.16 Wallace Stevens

Psychologist Lucy Pollard-Gott found that the pattern of repetition of some words in some poems of Wallace Stevens suggests fractal structures. To detect these patterns, first Pollard-Gott represented a poem with a linear array of boxes, one for each word of the poem. Next, she selected a *dust particle*, a word important to the theme and repeated in the poem. Then she shaded each box corresponding to that dust particle and looked for geometric patterns in the resulting sequence. Poems "Flock" by Caroline Sydney and "Expressions about the Time and the Weather" by Caroline Kanner illustrate this construction; in both, words representing animals are dust particles – in boldface here.

Flock

In time, **partridges** (also **hens**) bracingly
splinter into **chicks**
pecking **insects**. Day takes hold, making a pecking thing
a waiting one.
Dove:
the **bird** to end waiting;
chordate, specifically **columbidae**;
something promised.

Expressions about the Time and the Weather

Such phrases contain
animals, often – **dogs**, especially.

Take: entre **chien** et **loup**.
This, to the French, means dusk —
the hour when **dogs**
become **wolves**?

Dusk, and also
rainstorms: **cats** and **dogs**.

Applying Pollard-Gott's method to these poems reveals good approximations of Cantor sets, signaling repetition across scales.

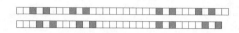

Figure 4.14: The graphs for these poems, first "Flock," then "Expressions."

An interesting pattern emerges involving the repetition of the dust particle **know** in Wallace Stevens' "The Sail of Ulysses (Canto I)." Graphed, the pattern is the top line of Fig. 4.15. We see it is a rough copy of the simple Cantor set shown below it.

Figure 4.15: Top: Dust particle pattern in "The Sail of Ulysses (Canto I)." Bottom: An approximate match with a Cantor set.

Certainly, longer works can reveal more levels of hierarchy. In addition to words, dust particles can be sounds or ideas. Units – the literary element represented by a box – can be words (the case at hand) or, for longer works, half-lines or lines. A pattern of sounds repeated across several scales can add to the perception of infinite depth: the pattern could repeat forever, on ever-smaller scales. A small collection of words can give a window on another world, invisible levels hinted by the scaling of repeated words. In his book *Wallace Stevens: The Making of the Poem*, Frank Doggett cites the feeling of inexhaustibility, of unending depth, often associated with reading Stevens.

Poems could be written specifically to include fractal scaling, but how much more interesting, how much more compelling, is the discovery of this scaling, when it occurs unknown to the poet.

4.8.17 Tom Stoppard, again

In addition to using fractals and chaos as subjects, Stoppard has built these concepts into the structure of his play *Arcadia*. In Scene 7, the longest scene and the dramatic climax of the play, characters from both time periods are present on stage together (though characters from one

4.8. LITERATURE

time cannot see or interact with those belonging to the other). Occasionally people from one time period have long discussions while people from the other time period remain silent, but more often the conversations among people from the two time periods are intermingled. They talk over and through one another. This led Josie Rodberg, a student in the older author's fractals class some years ago, to plot the changes of time periods of the speakers, as shown in Fig. 4.16. Some details of how these plots were generated are given in A.44.

Figure 4.16: Top: Lines from the present (dark) and the past (light). Bottom: Locations of the time period changes.

A signature of fractality is the precise way gap size is related to the number of gaps. Visually, these graphs have the right form; mathematically, they exhibit a power law scaling. Did Stoppard design this scaling distribution? Probably not. Perhaps an echo of natural fractals motivated this choice.

Another example is sensitivity to initial conditions, illustrated several places in the play. Perhaps the most glaring example is Valentine's surprise when he finds Thomasina's formulas for the Mandelbrot set. Certainly, Valentine is familiar with the main constructions of fractals and chaos, so why was the Coverly set, the thinly veiled Mandelbrot set, the best-known image associated with fractals, unfamiliar to him? How could someone who uses computers and knows some detail about chaos and fractals, not recognize the Mandelbrot set?

Our best answer is a conjecture, a guess really, about another facet of fractals and chaos, shown by the contingency of some historical developments. It's okay, we promise – we make a lot of home-grown musings. Some developments in science resulted from the combined efforts of many people but probably would have gone forward without any particular individual. For example, quantum mechanics in some form would have been developed without Heisenberg, or without Schrödinger, or without Pauli. Perhaps not in the same way we know it now, but still.

However, would fractal geometry have arisen without Mandelbrot? Examples of self-similar sets had been known to mathematicians for about a century when Benoit began the investigations that would occupy much of the rest of his life. Without his particular genius, who would have applied these ideas to develop a geometry of nature, a quantification of roughness? Though fractal geometry has grown to include very deep mathematics, high school geometry and algebra suffice to take anyone well into the field. Thomasina Coverly, whose ever-curious mind offers a good picture of how Benoit thought, could have discovered the basics of fractal geometry. We think Stoppard has imagined a history without Mandelbrot – maybe he was killed as a boy in occupied France, which could have happened – and in which fractal geometry was glimpsed by an English prodigy. But then she died before she could develop these ideas. Septimus, Thomasina's tutor, spent the rest of his life in isolation trying and failing to complete her work.

Perhaps Stoppard is presenting contingent history only for the example of fractal geometry. Bernard Nightingale's misreading, hilarious for us, not so much for him, of Lord Byron's activities at Sidley Park suggests – to us, at least – that Stoppard believes contingency to be thoroughly mixed up in all the world. How very delicately balanced on a knife's edge are so many of the choices in our lives. Think of how you happen to be reading this particular book right now.

The world is filled with complicated wonders.

Finally, we push our search for fractality to an even larger scale, *Metastructural fractals*.

In her analysis of Wallace Stevens, Pollard-Gott eventually zooms out, investigating fractal patterns over the longer scale of Stevens' works as a whole. A longer text can reveal more levels of a hierarchy, though the patterns can be harder to see. Nevertheless, we can intuit a larger shape.

Certainly, this is interesting, but let's zoom further out. Could the oeuvre of an author possess patterns exhibited by individual works? Author David Mitchell, perhaps best known for the meta-novel *Cloud Atlas*, sometimes allows his characters to wander from book to book. He has stated that he views all of his novels as "individual chapters in a mega-novel."

Taken literally, this scaling across an author's whole life would be remarkable: surely later works were not already in the author's mind when the first sentences of the first story are crafted.[1] But what if they were? Not consciously, to be sure, but reflecting, responding to, the slow rhythms of life. Perhaps first stories are generally bright, energetic, less experienced: *Look at me! I'm a writer!* Having had more experiences, learned more of how people live, the author's middle stories are mature, measured, more subtle. Informed by the collection of earlier works, the author's last stories are what? Satisfied? Melancholy? Perhaps enjoying the trip through the twilight. Looking into the forest that at dawn was cheerful and full of promise, and at noon was warm, peaceful, familiar, but now becomes dark and mysterious. *I used to know, or thought I knew, what was behind those trees. Now I'm not so sure.* A good ending for a poem, a book, a life's work.

---

[1] Although in 1856 Charles Dickens wrote, "when I was a very odd little child with the first faint shadows of all my books, in my head"

Is our approach to life – how curious we are, how much importance we place on helping others, that sort of thing – established early on? If it is, will this inform how we grow? Will we unconsciously build these same arcs into our writing? The characters and circumstances are introduced in the early morning of the story. Their development and interaction is the mature middle section. Even if pyrotechnical, the conclusion is when the author knows the curtain is coming down and the house lights coming up.

In a 1981 interview in *The Paris Review*, Gabriel García Márquez said,

> One of the most difficult things is the first paragraph. I have spent many months on a first paragraph and once I get it, the rest just comes out very easily. In the first paragraph you solve most of the problems with your book. The theme is defined, the style, the tone. At least in my case, the first paragraph is a kind of sample of what the rest of the book is going to be.

Does this pattern occur on other levels? Rereading *One Hundred Years of Solitude* with this question in mind will be our reward for shipping the final version of this book to our publisher.

Or maybe our lives sculpt our reading: The beginning of most books we've read have been mysterious, full of bright promise, because we don't know the characters or circumstances. By the middle we think we know the players and themes, maybe can recognize some of the subtle points the author includes. By the end, we're saying goodbye. Even if these characters continue in other volumes, the experiences contained in that story are coming to a close.

Zoom out further still. Does the literature of a culture or a period itself have an arc like that of an author – Fenimore Cooper to Theodore Dreiser to Don DeLillo? In *The Landscape of History. How Historians Map the Past*, John Lewis Gaddis wrote about a fractal geometry of terror, that Joseph Stalin's acts of violence and terror occurred at scales from individuals to millions of individuals. This is very gloomy, but does suggest a direction. Is there also a fractal geometry of narrative, in which stories reach out past individuals and become myth?

4.9 Visual art

Unlike music and literature, the experience of visual art is reliant on the viewer's choice of where to look first – whether to focus on a tiny detail in a corner of a painting or to step back to observe the flow of color and shape. Depending on whether we start on a small scale or on a large scale, fractal aspects of a painting must be extrapolated outward or interpolated inward to see all the scales. Could a delicate gradation in shade, or in shades, give a fractal pattern? Could a clustering of like shapes, or maybe even of shapes of similar narrative impact? So many possibilities.

What about kinetic art, sculptures moving by wind or machine, casting shadows that sometimes are fractal? Or optical sculpture, linked video monitors or something with flashing neon or reflected laser light? While fractals in some visual arts are perceived more easily than those in literature, others may exhibit similar indirectness. This arena is not so obvious as it might have at first appeared. We sample instances; as usual, they are just a few from among many.

4.9.1 T. E. Breitenbach

American artist T. E. Breitenbach gave this answer to our question about whether the evolution of an artist's work might reflect the evolution of art over some time period.

> Since every artist is influenced by someone else in the past or present, you could identify connecting strands of influence or relatedness. It would look more like a complex set of nets with strings reaching back in time, and crossing on the perpendicular in the present (and at all stages in time – the past presents). The nets would in some places be connected with one another or maybe be woven through each other in a tangle. At some places, the stitching of the nets would be excessively tight, dense, thick and swollen, swirling around the various larger "movements" of art; the Egyptian period, Medieval religious art, Dadaism, etc.
>
> I could envision a particular art Movement as the beginning formula for a fractal, which, then as you zoom into the edges representing the passage of time, some of these patterns continue, but they are gradually influenced or blended with fractal formulas representing neighboring or future Movements.

Art across time scales. Fractals in time, not in space, on canvas. Perhaps our practice in recognizing sequential fractals in music may help us perceive fractal patterns in the history of art.

4.9.2 Alexander Cozens

In his masterpiece *History of Art*, Russian-American art historian H. W. Janson recounts that British landscape painter Alexander Cozens felt that using actual landscapes as models "could produce only stereotyped variations on an established theme" and "did not supply the the imaginative, poetic quality that for him constituted the essence of landscape painting." Cozens' method was to crumple a piece of paper, flatten it out, and then blot ink on it. Some of the resulting patterns become the basis for an imaginative landscape. This works because crumpled paper

4.9. VISUAL ART

and landscapes exhibit similar (fractal) distributions of folds. Although Cozens did not know the language of fractals, he understood the concept.

This notion can be traced even further back. Janson recounts that in his *Treatise on Painting*, da Vinci suggested that artists can find inspiration for painting mountain ranges in stains on old walls. Inglorious natural patterns mimic patterns in landscape and inspire patterns in art. This is an early, if extreme, instance of scale ambiguity.

Often, Benoit commented that artists, and small children, understood the lesson of fractals without seeing the math.

4.9.3 Salvador Dalí

In Salvador Dalí's *Visage of War* (1940), a grimacing skeletal face's eyes and mouth each contain a face, whose eyes and mouth each contain a face, whose eyes and mouth each contain a face – a clear, if gruesome, example of fractals in art.

In 1940, Dalí was thinking of the Spanish Civil War, a source of images he found deeply frightening. In his book *Dalí*, Robert Descharnes described the painting as having "eyes filled with infinite death," referring to the recursive effect set in motion by self-similarity. The pattern of faces within faces is mapped by a kind of Sierpinski gasket.

Also in Descharnes' book is a preliminary study of the painting. Here only the mouth contains a face with smaller faces in eyes and mouth. One eye of the largest face is filled with rings of a tree trunk, the other with cells of a honeycomb. Although certainly disturbing, Dalí rejected this version because he recognized that it does not capture the notion of infinity so clearly implied by the self-similarity of the painting.

Benoit and the older author spent hours discussing this painting. Long before the language of fractal geometry was developed, Dalí had discovered the visual power of self-similarity. When the older author showed him the preliminary study, Benoit was silent for almost a minute, ages at the speed he thought. Finally, he said, "Michael, today you have earned your pay." For you see, Dalí did not adapt the pattern of a picture of the Sierpinski gasket he had seen. No, in thinking about infinity, he recognized that

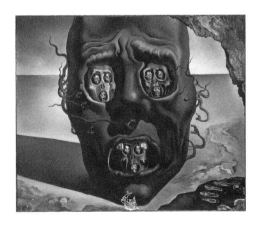

Figure 4.17: Dalí's *Visage of War*.

repeating the same pattern across scales, something we see so often in nature, is the best way to go. Benoit thought *Visage of War* is one of the clearest expressions of an artist recognizing self-similarity in nature.

4.9.4 Max Ernst and others

Decalcomania is best illustrated by a simple experiment. Put some fairly viscous paint – finger paint or oil paint from a tube work well – on a piece of stiff paper or canvas fastened to a table. Cover this surface with another surface, a piece of paper or canvas, glass or metal. Flatten and spread the paint by applying pressure to the top surface. When you are ready, pull the surfaces apart. The result, illustrated in Fig. 4.18, is not the simple smear you might expect, but a branching pattern of surprising (though perhaps not for the reader who has read this far) complexity.

Figure 4.18: Decalcomania.

If you were to repeat this experiment in slow motion, you would see that paint adheres to both the top and bottom surfaces as they are pulled apart, forming ridges between the surfaces. With increasing distance between the surfaces, the paint ridges coalesce across a range of scales and a branching pattern appears. As more and more ridges come together, a dendritic fractal forms.

Around 1935 this process was rediscovered and named decalcomania by Spanish surrealist painter Óscar Domínguez. Domínguez and Ukranian-American artist Boris Margo used paper. German artist Hans Bellmer, French surrealist Marcel Jean, American painter Enrico Donati, and French artist (not the economist) André Masson used canvas. French mathematician, poet, and collagist Max Bucaille created his decalcomanias on glass.

Perhaps the best-known examples are those of Max Ernst. Ernst used decalcomania to obtain richly textured images, evoking a dream-like atmosphere as a background for his haunting figures. Examples include *Mythological Figure - Woman* (1940), *Three Well-Tempered Cypresses* (1949), *Europe after the Rain* (1940-42), and *Blue Mountain and Yellow Sky* (1959). These surreal landscapes look like underwater vistas of rich, strange corals, all the more strange for their naturalism in a world of fantastical figures.

Next, here are examples from three student projects for the older author's fractal geometry class.

4.9. VISUAL ART

Natalie Eve Garrett

In her autumn 1998 project *Aesthetics in Visual Art and Fractals*, Natalie Eve Garrett put acrylic paint between pieces of gessoed paper. To (literally) include the artist's handprint in the final images, Natalie pressed her fingertips on the top sheet

> so that small explosions were visible in the fractal pattern. ... On the one hand, this plate remains objective in that it is governed by fractal structures. ... Yet it was important to me, as an artist, to recognize my role in the creative process and reference my hand in the work.

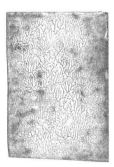

Figure 4.19: Decalcomania by Garrett, Geis, and Miller and Norris.

Tanja Geis

Tanja Geis devised another variation in her autumn 2001 project *Decalcomania*. She explored how variations in the pressure on the top paper altered the branching patterns. Removing, rotating by 90°, and replacing the top paper, perhaps with several repetitions, had the effect of breaking the long branches and randomizing the orientations of the smaller branches.

The next effect, pictured in the middle of Fig. 4.19, surprised the older author. After she applied pressure to the upper paper, Tanja scored the paper with the blunt edge of a knife. In every example, the thinnest ridges bent to intersect perpendicularly the pressure scorings. These results were found by play. *Here's the basic set up. I wonder what happens if I do this?*

Claire Miller and Cara Norris

In their autumn 1999 project *Fractal Painting, au Natural*, Claire Miller and Cara Norris used fingerpaint and glossy fingerpaint paper. They applied paint to both pieces of paper before placing them together, without pressure, with only gravity settling the top piece onto the bottom. Because the top sheet is lifted off gently, we find delicate, multiple branching throughout the picture. In an interesting variation, sometimes they replaced the top paper on the bottom, as many as 15 times, obtaining a more uniform branching pattern. Basic decalcomania is quite simple, but the range of variations is substantial.

As the boundaries of science have pushed further from the scale of people, the amazing work of the Curiosity rover on Mars, the Hubble Space Telescope, the Large Hadron Collider, have made science a very serious business. We've lost the playful aspect of science, moments of unfettered imagination, curiosity (not on Mars, but in the space behind our eyes) – our most important emotion. Because fractals span so many scales, they are found at human size, and in ways our minds have not yet learned to recognize easily. They hold a surprise around every corner. Better: between every pair of surprises, another waits.

4.9.5 Augusto Giacometti

One of Benoit's favorite artists, Augusto Giacometti, studied art in Zurich, Paris, and Florence. Of particular interest are his paintings inspired by fields of alpine flowers pushing through the snow. Unlike the translational symmetry of a field of daisies or phlox, these alpine flowers exhibit scaling symmetry, floral fractals. Giacometti abandoned traditional representationalism, instead applying thick blobs of paint to the canvas, producing paintings that consist of only a distribution of large and small patches of color. These and a series of similar pastels, some painted over a decade before Wassily Kandinsky's first abstract works, are the basis of some art historians' claim that Giacometti was the first abstract painter. Giacometti certainly had discovered – and expressed – the view that natural scenes are made of ever more copies of ever-smaller details, without using the language of scaling relations.

4.9.6 Katsushika Hokusai

Perhaps best known in the West for *In the Hollow of a Wave off the Coast of Kanagawa*, the Japanese artist Katsushika Hokusai made wonderful woodblock prints that present fractal aspects of nature with a sophistication rarely matched even today. Benoit was very impressed with Hokusai's work. He often cited Hokusai as someone with a careful eye noticing the salient features of fractals, without being familiar with their mathematics.

In Fig. 4.20 we see *Fuji from the Seashore*. Note the roughness in many places and across many levels, not simply irregular curves, but hierarchies of irregularities in the froth of the wave, in the frilled tangles of seaweed on the beach.

4.9. VISUAL ART

This complexity, the result of Hokusai's careful eye and hand, reminds us of a filled-in Julia set, a construction we will explore in Sect. 5.3.

In Fig. 4.21 is Hokusai's *Fuji in a Thunderstorm*. Again we see wiggles upon smaller wiggles upon still smaller wiggles, in the clouds and profiles of the trees, rocks and shrubs, even in the distribution of spot sizes. Interestingly, the broken lines on the left panel represent lightning. Compare these clumsy strokes to the wonderful observational detail in other parts of this picture. Of course, Hokusai worked before photography had been developed, so he had only the briefest of glimpses of lightning to guide his drawing. The fractal nature of lightning bolts may not have been noticed by Hokusai.

Figure 4.20: Hokusai's *Fuji from the Seashore*.

Or maybe not. Even a glimpse of lightning reveals something more complicated than straight lines, even if it's not clear how magnificently complex lightning really is. Compare the lightning to the simplistically drawn mountain, however, and there is a sense of artistic intent – the complicated detail of nearby life and the abstraction into essential forms of that which is far away .

We are happier with this explanation. It is consistent with Benoit's assessment that "Hokusai got fractals."

Figure 4.21: Hokusai's *Fuji in a Thunderstorm*.

4.9.7 Jackson Pollock

Created between 1943 and 1952, Jackson Pollock's drip paintings were an important step in the development of modern art. Placing a large canvas on the floor of his barn, Pollock dripped and drizzled paint onto the canvas while moving around it. In *History of Art* Janson writes this about Pollock: "He is himself the ultimate source of energy for these forces, and he 'rides' them as a cowboy might ride a horse, in a frenzy of psychophysical action." The drip paintings are complex, intricate, layered. Are they fractal?

In their 1999 *Nature* paper, Richard Taylor, Adam Micolich, and David Jonas reported on their quantification of the complexity of some of Pollock's drip paintings through a fractal dimension called the box-counting dimension. We'll study this in Chapter 6, but for now will say just that to compute this dimension we cover the shape with boxes and seek a power law relation in how the number of boxes grows as the box size shrinks. For drip paintings presenting layers of different colors, the color layers are digitally separated and the box-counting dimension of each is measured. Combining all layers gives the dimensions of the whole painting.

The calculations of Taylor and his coworkers revealed two power law scalings, suggesting separate dimensions in each regime. That for smaller boxes, determined by the dynamics of paint dripping, is the *drip dimension*; that for larger boxes, determined by Pollock's motion around the canvas, is the *Lévy dimension*, reflecting the resemblance of the artist's motion to a Lévy flight, a type of random fractal we'll describe in A.10. For now, foraging paths of deer are a good example of Lévy flights.

Pollock's drip paintings are divided into three periods: preliminary (1943–45), transitional (1945–47), and classic (1948–52). Paintings from the preliminary period have lower drip dimensions, around 1.1, meaning his paintings are only slightly more mathematically complicated than smooth curves. Drip dimension rises during the transitional period, achieving a maximum in the classical period. For example, *Autumn Rhythm: Number 30* (1950), has a drip dimension of about 1.7.

Pollock was not so well known when he began developing his drip painting technique. He does not appear to have kept careful records of his early work, and rumors suggest that he may have sold some of these earlier paintings. So if a previously uncatalogued drip painting is found when cleaning out an old house on Long Island, the question of whether this is an early Pollock must be taken seriously. Taylor, Micolich, Jonas, and coworkers proposed that their dimension analysis be added to the toolbox of those seeking to authenticate Pollock paintings.

To test how well the observed fractality of Pollock's drip paintings characterizes Pollock's style, 37 University of Oregon undergraduates produced drip paintings in an attempt to match Pollock's work, but none of their works matched the dimensions calculated for his paintings. Analysis of 14 drip paintings by unknown artists gave similar results.

However, the approach of Taylor's group was not embraced universally. For example, in a 2006 paper in *Nature*, Katherine Jones-Smith and Harsh Mathur criticized Taylor's analysis because not enough levels of scaling were found in either the drip dimension regime or the Lévy dimension regime, even though box-counting was applied over three orders of magnitude. In addition, Jones-Smith objected that applying box-counting to a simple drawing of five-pointed stars gave power laws having two regimes. Taylor responded by showing that other simple drawings of stars give no straightforward power laws at all. Response and response to response followed, the arguments ever more intricate. We have not found a claim by Taylor that fractal analysis should be the only, or even the most important, tool in authenticating Pollock's drip paintings, so we agree with

4.9. VISUAL ART

Benoit's statement that his is an interesting direction that "deserves to be pursued further."

4.9.8 Rhonda Roland Shearer and others

Several scholars have noted that the two greatest revolutions in art, the Renaissance and the birth of modern art, were catalyzed by artists thinking about new geometries: perspective geometry for the former, non-Euclidean and higher-dimensional geometry for the latter.

Williams College art historian Samuel Edgerton makes a convincing argument about the open relation between Renaissance painters and mathematicians. Some Renaissance artists – Filippo Brunelleschi and Leon Battista Alberti, for example – wrote about perspective painting, but the lead came from geometry. University of Texas at Austin art historian Linda Dalrymple Henderson thoroughly explores the influence of non-Euclidean and higher-dimensional geometries on the birth of modern art. Indeed, Henderson states that these geometries, and not relativity, were the main scientific influence on modern art. Dalí's *The Crucifixion (Corpus Hypercubicus)* (1954) uses an unfolded hypercube for the cross.

In turn, fractal geometry may offer a new set of vistas for art to explore. Artist Rhonda Roland Shearer uses her own sculpture to combine Euclidean geometric objects with the natural fractal forms of plants, drawing attention to different registers of artifice. In her essay "Chaos Theory and Fractal Geometry: Their Potential Impact on the Future of Art," Shearer notes that fractal geometry fits many of Thomas Kuhn's eight traits of scientific revolution. For example, fractal geometry is a new language for the irregularity of nature, gives a new perspective on part/whole relationships (through self-similarity), and is a powerful new cultural icon. If you doubt this last point, consider the number of book jackets, and biker biceps, sporting the Mandelbrot set.

Shearer, Breitenbach, the artist Javier Barrallo, and others claim something stronger than isolated artists using fractals, consciously or not. They see a relation between part and whole on a level more subtle than how the small branches of decalcomania resemble the large branches. This pattern involves many artists and many time scales. It can be appreciated only after some study.

Once we learn to recognize obvious physical fractals, the next challenge is to train the eye and mind to notice patterns in pieces separated in time, in space, and by artist. Music is one place to sharpen these skills; visual art, we see, is another.

4.10 Building fractals

So much of our manufacture rests on lines, planes, circles, and spheres, shapes seen as simple in Euclidean geometry. Geometry, a word that translates literally as "the measurement of earth," grew from practical issues like surveying rich farmlands after the annual flooding of the Nile. Except for variations imposed by terrain, boundaries were straight line segments, farms were the insides of polygons. Greek geometry abstracted these, focusing on archetypes inhabiting a Platonic universe. Nature was these Euclid archetypes, locally roughed up.

Much of nature does reinforce this: sun and moon appear to be circles, a honeycomb is a grid of hexagons, the trunk of a tree is almost a cylinder. Close inspection reveals variations, but that's built into this worldview. With the geometry of Euclid as our language, it is little wonder that manufacture focused on lines and circles. Indeed, building anything else requires some effort, but is this such a surprise, given the geometry on which industry is built?

But all this might be about to change, because 3D printing, programmable and working on small scales, can produce fractals easily. As 3D printing technology evolves to accommodate more forms of manufacture, fractals may become a more common part of the built world.

How would the world have been different if the first observers of nature understood detail as well as general trends? If the first description of a tree had been "It's made up of little trees," instead of "It's a cylinder with smaller cylinders sticking off it," would our world be different now? If magnification symmetry were the first symmetry we learned, would our manufacture be based on scaling? Would fractal antennas and chemical mixers be obvious and old tech? Would the structure and growing instructions of the lungs be a short chapter in high school biology texts?

If you think the ways we learned to manipulate the world were imposed by immutable historical forces, that nothing of consequence could have been different, look at the arrangement of keys of your laptop.

Chapter 5

The Mandelbrot set: infinite complexity from a simple formula

The first pictures of the Mandelbrot set were introduced to the general public in August 1985, in A. K. Dewdney's "Computer Recreations" column of *Scientific American*. The result was a tectonic shift in geometry, in computer graphics, and in popular culture. The images are breath taking, complex without apparent end, baroque, rococo, spirals made of spirals made of spirals, whorls within whorls within whorls, continuing to the infinitesimally small; richly varied, still each shape hints at some common structure. Many have assumed that this complexity must come from complicated formulas. In fact, the formula is one very short line, the computer code not much longer. Anyone familiar with loops can write a Mandelbrot set program, and many have, from school children working on computers in their parents' kitchens, to mathematicians commanding powerful parallel processors at large universities. Images generated by these programs have led to hard, hard problems in pure mathematics, which continue to fascinate top mathematicians working today. Perhaps even more importantly, mathematics – and interesting, beautiful mathematics at that – was brought into people's homes. These pictures built a bridge between MathWorld and the world of the curious public. This is an immensely important – some have argued, the most important – consequence of fractal geometry. When asked later if he had foreseen this, Benoit replied, "Well, I had hoped."

In this chapter, we will survey pictures of Julia sets and the Mandelbrot set and find some of their basic properties. We will discuss conjectures that have been proved, as well as those that have withstood the efforts of brilliant mathematicians. How can there still be questions involving the familiar arithmetic of complex numbers that no one knows how to answer? We shall see.

5.1 Some pictures

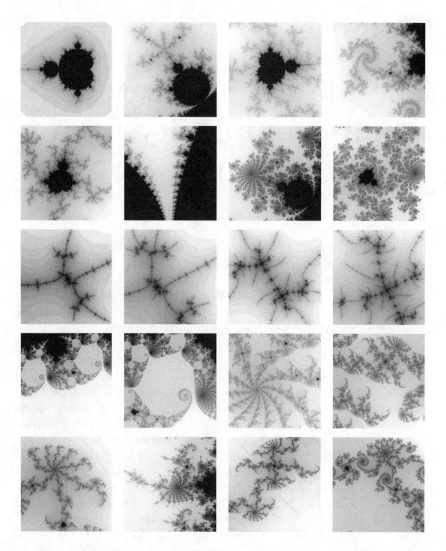

Figure 5.1: The Mandelbrot set and some magnifications. The c-values are in A.45.

The black parts of the images of Fig. 5.1 form approximations of parts of the Mandelbrot set. The first image of the top row shows the whole Mandelbrot set and contains all the other images. Or, more precisely, all the other images are magnifications of parts of this first image. Early on in the exploration of the Mandelbrot set, the choice of which parts to magnify was guided by chance and a fuzzy notion of what looked interesting.

Key features are what appear to be island copies of the Mandelbrot set nestled amongst the swirling patterns. Some look like perfect copies of the whole set, while others (first image of the second row) are quite distorted. All the little copies, even very dinky ones, are inevitably surrounded by a halo of still dinkier copies of the Mandelbrot set, and on and on forever.

To find these small copies we must look very carefully. From the images of the third row we might think the Mandelbrot set is behaving like the Cheshire cat, vanishing except for its grin. Look closely enough and we'll see that the grin is made of little Mandelbrot sets.

The pictures suggest that these island copies are scattered around the main body of the Mandelbrot set, spread infinitely thin as a fractal dust. But Benoit conjectured, and Adrien Douady and John Hubbard proved, that each island is attached to the main body by an isthmus of still smaller Mandelbrot sets.

Since the interior of each black region is without substructure, the lion's share of visual interest is concentrated around the boundary of the Mandelbrot set. Some boundary regions (row 3 again) are sparsely filled; in others (fourth image of row 2) the boundary is so crumpled that it resembles the wash of foam on a beach after a violent storm. The nearer we look at the body of the Mandelbrot set, the more crowded the picture becomes. In fact, Benoit conjectured, and Mitsuhiro Shishikura proved, that the boundary of the Mandelbrot set wiggles around so much it is 2-dimensional, unlike the smooth 1-dimensional boundaries of familiar 2-dimensional shapes. If you look closely enough at the boundary of the Mandelbrot set, you'll bop into tiny copies of the Mandelbrot set, and into the thickly branched spirals that support them.

All these pictures, and infinitely many more, each different from all the others, come from a simple formula. Let's see how.

5.2 The algorithm

In middle school we learn to solve quadratic equations of the form $ax^2 + bx + c = 0$ by using the quadratic formula

$$x = \frac{-b \pm \sqrt{b^2 - 4ac}}{2a}$$

At least early in our mathematics education we are told there is no solution if $b^2 - 4ac < 0$, because negative numbers have no square roots. The more complete statement is that negative numbers have no square roots that are real numbers. Around 1545, Italian mathematician Gerolamo Cardano introduced the concept of complex numbers in order to solve quadratic equations with $b^2 - 4ac < 0$. Not much later Rafael Bombelli developed the rules for complex number addition, subtraction, multiplication, and division.

Complex numbers have the form $a + ib$, where a and b are real numbers and $i^2 = -1$. The number a is called the *real part* of the complex number, and b (not ib) the *imaginary part*. All quadratic equations can be solved using complex numbers.

Of course, complex numbers are interesting in their own right. The Mandelbrot set, and its cousins, the Julia sets, are generated by asking some fundamental questions about the dynamics of complex numbers. In fact, one of the simplest complex number formulas is all we need to generate Julia set and Mandelbrot set images.

To add complex numbers, first add the real parts, then add the imaginary parts. That is,

$$(a + ib) + (c + id) = (a + c) + i(b + d)$$

To multiply complex numbers, distribute (or apply FOIL, "first, outer, inner, last," an acronym unfamiliar to the older author, but learned by the younger author in middle school) and simplify, recalling $i^2 = -1$. That is,

$$(a + ib) \cdot (c + id) = (ac - bd) + i(bc + ad)$$

where

$$ib \cdot id = i^2 bd = -bd.$$

The usual notation for the formula to generate images of Julia sets and the Mandelbrot set is

$$z \to z^2 + c$$

with z and c both standing for complex numbers. Unpacking this formula, we see that the arrow indicates a process repeated again and again to yield a sequence in which the output of one step is the input of the next step. Starting with a number z_0, this process generates a sequence of numbers $z_1, z_2, z_3, \ldots$

$$z_1 = z_0^2 + c$$
$$z_2 = z_1^2 + c$$
$$z_3 = z_2^2 + c$$
$$\ldots$$
$$z_{k+1} = z_k^2 + c$$
$$\ldots$$

We can collapse this sequence back down into a single formula, the *Mandelbrot formula*:

$$z_{n+1} = z_n^2 + c$$

5.2. THE ALGORITHM

We can dissect this formula using what we know about adding and multiplying complex numbers, splitting each component into its real and imaginary parts:

$$c = a + ib, \quad z_n = x_n + iy_n, \quad \text{and} \quad z_{n+1} = x_{n+1} + iy_{n+1}$$

Using these and the complex arithmetic rules, we can find individual formulas for the real and imaginary parts of the Mandelbrot formula:

$$x_{n+1} = x_n^2 - y_n^2 + a \quad \text{and} \quad y_{n+1} = 2x_n y_n + b$$

To produce pictures of Julia sets and of the Mandelbrot set, the fundamental question is this: for a given starting point z_0 and a given value of c, will the sequence $z_1, z_2, z_3, \ldots$ gotten by iterating the Mandelbrot formula stay bounded or run off to infinity? What, precisely, do we mean by "run off to infinity"? The distance to the origin from a complex number $z_n = x_n + iy_n$ is

$$|z_n| = |x_n + iy_n| = \sqrt{x_n^2 + y_n^2}$$

Then running off to infinity means that $|z_n|$ grows without bound as n increases.

Let's look at two examples, one for $c = i$, one for $c = 1$, and find the first few iterates of the Mandelbrot formula $z_{n+1} = z_n^2 + c$, starting from $z_0 = 0$.

$c = i$	$c = 1$
$z_0 = 0$	$z_0 = 0$
$z_1 = 0^2 + i = i$	$z_1 = 0^2 + 1 = 1$
$z_2 = i^2 + i = -1 + i$	$z_2 = 1^2 + 1 = 2$
$z_3 = (-1+i)^2 + i = -i$	$z_3 = 2^2 + 1 = 5$
$z_4 = (-i)^2 + i = -1 + i$	$z_4 = 5^2 + 1 = 26$

For $c = i$ the iterates enter a repeating pattern: $z_3 = -i$, $z_4 = -1 + i$, $z_5 = -i$, $z_6 = -1 + i$, and back and forth, forever. In fact, if we ever see two iterates that are equal – here it was $z_2 = z_4 = -1 + i$ – then successive iterates will repeat forever.

For $c = 1$ the iterates appear to grow without bound, a belief reinforced by a few additional calculations: $z_5 = 677$ and $z_6 = 458,330$. In A.46 we'll show that if the distance of an iterate from the origin ever exceeds 2, then later iterates will run away to infinity.

Now we are ready for the definitions of Julia sets and the Mandelbrot set.

5.3 Julia sets

To understand the Mandelbrot set, we should begin with its historical and mathematical predecessors, Julia sets. Discovered by the French mathematician Gaston Julia early in the 20th century, Julia sets are a family of shapes in the plane determined by the iteration of the Mandelbrot function. Here's how.

Choose a complex number c. For this c we'll test every complex number z_0, iterating the Mandelbrot formula. The z_0 whose iterates never get farther than 2 from the origin – that is, whose iterates do not run away to infinity – constitute the *filled-in Julia set* K_c of the complex number c.

But this leads to another question: if $|z_n| \leq 2$ for $n \leq 100$, say, can we conclude that $|z_n| \leq 2$ for all n? Not necessarily, and we lack the infinite expanse of time needed to test if $|z_n| \leq 2$ for all n. So we choose the maximum number of iterations, called the *dwell*, that we are willing to check. If $|z_n| \leq 2$ for all n up to the dwell, we assume $|z_n| \leq 2$ for all n, and consequently the starting point z_0 belongs to K_c. We paint the pixel containing z_0 black, indicating it belongs to the filled-in Julia set. If for some $n \leq$ dwell, $|z_n| > 2$, then z_0 cannot belong to K_c and the pixel containing z_0 is painted a color (or shade of gray) assigned from a list that pairs a color with this value of n. Usually brighter colors (or lighter shades of gray) signal more rapid escape, so that the boundary of the set displays a gradient of colors indicating how quickly the iterates of that point fall away from the set.

As the dwell increases, more detail can be seen around the edge of K_c. In Fig. 5.2 we see four renderings of the same Julia set with dwell 10, 15, 20, and 100. The higher dwell gives the most accurate picture: points which had stayed bounded up to a smaller dwell may run away after more iterations, thus carving out the intricate swirls of the Julia set boundary.

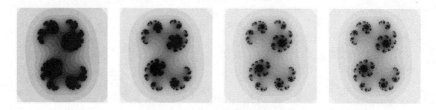

Figure 5.2: The $c = 0.4 + 0.1i$ filled-in Julia set with dwell 10, 15, 20, and 100.

But before looking at additional pictures, we must clear up one more point. So far, all the sets described are filled-in Julia sets, denoted as K_c. For each complex number c, the *Julia set* J_c is the boundary, or edge, of the filled-in Julia set K_c. When K_c is a solid black region, J_c is the (usually very crinkly) edge of the solid region. When K_c is a fragmented

Cantor set, it has no solid black regions and $J_c = K_c$.

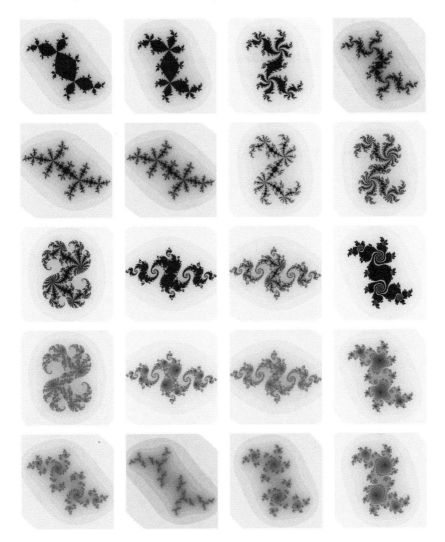

Figure 5.3: Filled-in Julia sets K_c for c values in A.47.

Now we're ready for some pictures. In Fig. 5.3 we see 20 examples of filled-in Julia sets. These pictures tell several stories. First, the Julia sets can be sorted into two main types. Each Julia set in the top row clearly consists of only one piece, a solid black region pinched and swirled into an elaborate but connected shape; each Julia set in the bottom row is a dust, that is, a Cantor set split into infinitely many pieces. Of course, these are not copies of the familiar Cantor middle-thirds set that lies in a straight line. Rather, these Cantor sets involve rotations and scaling factors that vary with location. What characterizes these as Cantor sets is their repeated subdivision into two distinct pieces. This dichotomy is not so clear for all of the middle Julia sets. Some look as though they might contain solid patches separated from one another. Nevertheless, if you can find one point of separation, you can be sure that the set is endlessly subdivided; and if the set contains a solid region, then the apparent islands are bridged by a narrow isthmus. This is because early in the 20th century, Julia and another French mathematician, Pierre Fatou, proved that this dichotomy is mathematically true: every Julia set generated by iterating the Mandelbrot formula $z_{n+1} = z_n^2 + c$ is either connected (one solid piece) or a dust (a Cantor set). There is no middle ground. We give some detail about this in A.48. Look closely: the Julia sets of the first three rows are connected; those of the last two rows are dusts.

The second story is illustrated by the Julia sets of Fig. 5.4. In broad outline as well as detailed pattern, these Julia sets are much like one another. The main difference is that the left Julia set is a Cantor dust, while the right Julia set has a solid black region. The magnifications of the centers of these Julia sets make their difference crystal clear. With such similar patterns, we expect the c values of these Julia sets to be close together, and they are.

The disconnected Julia set is for $c = -0.78517 + 0.14i$, the connected Julia set is for $-0.78541 + 0.13713i$; this small change in c gives rise to a large change in the geometry of the Julia set – connected or Cantor set. Visually distinguishing connected sets from Cantor sets can be challenging, so we are lucky there is a simpler way. In A.48 we describe a theorem of Julia and Fatou that implies that a Julia set is connected if the iterates of $z_0 = 0$ do not run away to infinity, and is a Cantor set if the iterates of $z_0 = 0$ do run away to infinity. For the first Julia set of Fig. 5.4, the iterates of

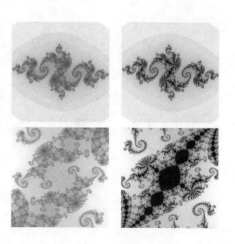

Figure 5.4: Top: A Julia dust and a nearby connected Julia set. Bottom: Magnifications.

$z_0 = 0$ run away to infinity (the center of the magnification, the point $z_0 = 0$, is not black); for the second Julia set, they don't (the center of the magnification is black). In the next section we will see that this difference is the basis of the definition of the Mandelbrot set.

5.3. JULIA SETS

  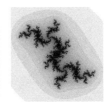

Figure 5.5: Connected Julia sets. We focus on the number of lobes that meet at each branch point; the c values are in A.47.

Figure 5.6: A magnification of the fourth image of Fig. 5.5

A third story involves the images with solid black regions of connected Julia sets. Some appear to consist of pieces, or lobes, that meet at points called branch points, a concept introduced in Sect. 2.2 and illustrated in Fig. 2.10. In the first Julia set of Fig. 5.5, each branch point is the meeting point of three lobes, down to the tiniest detail we can see. For the second Julia set, four lobes meet at each branch point; for the next after that, five lobes meet at each branch point. But for the fourth Julia set, three lobes meet at some branch points, and four lobes meet at others. The three are obvious, but the four might be hard to see. The magnification in Fig. 5.6 should establish the presence of four lobes meeting at a point. When we study the Mandelbrot set, we'll see that the number of lobes meeting at a branch point tells us something about the way in which the iterates of $z_0 = 0$ avoid running off to infinity. Also, the number of lobes meeting at each branch point provides a roadmap for locating the part of the Mandelbrot set where the point c lies. If the same number of lobes meet at each branch point of K_c, the roadmap locating c in the Mandelbrot set is (usually) fairly simple; if different numbers of lobes meet at some branch points, the path to locate c in the Mandelbrot set is less direct, more winding, but can be read with some practice, and some patience.

A final story ties Julia sets to chaos. Points outside K_c iterate to infinity, which is a kind of fixed point, although seeing that does require a bit of work, sketched in A.49. Points inside K_c (by which we mean points in K_c and not in J_c, so K_c must have a filled-in region) iterate to some cycle, a repeating pattern of points. The points of J_c do not run away to infinity and do not converge to a cycle. Rather, they twirl in intricate dances, and almost all of these dances never repeat. In fact, iteration on the Julia set is chaotic, in the modern, technical sense sketched in A.50. Using the expression of complex number multiplication in polar coordinates (A.51), in A.52 we show that iteration of the Mandelbrot function is chaotic on the $c = 0$ Julia set. The argument for other Julia sets has more technical complications, but the $c = 0$ case illustrates the main points.

So we see that the Julia set J_c is the border separating those points that run away to infinity from those that run to a cycle. The most interesting things usually are found on borders, so it is no wonder that Julia sets are very interesting.

5.4 The Mandelbrot set

At the urging of his uncle Szolem Mandelbrojt, a noted mathematician at the Collège de France, young Benoit had read the papers of Fatou and Julia. Though he saw nothing he could add to them at the time, these ideas stayed with him. Many years later when he had the use of a programmer and a computer with crude graphics capabilities, Benoit asked this question: because each Julia set J_c is either connected or a dust, can we draw a map of the c for which J_c is connected?

The result was the Mandelbrot set, a roadmap giving the locations of all the connected Julia sets. For every point c in the Mandelbrot set, the Julia set J_c is connected. For every point c not in the Mandelbrot set, the Julia set J_c is a dust. But the Mandelbrot set is a subtle map. Read properly, it encodes so much more information than just the connected/dust dichotomy. Learning to read some of these nuances is one of our goals.

It is important for us to note that the Mandelbrot set and the Julia sets live in different spaces: the space of c values for the Mandelbrot set, the space of z_0 values for Julia sets. Some authors emphasize this difference by saying the Mandelbrot set lives in the *parameter plane*, Julia sets in the *dynamical plane*. An analogy, certainly imperfect, is to think of scanning across points c in the Mandelbrot set as similar to adjusting the tuner on a radio, and to think of the Julia set J_c as the sound that comes out of the radio tuned to that frequency.

A bad – or at least, extremely tiring – way to answer Benoit's question is to generate a picture of J_c for every value of c and then try to decide if J_c is connected or is a Cantor set. Look at the examples in Fig. 5.3. For some c, the answer to "connected or dust " is clear. But for others, it is much murkier. The problem becomes tractable with the theorem of Fatou and Julia, mentioned in the last section and in A.48. This theorem implies that the Julia set J_c is connected if and only if the iterates of $z_0 = 0$ remain bounded.

The window we use for the Mandelbrot set represents c values. For each pixel in the window, take c to be the point at the middle of that pixel. Then for that c value, starting with $z_0 = 0$ (see A.53), generate the sequence $z_1, z_2, z_3, \ldots$ by the Mandelbrot formula $z_{n+1} = z_n^2 + c$. If for each $n \leq$ dwell, the distance of z_n from the origin does not exceed 2, we assume the iterates do not run off to infinity, so by the theorem of Julia and Fatou, J_c is connected and c belongs to the Mandelbrot set M. Paint

5.4. THE MANDELBROT SET

the pixel containing c black. If for some $n \leq$ dwell, the distance of z_n from the origin does exceed 2, then J_c is a dust, c does not belong to the Mandelbrot set, and we paint the pixel containing c a color or shade of gray assigned by the number of iterates needed to get farther than 2 from the origin.

The examples of Fig. 5.7 emphasize how the filled-in Julia set K_c depends on whether c does or does not belong to the Mandelbrot set. In both Mandelbrot set pictures, the gray cross marks the point c, $-0.5 - 0.6i$ (left) and $-0.5 - 0.65i$ (right), determining the Julia set K_c in the picture below it. Comparing these pictures, notice that if the points sampled in the Mandelbrot set are nearby, then the Julia sets are very similar, at least in rough outline. However, as soon as the points stray outside the Mandelbrot set, the Julia set breaks apart into a Cantor dust.

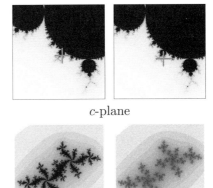

Figure 5.7: Top: crosses indicate points c in or near the Mandelbrot set. Bottom: the corresponding Julia sets K_c.

Let's think a bit more about the condition defining membership in the Mandelbrot set. One way the *Mandelbrot sequence*,

$$z_0 = 0, z_1 = z_0^2 + c, z_2 = z_1^2 + c, z_3 = z_2^2 + c, z_4 = z_3^2 + c, z_5 = z_4^2 + c, \ldots$$

the iterates of $z_0 = 0$ by the Mandelbrot formula, stay bounded is by converging to a repeating pattern, either a fixed point or a cycle. We are most interested in patterns that are stable, so iterates of points near $z_0 = 0$ converge to the same pattern. (How to test stability is described in A.54.) For example, for any c in the largest cardioid of the Mandelbrot set, called the *Main cardioid*, the Mandelbrot sequence converges to a single point, different for each c in the Main cardioid. The reason for this is described in A.55. For any c in the large disc to the left of the Main cardioid, the Mandelbrot sequence converges to a pair of points, hopping back and forth between them with every iteration. We call this a 2-cycle. In A.56 we show how to locate the disc of these 2-cycles. For any c in the largest disc near the bottom of the Main cardioid, the Mandelbrot sequence converges to a 3-cycle. Fig. 5.8 illustrates this, with the point c indicated on the Mandelbrot set and the corresponding iterates of $z_0 = 0$ plotted in the dynamical plane.

In A.57 we derive a way to count the number of discs and cardioids corresponding to each cycle length. In a sample calculation there, we

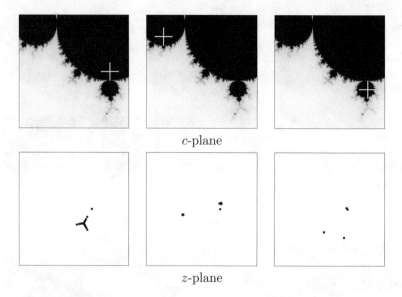

Figure 5.8: Top: three points in the Mandelbrot set, lying in the parameter plane. Bottom: iterates converging to a fixed point, 2-cycle, and 3-cycle in the dynamical plane.

locate the c value of the "center" of both 3-cycle discs attached to the Main cardioid.

We can label each disc and cardioid of the Mandelbrot set with the length of the cycle the Mandelbrot sequence produces for any c in that disc or cardioid. The relative locations of all these cycles have been understood for some time now. To start, we label the Main cardioid with a 1, because for any c in the Main cardioid, the iterates of $z_0 = 0$ converge to a fixed point, and we can think of fixed points as cycles of length 1.

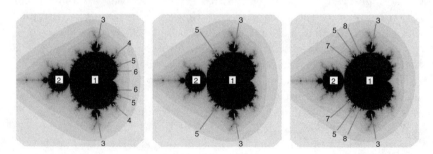

Figure 5.9: First: Part of the principal sequence. Second and third: The beginning of the Farey sequence.

On the first image of Fig. 5.9 we have labeled some discs of the *principal*

5.4. THE MANDELBROT SET

sequence: from the 2-cycle disc, going around the Main cardioid, either counterclockwse or clockwise, the largest disc we encounter is a 3-cycle disc. Continuing in the same direction, the next largest is a 4-cycle disc, then a 5-cycle disc, and so on, forever, though we stopped labeling the discs at 6. Deep in the cusp of the Main cardioid are tucked tiny million-cycle discs, tinier still billion-cycle discs, and on and on, ever-longer cycles for ever-smaller discs.

The cycle numbers of all the other discs attached to the Main cardioid are arranged according to the *Farey sequence*. We know that consecutive principal sequence discs have consecutive cycle numbers, say n and $n+1$. The Farey sequence tells us that the largest disc (which has the smallest cycle number) between these two discs will have cycle number $n+(n+1)$. For example, between the 2- and 3-cycle discs, the largest is a 5-cycle disc. See the second image of Fig. 5.9.

Finding the cycle numbers of the discs attached to the Main cardioid follows a certain order. First, the discs of the principal sequence are numbered. Next, between consecutive principal sequence discs, the largest disc has cycle number given by the Farey sequence. These are the two steps we have done already. The next step is this: between any consecutive pair of discs already numbered, the cycle number of the largest disc is given by the Farey sequence. For example, between the 2- and 5-cycle discs already numbered, the largest disc is a 7-cycle disc. Between the 5- and 3-cycle discs already numbered, the largest disc is an 8-cycle disc. See the third image of Fig. 5.9. Is the pattern clear?

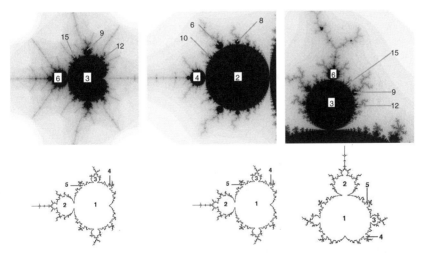

Figure 5.10: Top: the multiplier rule for a small cardioid and two discs. Bottom: Corresponding features of the Mandelbrot set, each in the orientation of the image above.

Repeat this again and again, forever, using the Farey sequence to number the largest disc between two consecutive discs already numbered. This gives the cycle numbers of all the discs attached to the Main cardioid.

The Mandelbrot set has other cardioids besides the Main, and also discs attached to discs and to these other cardioids, and discs attached to discs attached to discs, and so on. The pattern of cycle numbers for all these discs can be obtained by a simple extension of the principal sequence and the Farey sequence.

Fig. 5.10 illustrates this extension, called the *multiplier rule*. On the left is a small copy of the Mandelbrot set. We happen to know (and will explain how in A.57) that for every c in the largest cardioid of this copy, the Mandelbrot sequence converges to a 3-cycle. The multiplier rule works this way. Map the discs around this little cardioid to the corresponding discs around the Main cardioid. The cycle number of each disc around this little cardioid is 3 times the cycle number of the corresponding disc around the Main cardioid. For example, corresponding to the 2-, 3- and 4-cycle discs of the principal sequence around the Main cardioid, we have 6-, 9-, and 12-cycle discs around the little cardioid. Corresponding to the 5-cycle discs of the Farey sequence on the Main cardioid, we have 15-cycle discs on the little cardioid, and so on.

This same rule can be applied to smaller discs attached to discs or cardioids, once we see the correspondence of these smaller discs with discs attached

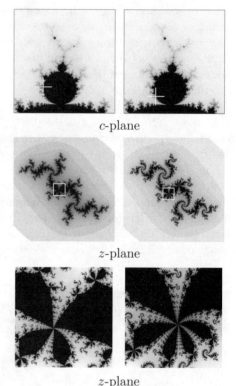

Figure 5.11: Some filled-in Julia sets K_c, with the point c located relative to a 3-cycle disc of the Mandelbrot set, illustrating how the multiplier rule is seen in the number of lobes meeting at Julia set branch points.

to the Main cardioid. In the middle image of Fig. 5.10 is the 2-cycle disc of the principal sequence around the Main cardioid. Think of the point where this disc attaches to the Main cardioid as corresponding to the cusp of the Main cardioid, and the big disc opposite this as corresponding to the 2-cycle disc attached to the Main cardioid. This establishes the orien-

5.4. THE MANDELBROT SET

tation of the attached discs. The biggest disc near the top and bottom of this disc corresponds to the 3-cycle discs attached to the Main cardioid, and so on. Note that each cycle number is 2 times the cycle number of the corresponding disc attached to the Main cardioid.

On the right side of Fig. 5.10 is a 3-cycle disc of the principal sequence around the Main cardioid. Though the orientation is different from the traditional view of the Main cardioid, the same pattern can be used to find the cycles when we map the attachment site to the Main cardioid cusp.

Now we can decode the meaning behind the number of lobes meeting at Julia set branch points. For a c from any disc attached to the Main cardioid, the branch points of J_c feature the number of lobes equal to the cycle number of that disc. For discs attached to discs or to other cardioids, the multiplier rule determines some aspects of the shape of the Julia set. For example, on the left side of Fig. 5.11 we see a 12-cycle Julia set for $c = -0.226 + 0.752i$, a point in a 12-cycle disc attached to a 3-cycle disc. In this Julia set, the most obvious branch point features three lobes (because the 12-cycle disc is attached to the 3-cycle disc), though elsewhere we see four-lobed branch points, signaling that the 12-cycle disc is attached to the 3-cycle disc in the position of the 4-cycle disc around the Main cardioid. The magnification below the Julia set picture makes these four lobes clear. On the right side of Fig. 5.11 we see an 18-cycle Julia set for $c = -0.212 + 0.699i$, a point in an 18-cycle disc attached to the same 3-cycle disc. Here again we find branch points where three larger lobes meet, and other branch points where six smaller lobes meet. Reading the sequence of lobe numbers by their relative sizes, we see this c lies in an 18-cycle disc in the 6-cycle position relative to the 3-cycle disc.

With the principal sequence, the Farey sequence, and the multiplier rule, we can locate – at least relative to one another – every disc of the Mandelbrot set. This leaves only the small Mandelbrot set copies, which can be located by a method called *Lavaurs' algorithm*. Lavaurs' algorithm tells us which parts of the Mandelbrot set are attached to which parts. Not for the last time, we remind you that all this comes from $z_{n+1} = z_n^2 + c$. Although its proof is subtle, Lavaurs' algorithm can be described and applied relatively easily.

Lavaurs' algorithm

To set up the basic structure, we draw a unit circle, labeling some points with the fraction of the circle they mark off. We are interested in the rationals with denominators of the form $2^k - 1$. Why is this? If z is very large, $z^2 + c$ is close to z^2. In A.51 we see that squaring z doubles its argument, so as a model for this beh (the angle of z in its polar representation)avior, Lavaurs studied the angle-doubling function on the unit circle. The fractions with denominators $2^k - 1$ belong to k-cycles for the angle-doubling function. We'll see why in A.59. (By a simple application of the doubling map, in A.60 we prove Fermat's Little Theorem, a result in number theory, just by counting cycles of the doubling

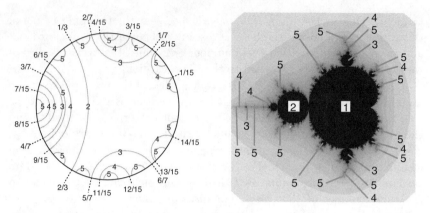

Figure 5.12: First: Lavaurs' algorithm through period 5. Second: The Mandelbrot set showing the components through period 5.

map.) The *abstract Mandelbrot set* is obtained by drawing arcs between certain pairs and pinching each arc to a point. Which pairs to connect boils down to two basic steps.

- Connect 1/3 and 2/3, the numbers of period 2. (For $k = 2$, $2^k - 1 = 2^2 - 1 = 3$.)

- Now, having taken care of the numbers with $k = 2$, continue with the next k up, $k = 3$, so the denominators are $2^3 - 1 = 7$. Connect the smallest unconnected fraction with the next smallest unconnected fraction. In this case, the first arc is drawn between 1/7 and 2/7, the next between 3/7 and 4/7, and the last between 5/7 and 6/7. Always make sure no connecting arc crosses any previously drawn arc (sometimes this will force us to connect a fraction to one that is not its nearest neighbor in the list), and no fraction is connected to more than one other fraction. Continue in this way, one k at a time, for $k \geq 4$.

The first image of Fig. 5.12 illustrates Lavaurs' algorithm for features through period 5 ($2^5 - 1 = 31$). Near the arcs, we have written the period of the corresponding cycle, and thus the period of the component of the abstract Mandelbrot set formed when each arc is pinched to a point. To read Lavaurs' graph, look at the 3-cycle arc connecting 1/7 to 2/7. Everything above this arc is connected to the 3-cycle disc attached to the Main cardioid. Between these, we see a 4-cycle arc connecting 3/15 to 4/15, and two 5-cycle arcs, one connecting 5/31 to 6/31, and one connecting 7/31 and 8/31. By the multiplier rule, the 4-cycle cannot be a disc attached to the 3-cycle disc, because 4 is not a multiple of 3. Therefore this 4-cycle must be the cardioid of a small Mandelbrot set attached back to the 3-cycle

5.4. THE MANDELBROT SET

disc. Similarly, the two 5-cycle component arcs must correspond to small Mandelbrot sets, because 5 is not a multiple of either 3 or 4. In the second image of Fig. 5.12 we can find these 4- and 5-cycle small Mandelbrot sets swimming in the decorations above the 3-cycle disc.

These rules find all the discs and cardioids relative to one another, but they tell us nothing about where these features are located in the c-plane. One approach, sketched in A.57, involves finding the roots of a family of polynomials. Although the polynomials of this family are fairly simple and are built by iteration, this approach does not add to our intuition about the Mandelbrot set. Rather, to show how remarkable it is that we can learn so much about the Mandelbrot set, we'll sketch the main steps of two results that elevate the complexity of the Mandelbrot set to another level altogether. These are Tan Lei's mapping of parts of the Mandelbrot set to parts of some Julia sets, and Shishikura's proof that the Mandelbrot set boundary is so complicated that it has the highest possible dimension of any shape in the plane. Before Shishikura's proof, evidence of the complexity of the Mandelbrot set boundary was accumulating. John Milnor provided especially clever examples.

c-plane

z-plane

Figure 5.13: An illustration of Tan Lei's theorem. Top: part of the Mandelbrot set with a Misiurewicz point c indicated, and a magnification about that point. Bottom: the Julia set for that c, and a magnification about that c.

Tan Lei's theorem applies to points called *Misiurewicz points*. In the Mandelbrot set, Misiurewicz points occur at the tips of branches and centers of spirals. Each Misiurewicz point c belongs to the boundary of the Mandelbrot set, and in A.61 we see that this same number c belongs to the Julia set J_c. For example, the complex number i in the parameter plane belongs to the Mandelbrot set, and, as we'll see in the next paragraph, the complex number i in the dynamical plane belongs to the Julia set J_i. Tan Lei showed that for every Misiurewicz point c, as we magnify the Mandelbrot set at c and magnify the Julia set J_c at c by an appropriate factor (described in A.62), the two become indistinguishable. See Fig. 5.13. Long before Tan Lei's theorem was proved, observing this similarity for many examples of connected Julia sets was what led Benoit to ask if the Mandelbrot set is connected.

To understand this result, we must know how to recognize Misiurewicz points. *Misiurewicz points* are those c for which $z_0 = 0$ iterates to a cycle, but where $z_0 = 0$ itself does not belong to a cycle. For example, recall the $c = i$ calculation from Sect. 5.2:

$$z_0 = 0,$$
$$z_1 = z_0^2 + i = 0^2 + i = i,$$
$$z_2 = z_1^2 + i = i^2 + i = -1 + i,$$
$$z_3 = z_2^2 + i = (-1+i)^2 + i = -i,$$
$$z_4 = z_3^2 + i = (-i)^2 + i = -1 + i,$$

and so on. That is, $z_0 = 0$ iterates to a 2-cycle $\{-1+i, -i\}$, so $c = i$ is a Misiurewicz point.

Tan Lei's result would not be so important if Misiurewicz points were rare or spread sparsely across the boundary of the Mandelbrot set. But this is not the case: for every point c_0 of the Mandelbrot set boundary, every disc, no matter how tiny, centered at c_0 contains infinitely many Misiurewicz points. This is the basis for the (slightly exaggerated) statement that the Mandelbrot set boundary is a visual dictionary of Julia sets.

In 1983 Benoit conjectured that the boundary of the Mandelbrot set is so complicated that it is 2-dimensional. In 1991, Shishikura proved this by finding in the cusp of the Main cardioid a sequence of c for which the dimension of the Julia sets J_c get as close as we like to 2. Fig. 5.14 illustrates this: the first disc lies more deeply in the cusp than does the second disc; the left Julia set fills space more fully than the right. Tan Lei's theorem shows that at least near some points, the boundary of the Mandelbrot set resembles these Julia sets and so is 2-dimensional. In fact, the whole boundary of the Mandelbrot set is 2-dimensional, and everywhere we look around the boundary of the Mandelbrot set,

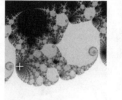 

c-plane

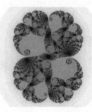 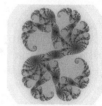

z-plane

Figure 5.14: An illustration of Shishikura's theorem. Top: discs deep in the cardioid cusp. Bottom: Julia sets at the indicated c values.

we find c values with 2-dimensional Julia sets J_c. In this way, Shishikura proved the Mandelbrot set boundary has dimension 2.

In its appearance and in its mathematics, the Mandelbrot set boundary is so complicated, so beautiful, so surprising. Worlds mysterious and unknown are in every dust mote swirling around the edge.

5.5 Other Mandelbrot sets

Everything we've seen so far indeed does come from the single function $z_{n+1} = z_n^2 + c$. With other functions, we can get complicated shapes that are different, and yet ... some familiar aspects remain.

The simplest variation is to replace $z^2 + c$ by $z^n + c$ for integer values of $n > 2$. Fig. 5.15 shows the Mandelbrot set and a small copy of that Mandelbrot set for $n = 2, 3, 4$, and 5. (The locations of the small copies are given in A.63.) The small copies are copies of the whole set of which they are a part, many with distortions familiar from the quadratic Mandelbrot set.

Applied to these functions $z^n + c$, the theorem of Fatou and Julia (A.48) implies that the Julia set for $z^n + c$ is either connected or a dust, depending on whether the iterates of $z_0 = 0$ remain bounded or escape to infinity. Consequently, we say c belongs to the $z^n + c$ Mandelbrot set if the iterates of $z_0 = 0$ remain no farther than $2^{1/(n-1)}$ from the origin for iterations up to the dwell. (In A.64, we see why the condition $|z_k| > 2$ is replaced by $|z_k| > 2^{1/(n-1)}$ in this case.)

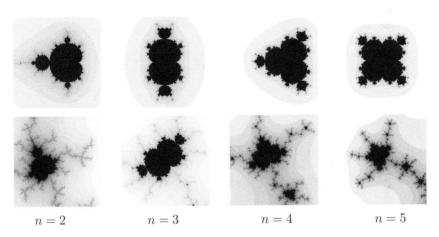

$n = 2$ $\qquad\qquad$ $n = 3$ $\qquad\qquad$ $n = 4$ $\qquad\qquad$ $n = 5$

Figure 5.15: Top: The Mandelbrot set for $z^2 + c$, $z^3 + c$, $z^4 + c$, and $z^5 + c$. Bottom: Magnifications of the image above, showing small copies of the corresponding Mandelbrot sets.

Another way to vary the original Mandelbrot set construction is to replace $z^2 + c$ by a more complicated function $g(z)$ for which the connectedness of the Julia set depends on the iterates of several initial points z_0, the "critical points" of $g(z)$ (A.48). For example, suppose $g(z)$ has two critical points, $z_a = 0$ and $z_b = -1$. Having two critical points opens up the possibility that the iterates of one critical point run away to infinity, while the iterates of the other do not. This produces a richer variety of structures for Mandelbrot and Julia sets. To illustrate these possibilities, we take $g(z) = z^3/3 + z^2/2 + c$, which in A.65 we see has critical points $z_a = 0$ and $z_b = -1$. Then the definition of the Mandelbrot set has three variations: plot those c for which the iterates of z_a remain bounded (first image of Fig. 5.16), plot those c for which the iterates of z_b remain bounded (the second image of Fig. 5.16), and plot those c for which the iterates of both z_a and z_b remain bounded (the third image of Fig. 5.16). Not surprisingly, the black region of this last picture consists of the points in common in those of the two images to its left.

Figure 5.16: The Mandelbrot set for the iterates of z_a (first image), for the iterates of z_b (second), for both (third), and a magnification (fourth).

What is surprising is the right-most image of Fig. 5.16, a magnification of this third set. Looking closely, you can see that the $z^2 + c$ Mandelbrot set has returned. Although we did not expect to find copies of the quadratic Mandelbrot set in other, more complicated functions, we are starting to see this is quite common, once we know where and how to look.

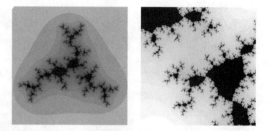

Figure 5.17: A filled-in Julia for $g(z)$ for some c where one critical point iterates to infinity and the other remains bounded. A magnification: disconnected pieces with positive area.

Before we get to that, we should point out that two critical points give a new possibility for Julia sets. Recall that filled-in Julia sets generated by iterating $z^2 + c$ are either dusts or connected, depending on whether the single critical point $z_0 = 0$ iterates to infinity or remains bounded. If the function has more than one critical point, the theorem of Fatou and Julia (A.48) says that the Julia set is connected if and only if the iterates of all critical points remain bounded, and the Julia set is a dust if and only if the iterates of all critical points run to infinity. However, if the iterates of one critical point remain bounded, while the iterates of another run to infinity, the Julia set is neither connected nor a dust. Because only a connected set is all one piece, that the Julia set isn't connected means it has at least two pieces. Then self-similarity guarantees it has infinitely many pieces. But as we see in Fig. 5.17, unlike a Cantor dust, each piece has positive area.

This Julia set is an example of the middle ground – not connected and not a dust – impossible for $z^2 + c$. Our first examples of Julia sets were quite complicated, but FractalWorld is rich in structures more intricate still. Perhaps this pattern of surprises upon surprises continues without bound, but we don't know.

5.6 The universality of the Mandelbrot set

Many stories about mathematics begin with Newton, and rightly so: he gave mathematics a new language for expressing itself. We still make recourse to Newton when trying to find the *roots* of the function f, those x for which $f(x) = 0$. Yes, this is an important problem, but why are we mentioning it here? Because the Mandelbrot set makes a surprise appearance in some versions of root finding.

Finding the roots of an equation may seem like a simple-enough task. Just graph the function and look at where the graph crosses the x-axis. The problem is that if we need to know the value of the root with some accuracy, we have to do a lot of work to draw a sufficiently detailed graph. To find a faster way, Newton used his recently invented mathematical tool, calculus. Oddly enough, we find that the easiest way to explain Newton's graphless method is to illustrate it first with graphs.

Start with a graph of some curve $y = f(x)$. Pick a point x_0 on the x-axis, our guess at a root of the function f. Unless we are very lucky, $f(x_0) \neq 0$. So we go to the point $(x_0, f(x_0))$ on the graph of f and draw the line tangent to the graph at that point. If this tangent line is not horizontal, it crosses the x-axis at some point. We call it x_1. (If the tangent line is horizontal, it never crosses the x-axis and there is no next point x_1.)

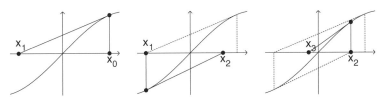

Figure 5.18: Three iterates of Newton's method.

If $f(x_1) = 0$, we've found a root. If not, go to the point $(x_1, f(x_1))$ on the graph, and again draw the line tangent to the graph at that point. If this tangent line is not horizontal, it intersects the x-axis at some point. We call it x_2.

If $f(x_2) = 0$, we've found a root. If not, ...you see what to do: draw another line vertically to the graph, then draw the tangent line to where it crosses the x-axis. Repeat, generating a sequence of points $x_0, x_1, x_2, x_3, \ldots$. Continue until the difference $|x_{n+1} - x_n|$ is very small. A root of $f(x)$ is near the last point in the sequence we've generated. Fig. 5.18 illustrates the geometry of Newton's method.

Once derived, this method can be used without plotting a single point. Using the geometry of how x_1 is obtained from x_0, in A.66 we find the formula for Newton's method:

$$x_{i+1} = x_i - \frac{f(x_i)}{f'(x_i)}$$

where $f'(x_i)$ is the derivative of $f(x)$ at $x = x_i$, that is, the slope of the tangent line to $f(x)$ at $x = x_i$. Because the derivative can be calculated easily from the formula for $f(x)$ by familiar rules of calculus, no graphs required.

Once we understand how to use it, Newton's method seems straightforward. But there are a couple of questions lurking right below the surface. For instance, if there are several roots, which initial guesses lead to which root? In Fig. 5.19 we see that using Newton's method and starting from nearby points can produce a sequence converging to one root (the solid gray lines) or another (dashed lines).

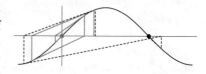

Figure 5.19: Nearby initial guesses can converge to different roots.

This isn't a surprise: if there are two roots and if (as it turns out, almost) every initial guess converges to some root, there must be an abrupt transition of the root to which an initial guess converges.[1] A natural problem, then, is to draw a map of which initial guess converges to which root.

To appreciate just how subtle this problem can be, we look back into the pre-computer history of Newton's method. In 1879, Arthur Cayley adapted Newton's method to finding roots of complex equations,

$$z_{n+1} = z_n - \frac{f(z_n)}{f'(z_n)}$$

and asked this question: for each of the roots $z_a, z_b, \ldots, z_d$ we already know, what is the collection of initial guesses z_0 from which Newton's method will converge to that root? This collection of points makes up the *basin of attraction* of the root, written $A(z_a)$, $A(z_b)$, and so on.

For $f(z) = z^2 - 1$ the roots are $z_a = -1$ and $z_b = +1$. With a short, but very clever, argument (recounted in A.67), Cayley showed that from

---

[1] Sounds plausible, doesn't it? But some plausible notions turn out to be dead wrong.

5.6. THE UNIVERSALITY OF THE MANDELBROT SET

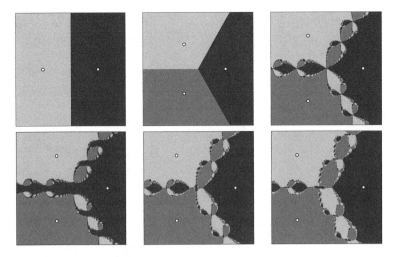

Figure 5.20: Top: Basins of attraction for $z^2 - 1$, a guess for $z^3 - 1$, and the true basins for $z^3 - 1$. Bottom: Basins of attraction for three other cubics.

any z_0 with positive real part, Newton's method converges to $z_b = +1$, and from any z_0 with negative real part, Newton's method converges to $z_a = -1$. A map of these basins is the simplest picture (top row, first image) in Fig. 5.20.

What about starting points z_0 on the imaginary axis? In some rough sense, these points should be attracted equally to both roots, so we expect they will converge to neither. In fact, if z_0 is imaginary, then by the Newton's method formula, all the z_n are imaginary. Although he didn't have this language, Cayley's argument shows that Newton's method is chaotic on the imaginary axis. This, too, we show in A.67.

At the end of this paper, Cayley wrote that next he would solve the corresponding problem for $z^3 - 1$. This function has three roots, points spaced 120° apart on the unit circle. The top row, second picture of Fig. 5.20 shows the obvious guess; the top third picture of Fig. 5.20 shows the real basins, discovered by Hubbard a century after Cayley. Though the majority of each basin falls where we expect it, the boundary offers a complicated interlacing of smaller and smaller sub-basins for all three roots. Little surprise that Cayley was unable to find the basins for this cubic with the tools available in his simpler universe.

The second row of pictures in Fig. 5.20 shows the basins of attraction for three other cubic polynomials, whose roots are given in A.68. All these pictures have a peculiar property. Pick any point of one basin and any point of another basin. Then the line connecting them contains a point of the third basin, no matter how close together the first two points are, and continuing on for ever-smaller scales ad infinitum.[2] This is called the *Wada property*, described in A.69. Three sets exhibit the Wada property if all three have a common boundary. Common sense may suggest that this is impossible, but we see this in the basins of attraction of Newton's method for roots of cubic (and higher) polynomials and in PhysicalWorld nested reflections, as in the familiar gazing globe lawn ornaments shown in Fig. 5.21. Recall the set up of the globes is shown in Fig. 1.14.

These Newton basin pictures are pretty, but more importantly, we are only one step away from stumbling across the Mandelbrot set again. This is related to another question about Newton's method. We've seen that if the graph of f has a horizontal tangent at some x_i, then Newton's method will not produce x_{i+1}. Then we say Newton's method *fails to converge* to a root of f. Are there other, perhaps more pernicious, ways Newton's method can fail to converge to a root?

Figure 5.21: The Wada property for shiny spheres.

Indeed there are. Another problem, illustrated in Fig. 5.22, is that Newton's method might get stuck in a cycle, a repeating pattern. In the first image of Fig. 5.22 we see a 2-cycle: applied to x_0, Newton's method gives x_1; applied to x_1,

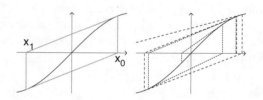

Figure 5.22: A Newton's method cycle. Nearby points don't cycle.

Newton's method gives x_0. And on it goes, hopping between these two points, forever. But the second image of the figure shows why we don't so much care about this cycle. If we're even a tiny bit off from the cycle, Newton's method carries us away from the cycle, inward (small dashes) or outward (large dashes). This is called an *unstable cycle*, because nearby points iterate away from it. Unstable cycles are (almost always) invisible to computers.

In 1983, James Curry, Lucy Garnett, and Dennis Sullivan wondered if Newton's method can have *stable* cycles, points that repeat a pattern, and whole neighborhoods of nearby points that repeat the same pattern. They used this family of polynomials:

$$f(z) = z^3 + (c-1)z - c$$

---

[2] This is why the abrupt transition mentioned a few paragraphs earlier does not occur.

5.7. THE MANDELBROT SET IN FOUR DIMENSIONS

Here c is a complex parameter analogous to the c in $z^2 + c$. They plotted those c for which Newton's method for this f has a stable cycle. As explained in A.70, a stable cycle must attract the point $z_0 = 0$.

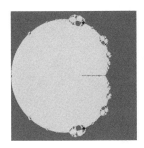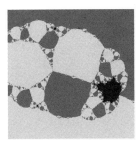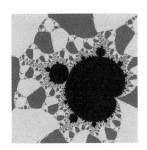

Figure 5.23: The Curry-Garnett-Sullivan graph and two successive magnifications, revealing a familiar shape. From this main cardioid, every c gives a 2-cycle. The multiplier rule gives the cycles for all other components.

Fig. 5.23 shows the result. For points painted dark gray, the iterates of $z_0 = 0$ converge to $z = 1$, a root of $f(z)$ for all c. For points painted light gray, the iterates of $z_0 = 0$ converge to some other root of f. For points painted black, the iterates of $z_0 = 0$ converge to some cycle. Successive magnifications reveal a black $z^2 + c$ Mandelbrot set. For c in the main cardioid of this Mandelbrot set, iterating Newton's method from $z_0 = 0$ converges to a 2-cycle. Then the multiplier rule, principal sequence, and Farey sequence give the cycle numbers of the other features of this Mandelbrot set. Keep in mind that the Newton function for these cubic polynomials f is a ratio of cubic/quadratic, and looks nothing like $z^2 + c$. That we again find the $z^2 + c$ Mandelbrot set seems unlikely.

But maybe not. Douady and Hubbard, and Curtis McMullen, have shown that we can find tiny $z^n + c$ Mandelbrot sets in the parameter planes of many, many other functions. So not only is the Mandelbrot set generated by iterating a very simple function, it can be generated by iterating many other functions as well. This beautiful, complicated object has been hidden behind the folds of traditional mathematical analysis, waiting for Benoit and others to ask the right questions and have access to computers and programmers to generate the pictures. Other, perhaps many other, similar miracles may remain just out of sight. For now.

5.7 The Mandelbrot set in four dimensions

So far, we have looked at functions with only a single complex parameter, the c of $z^2 + c$, for example. But we can add more parameters, and expect even richer structure, provided that we add new parameters in a sensible way. To understand how to do this, let's look more closely at

$z^2 + c$.

Why do we iterate this particular quadratic, $z^2 + c$, and not the more general $a_2 z^2 + a_1 z + a_0$? The reason is that by a straightforward renaming of the variable, the quadratic $a_2 z^2 + a_1 z + a_0$ becomes $z^2 + c$, for some constant c. We derive this in A.71. Similarly, the most general cubic can be converted into

$$z^3 + a_1 z + a_0 = z^3 + (a + ib)z + (c + id)$$

We have written the coefficients $a_1 = a + ib$ and $a_0 = c + id$ to emphasize that they are complex. That is, the cubic analog of the Mandelbrot set is determined by two complex parameters, $a+ib$ and $c+id$, or equivalently, by four real parameters, a, b, c, and d. Each complex parameter gives two real parameters, so Mandelbrot sets always live in even (real) dimensional spaces.

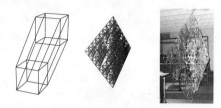

Figure 5.24: Hypercube and Sierpinski hypertetrahedron, computer-generated and physical.

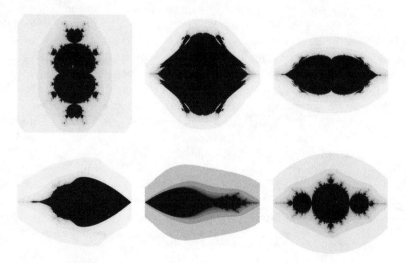

Figure 5.25: Cross-sections of the cubic Mandelbrot set, each formed by intersecting the 4-dimensional set with the plane indicated in the corresponding location of Table 5.1. Fig. 5.27 shows a magnification of the lower right section.

5.7. THE MANDELBROT SET IN FOUR DIMENSIONS

$a = b = 0$	$a = c = 0$	$a = d = 0$
$b = c = 0$	$b = d = 0$	$c = d = 0$

Table 5.1: Cross-section planes for Fig. 5.25.

How can we understand a Mandelbrot set, or any shape, in 4 dimensions? It turns out that with a bit of practice, we can learn to sketch simple shapes, cubes for example, in 4 dimensions (Fig. 5.24, first, and A.72). With a bit more practice we can sketch the analog of the Sierpinski gasket in 4 dimensions, or even build a model from pipe cleaners (Fig. 5.24, second and third, and A.73).

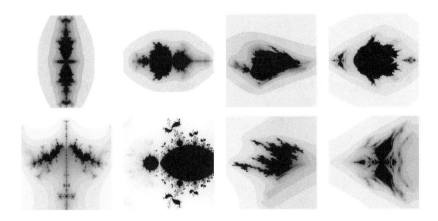

Figure 5.26: Top: Some cross-sections of the cubic Mandelbrot set. Bottom: Magnifications.

But trying to sketch the Mandelbrot set even in only 2 dimensions is discouraging. It is far too complicated to capture anything other than the simplest features. Trying to sketch the 4-dimensional Mandelbrot set is certain to end in a frustrated artist and an incomprehensible picture. So here, we will have to be content with cross-sections. Hold two of a, b, c, and d constant, and plot how the Mandelbrot set intersects the resulting plane. Fig. 5.25 shows some examples of the cubic Mandelbrot set crossing the six coordinate planes, indicated by the corresponding positions in Table 5.1. A systematic exploration of this fractal in 4 dimensions would be challenging

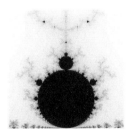

Figure 5.27: A magnification of the $c = 0$, $d = 0$ section, $-0.25 \leq a \leq 0.25$, $1 \leq b \leq 1.5$.

and lengthy. Instead, we will be satisfied, for the time being, with the rich variations and textures of this set, derived, once again, from the simplest of formulas.

In Fig. 5.26 we show the intersection of this Mandelbrot set with four planes offset from the origin, each with a magnification of an interesting part shown in the image below. Parameters are given in A.74. Some of these are symmetric and lovely, others look moth-eaten. Recall that this Mandelbrot set lives in four real dimensions, and we are viewing 2-dimensional slices through this set. Because these slices do not cut through the origin, they need not be symmetrical.

These pictures are a very few, selected randomly without much forethought, and yet they yield the complexity we have come to expect from Mandelbrot iterations. If our random sampling can turn up such a wealth of interesting questions, you can only imagine what a more rigorous study might reveal. Other planes, other magnifications, may show whole new categories of complications.

5.8 Some things we know; some things we don't know

In MathWorld asking, "What do we know?" means "What can we prove?" For example, mathematicians have proved

- The Mandelbrot set is connected. (Douady and Hubbard)

- The boundary of the Mandelbrot set is filled with little copies of the Mandelbrot set.

- The Mandelbrot set boundary has dimension 2. (Shishikura)

- The Mandelbrot set boundary is filled with bits of Julia sets. (Tan)

- The Mandelbrot set is universal. (Douady, Hubbard, McMullen)

- The n-cycle discs attached to the big cardioid have radius about $1/n^2$. (John Guckenheimer and Richard McGehee. A variation on this appears in A.75.)

In addition to these specific points, mathematicians have proved many very technical results involving intricate and surprising detail of the deep math of the Mandelbrot set.

However, this list is dwarfed by its complement, the list of things we don't know. The main conjecture, yet to be proved or disproved, is that the Mandelbrot set is *locally connected*. That is, for every point p in the set and every small-enough disc D centered at p, the part of the set inside D is all one piece.

5.8. UNANSWERED QUESTIONS

For example, in the first image of Fig. 5.28 the set is a bent comb with many missing teeth, and with a row of teeth crowding together on the left. This set is not locally connected, because for all small-enough discs D we draw centered on any point p of the left edge of the comb, the part of the comb in this disc contains a collection of parts of teeth, separated from one another.

However, if we remove the left edge of the comb, everything that remains is locally connected. Do you see why? (Hint: after removing the left edge of the comb, any two comb teeth are a separated by some positive distance.)

Douady and Hubbard conjectured that the Mandelbrot set is locally connected, but so far proving this has eluded the efforts of brilliant mathematicians, including Fields Medalists Jean-Christoph Yoccoz and Curtis McMullen, though local connectivity has been established at some points.

On the other hand, the general cubic Mandelbrot set (Sect. 5.7) is *not* locally connected. In the second and third pictures of Fig. 5.28 ($0.5 \leq a \leq 0.9$ and $-0.2 \leq c \leq 0.2$ [second], and $0.55 \leq a \leq 0.75$ and $0 \leq c \leq 0.2$ [third]), we see a portion of the real part of the cubic Mandelbrot set, that is, points with coordinates (a, c), where the cubic Mandelbrot formula is

$$z_{n+1} = z_n^3 + (a + ib)z + (c + id)$$

Of course, this picture is not proof by itself, because the whole cubic Mandelbrot set is 4-dimensional, so these thin diagonal peninsulas that appear to be accumulating along the left side may be slices of a higher-dimensional shape that is locally connected. But in fact, that the cubic Mandelbrot set is not locally connected has already been proved by Lavaurs.

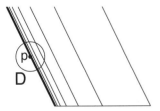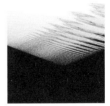

Figure 5.28: First: A simple set that is not locally connected. Second: A portion of the real part of the cubic Mandelbrot set. Third: A magnification, suggesting that the cubic Mandelbrot set is not locally connected.

Other than the fact that it is very hard to prove – itself surprising, because the local connectivity question is answered in higher dimensions, but not in one complex dimension – why do we care about the local connectivity of the Mandelbrot set? If proven, this conjecture would unlock most of the remaining problems with the Mandelbrot set. For example, local connectivity implies that Lavaurs' method for locating the discs and cardioids of the Mandelbrot set successfully finds all these components. Also, it implies the *hyperbolicity conjecture*: that every solid piece of the Mandelbrot set corresponds to some cycle, that chaotic iteration is restricted to some points on the boundary of the Mandelbrot set.

Very deep, very subtle, very beautiful math comes attached to the Mandelbrot set. And again, remember that all this comes from iterating $z_{n+1} = z_n^2 + c$.

Stepping back a bit, we encounter an epistemological issue. A complete definition of the Mandelbrot set is this: the collection of all complex numbers c for which the iterates of $z_0 = 0$ under $z_{n+1} = z_n^2 + c$ remain within the circle of radius 2 for all n. That's all. Every spiral, every branch, constellations of small copies of the Mandelbrot set, each different from every one of its siblings, all of it comes from this short definition. So we're left with a quandary. Is the Mandelbrot set really simple, or is $z_{n+1} = z_n^2 + c$ really complicated, or can a complicated shape have a simple cause? Surely the third option is the right one.

The interplay of complexity and simplicity is maybe the most beautiful and stimulating quality of fractals, a quality which never ceases to amaze, excite and dazzle those who come across it. The Mandelbrot set is our best example of this.

More is going on than just a simple formula. It's a simple formula that is iterated many, many times. A small change, perhaps, but one that makes all the difference. Benoit said this perfectly in the last line of his July 6, 2010, TED talk, delivered three months before he died. We think this is the best one-line summary of fractal geometry.

> Bottomless wonder springs from simple rules which are repeated without end.

Chapter 6

Quantifying fractals: What is fractal dimension?

Some ideas pop into our heads quickly, others grow slowly over years. A long time after Benoit had developed the basics of fractal geometry, he had the gradual realization that a tool he had been using all along, fractal dimension, was the first reliable measure of roughness. Though we could measure temperature by degrees, color by wavelength, and loudness by decibels, roughness has been an unquantifiable puzzle. An early mention of this puzzle is found in the *Timeaus* where Plato wrote that roughness is a combination of hardness and irregularity.

Though the most familiar meaning of the word "dimension" is the number of independent directions in which we can move in space, the interpretation of dimension we make with fractals has to do with the relative amount of space a complicated shape occupies. As with our introducton to self-similarity, first we will define and compute dimensions for exact mathematical fractals. After establishing some familiarity with this construction, we will be able to move on to calculating and interpreting dimensions of physical objects.

To compute the fractal dimension, some familiarity with logarithms is needed. In fact, we only need to know three things: (1) logs turn products into sums, (2) logs turn exponents into factors, and (3) we compute logs by pressing the log button on our calculators.

So let's see how a shape can have a dimension between 1 and 2.

6.1 Similarity dimension

Instead of thinking of dimension in terms of polarized movie glasses turning flat worlds into 3D ones, we are going to think of it as a gradient. A smooth curve has a relatively low dimension, 1, and a crooked curve such as a Koch curve has a slightly higher dimension, though still less than 2, which is the dimension of a filled-in square. We can start to explain this by investigating what happens when we slice some simple geometric shapes into pieces.

First, try dividing a line segment of length L into segments of length rL for $r = 1/2$. (A good question would be: why not just divide a segment of length L into segments of length $L/2$? But we want to look for a more general pattern, so we write the number N of these smaller segments in terms of r.) Dividing this line segment gives

$$N = 2 = 2^1 = \left(\frac{1}{1/2}\right)^1 = \left(\frac{1}{r}\right)^1$$

Try the analogous division of a filled-in square with side length L into N squares of side length rL for $r = 1/2$:

$$N = 4 = 2^2 = \left(\frac{1}{1/2}\right)^2 = \left(\frac{1}{r}\right)^2$$

Look at the exponents. See the pattern yet?

Try a filled-in cube. Using the same steps, we find that the number N of little cubes of side length rL is

$$N = \left(\frac{1}{r}\right)^3$$

Fig. 6.1 illustrates these subdivisions.

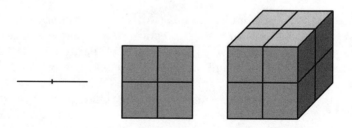

Figure 6.1: Subdividing a line segment, a square, and cube.

The exponent gives the relation between the size and the number of small copies making up a shape. In other words, for any evenly subdivided shape, the dimension, d, can be found by solving

$$N = (1/r)^d$$

6.1. SIMILARITY DIMENSION

This rule works well enough to start evaluating shapes made up of smaller copies of themselves, which means not just Euclidean cubes and line segments but also fractals.

Again, the gasket can be our guide. In Fig. 6.2 we see the gasket divided into $N = 3$ smaller gaskets, each scaled by a factor of $r = 1/2$. By plugging these values into $N = (1/r)^d$, we obtain

$$3 = 2^d$$

This is slightly more complicated than square and cube roots. To get at the exponent, we will need logarithms. Since we know how exponents behave when numbers are multiplied or raised to a power,

$$10^a \cdot 10^b = 10^{a+b}, \quad (10^a)^b = 10^{ab}$$

we can deduce the product rule and the exponent rule for logarithms,

Figure 6.2: The Sierpinski gasket, and the same gasket outlined into three smaller copies of itself.

$$\log(a \cdot b) = \log(a) + \log(b), \quad \log(a^b) = b \cdot \log(a)$$

Applying the exponent rule of logarithms to the gasket's $3 = 2^d$, we see

$$\log(3) = \log(2^d) = d \log(2)$$

We isolate d with some algebra, then press, twice, the log button on a calculator and divide to find the value of the gasket's dimension:

$$d = \log(3)/\log(2) \approx 1.58496$$

We've already agreed that we can define dimension by the exponent of the scaling relation between the number and the size of small copies making up a shape, which made perfect sense for line segments, squares, and cubes. Arguing similarly, we find the gasket has a dimension between 1 and 2. Though the gasket irrefutably lives in two spatial dimensions, its similarity dimension is a mathematical measurement, not a physical one. As an exponent in a scaling relation, noninteger values make no less sense than integer values. Think of it like this. The gasket is more complicated, more rough, than a line segment with dimension 1, but has more empty space than the filled-in square with dimension 2.

Broadening this rule, we can see that for every self-similar shape made of N copies of itself, each scaled by a factor of r, the dimension d_s is given by

$$d_s = \frac{\log(N)}{\log(1/r)}$$

Because this equation is derived from the scaling relation for exactly self-similar shapes, d_s is called the *similarity dimension*, and this equation is the *basic similarity dimension formula*.

Before continuing, let's clear up two questions about this formula. The first is what's the relation between the similarity dimension scaling factor r and the r of an IFS table? The answer is that with a caveat, they're the same. The caveat is that this dimension formula works only for self-similar fractals. o if you were wondering whether to choose r or s if they differ, don't bother. For fractals that are self-affine ($r \neq s$), finding the dimension is *much, much* more difficult, and no one knows how to do this in general. If you're wondering which r to use in the dimension formula if an IFS table has different r values, but each $r_i = s_i$, be patient. This we'll handle with the Moran equation.

The second question involves the fact that the gasket self-similarity gives rise to infinitely many decompositions: 9 copies each scaled by 1/4, 27 copies each scaled by 1/8, and on and on forever. How do we know which to use in the basic similarity dimension formula? The answer is that all are correct decompositions, and all give the same dimension. For example,

$$\frac{\log(9)}{\log(4)} = \frac{\log(3^2)}{\log(2^2)} = \frac{2\log(3)}{2\log(2)} = \frac{\log(3)}{\log(2)}$$

In fact, any decomposition painted by the self-similarity of a shape will give the same dimension for the shape.

In Fig. 6.3 we illustrate the ease of the computing similarity dimension once the pattern of decomposition is found. Fractal (a) consists of $N = 6$ pieces, each scaled by $r = 1/3$, so $d_s = \log(6)/\log(3) \approx 1.63093$. We don't even have to pay attention to the arrangement of the pieces. Fractal (b) has $N = 4$ pieces, each scaled by $r = 1/4$, so $d_s = \log(4)/\log(4) = 1$. In this case, the square Cantor set is as space-filling as a straight line, though its pieces are arranged differently in the plane. This example, clearly and manifestly fractal, is a counterexample to the often-repeated, but completely incorrect, statement that fractals are shapes with noninteger dimensions. When facing undemonstrated generalizations, readers should be skeptical. Don't believe something just because we write it. In math, everything should make sense, if you're willing to think hard.

Fractal (c) is the Sierpinski tetrahedron, made of $N = 4$ pieces, each scaled by $r = 1/2$, and so $d_s = \log(4)/\log(2) = 2$. This is another fractal with an integer dimension. Though the tetrahedron lives in three spatial dimensions, it has enough empty space that it is effectively 2-dimensional. Fractals (d) and (e) consist of $N = 3$ pieces, each scaled by $r = 1/2$, so they have dimension $d_s = \log(3)/\log(2)$, the same as the Sierpinski gasket. However, neither looks much like the gasket, nor do they look like one another, proving that the similarity dimension does not fully characterize the fractal. This is hardly a surprise: how could one number completely describe such complicated shapes?

6.1. SIMILARITY DIMENSION

Figure 6.3: Examples for computing the similarity dimension.

Another question: can we still find the dimension if some pieces are scaled by different factors? There is no obvious way to put more than one scaling factor into the basic similarity dimension formula, so some reformatting is necessary. Suppose a fractal consists of N copies, scaled by $r_1, r_2, \ldots, r_N$. First, rewrite the basic similarity dimension formula as

$$d_s \log(1/r) = \log(N)$$

Using the exponent rule in reverse, we pull d_s inside the log,

$$\log(1/r^{d_s}) = \log(N)$$

Because both logs are base 10, we can transform the equation by raising 10 to each side, dropping the logs because $10^{\log(x)} = x$ for all positive x. Thus transformed, the equation is:

$$1/r^{d_s} = N$$

Multiply both sides by r^{d_s}, obtaining

$$1 = N \cdot r^{d_s}, \quad \text{or} \quad 1 = r^{d_s} + r^{d_s} + \cdots + r^{d_s} \quad \text{with } N \text{ terms}$$

Now each of the N terms r^{d_s} can be replaced by $r_1^{d_s}, r_2^{d_s}, \ldots, r_N^{d_s}$, each with its own scaling factor. This gives the *Moran equation*,

$$r_1^{d_s} + r_2^{d_s} + \cdots + r_N^{d_s} = 1$$

Often we can solve this equation fairly easily, using one of the patterns illustrated in Fig. 6.4.

Fractal (a) is fairly straightforward, involving three pieces scaled by $r = 1/2$ and two by $r = 1/4$. For this fractal the Moran equation looks like this:

$$(1/2)^d + (1/2)^d + (1/2)^d + (1/4)^d + (1/4)^d = 1$$

That is,

$$3 \cdot (1/2)^d + 2 \cdot (1/4)^d = 1$$

Solving for d may appear challenging – after all, taking logs of both sides won't help this time. However, because the scaling factors $r = 1/2$ and $r = 1/4$ both are powers of a common factor, $1/2$,

$$1/2 = (1/2)^1 \quad \text{and} \quad 1/4 = (1/2)^2$$

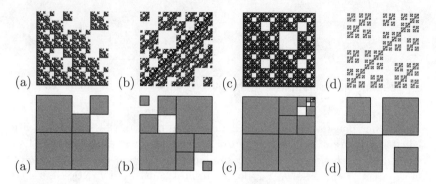

Figure 6.4: Top: Examples for computing the similarity dimension by the Moran equation. Bottom: Guides to decomposing these fractals into pieces. The box side lengths are 1/2, 1/4, and 1/8 that of the unit square.

we see that the terms of the Moran equation, $(1/2)^d$ and $(1/4)^d$, can be written as,

$$(1/2)^d = x \text{ and } (1/4)^d = ((1/2)^2)^d = ((1/2)^d)^2 = x^2$$

With these substitutions, the Moran equation becomes the quadratic equation

$$3x + 2x^2 = 1$$

which has solutions $x = (-3 \pm \sqrt{17})/4$. The Moran equation always has exactly one solution (We'll prove this in A.76), so now we must choose. Since x is a positive number raised to a positive power, the value must be positive, and we can discard the negative solution. That is,

$$(1/2)^d = x = (-3 + \sqrt{17})/4$$

Now take the logarithm of both sides and solve for d:

$$d_s = \frac{\log((-3+\sqrt{17})/4)}{\log(1/2)} \approx 1.83251$$

Of course, our calculators are smart enough to solve $3 \cdot 0.5^d + 2 \cdot 0.25^d = 1$ numerically, but this gives just an approximate value. So long as all the scaling factors are r or r^2 for some $r < 1$, the Moran equation can be solved easily and exactly by the quadratic formula. If the r values are not so neatly related, we let our machines solve it for us.

At first glance, fractal (b) seems to fit this template:

$$r_1 = r_2 = 1/2, r_3 = \cdots = r_7 = 1/4, \text{ and } r_8 = r_9 = 1/8.$$

Then the Moran equation is

$$2 \cdot (1/2)^d + 5 \cdot (1/4)^d + 2 \cdot (1/8)^d = 1$$

6.1. SIMILARITY DIMENSION

Again, all three r terms are still related in a way that lets us do some simplifying. As in the first example, we take $x = (1/2)^d$, but here the Moran equation becomes the cubic

$$2x + 5x^2 + 2x^3 = 1, \text{ which we rewrite as } 2x^3 + 5x^2 + 2x - 1 = 0$$

There is a cubic formula, the analog of the quadratic, though much longer, but because the linear term is 1, perhaps we don't need the cubic formula. One trick to simplify the cubic is to try dividing it by $x + 1$ or $x - 1$. In this case, $x + 1$ works, giving

$$2x^3 + 5x^2 + 2x - 1 = (x + 1)(2x^2 + 3x - 1)$$

Solving for x, we find the solutions are

$$x = -1 \quad \text{and} \quad x = (-3 \pm \sqrt{17})/4$$

Again, we discard the negative solutions and find that this fractal has $d_s = (\log((-3 + \sqrt{17})/4))/(\log(1/2))$, the same as example (a).

Fractal (c) decomposes into three copies scaled by $1/2$, two copies scaled by $1/4$, two scaled by $1/8$, two scaled by $1/16$, and so on, to infinity. This time the left-hand side of the Moran equation becomes an infinite series, the *infinite Moran equation*

$$3 \cdot (1/2)^d + 2 \cdot (1/4)^d + 2 \cdot (1/8)^d + 2 \cdot (1/16)^d + \cdots = 1$$

How can we deal with this infinite series? Once again, all the scaling factors are powers of $1/2$, so we can follow the pattern for the quadratic and cubic Moran equations and take $x = (1/2)^d$:

$$3x + 2x^2 + 2x^3 + 2x^4 + \cdots = 1$$

It's still an infinite series, so we should look for other simplifications. We can start by breaking the $3x$ term into $x + 2x$, then factor $2x$ from all the terms except x.

$$1 = x + 2x + 2x^2 + 2x^3 + 2x^4 + \cdots$$
$$1 = x + 2x(1 + x + x^2 + x^3 + \cdots)$$
$$1 = x + \frac{2x}{1 - x}$$

where we have used the fact that the geometric series $1 + x + x^2 + x^3 + \cdots$ has the sum $1/(1-x)$, provided $|x| < 1$. The reason for this is reviewed in Sect. A.2, but keep an eye out in case this series turns up again. Multiplying both sides by $1 - x$, the infinite Moran equation gives rise to another quadratic equation,

$$x^2 - 4x + 1 = 0$$

with solutions
$$x = 2 \pm \sqrt{3}$$
Another hitch: both of these are positive, so how do we know which to use? Since summing the geometric series required $|x| < 1$, we must take $x = 2 - \sqrt{3}$, giving
$$d_s = \frac{\log(2 - \sqrt{3})}{\log(1/2)} \approx 1.89997$$

In A.77 we'll justify extending the Moran equation to infinite series; in A.78 we'll work through two more examples.

Finally, fractal (d) has $r_1 = r_2 = 1/2$ and $r_3 = r_4 = 1/3$. Our trick won't work here, because $1/2$ and $1/3$ are not simple powers of a common number. Instead, we solve the Moran equation
$$2 \cdot (1/2)^d + 2 \cdot (1/3)^d = 1$$
numerically with the "solve" button on a graphing calculator, obtaining the approximate value of $d_s \approx 1.6055$.

So far, we have always split the fractal into pieces that do not overlap except at isolated points or, at most, along edges. Any more substantial overlap can produce paradoxical results. How would we know that a dimension calculation has given a paradoxical result? Here, common sense can tell us a lot. For example, because the plane is 2-dimensional, the dimension of any shape lying in the plane cannot exceed 2. Yet, according to the basic similarity dimension formula, a fractal made of $N = 5$ pieces, each scaled by a factor of $r = 1/2$, should have dimension
$$d_s = \frac{\log(N)}{\log(1/r)} = \frac{\log(5)}{\log(2)} \approx 2.32193,$$
certainly impossible for a fractal lying in the plane. The point: when applying the similarity dimension formula, be careful that the pieces of the decomposition do not overlap too much. Specifying precisely how much overlap is too much is something we'll discuss in Sect. 6.5.

Finally, to help build some intuition that the similarity dimension is a measure of complexity, let's look at some relatives of the Koch curve, all made of $N = 4$ pieces, with scaling factor r increasing from $r = 1/4$ to $r = 1/2$, shown in Fig. 6.5. The relation between r and d_s is given by the basic similarity dimension formula for these fractals: $d_s = \log(4)/\log(1/r)$, or if we want r in terms of d_s, $r = 4^{-1/d_s}$. Not surprisingly, we find a straight line for $r = 1/4$ ($d_s = 1$) and a filled-in triangle for $r = 1/2$ ($d_s = 2$).

Exercises

To practice your familiarity with these dimension calculations, try your hand at finding the similarity dimensions of the fractals in Fig. 6.6.

6.2. BOX-COUNTING DIMENSION

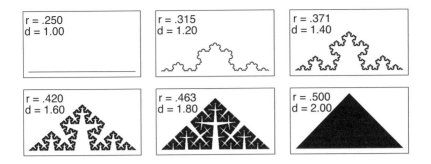

Figure 6.5: Relatives of the Koch curve, with dimensions increasing from 1 to 2 and the corresponding r values indicated.

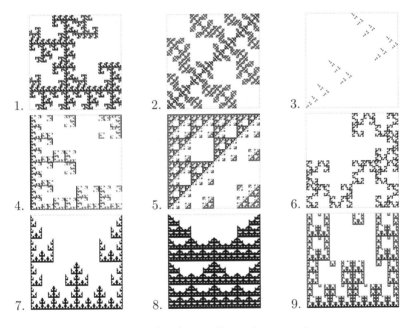

Figure 6.6: Similarity dimension exercises.

6.2 Box-counting dimension

The similarity dimension works just fine for exactly self-similar fractals, but nothing in nature is exactly self-similar. To deal with this we use two variations, the box-counting dimension d_b and the mass dimension d_m, the subject of the next section. Just as similarity dimension d_s is an exponent, both d_b and d_m are the exponents in power law relations.

If some complicated irregular shape isn't made up of small copies of itself, we use d_b and d_m to measure its complexity and to test for the presence of a power law. Box-counting involves covering the shape with the minimum number of boxes of side length ϵ and finding how the number of boxes $N(\epsilon)$ scales with ϵ. The *scaling hypothesis* is that $N(\epsilon)$ is related to ϵ by a power law,

$$N(\epsilon) \approx k \cdot (1/\epsilon)^{d_b}$$

We write $1/\epsilon$ instead of ϵ because we'll need more boxes as the side length decreases. Though the presence of a power law scaling in any shape starts as hypothesis, the steps we'll take to compute the box-counting dimension automatically tests the validity of this hypothesis. We don't need a separate test of the scaling hypothesis.

There are two avenues for computing d_b, depending on the kind of information we have about $N(\epsilon)$. For some mathematical objects, self-similar or not, we can find an exact formula for $N(\epsilon)$. In this case, d_b is computed by taking a limit of the appropriate expression. For physical fractals and random mathematical fractals, we are unlikely to find an exact formula for $N(\epsilon)$. In these cases, d_b can be computed by measuring the slope of a graph.

6.2.1 Box-counting: the limit approach

For mathematical fractals, we can often find an exact formula for $N(\epsilon)$. Then, using the scaling hypothesis as our guide, we compute the box-counting dimension by the *limit formula for d_b*,

$$d_b = \lim_{\epsilon \to 0} \frac{\log(N(\epsilon))}{\log(1/\epsilon)}$$

(If this seems like a bit of a leap, you can see how to derive the limit formula from the scaling hypothesis in A.79.) To use the limit formula we need not have a formula for $N(\epsilon)$ for *all* $\epsilon > 0$, but just for a sequence of box sizes $\epsilon_n \to 0$ as $n \to \infty$. Specifically,

$$\lim_{n \to \infty} \frac{\log(N(\epsilon_n))}{\log(1/\epsilon_n)} = \lim_{\epsilon \to 0} \frac{\log(N(\epsilon))}{\log(1/\epsilon)}$$

Provided the ϵ_n don't go to 0 too rapidly (that is to say, the rate of decrease is no faster than a fixed fraction), we can simplify the calculation of d_b because if

$$\lim_{n \to \infty} \frac{\log(N(\epsilon_n))}{\log(1/\epsilon_n)} \quad \text{exists, then} \quad \lim_{\epsilon \to 0} \frac{\log(N(\epsilon))}{\log(1/\epsilon)} \quad \text{must also exist,}$$

and the limits are equal. We told you not to take our word for anything, so the proof is in A.80. Recognizing this equivalence, we end up with the *sequential formula for d_b*:

$$d_b = \lim_{n \to \infty} \frac{\log(N(\epsilon_n))}{\log(1/\epsilon_n)}$$

6.2. BOX-COUNTING DIMENSION

Yet again, and probably not for the last time, the gasket is our guide. In Fig. 6.7 the gasket is covered with boxes of side length $\epsilon_1 = 1/2$, $\epsilon_2 = 1/4$, $\epsilon_3 = 1/8$, and $\epsilon_4 = 1/16$. We see

$$N(1/2) = 3, \quad N(1/2^2) = 3^2, \quad N(1/2^3) = 3^3, \ldots, \text{ and so on.}$$

From this, we can see the relationship is $N(1/2^n) = 3^n$. Then by the sequential formula, we can calculate d_b of the gasket.

$$\begin{aligned} d_b &= \lim_{n \to \infty} \frac{\log(N(1/2^n))}{\log(1/(1/2^n))} = \lim_{n \to \infty} \frac{\log(3^n)}{\log(2^n)} \\ &= \lim_{n \to \infty} \frac{n \log(3)}{n \log(2)} = \frac{\log(3)}{\log(2)} \end{aligned}$$

As you can see, since the gasket exhibits exact self-similarity, it has a similarity dimension, which we have already computed. For the gasket, the similarity dimension and the box-counting

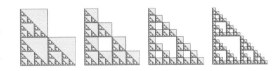

Figure 6.7: Covering the gasket with boxes.

dimension agree, but be careful, because for some shapes they differ.

For this calculation we took $\epsilon_n = 1/2^n$ to use the 1/2 scaling of the gasket. In fact, the sequential formula will give the same value of d_b for every sequence $\epsilon_n \to 0$, but finding $N(\epsilon_n)$ can be substantially more difficult if the sequence of ϵ_n does not utilize the scaling symmetry of the fractal. Since all sequences will give the same answer, we might as well select the sequence giving the simplest calculation.

Before introducing the log-log approach and applying it to some physical fractals, let's look at two additional mathematical examples.

On the left side of Fig. 6.8 we see a construction called a *product* of a Cantor middle-thirds set C and a (unit length) line segment I. We denote this construction by $C \times I$. The Cantor set is drawn along the x-axis, the line segment along the y-axis. We can visualize this by putting a vertical line seg-

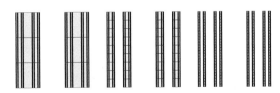

Figure 6.8: Covering the product of a Cantor set and interval with boxes of side length 1/3, 1/9, and 1/27. The first image demonstrates that $C \times I$ is made of six copies scaled by 1/3.

ment above every point of the Cantor set or else by putting a horizontal Cantor set beside every point of the line segment.

What size boxes should we use to cover this shape? The line segment can be covered with boxes of any size $1/m$. However, the scaling symmetry of the Cantor set suggests covering it with boxes of side length $1/3$, $1/9$, $1/27$, and so on. Perusing the three images of Fig. 6.8, we see

$$N(1/3) = 3 \cdot 2 = 6, \quad N(1/9) = 9 \cdot 4 = 36, \quad N(1/27) = 27 \cdot 8 = 216,$$

and in general,

$$N(1/3^n) = 6^n$$

Then by the sequential formula for d_b we find

$$d_b = \lim_{n \to \infty} \frac{\log(6^n)}{\log(1/(1/3^n))} = \lim_{n \to \infty} \frac{n \log(6)}{n \log(3))}$$
$$= \frac{\log(6)}{\log(3)} = \frac{\log(2)}{\log(3)} + \frac{\log(3)}{\log(3)} = \frac{\log(2)}{\log(3)} + 1$$

It turns out that we can also calculate the same dimension of $C \times I$ using the basic similarity dimension formula, once we recognize its exact self-similarity.

Our last example – for now – is the limit computation of d_b for the shape seen in Fig. 6.9. This construction is called the *union* of a gasket G and a (unit length) line segment I. We write this union as $G \cup I$. Basically, the two

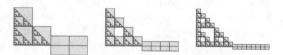

Figure 6.9: The union of a gasket and a line segment, covered with boxes of side length $1/2$, $1/4$, and $1/8$.

shapes are pasted together in the same space. In this example, the gasket is built from the triangle with vertices $(0,0)$, $(1,0)$, and $(0,1)$; the line segment has endpoints $(1,0)$ and $(2,0)$.

What size boxes should we use? Again, the line segment can be covered with boxes of any size $1/m$. The scaling symmetry of the gasket suggests covering it with boxes of side length $1/2$, $1/4$, $1/8$, and so on. From this, we find

$$N(1/2) = 3 + 2 = 5, \quad N(1/4) = 9 + 4 = 13, \quad N(1/8) = 27 + 8 = 35.$$

There's no obvious pattern to $5, 13, 35, \ldots$, but we can find one by noting $5 = 3 + 2$, $13 = 9 + 4$, $35 = 27 + 8$. That is, we keep track of how many boxes cover the gasket and how many cover the line segment. Sometimes the source of the sum is more instructive than the sum itself.

With boxes of these sizes, for the union of a gasket and a line segment we see

$$N(1/2^n) = 3^n + 2^n$$

6.2. BOX-COUNTING DIMENSION

Then by the sequential formula for d_b,

$$d_b = \lim_{n\to\infty} \frac{\log(3^n + 2^n)}{\log(1/(1/2^n))}$$

Now we're in some trouble, because log doesn't behave nicely with respect to sums. But log does behave nicely with respect to products, and it turns out that any sum can be written as a product by factoring out the larger term:

$$3^n + 2^n = 3^n\left(1 + \frac{2^n}{3^n}\right)$$

Then

$$d_b = \lim_{n\to\infty} \frac{\log\left(3^n\left(1 + \frac{2^n}{3^n}\right)\right)}{\log(2^n)} = \lim_{n\to\infty} \left(\frac{\log(3^n)}{\log(2^n)} + \frac{\log\left(1 + \frac{2^n}{3^n}\right)}{\log(2^n)}\right)$$

$$= \lim_{n\to\infty} \frac{\log(3^n)}{\log(2^n)} + \lim_{n\to\infty} \left(\frac{\log\left(1 + \frac{2^n}{3^n}\right)}{\log(2^n)}\right)$$

To deal with the last term, observe that as $n \to \infty$,

$$\log\left(1 + \frac{2^n}{3^n}\right) = \log\left(1 + \left(\frac{2}{3}\right)^n\right) \to \log(1) = 0$$

and so for the union of the gasket and line segment

$$d_b = \lim_{n\to\infty} \frac{\log(3^n)}{\log(2^n)} = \lim_{n\to\infty} \frac{n\log(3)}{n\log(2)} = \frac{\log(3)}{\log(2)}$$

In A.81 we list some other choices of $N(\epsilon)$ that can be used to calculate the box-counting dimension.

6.2.2 Box-counting: the log-log approach

Now suppose we have some rough or complicated shape: a Google Maps view of a coastline or a river, or sutures between bones of a deer skull (Fig. 6.10), for example. We won't bother with trying to find a formula for $N(\epsilon_i)$ for each box size ϵ_i. Certainly this is a hopeless exercise. Rather, we'll go back to the scaling hypothesis, $N(\epsilon) \approx k \cdot (1/\epsilon)^{d_b}$, and unfold it differently. Suppose we measure $N(\epsilon_i)$ for a sequence $\epsilon_1 > \epsilon_2 > \cdots > \epsilon_i > \cdots > \epsilon_n$ of box sizes. Then we can take the log of both sides of the scaling hypothesis formula using the product and exponent rules for logarithms to break it down. Thus we obtain

$$\log(N(\epsilon_i)) = \log(k \cdot (1/\epsilon_i)^{d_b})$$
$$= \log(k) + \log((1/\epsilon_i)^{d_b})$$

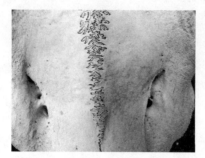

Figure 6.10: Remarkable deer skull sutures: fibrous joints that connect bones of the skull, allowing the skull some elasticity.

That is,
$$\log(N(\epsilon_i)) = d_b \log(1/\epsilon_i) + \log(k)$$
This equation is the basis of the *log-log approach* for finding d_b. The components of this equation match up with those of the familiar equation of a straight line,
$$y = mx + b$$
If the scaling hypothesis holds, the points $(\log(1/\epsilon_i), \log(N(\epsilon_i)))$ should lie roughly on a straight line with slope d_b. Since we know that no object in nature is grown or sculpted in isolation from its surroundings, we expect environmental perturbations will cause some variability in each count $N(\epsilon_i)$. We can use well-established statistical methods to estimate the likelihood that the scattering of points in a log-log plot are a sample of data that lie on a straight line, that is, the likelihood that the plotted points support the scaling hypothesis.

The other issue is the *scaling range*, the range of ϵ values over which the scaling hypothesis holds. Unlike the laws of geometry, the physical forces that build an object work over only a limited range. We know that our lungs branch and branch again, for an average of 23 levels, with the branches ranging in length from a few tens of centimeters to a few tenths of a millimeter, but ultimately the cells making up the lungs do not look like little lungs. How can we account for this finite range of scales? How many levels are enough?

Figure 6.11: Fractality. First: Not enough levels. Second: Closer to enough levels.

Fractals are defined by repetition across scales, but if there aren't enough scales, the repetition doesn't capture an important feature of the

6.2. BOX-COUNTING DIMENSION

shape. The first image of Fig. 6.11 does not exhibit any repeated pattern; the second possibly does; the lungs certainly do. A good rule of thumb is that the smallest scale exhibiting the pattern should be smaller than 1/100th the size of the whole shape. Truly, to establish fractality, more is more.

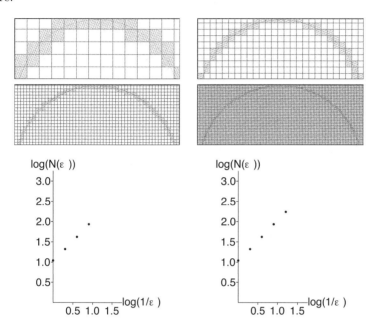

Figure 6.12: Approximating the box-counting dimension of a semicircle. The covering by the largest boxes is not shown.

A few mathematical examples should give us a sense of how box-counting works. This method is illustrated in Fig. 6.12, where we have covered a semicircle with boxes of several sizes. The smaller the boxes, the more closely the collection of boxes matches the shape of the semicircle. With all but the smallest boxes (left graph), the four points lie on a line of slope $d_b = 0.99$, pretty close to the expected value of $d_b = 1$. With all the sizes of boxes (right graph), the five points lie on a line with slope $d_b = 1$. This should not surprise us, because the semicircle is definitely not a complicated curve. Here the difference between the left graph and the right graph is small. We bother to mention this because in some instances – the Koch curve will be one – smaller boxes are needed to obtain an accurate estimate of d_b.

In Fig. 6.13 we see a Koch curve covered with squares of four sizes. We know the similarity dimension of the Koch curve is $d_s = \log(4)/\log(3) \approx 1.26$, and we expect the same for d_b, since this fractal is exactly self-similar. Computing the dimension using the first four box sizes gives the

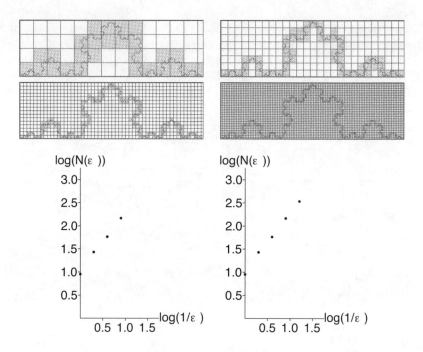

Figure 6.13: Approximating the box-counting dimension of a Koch curve.

estimate $d_b \approx 1.319$. Using all five levels of box size gives $d_b \approx 1.294$. More may be more, and smaller boxes detect more detail, but here the smallest boxes are only $1/16^{\text{th}}$ the size of the largest. We would need probably two more levels of smaller boxes to get $d_b \approx 1.27$. When asked about the slow convergence of box-counting dimension calculations, Benoit remarked, "Asymptotia can be very far away."

When it comes to PhysicalWorld, we have to be more careful. In Fig. 6.14 we see a portion of the suture in a deer skull, covered by boxes of side length 1.6 cm at the largest and 1 mm at the smallest. The common practice, which we use, is to measure lengths in terms of the largest boxes. Thus the 1.6 cm box is taken to have side length 1, the 8 mm box to have side length $1/2$, and so on. If we consider these, as well as the three intermediate sizes, we can start to see a pattern in the log-log plot. Though still not a sufficient range to take the dimension calculation seriously, this does illustrate some of the challenges to implementing the computation of the box-counting dimension. For these sizes, the number of boxes is 20, 52, 165, 448, and 1080. In the log-log plot, the best-fitting line has a slope about 1.462. Indeed, this suture does look quite a bit more wiggly than the Koch curve's 1.26, reinforcing the observation that higher dimensions come closer to filling 2-dimensional space.

6.2. BOX-COUNTING DIMENSION

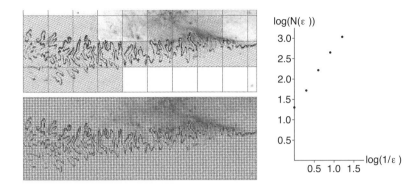

Figure 6.14: Approximating the box-counting dimension of a skull suture.

6.2.3 Mistakes in the computation of the box-counting dimension

We know that when common words are used as technical terms, sometimes the precise definition is abandoned for a homemade twist on the original meaning. Introducing a new word should solve this problem, but only if the root of the new word is understood. For example, people who think that fractal derives from "fraction" assert that a fractal is any shape with a fractional dimension. But we know of some fractals which have integer dimension. For example, the product of two Cantor middle-halves sets consists of four copies each scaled by $r = 1/4$, and so has dimension $d_s = 1$. And Cantor sets, and their products, are very definitely fractal.

Other mistakes derive from misunderstandings of the box-counting dimension. The most common is to make deductions based on too small a range of box sizes. Especially if an example relies on boxes laid out in a grid, with no effort to find the minimum number of boxes needed, it's very important to use small-enough boxes to get enough data points. While it is true that either the minimum number of boxes or the number of boxes in a fixed grid will eventually reach the same limit as the box size goes to 0, this limiting value may be approached only with boxes very much smaller than those commonly used.

We can illustrate this with an example from a real website, though we'll use fake, but similar, data. The goal of this analysis was to show, using boxes of sizes $\epsilon_1 = 1$, $\epsilon_2 = 1/2$, and $\epsilon_3 = 1/4$, that the side elevation of a house is a fractal.

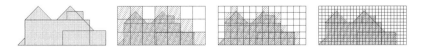

Figure 6.15: A side elevation of a house and several levels of box-counting.

Covering our own side elevation sketch with boxes, we find $N(\epsilon_1) = 17$, $N(\epsilon_2) = 53$, and $N(\epsilon_3) = 183$. Plotted in Fig. 6.16, the points appear to lie along a straight line with slope about 1.7. The website author asserts that because the side elevation of the house has a noninteger dimension, it must be a fractal.

This is absurd: the side elevation consists of filled-in rectangles and triangles, which are manifestly 2-dimensional. Many more levels of box sizes would be needed to get the right result, but it would have been achieved eventually. Be careful about believing what you read on the web. The very first question should be this: based on what I know from other sources, does this make any sense?

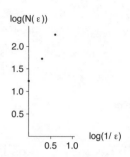

Figure 6.16: A log-log plot of the house side elevation data.

We recall another site that used two data points to calculate the dimension of clumped crystals for the very reason that when three points were used, they did not lie along a straight line. Certainly no credence should be given to any calculation of dimension based on only two box sizes. Two data points *always* lie on a line, and can give no evidence of a linear trend in a log-log plot.

We'll repeat the rule of thumb, that the data should span a range of sizes covering at least two orders of magnitude (or two decades, using the terminology of scientific literature). That is, the smallest boxes should have sides no larger than 1/100 times the sides of the largest boxes. This is a *minimum* requirement in order to claim that fractality is an important characteristic of an object.

Respecting these cautions, scientists have applied box-counting correctly to quantify parts of our DNA, earthquake fault lines, the roughness of metal fractures, coastlines, and watershed boundaries, to name a few. So much of rough nature turns out to be rough in ways we can measure.

However, nature usually has the last laugh. Often the amount of roughness is itself roughly distributed. But Benoit found a way – multifractals – to measure this roughness of roughness. We'll talk about multifractals in Chapter 7, but at this point we can't keep ourselves from asking, Is this all, or are there multi-multifractals? Is there whole cascade of modulations upon modulations of roughness measures? Or an entirely new category of fractality that encompasses all of these? Probably nature still is laughing at us.

Exercises

To practice your familiarity with these calculations, find the box-counting dimensions of the shapes in Fig. 6.17. These are

1. The union of a Cantor middle-thirds set and a line segment.

6.3. MASS DIMENSION

2. The union of a Cantor middle-thirds set and another line segment.

3. The union of two gaskets, each with base and altitude of length 1.

4. The product of two Cantor middle-thirds sets.

5. The product of a gasket and a line segment.

6. The product of a gasket and a Cantor middle-halves set.

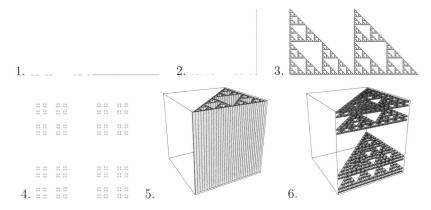

Figure 6.17: Images for the box-counting dimension exercises.

Note that changing the Cantor set of problem 6 to the middle-thirds set makes it much more challenging: the gasket is covered by boxes of side length $1/2^n$ and the Cantor set by boxes of side length $1/3^n$. There is no obvious efficient cover for both. Nevertheless, in Sect. 6.5 we will find a straightforward approach for computing this dimension. That is, don't be distressed if you can't figure out the dimension of this modified problem 6 with the tools we have so far. Sometimes problems we can't solve with one approach lead us to find new techniques.

6.3 Mass dimension

For fractals living in three dimensions – for instance, sea sponges and dust clumps, objects with a power law distribution of holes or gaps – finding the number of boxes needed to cover the interior regions of the object would require a CAT scan. In fact, it can be just as effective, and much simpler, to weigh similar objects of different sizes and see how the mass varies with size. With $M(\delta)$ as the mass of the object of diameter δ, the power law relation is

$$M(\delta) = k \cdot \delta^{d_m}$$

If this power law holds true over a reasonable scaling range, the exponent d_m is called the *mass dimension* of the object. Finding a formula for $M(\delta)$ is unlikely, except for purely mathematical objects. For physical examples we rely again on the log-log approach, but we note one difference from box-counting. While box-counting saw the number of boxes increasing as the box size decreased, calculations of the mass dimension have mass and size of the object increasing together. For d_m we plot the points $(\log(\delta_i), \log(M(\delta_i)))$ for $\delta_1 < \delta_2 < \cdots < \delta_n$. If these points lie approximately along a line, then these measurements support a power law relation between mass and size, and the slope of that line is our estimate of the mass dimension. Here we'll give two examples, crumpled paper and clusters of chick peas. Some other examples are in Sect. 6.6.

6.3.1 Crumpled paper

Possibly the simplest example of fractals made by hand is no farther away than the wastepaper basket. Crumpling paper is done easily, without purposeful thought. Hold the paper between your hands and push your hands together while crushing the paper with your fingers. That such a simple process, done easily and often, can produce fractals may be a surprise. But the process of crumpling introduces creases and folds and cavities of all sizes, and it's at least plausible that the relation between the number of cavities and the size of the cavities would follow a power law. Because the cavities occupy volume without contributing weight, we would expect to find a power law relation between weight and size of paper crumples.

Figure 6.18: First: Some paper crumples. Second: A paper crumple sliced open, revealing cavities of many sizes.

In the first image of Fig. 6.18 we see some crumpled sheets of paper. The filled-in dots in the log-log plot of the graph in Fig. 6.19 fall almost along a line of slope 2.5. The second image of Fig. 6.18 shows a paper crumple sliced open to reveal cavities of many sizes: the smaller the cavity, the more of them we tend to find. This distribution of cavity sizes is a characteristic of the fractality of crumpled paper.

6.3.2 Clusters of peas

For contrast, we will weigh collections of different numbers of peas. Depositing each bunch of peas in a piece of plastic wrap and pulling it tight gives an approximately spherical cluster. We see some examples in the photograph of Fig. 6.19. The circles in the log-log plot of the graph in Fig. 6.19 fall almost along a line of slope 3, suggesting that the pea cluster is a solid 3-dimensional object and certainly not a fractal.

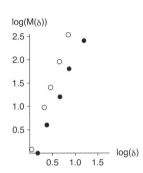

Figure 6.19: First: Some pea clusters. Second: A log-log plots of paper crumples (dots) and pea cluster (circles).

But how can we be certain the pea clusters aren't fractal? After all, there are fractals that are 3-dimensional. The real indication of fractality is a hierarchical distribution, as with the gaps we saw in the sliced open crumpled paper. Unless we freeze a pea cluster, we can't slice it in half without the peas rolling across the table. But we don't need to slice the pea cluster, because we know that gathering the peas into a cluster by pulling the plastic wrap tight packs the peas closely together, which leaves gaps all of about the same size. As far as the mass-diameter power law relation is concerned, a cluster of peas acts like a solid, with neither the gaps nor the peas varying in size at any scale.

Mass dimension has been applied to quantify the behavior of cosmic dust clumps that catalyze planet formation, the distribution of clusters of galaxies, and also many things in between. Mass dimension is a familiar implement in the toolboxes of many scientists, and now it is in yours, too.

6.4 Dimensions of more complicated fractals: random, IFS with memory, and nonlinear

Of course, we can use these same basic tools to take on some trickier examples, including random fractals, those generated by IFS with memory, and (albeit with a lot more work) some nonlinear fractals. In these cases, the Moran equation can be extended to help us calculate the dimension of each fractal. Simplest are the random fractals, so let's start there.

6.4.1 Dimensions of some random fractals

What can we measure about random sets? Leaving aside painstaking box-counting approximations, the first thing that should come to mind is the average, also called the expected value. We use the notation $\mathbb{E}(x)$ for the *expected value* of x. For example, suppose x can take only two values, $1/2$ and $1/3$. If we measure the value of x at different times, we might get a sequence that looks like this:

$$1/2, 1/2, 1/3, 1/2, 1/3, 1/3, 1/3, 1/3, 1/2, 1/3, 1/3, 1/3, 1/3, 1/2, 1/2$$

The expected value is the sum of all these measurements (6), divided by the total number of measurements (15). If we have an even larger list of measurements, it is simpler to group them into their common values, in this case $1/2$ and $1/3$, and count the number of each. Then the expected value is the sum of the number of each measurement times the measurement value, divided by the total number of measurements. For the example at hand,

$$\mathbb{E}(x) = \frac{6 \cdot (1/2) + 9 \cdot (1/3)}{15} = \frac{6}{15}$$

We can also write this as

$$\mathbb{E}(x) = \frac{6}{15} \cdot \frac{1}{2} + \frac{9}{15} \cdot \frac{1}{3}$$

This example contains the main idea of the expected value formula which we'll state in the next paragraph. Assuming that the number of times $1/2$ and $1/3$ occurred is typical for the data set being sampled – an assumption more easily believed for longer strings of measurements – then $6/15$ and $9/15$ are the *relative frequencies*, or probabilities, of obtaining each value. In this case, we see that $1/3$ occurs approximately 50% more often than $1/2$.

The general result is that we can write an expected value formula for any set of measurements. If a measurement x can take the values $x_1, x_2, \ldots, x_N$, with corresponding probabilities $p_1, p_2, \ldots, p_N$, then the *expected value formula* is

$$\mathbb{E}(x) = p_1 \cdot x_1 + p_2 \cdot x_2 + \cdots + p_N \cdot x_N$$

The expected value is the tool we need to extend the Moran equation to some kinds of random fractals. Suppose a random fractal is made of N pieces with scaling factors $r_1, \ldots, r_N$ selected randomly. Then the dimension d of this fractal is the solution of the *random Moran equation*,

$$\mathbb{E}(r_1^d + r_2^d + \cdots + r_N^d) = 1$$

6.4. RANDOM, WITH MEMORY, AND NONLINEAR

Because $\mathbb{E}(x+y) = \mathbb{E}(x) + \mathbb{E}(y)$, the random Moran equation can be separated out into independent terms:

$$\mathbb{E}(r_1^d) + \mathbb{E}(r_2^d) + \cdots + \mathbb{E}(r_N^d) = 1$$

Just what dimension are we computing here? It can't be the similarity dimension, because almost certainly these random fractal constructions are not self-similar. We'll discuss this a bit at the end of this section. For now, we'll be content with saying that we're computing the fractal dimension.

For example, suppose we build a fractal with four pieces, each having scaling factor $r = 1/2$ with probability $p = 1/2$ and $r = 1/4$ with probability $p = 1/2$. We can reinsert the expected value formula into each piece of the random Moran equation.

$$\mathbb{E}(r^d) = \frac{1}{2} \cdot \left(\frac{1}{2}\right)^d + \frac{1}{2} \cdot \left(\frac{1}{4}\right)^d$$

The equation for all four pieces is

$$4 \cdot \left(\frac{1}{2} \cdot \left(\frac{1}{2}\right)^d + \frac{1}{2} \cdot \left(\frac{1}{4}\right)^d\right) = 1$$

Now we can solve with our usual quadratic equation approach. If we take $x = (1/2)^d$, this becomes the quadratic equation $2x + 2x^2 = 1$. Solving for d gives

$$d = \frac{\log((-1+\sqrt{3})/2)}{\log(1/2)} \approx 1.44998$$

The first and second images of Fig. 6.20 show versions of this random construction, with different sequences of $r = 1/2$ and $r = 1/4$, both with probability $p = 1/2$. We expect that both have the same dimension, though of course the values may differ a bit depending on the random sequences.

The third image of Fig. 6.20 has $r = 1/2$ with probability $p = 1/4$ and $r = 1/4$ with $p = 3/4$. Note that this fractal looks more sparse than the first and the second, and the dimension calculation bears out this observation.

$$4 \cdot \left(\frac{1}{4} \cdot \left(\frac{1}{2}\right)^d + \frac{3}{4} \cdot \left(\frac{1}{4}\right)^d\right) = 1$$

gives

$$d = \frac{\log((-1+\sqrt{13})/6)}{\log(1/2)} \approx 1.20337$$

To reiterate, the more sparse the fractal, the lower its dimension.

For the fourth image of Fig. 6.20, the probabilities of the scaling factors vary with the pieces. Specifically, the lower two pieces have $r = 1/2$ with

Figure 6.20: Four random fractals, with scaling factors 1/4 and 1/2.

$p = 1/2$ and $r = 1/4$ with $p = 1/2$, and the upper two pieces have $r = 1/2$ with $p = 1/4$ and $r = 1/4$ with $p = 3/4$. Here the random Moran equation looks a little more complicated but is in fact just as easily solved as our other Moran equations that involve only $r = 1/2$ and $r = 1/4$.

$$2 \cdot \left(\frac{1}{2} \cdot \left(\frac{1}{2}\right)^d + \frac{1}{2} \cdot \left(\frac{1}{4}\right)^d \right) + 2 \cdot \left(\frac{1}{4} \cdot \left(\frac{1}{2}\right)^d + \frac{3}{4} \cdot \left(\frac{1}{4}\right)^d \right) = 1$$

With our usual $x = (1/2)^d$, we find that $x = 2/5$ and so

$$d = \log(2/5)/\log(1/2) \approx 1.32193.$$

There are more general random extensions of the Moran equation. For example, suppose instead of taking on only a small number of values, the scaling factors can be continuously distributed in some range. Check A.82 for an example of this.

6.4.2 Dimensions of memory IFS

For an IFS consisting of similarity transformations $T_1, \ldots, T_N$, with scaling factors $r_1, \ldots, r_N$, if each transformation can be applied after each transformation, then we know this IFS generates a fractal with dimension d, the solution of the Moran equation, $r_1^d + \cdots r_N^d = 1$. We can think of these IFS as being memoryless: the transformations can be applied in all combinations. What has been done before has no influence on what may be done next. The simplest way to include memory in an IFS is to allow some combinations of the T_i and forbid others. Forbidding some combinations of transformations will remove some regions of the fractal, making is sparser and lowering its dimension. But even here, we can modify the Moran equation to find the dimension. Let's see how.

For illustration, we'll use the four transformations of the square IFS of Sect. 2.5. For memory IFS we used the transition graph to encode which combinations of the T_i are allowed. To compute the dimension, we reformulate this graph as the *transition matrix*, a square array of numbers

$$[m_{ij}] = \begin{bmatrix} m_{11} & m_{12} & m_{13} & m_{14} \\ m_{21} & m_{22} & m_{23} & m_{24} \\ m_{31} & m_{32} & m_{33} & m_{34} \\ m_{41} & m_{42} & m_{43} & m_{44} \end{bmatrix}$$

6.4. RANDOM, WITH MEMORY, AND NONLINEAR

For the matrix entry m_{ij}, the number i indicates the row in which the entry lies, j represents the column.

The entries of the transition matrix are 0 or 1:

$$m_{ij} = \begin{cases} 1 \text{ if address } ij \text{ is occupied in the fractal, that is, if the} \\ \quad \text{composition } T_i \circ T_j \text{ is allowed, and} \\ 0 \text{ if address } ij \text{ is unoccupied in the fractal, that is, if the} \\ \quad \text{composition } T_i \circ T_j \text{ is forbidden.} \end{cases}$$

To help understand what the transition matrix tells us, observe that if $m_{ij} = 1$, then we know T_i is applied after T_j has been applied. That is, $m_{ij} = 1$ takes the part of the fractal with address j, makes a copy of that part shrunk by r_i, the scaling factor of T_i, and places that copy in address ij. For example, in Fig. 6.21 we see a fractal produced by a memory IFS and

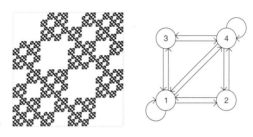

Figure 6.21: A fractal generated by an IFS with memory, along with its transition graph.

the corresponding transition graph. This fractal has empty length-2 addresses 22, 23, 32, and 33. The corresponding forbidden compositions are $2 \to 2$, $3 \to 2$, $2 \to 3$, and $3 \to 3$. This means the transition matrix entries m_{22}, m_{23}, m_{32}, and m_{33} are 0, and the transition matrix looks like this:

$$m = \begin{bmatrix} 1 & 1 & 1 & 1 \\ 1 & 0 & 0 & 1 \\ 1 & 0 & 0 & 1 \\ 1 & 1 & 1 & 1 \end{bmatrix}$$

One thing we see is that all the entries of the first and fourth rows are 1, which means that addresses 1 and 4 each contain a copy of the whole fractal, scaled by 1/2. We know this also because the transition graph shows 1 and 4 are romes.

The dimension of a memory IFS fractal is the solution d of the *memory Moran equation*

$$\rho[m_{ij}r_i^d] = 1$$

where for any square (number of rows = number of columns) matrix M, $\rho[M]$ is the largest eigenvalue of M. In general, the magnitude of the largest eigenvalue of M is called the *spectral radius* of M. The definition of eigenvalues is developed briefly in A.83; very roughly, the eigenvalues give the growth rate, in particular directions, under a system described by the matrix. Typically, we do not compute these by hand, but use a computer algebra system, like Mathematica. In A.84 we describe how to get Mathematica to calculate the eigenvalues of a matrix. All computer algebra systems, and some graphing calculators, can compute eigenvalues.

The proof of this extension of the Moran equation is subtle, but at least we can show it is plausible by observing that in the case of an IFS with all $m_{ij} = 1$, that is, all combinations of all transformations allowed, then the eigenvalues of the matrix $[m_{ij}r_i^d]$ are

$$\underbrace{0, 0, \ldots, 0}_{n-1}, r_1^d + r_2^d + \cdots + r_n^d$$

The largest eigenvalue is $r_1^d + r_2^d + \cdots + r_n^d$, which should look familiar. In this circumstance, the memory Moran equation $\rho[m_{ij}r_i^d] = 1$ is just the familiar Moran equation $r_1^d + r_2^d + \cdots + r_n^d = 1$.

For the example of Fig. 6.21, the eigenvalues of the transition matrix are $0, 0, 1 \pm \sqrt{5}$. In the case that every r_i takes on the same value r, the memory Moran equation simplifies to

$$r^d \rho[m_{ij}] = 1$$

In this example, every $r_i = 1/2$ and the memory Moran equation becomes

$$(1/2)^d(1 + \sqrt{5}) = 1,$$

so $d = \log(1 + \sqrt{5})/\log(2) \approx 1.69424$.

We can check this dimension calculation, since we can see that the fractal consists of two copies (recall addresses 1 and 4 are romes) scaled by $1/2$ and four copies (addresses 21, 24, 31, and 34) scaled by $1/4$. Solving the standard Moran equation

$$2 \cdot (1/2)^d + 4 \cdot (1/4)^d = 1$$

gives $d = \log((-1 + \sqrt{5})/4)/\log(1/2)$. With a bit of algebra, we can see this is identical to the previously calculated value of the dimension.

A final check: in A.85 we compare the memory Moran equation calculation of the dimension of Fig. 6.4's fractal (c) with the infinite Moran equation computation. Naturally, these results are the same.

We can apply the memory Moran equation to some examples from Sect. 2.5, specifically, the fractals of Fig. 2.34, shown again in Fig. 6.22. We've already seen that the dimension of (a) can be found by the Moran equation, and the dimension of (b) by the extension of the Moran equation to infinite series, but the dimensions of (c) and (d), and many others besides, cannot be found this way. Now we can compute the dimensions of these by finding the eigenvalues of the transition matrices.

6.4. RANDOM, WITH MEMORY, AND NONLINEAR

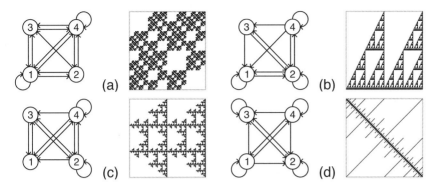

Figure 6.22: Examples of IFS with memory with their transition graphs.

The transition matrices are

$$(a) \begin{bmatrix} 1 & 1 & 1 & 1 \\ 1 & 0 & 0 & 1 \\ 1 & 1 & 0 & 1 \\ 1 & 1 & 1 & 1 \end{bmatrix} \quad (b) \begin{bmatrix} 1 & 1 & 1 & 1 \\ 1 & 1 & 1 \\ 0 & 1 & 0 & 1 \\ 0 & 1 & 0 & 1 \end{bmatrix}$$

$$(c) \begin{bmatrix} 0 & 1 & 1 & 1 \\ 0 & 1 & 1 & 1 \\ 1 & 1 & 0 & 1 \\ 1 & 1 & 0 & 1 \end{bmatrix} \quad (d) \begin{bmatrix} 1 & 0 & 0 & 1 \\ 1 & 1 & 1 & 1 \\ 1 & 1 & 1 & 1 \\ 1 & 0 & 0 & 1 \end{bmatrix},$$

and the corresponding eigenvalues are

(a) $0, 3.365, -0.683 \pm 0.358i$ (b) $0, 0, 0, 3$
(c) $-1, 0, 0, 3$ (d) $0, 0, 2, 2$

Running these through the memory Moran equation, we find the dimensions are

(a) $\log(3.365)/\log(2) \approx 1.750$ (b) $\log(3)/\log(2) \approx 1.585$
(c) $\log(3)/\log(2) \approx 1.585$ (d) $\log(2)/\log(2) = 1$

The Moran equation can do all this and more. Before we get to its multifractal applications in the next chapter, let's look at some nonlinear fractals.

6.4.3 Dimensions of some nonlinear fractals

So far, we have used IFS consisting of linear functions: the variables x and y appear only as x and y, and there is no x^2, $\sqrt{y}$, or anything more complicated. But we can also build IFS with nonlinear functions. For example, in Fig. 6.23 we see two nonlinear gaskets, defined by the IFS

$$T_i(x,y) = (.5x^c, .5y^c) + (e_i, f_i)$$

where

$$(e_1, f_1) = (0,0), \ (e_2, f_2) = (.5, 0), \text{ and } (e_3, f_3) = (0, .5)$$

and $c = 0.9$ for the first gasket, 1.1 for the second gasket.

By changing the power to which the variables x and y are raised, we can make different pictures, different wobbly versions of the gasket. The values of e_i and f_i give the coordinates of the lower left corners of the three pieces of these gaskets.

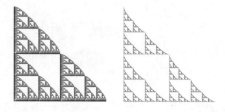

Figure 6.23: Two nonlinear gaskets.

For self-similar fractals, relative scalings are independent of how deeply in the fractal the piece lies. Certainly, this isn't the case for these nonlinear gaskets. Here, the amount of contraction depends on location, which is how the hypotenuse of the triangle turns into curves that are increasingly crinkled the closer you look.

Because the scalings depend on location, a nonlinear transformation cannot be described by a single scaling factor. Instead, we'll look at the *diameter* $\|A\|$ of a set A, defined as the greatest distance between pairs of points in A, and note how nonlinear transformations alter diameter. As usual, $A_{i_1 \ldots i_n}$ denotes a region with address $i_1, \ldots, i_n$. Then the *nonlinear Moran equation* is

$$\lim_{n \to \infty} \left(\sum_{i_1, \ldots, i_n} \|A_{i_1 \ldots i_n}\|^d \right)^{1/n} = 1$$

In general, evaluating this limit is challenging. Instead, we approximate the dimension by solving numerically

$$\left(\sum_{i_1, \ldots, i_n} \|A_{i_1 \ldots i_n}\|^d \right)^{1/n} = 1$$

for $n = 1, 2, \ldots, 10$. That is, sum the dth power of the diameters of all the length-n address regions, take the nth root, set this equal to 1, and solve for d. Then we graph the results and look for any visible trends. Fig. 6.24 suggests that $d \approx 1.9$ for the first gasket and $d \approx 1.5$ for the second. Since we only looked through $n = 10$, we can draw only a shaky conclusion about the exact value of d. Still, without looking any further, we see that the dimension of the second gasket is appropriately lower than the first.

In some cases, including many circle inversion fractals, we can use an easier approach. These are called *cut-out sets*. Starting with some set in the plane, we remove discs of radii $r_1 \geq r_2 \geq \cdots$, with $r_k \to 0$, so that the

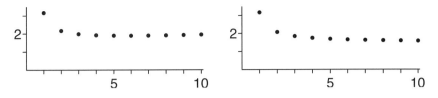

Figure 6.24: Estimating the dimensions of the nonlinear gaskets.

sum of the areas of the removed discs equals the area of the original set. If
$$\lim_{k \to \infty} \frac{\log(r_k)}{\log(k)}$$
exists, call it B. Then the box-counting dimension of the cut-out set is, simply,
$$d_b = -\frac{1}{B}$$
The first image of Fig. 6.25 is a circle inversion fractal which is also a cut-out set. The second image is the log-log plot of r_k vs. k, revealing that these points lie (almost) on a line with slope -0.752. If we take this value for B, we get the dimension
$$d_b \approx -\frac{1}{-0.752} \approx 1.328.$$

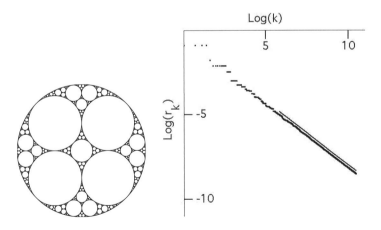

Figure 6.25: A circle inversion cut-out set and a graph used to estimate its dimension.

Dimensions can be computed by many other approaches, of course. We have presented only a few methods for a few kinds of fractals. Other methods may be more useful depending on the shape itself. For instance, the method of Hermann Minkowski and Georges Bouligand finds d_b by measuring how quickly a shape fills up the space around it when all parts of the shape are thickened.

When we talk about, say, the dimension of random fractals with only statistical self-similarity, which dimension are we talking about? Not similarity, because these shapes aren't exactly self-similar. Probably not box-counting, because unless we are very lucky, we'll never find a pattern in the number of boxes. In fact, what we are measuring for random fractals and for nonlinear fractals is the *Hausdorff dimension*, a mathematical construct of greater logical depth than that of the other ideas we have encountered. In Sect. A.99 we'll give the briefest of sketches of how the Hausdorff dimension works.

The Hausdorff dimension is a favorite of mathematicians because its definition is general enough that theorems can be proved about it. But this same generality makes computing it, even in simple cases, quite difficult. Worked out directly, the proof that the Cantor middle-thirds set has Hausdorff dimension $\log(2)/\log(3)$ covers several pages of fairly challenging math. Given this, it may be surprising that McMullen and Kenneth Falconer, and a few others, have computed the Hausdorff dimensions for some self-affine fractals, fractals having different scalings in different directions. In general, the dimensions of self-affine fractals are very difficult, and in some cases apparently impossible, to compute. An indication of just how peculiar these shapes are is this: the dimension of a self-similar fractal depends on only the number of pieces and the sizes of the pieces, while for some self-affine fractals the dimension changes just by moving the pieces around.

And this is only the start. Dimension is a beautiful, deep, and very intricate branch of mathematics.

Exercises

1. Suppose a random fractal consists of two pieces with scaling factors

$$r_1 = \begin{cases} 1/2 & \text{with } p = 1/2 \\ 1/4 & \text{with } p = 1/2 \end{cases} \quad \text{and} \quad r_2 = \begin{cases} 1/2 & \text{with } p = 1/4 \\ 1/4 & \text{with } p = 3/4 \end{cases}$$

(a) Find the expected value of the dimension of this random fractal.

(b) Find the maximum possible dimension of this random fractal.

(c) Find the minimum possible dimension of this random fractal.

(d) Is the expected value closer to the minimum or the maximum? (A numerical approximation of the expected value may be useful here.) Can you explain this result?

2. Suppose a random fractal consists of two pieces with scaling factors

$$r_1 = 1/2 \quad \text{and} \quad r_2 = \begin{cases} 1/2 & \text{with } p = 1/2 \\ 1/4 & \text{with } p = 1/2 \end{cases}$$

6.5. DIMENSION RULES

(a) Find the expected value of the dimension of this random fractal.
(b) Find the maximum possible dimension of this random fractal.
(c) Find the minimum possible dimension of this random fractal.

3. Suppose a random fractal consists of two pieces with scaling factors

$$r_1 = \begin{cases} 1/2 & \text{with prob } p \\ 1/4 & \text{with prob } 1-p \end{cases} \quad \text{and} \quad r_2 = \begin{cases} 1/2 & \text{with prob } p \\ 1/4 & \text{with prob } 1-p \end{cases}$$

(a) Show the maximum and minimum values of the dimension of this random fractal are 1 and 1/2.
(b) Find the exact value of p so the expected value of the dimension is 3/4.

4. Find the dimension of the fractal determined by each of these transition graphs. If you can find the dimension by the Moran equation or the infinite Moran equation, do that and compare the answer with that you obtained using the memory Moran equation.

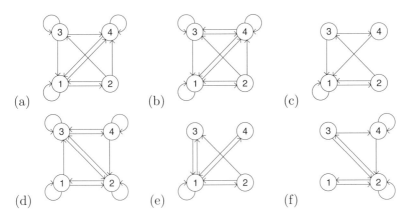

Figure 6.26: Transition graphs for Exercise 4.

5. Draw transition graphs that give rise to these Moran equations.
(a) $x + 3x^2 = 1$.
(b) $x + 3x^2 + x^3 = 1$.
(c) $2x + x^2 + x^3 = 1$.

6.5 Some rules dimensions must obey

In all the examples we have seen so far, computing the dimension has been pretty straightforward. However, some shapes require subtle attention to find the dimension, especially when they are made up of combinations of other shapes. In this section we will present some rules that dimensions obey and examples of applying these rules to compute dimensions of some complicated shapes. Surprisingly, a few of these calculations are easier than anything we have done so far.

Despite the general difficulty of computing the Hausdorff dimension, in some cases – specifically, when the pieces do not overlap too much – the Hausdorff dimension of self-similar fractals is the solution of the Moran equation. The specific parameters of "not overlapping too much" were given by Hutchinson in his *open set condition*, described in A.86. Even without the proof, the idea is simple enough to intuit: for fractals lying in the plane, the pieces can overlap only at points or along line segments. Any more substantial overlap would cause regions to be double counted in the Moran equation, throwing off the result.

Although there are some other dimensions – mass dimension d_m, which we have seen, something called packing dimension, and still others – we will focus on three kinds of dimension, $d_s, d_b,$ and d_h: similarity, box-counting, and Hausdorff. All three are different ways of approaching fractal dimension, and each can yield a slightly different answer. Nevertheless, there are some general relations between these. For example, typically,

$$d_h \leq d_b \quad \text{and} \quad d_h \leq d_s$$

Often, but not always, they are equal. For example, if the open set condition holds, then $d_h = d_s$. We will express the remaining results of this section in terms of d_b and d_h, but think of d_h as a generalization of d_s.

These dimensions satisfy five rules: monotonicity, stability, invariance, the product rule, and the intersection rule.

1. *Monotonicity* The dimension of a part cannot exceed the dimension of the whole. In symbols we write $A \subseteq B$ to indicate A is contained in B. Then the monotonicity rule is

$$A \subseteq B \text{ imples } d_b(A) \leq d_b(B) \text{ and } d_h(A) \leq d_h(B).$$

This makes sense. A square cannot contain a cube.

2. *Stability* For a shape made of several pieces, the dimension of the whole shape is the largest of the dimensions of the pieces. We write $A \cup B$ for the *union* of A and B, that is, everything that belongs to A or to B, or to both. Then the stability rule is

$$d_b(A \cup B) = \max\{d_b(A), d_b(B)\} \text{ and } d_h(A \cup B) = \max\{d_h(A), d_h(B)\}.$$

For instance, suppose $d_b(A) > d_b(B)$. Then the power law for the number of boxes that cover A has a larger exponent than that for the number of boxes covering B. The larger exponent determines the power law of the whole shape.

In Fig. 6.27 you'll see three familiar box-counting dimension exercises. Watch how easily the calculations go when we use the stability rule. Since

6.5. DIMENSION RULES

Figure 6.27: Stability of dimension examples.

all three are made of pieces that are self-similar without substantial overlaps, we use d to stand for both d_b and d_h, which are equal for all three.

(a) This is the union of a Cantor middle-thirds set and a line segment, so
$$d = \max\{\log(2)/\log(3), 1\} = 1$$

(b) This is the same thing as (a), just with a different arrangement of the pieces, so again $d = 1$.

(c) This is a union of two gaskets, so the dimension is the same as that of a single gasket.
$$d = \max\{\log(3)/\log(2), \log(3)/\log(2)\} = \log(3)/\log(2)$$

3. *Invariance* If we transform a space without too much stretching and twisting, dimension is preserved. The simplest case is that of a similarity transformation f. Then the dimension of the transformed set is equal to that of the original set. In symbols,
$$d_b(f(A)) = d_b(A) \text{ and } d_h(f(A)) = d_h(A).$$

This is no surprise: $f(A)$ is just a scaled copy of A, and scaling preserves the relation between whole and part, hence does not change dimension. But in fact, invariance holds for much more general functions. For example, suppose for every pair of points p and q in A, the distance between $f(p)$ and $f(q)$ is bounded above and

Figure 6.28: Top: Cantor middle-thirds set. Bottom: A nonlinear image.

below by multiples of the distance between p and q. That is, the distortion of distance by f lies in the Goldilocks zone: the distance becomes neither too large nor too small. Then f preserves both the box-counting and the Hausdorff dimension.

Fig. 6.28 illustrates such a function. There the upper fractal is the Cantor middle-thirds set, while the lower is a version of this Cantor set, with each point mapped by the function $f(x) = x + x^2$. For example, the left endpoint of the upper Cantor set, $x = 0$ is mapped to the left point, $f(0) = 0$, in the lower Cantor set, and the right endpoint $x = 1$ is mapped to $f(1) = 2$. In A.87 we'll prove that the distortion of this function f lies in the Goldilocks zone and conclude that both these fractals have the same dimension.

4. *The product rule* To visualize the product, $A \times B$, of two shapes A and B, imagine that the shapes lie in perpendicular spaces, and above every point of A lies a whole copy of B. When dealing with products, finding the dimension is almost always a matter of adding up the dimensions of the factors. In symbols, typically,

$$d_b(A \times B) = d_b(A) + d_b(B) \text{ and } d_h(A \times B) = d_h(A) + d_h(B)$$

For example, the dimension of a filled-in cube, which is the product of a line segment and a filled-in square, is the sum of the dimensions of the line segment and the square, $3 = 1 + 2$. Watch how easily the calculations go when we use the product rule on the examples of Fig. 6.29. Again, here d denotes both d_b and d_h, since the values are equal.

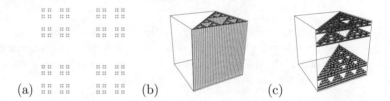

Figure 6.29: Examples of the product rule for dimensions.

(a) This is the product of two Cantor middle-thirds sets, whose dimension we already know, so

$$d = \frac{\log(2)}{\log(3)} + \frac{\log(2)}{\log(3)} = \frac{2\log(2)}{\log(3)} = \frac{\log(4)}{\log(3)}$$

(b) This is a product of a gasket and a line segment, so

$$d = \frac{\log(3)}{\log(2)} + 1 = \frac{\log(3) + \log(2)}{\log(2)} = \frac{\log(6)}{\log(2)}$$

(c) This is a product of a gasket and a Cantor middle-halves set, so

$$d = \frac{\log(3)}{\log(2)} + \frac{\log(2)}{\log(4)} = \frac{\log(3)}{\log(2)} + \frac{1}{2}$$

6.5. DIMENSION RULES

The product rule can be applied even to fractals that have different scalings in different directions. In general, computing the dimension of one of these self-affine fractals is extraordinarily difficult. However, if a self-affine fractal is a product of two self-similar fractals, the product rule can find the dimension much more quickly and painlessly.

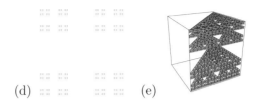

Figure 6.30: More product rule examples.

(d) This squashed-looking fractal is actually the product of a Cantor middle-thirds set and a Cantor middle-halves set, so

$$d = \frac{\log(2)}{\log(3)} + \frac{\log(2)}{\log(4)} = \frac{\log(2)}{\log(3)} + \frac{1}{2}$$

(e) Recall that when we saw this example before, we could not do the calculation by box-counting, because the gasket suggests scaling factors that are powers of $1/2$, while the Cantor middle-thirds set suggests scaling factors that are powers of $1/3$. Since no happy single scaling could be agreed upon, it would be very difficult to find a formula for $N(\epsilon)$. By using the product rule, now we can see that

$$d = \frac{\log(3)}{\log(2)} + \frac{\log(2)}{\log(3)}$$

5. *The intersection rule* The intersection of two sets A and B, denoted by $A \cap B$, consists of everything that belongs to both. The rule for the dimension of the intersection is the most complicated, because $A \cap B$ depends on the relative positions of A and B.

For example, think of two lines in the plane. If they are parallel, then either they coincide or they never meet. Far more common is for their slopes to be different, in which case the lines intersect at a single point, which has dimension 0. Because the dimension of the plane is 2, we can surmise that the general rule is the sum of the dimensions of the two sets, minus the dimension n of the space in which they lie. That is,

$$d(A \cap B) = d(A) + d(B) - n$$

A negative result for the intersection formula is interpreted as an empty intersection: that is, A and B have no points in common. However, it is worth mentioning that one of Benoit's last projects was to use negative values of the intersection formula to quantify degrees of emptiness. Though begun, this project remains unfinished.

To demonstrate just how tricky intersections can be, let's look at the three intersections of a line segment and a gasket illustrated in Fig. 6.31.

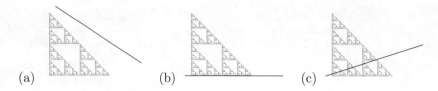

(a) (b) (c)

Figure 6.31: Three intersections of a gasket and a line segment.

(a) The line segment misses the gasket completely. So long as both shapes have finite size, empty intersections always can be achieved by moving the shapes far enough apart.

(b) The line segment is aligned exactly along a side of the gasket. The intersection is that side itself, a line segment, hence of dimension 1.

(c) We see a more typical intersection, which means we can use the intersection formula. Here, the intersection is a Cantor set having dimension

$$\frac{\log(3)}{\log(2)} + 1 - 2 = \frac{\log(3)}{\log(2)} - 1 \doteq \frac{\log(3/2)}{\log(2)}$$

These rules hold often, but they do not hold in absolutely every circumstance. Exceptions can be found, but typically these require very precise, finicky constructions of the shapes. In A.88 we'll spell out conditions guaranteeing that these rules hold.

By combining the five rules of this section, we can find the dimensions of some sets that would be a challenge to compute directly. For example, what is the dimension of the fractal X in the first image of Fig. 6.32? Looking closely, we can guess that fractal X is the portion of the product $C \times I$ of the Cantor middle-thirds set and the unit interval, lying under the parabola $y = x^2$.

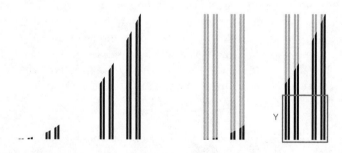

Figure 6.32: First: Find the dimension of this set. Second: How to do it.

In the second image of the figure, we find the key to the calculation. First, by adding in the gray segments we can see that indeed X is a subset of the product $C \times I$. By monotonicity and the product rule, we know

6.5. DIMENSION RULES

that this fractal's dimension must be less than or equal to the sum of the dimensions of C and I.

$$\dim(X) \leq \dim(C \times I) = \dim(C) + \dim(I) = \frac{\log(2)}{\log(3)} + 1$$

Next, the box on the right contains a scaled copy, Y, of $C \times I$. Then by invariance and monotonicity, we know that the dimension of the whole product must be the same as the dimension of Y, which is less than or equal to the dimension of X.

$$\log(2)/\log(3) + 1 = \dim(Y) \leq \dim(X)$$

Combining these two lines, we see that

$$\frac{\log(2)}{\log(3)} + 1 \leq d(X) \leq \frac{\log(2)}{log(3)} + 1,$$

where d stands for both the box-counting and the Hausdorff dimensions.

Before closing, let us mention that in A.89 we illustrate McMullen's calculations showing some of the subtle behavior that the dimension of self-affine fractals exhibit.

Exercises

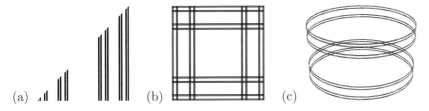

Figure 6.33: Exercises for the algebra of dimensions. Everything that looks like a Cantor set is a Cantor set, with a middle-thirds set for (a) and (c), and a middle-halves set for (b).

1. Find the dimension of the fractal in Fig. 6.33 (a) without using monotonicity.

2. Find the dimension of the fractal in Fig. 6.33 (b).

3. Find the dimension of the fractal in Fig. 6.33 (c), the product of a circle in the xy-plane and a Cantor set along the z-axis.

4. Find the scaling factor r of a Cantor set C consisting of two pieces, each scaled by r, so that the product $C \times C \times C$ has dimension 1. Think of $C \times C \times C$ as the product of a Cantor set along the x-axis, another along the y-axis, and a third along the z-axis.

5. Using the familiar square IFS rules

(a) find the dimension of the IFS with memory image produced by transition graph (i).

(b) Find the dimension of the IFS with memory image produced by transition graph (ii).

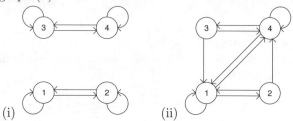

6.6 Are unearthly dimensions of earthly use?

Benoit's honors include the Japan Prize, the Steinmetz Award, the Wolf Prize, the Caltech Distinguished Alumni Award, a score of honorary doctorates, and many others, all in diverse categories, as we'd expect for a thinker of his power and range of interests. However, perhaps the most surprising of these came in 1988 when Benoit was made an honorary member of the Sudbury, Ontario, chapter of the United Mine Workers union. His work introducing dimension as a measure of roughness had led to the first reliable method for quantifying how dust particles stick to lung tissue, a topic of considerable interest to miners whose work-related risks include black lung. No child of priviledge, Benoit's early life included work on a horse farm, and as a toolmaker. He had great respect for physical labor, and was delighted, and proud, that his work improved the safety of miners.

Finding fractals in the wild is one thing, but finding ways to use the mathematical tools at our disposal to quantify and modify the world around us is even better. Dimension is one such tool.

6.6.1 Coastlines

In Sect. 3.3 we mentioned Richardson's work on measuring the lengths of coastlines by approximating them with segments of length ℓ. Denoting by $N(\ell)$ the number of segments needed, the length of the coastline can be approximated by $L(\ell) = N(\ell) \cdot \ell$. Richardson observed that $L(\ell)$ increases rapidly as ℓ decreases, instead of approaching a limiting value as it does for a smooth curve. For coastlines, $L(\ell)$ appears to grow without bound. Looking for a pattern underneath this

Figure 6.34: South Africa and west Britain coastlines.

6.6. ARE UNEARTHLY DIMENSIONS OF EARTHLY USE?

growth, Richardson plotted $\log(L(\ell))$ vs. $\log(\ell)$, obtaining points approximately along straight lines with slopes given in Table 6.1

-0.25	west coast of Britain
-0.15	land frontier of Germany
-0.14	land frontier of Portugal
-0.02	coast of South Africa

Table 6.1: Length power law exponents for some coastlines.

Richardson wrote that the slopes of the log-log plots "may be expected to have some positive correlation with one's immediate visual perception of the irregularity of the frontier." We can apply our knowledge of power laws to reformulate this in a more familiar relationship. Assuming $N(\ell) \approx (1/\ell)^d$, we find $L(\ell) = N(\ell) \cdot \ell = \ell^{1-d}$. That is, the slope of the line in the log-log plot is $1 - d$. Solving for d, we get the values in Table 6.2.

1.25	west coast of Britain
1.15	land frontier of Germany
1.14	land frontier of Portugal
1.02	coast of South Africa

Table 6.2: Dimensions of some coastlines.

Benoit interpreted this value of d as a dimension. Indeed, as seen in the same-scale Google Maps images of Fig. 6.34, the coast of South Africa is very smooth and the west coast of Britain is very rough. This was one of the first instances where physical objects were seen to have noninteger dimensions.

Ammonite sutures

Ammonites are extinct marine invertebrates that lived in coiled shells like those of the Nautilus. In Sect. 3.13, we discussed ammonite sutures, the wiggly boundary where new shell is added as the mollusk grows. The box-counting dimension of these sutures was used to quantify complexity, the results ranging from d only slightly greater than 1.0 to slightly greater than 1.6. A general observation that the dimension increases in descendant species suggested to some paleontologists that natural selection favors increased dimension, hence increased complexity.

In *Full House*, Gould used these dimension measurements to refute the notion that evolution favors increased complexity by observing that

- the majority of lineages of ammonites contain species with low-dimensional sutures throughout their histories,

- among the descendants of a given ancestor, no bias toward higher suture dimension was found,

- no correlation was observed between the suture dimension and the longevity of a species.

As we mentioned earlier, the drift toward higher dimensions was an instance of what Gould called a *left wall* effect. Since a curve cannot have dimension lower than 1, and since shell sutures are curves, random genetic drift must appear to push the dimension of ammonite shell sutures higher, when we consider ammonites as a group. Following the suture dimensions of individual genera does not present just a more nuanced view but leads to an entirely different conclusion.

Gould always looked for the bigger picture. He made a breath-taking extrapolation of his interpretation of ammonite suture complexity evolution, questioning the perceived drive toward greater complexity in all life. Specifically, you can't get much simpler than bacteria and remain alive, so for simple living systems the only direction for evolution toward explore is greater complexity. But analyzing the fossil record, Gould found no bias toward greater complexity in the descendants of more complex creatures. He deduced that there is no evolutionary drive toward greater complexity, which suggests that the evolution of intelligence need not be common, that we may be a "glorious accident." This is not where you expected to go from a discussion of the box-counting dimension of sutures of fossil shells, is it?

6.6.2 Pulmonary and circulatory systems

In his article "Mandelbrot's Fractals and the Geometry of Life," physiologist Ewald Weibel writes:

> When human lungs were studied morphometrically by light microscopy we had measured the internal surface area of an adult human lung at about $60 - 80 m^2$ whereas later, using the electron microscope with its higher resolving power, this estimate increased to $130 m^2$, the value now taken as real. This too is related to the fact that the lung's internal surface is a space-filling fractal surface whose dimension is estimated at 2.2.

For some time, Weibel had known that measuring at different resolutions gave different areas. Until he heard Benoit talk about Richardson's power law scaling for coastlines, Weibel had not thought of this difference between optical and electron measurements as a characteristic of the lungs having a dimension greater than 2.

In the same vein, W. Huang and coworkers found the pulmonary arterial tree has a dimension of about 2.71, and the pulmonary venous tree, of about 2.64. Interestingly, in *The Fractal Geometry of Nature* Benoit interpreted old physiological results to show that typical arterial trees will have a dimension of about 2.7. These numbers are all higher than the pulmonary system's 2.2, a fact which stems from the vascular system's job of delivering nutrients to every living cell of the body. As a result, the vascular system is much closer to space-filling, and so has a higher dimension, near 3. But the difference between "near 3" and "exactly 3" is important. Because our circulatory networks have dimensions less than 3, they occupy only a small fraction, about 3%, of the body's volume. Evolution does discover remarkable efficiencies.

6.6.3 DLA, BCCA, dust, planets, and the dark night sky

In Sect. 3.10 we saw that diffusion-limited aggregation (DLA) and ballistic cluster-cluster aggregation (BCCA) model the growth of dust clumps that may catalyze the formation of planets. Now we use the mass dimension to put some numbers behind this idea.

As described in physicist Jens Feder's *Fractals*, DLA clusters grown in two dimensions, by electrodeposition on the surface of a fluid, for example, have mass dimension $d_m \approx 1.71$, and DLA clusters grown in 3-dimensional space have $d_m \approx 2.5$. The mass of the cluster varies with its radius as

$$M(r) \approx r^{2.5}$$

while the volume enclosed by the cluster varies as

$$V(r) \approx r^3$$

so the density, the mass per unit volume, varies as

$$\frac{M(r)}{V(r)} = \frac{1}{r^{.5}}$$

In biophysicist Tamás Vicsek's *Fractal Growth Phenomena* we see that ballistic cluster-cluster aggregates grown in the plane have a dimension of about 1.55 and those clusters grown in space have a dimension of about 1.9. This gives

$$M(r) \approx r^{1.9}$$

so the density varies as

$$\frac{M(r)}{V(r)} = \frac{1}{r^{1.1}}$$

As these clusters grow in the solar nebula, their density continues to decrease, eventually matching that of the nebula. The motion of the cluster also slows to the rotation speed of the nebula. Collisions between these low-density clusters happen at relatively low speed. Because collisions collapse parts of the clusters, restructuring them by localized melting, both the density and the dimension of these clusters slowly increases.

Because diffusion-limited aggregation is so central to applications of fractal geometry, and also because this was a major focus of Benoit's work, we'll give a brief sketch of the mechanism of DLA.

All aggregation models begin with a seed particle. Moving particles travel in a direction chosen uniformly randomly from 0° to 360° and for a distance normally distributed – that is, following the bell curve – before changing direction and moving on. These changes are the effects of collisions with the particles of the surrounding medium (water, air, solar nebula) as they undergo random thermal motion. When the moving particle comes close enough to a point already in the cluster, it sticks.

As illustrated in the first image of Fig. 6.35, growth deep inside the fjords is very slow, because any randomly wandering particle must avoid contacting the outer parts of the cluster in order to travel all the way down a fjord. That is, the branches screen the interior from additional growth. For instance, the second image shows a small cluster, and the third image is the last 100 particles added, demonstrating that growth occurs on the periphery of the cluster. In the fourth image we see why side branches grow: additional growth of the side branch is more likely than is adding another side branch adjacent to this location. In DLA, branches tend to grow side branches, which in turn grow their own side branches. Fractality appears to be a natural consequence of this aggregation.

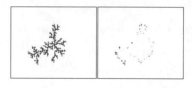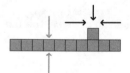

Figure 6.35: First: Screening of growth in the inner regions by the outer regions. Second: A small cluster. Third: the last 100 points added to a small cluster, illustrating peripheral growth. Fourth: A line of particles is unstable against side growth

But DLA is more complicated. Working with Benoit, physicist Alessandro Vespignani and programmer Henry Kaufman studied the statistical properties of million-particle DLA clusters. Their analysis was based on counting the number of cluster particles lying on a circle centered at the cluster seed particle. A circle is 1-dimensional (it's just a line segment with its endpoints glued together), so by the intersection formula of Sect.

6.5 applied to d_m,

$$\begin{aligned}d_m(\text{particles on the circle}) &= d_m(\text{DLA cluster}) + d_m(\text{circle}) - d_m(\text{plane}) \\ &= 1.71 + 1 - 2 \\ &= 0.71\end{aligned}$$

but careful numerical experiments give 0.65, a small, but noticeable, difference from 0.71. Benoit attributed this difference to lacunarity, a measure of the distribution of gaps in a fractal. Roughly, the higher the lacunarity, the more uneven the particle distribution. As the DLA cluster grows, it grows more arms and its lacunarity decreases. This remains an active area of study with many unanswered questions.

Finally, we can see how hierarchical clustering of galaxies can explain the dark night sky. From Sect. 3.12 recall the formulation of Olber's paradox: in a static universe infinite in space and time, our line of sight in every direction will encounter a star. So every part of the sky should look as bright as the surface of a star. Other than its hypotheses, what's wrong with this argument?

Suppose the large-scale distribution of galaxies is a fractal S of mass dimension d_m. Our line of sight L is just a line, of dimension 1. To find the likelihood that our line of sight intersects a star, we can use the intersection formula to find the likely dimension of the intersection of the line of sight with S.

$$\begin{aligned}\dim(L \cap S) &= \dim(L) + \dim(S) - 3 \\ &= d_m - 2\end{aligned}$$

So we see that if $d_m < 2$, then a typical line of sight will miss S, hence never will encounter a star. A sparse fractal distribution of galaxies explains the dark night sky. Of course, the expansion and finite age of the universe provide another explanation.

6.7 An irresponsible speculation: can we move in noninteger dimensions?

In his novel *Mr g*, Alan Lightman speculates on creating universes of different dimensions.

> In some, I even altered the number of dimensions of space: four, six, sixteen, to see what might happen. And why not try fractional dimensions, like 13.8?

Lightman doesn't describe the outcome of these experiments, so we'll take a moment to speculate on life in noninteger dimensions.

Suppose for a moment that we live on a plane and can move freely in 2 dimensions. The most obvious interpretation of this is that we are able to move in any of infinitely many directions from any point in the plane. Another possibility is that at some small scale space itself is quantized. Not infinitely divisible, space is grainy, a fine mesh, and at each point we can move in only four directions: left, right, forward, and back. If the mesh is fine enough, we might not notice the difference between moving on a grid and moving along a curve. This may sound impossibly strange, but loop quantum gravity (described in Lee Smolin's book *Three Roads to Quantum Gravity*, for example), postulates the quantization, or granularity, of space and time in an attempt to reconcile quantum mechanics and general relativity. Viewed at a small-enough scale, all paths may be zigzags rather than smooth curves.

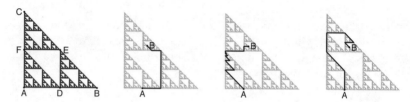

Figure 6.36: First: Gasket space with some points labeled. Others: Three paths between A and B.

Now suppose we live on a gasket. What would be different? For one thing, at some points the choice of directions varies. The first image of Fig. 6.36 shows a skeleton of gasket space. Real gasket space would contain many more, and much smaller, triangles. At the labeled points, the choices of directions are these:

- A: east and north only,
- B: west and northwest only,
- C: south and southeast only,
- D: east, north, northwest, and west,
- E: southeast, northwest, west, and south,
- F: east, north, south, and southeast.

All the other triangle vertices have one set of possibilities corresponding to D, E, or F. However, most points of the gasket are not triangle vertices. These are limits of sequences of triangle vertices, so the possible movement directions may be richer.

In Fig. 6.36 we see three paths, in the gasket, from point A to point B. Of course, we can find many more paths from A to B. If we lived in

6.7. A SPECULATION ABOUT DIMENSIONS

the gasket, how would these paths appear? Would we notice the lack of a straight line in the plane as we crossed the big empty middle of the gasket from A to B? No, because we wouldn't see the big empty middle of the gasket or for that matter, anything outside the gasket. Perhaps we can think of the line segments of the gasket as tunnels through which we must travel, our world a collection of subterranean passages like those of an ant colony or a prairie dog city. The only way we might notice the empty triangles is that in the straight path from D to E, all the side tunnels open toward the east, none toward the west. Over large-enough distances, we would find some constraints in the directions we could move.

Now the universe cannot actually be a gasket, because the inverse square law that gives stable planetary orbits, among other things, requires 3-dimensional space. But this does not prohibit fractal structure for space itself. We know the Sierpinski tetrahedron, consisting of $n = 4$ copies each scaled by $r = 1/2$, has $d_s = 2$ although it is a subset of 3-dimensional space. Logically, this can be extended to some 3-dimensional relative of a gasket, consisting of $N = 8$ pieces each scaled by $r = 1/2$, assembled in 4-dimensional space. We would see only the 3-dimensional insides, not the big gaps in hyperspace.

Certainly, this is a crazy idea, unsupported by any evidence available now. Better guides to the real possibilities of this idea can be found in Susie Vrobel's *Fractal Time* and Laurent Nottale's *Scale Relativity and Fractal Space-Time*; our little story is just an exercise in imagination. But imagination is so important in science, in literature, in life. Fractals give us many avenues for exercising our imaginations.

Chapter 7

Further developments

This chapter is necessarily speculative, and probably we will miss many important developments. After all, the original vision of John von Neumann, one of the architects of modern computers, was of a few large machines, run by armies of technicians, predicting the weather and managing the power transmission grid and the telephone system. No one imagined computers as household appliances, connecting us each to the Internet. For that matter, no one imagined the Internet. Mostly, our imaginations are far too conservative when it comes to the future of technology. Nevertheless, we'll share some guesses.

One is that the random IFS method introduced in Chapter 2 for generating fractals can be adapted to help visualize patterns in data. Another is multifractals, fractals too complex to be described by a single power law, whose roughness varies with location. Although this topic has many subtleties, the basics can be understood through a simple modification of IFS. Again, we will focus on the pictures, and the math that makes them. We'll include some applications of driven IFS and of multifractals and end with one other example, a favorite of ours.

7.1 Driven IFS and data analysis

We have seen how random IFS can fill in a shape gradually, but what happens if the random IFS algorithm is implemented with the transformations selected by some specific sequence of values? Can we learn something by looking at the IFS patterns produced by data?

British mathematician Ian Stewart explored some examples of this question for the IFS transformations of an equilateral gasket, with the order of the transformations selected by sequences derived from iterating functions, including the logistic map (defined in Sect. 7.2) and combinations of sines. Stewart obtained unevenly filled gaskets with some bits left empty. He concluded by commenting that these patterns appear to distinguish various types of correlation in chaotic time-series. Indeed they seem to be inordinately sensitive to such correlations, but in a rather unpredictable manner.

Can these phenomena be explained? If so, can the method be used to detect structure in "real" time-series? Maybe somebody would like to think about it.

And think about it is exactly what we're going to do.

Stewart's results must be read against the background of the gasket, which is not necessarily very easy, given the holes already in the gasket. We prefer to use the square IFS rules

$$T_i(x, y) = (x/2, y/2) + (e_i, f_i)$$

with translations

$$(e_i, f_i) = (0,0), \ (1/2, 0) \ (0, 1/2), \text{ and } (1/2, 1/2) \text{ for } i = 1, 2, 3, 4.$$

All our driven IFS plots start with $(1/2, 1/2)$, the point at the center of the square. The square IFS rules T_1, T_2, T_3, and T_4 can be expressed as moving halfway to corners $1, 2, 3$, and 4, that is, to the points $(0,0)$, $(1,0)$, $(0,1)$, and $(1,1)$.

Patterns in the underlying data are revealed as departures from a uniformly filled unit square. If 10,000 points are plotted by this IFS, each transformation having been applied independently with probability 0.25, the result will be a pretty uniform scattering of points, as we see in the first image of Fig. 7.1.

 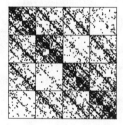

Figure 7.1: First: The square IFS with $p_1 = p_2 = p_3 = p_4 = 0.25$. Second: The square IFS with $p_1 = p_4 = 0.1$ and $p_2 = p_3 = 0.4$.

In the second image of the figure, we see 10,000 points, this time obtained by the probabilities $p_1 = p_4 = 0.1$ and $p_2 = p_3 = 0.4$. From these probabilities, we can make predictions about the pattern in the IFS, for example, the dense concentration of points along the 2–3 diagonal (the line between corners 2 and 3), the image of this diagonal line in squares 1 and 4, and the dearth of points along the 1–4 diagonal, the line between corners 1 and 4.

Let's try an example using values derived from nature instead of from our mathematical imaginations. Take the DNA sequence for amylase, a sequence of $3,957$ characters drawn from the alphabet $\{C, A, T, G\}$. If you read the sequence in order, applying T_1 when you encounter C, T_2 for A,

7.1. DRIVEN IFS

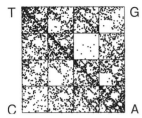
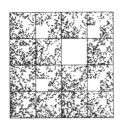
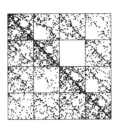

Figure 7.2: First: IFS driven by the amylase DNA sequence. Second and third: Surrogates of the amylase-driven IFS.

T_3 for T, and T_4 for G, you will get the first image of Fig. 7.2. This is the IFS *driven by* the amylase sequence.

Probably the most obvious feature of the amylase-driven IFS is that address 41 is almost empty. Since G immediately follows C only very rarely in the amylase genetic code, the corresponding transformations only lead to this square a handful of times. Because address 41 is almost empty, every address containing 41 must also be almost empty. We can see the most obvious examples of this at addresses 141, 241, 341, and 441, but many more squares, smaller yet, are almost empty because their addresses contain 41.

Just by forbidding the combination 41, with the points otherwise uniformly distributed, we can make a fairly believable copy of the amylase-driven IFS. However, comparing this forgery in the second image of Fig. 7.2 with the real deal, we can see that we are missing the strong diagonal between the lower right (A) and upper left (T) corners. The second image was constructed with about the same number of points in each of the four length-1 addresses, so we should adjust these probabilities to represent more accurately the actual distribution of bases in the amylase sequence. This sequence has 589 Cs, 1305 As, 1389 Ts, and 674 Gs. The third image of Fig. 7.2 forbids the composition $T_4 \circ T_1$ and has probabilities adjusted to $p_1 = 589/3957 \approx 0.149$, $p_2 = 1305/3957 \approx 0.330$, $p_3 = 1389/3957 \approx 0.351$, and $p_4 = 674/3957 \approx 0.170$. This produces a strong 2–3 diagonal quite similar to that of the amylase-driven IFS.

The point of Fig. 7.2 is cautionary. The left side of the figure, the IFS driven by the amylase sequence, looks complicated, rife with structures. But the right side shows just how much of this appearance can be obtained by matching the probabilities of the data and imposing a single exclusion. To make driven IFS an effective tool, we must learn to read more subtle patterns, to count and to think, as well as to look.

This is better done with a collection of examples that includes more than DNA sequences. To show the wider applicability of this method, now we'll see how to drive an IFS by time series. Recall this is a sequences of measurements ordered in time. Think of the intervals between beats of your heart, for instance. Or suppose we have a sequence of successive generations of a model insect population in an environment with limited resources. Rather than recording the actual numbers of insects in each generation, let us say that $x_1, x_2, \ldots, x_n$ are each generation's fraction of the *carrying capacity*, the maximum population supported by the environment. We'd like to turn this into a sequence of 1s, 2s, 3s, and 4s, instructions for driving the IFS. To do this, we *coarse-grain* or *bin* the data, that is, break the time series up into four chunks of measurements. How we break up the measurements depends on which features of the data we want to emphasize. Of the several avenues available, the main two are *equal-size bins* and *equal-weight bins*. (Another, *median-centered*, we'll mention a bit later.)

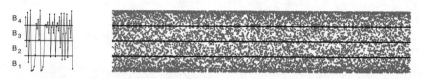

Figure 7.3: A time series example. First: The first 50 points. Second: 10,000 points.

For equal-size bins we take the range $R = \max\{x_i\} - \min\{x_i\}$ of the whole series and divide it into four intervals of equal length,

$$B_4 = [\min\{x_i\} + 3R/4, \max\{x_i\}]$$
$$B_3 = [\min\{x_i\} + R/2, \min\{x_i\} + 3R/4)$$
$$B_2 = [\min\{x_i\} + R/4, \min\{x_i\} + R/2)$$
$$B_1 = [\min\{x_i\}, \min\{x_i\} + R/4)$$

The first image of Fig. 7.3 shows the first few values of a time series from a model predicting generations of an insect population, together with horizontal lines separating the four bins. Successive values are connected by lines to emphasize the temporal ordering. The second image shows 10,000 points of this time series

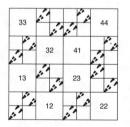
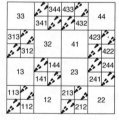

Figure 7.4: The IFS driven by the time series of Fig. 7.3. First: Empty length-2 addresses. Second: Empty length-3 addresses.

without the point-to-point lines. Whenever a data point lies in bin B_1, the lowest bin, in the driven IFS we apply transformation T_1. When a data point lies in B_2, wew apply T_2, and so on. In this way, the transformations are applied in the order determined by the data.

7.1. DRIVEN IFS

The IFS driven by this time series, with the data sorted into equal-size bins, is shown in Fig. 7.4. In the first image we label the empty length-2 addresses. In the second image we see that every empty length-3 address contains an empty length-2 address. In fact, every empty square in this IFS is the result of some forbidden pair, a length 2-address unoccupied, perhaps because of some restriction in the process generating the time series. (In a moment we'll begin a discussion of some causes of empty addresses.) At least in this example, once we know the forbidden pairs, we know every combination that cannot occur. In A.90 and A.91 we give a condition guaranteeing that every forbidden combination contains a forbidden pair, but as the next example shows, this relation does not hold for all driven IFS.

 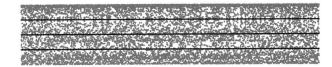

Figure 7.5: Another time series example.

In Fig. 7.5 we see a time series for another population, with only slightly different parameters. Again, we use equal-size bins; the driven IFS is shown in Fig. 7.6. Unlike what we saw with the previous example, in Fig. 7.6 we see some empty length-3 addresses (111, 141, and 334) that do not contain empty length-2 addresses. So in this example we must look at longer addresses, and consequently further into the past, to identify all the forbidden combinations.

A bit of thought reveals another problem. How can we tell whether address 141 is empty because it is the address of a combination excluded by the dynamics of the population or because we do not have a sufficiently long time series? In A.92 we give a calculation, based on the probabilities of the occupied

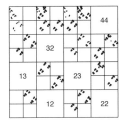 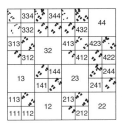

Figure 7.6: The IFS driven by the time series of Fig. 7.5. First: Empty length-2 addresses. Second: Empty length-3 addresses.

length-2 addresses and the length of the time series, for estimating the likelihood that this address is empty due to an exclusion in the dynamics.

Depending on the situation, it may make more sense to divide the data into equal-weight bins. Here the bin boundaries are placed so each bin contains about the same number of points, or roughly the same weight

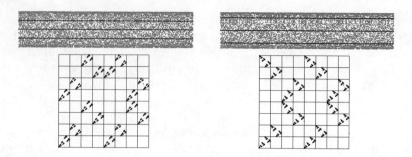

Figure 7.7: A time series with equal-size (first) and equal-weight (second) bins and their respective driven IFS.

pound for point. In the top of Fig. 7.7 we see the time series from Fig. 7.3 divided into equal-size bins in the first image, equal-weight bins in the second. The two IFS driven with these binnings are shown below the corresponding time series.

By visual inspection of the density of points in a region we can estimate the likelihood of the points landing in the combination of bins corresponding to the address. From this, we can estimate probabilities of certain behaviors a system may exhibit. Let's try this with some cardiac data, specifically, the durations of the intervals between successive heartbeats. In Fig. 7.8 we see a cardiac time series, with four different binnings. The top first has equal-size bins, and the top second has equal-weight bins. Both the two lower binnings are median-centered; that is, the boundary between bins B_2 and B_3 is the median value of the time series. The other two bin boundaries are 10% of the range of the values above and below the median for the first, and 5% above and below for the second. We use the median, or middle, value instead of the mean, or average, because the presence of a few extreme values can have a larger effect on the mean than on the median.

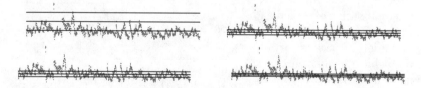

Figure 7.8: A cardiac time series, the top with equal-size bins first and equal-weight bins second. The bottom are two median-centered binnings.

From inspecting the first driven IFS in Fig. 7.9, we can quickly tell that the presence of extreme values may limit the usefulness of equal-size binning. Since most of the data points lie in B_1 and B_2, most of the driven IFS points in this first example fall on the lower edge of the square, hiding

7.1. DRIVEN IFS

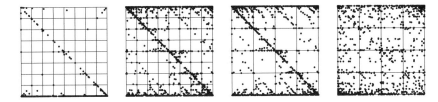

Figure 7.9: The driven IFS using the binnings of Fig. 7.8.

any further detail that may lie within that part of the sequence. Also, most of the points in B_3 are interspersed with points in B_2; the fairly dense collection of points on the 2–3 diagonal indicate long sequences of points falling in B_2 and B_3 in various combinations. There is not much else we can draw from this picture, since the majority of the graph is empty. Let's move on.

The equal-weight driven IFS, the second image of Fig. 7.9, shows much more structure, including a feature called the *backward Z*, often encountered in experimental data. The backward Z signals frequent movement of consecutive data points between adjacent bins. There are many data points visiting bins 1 and 2, sometimes several consecutive points falling in a single bin, switching bins, or alternating between them. This spreads driven IFS points along the bottom of the square. Points visiting bins 2 and 3 spread driven IFS points along the 2–3 diagonal; and data in bins 3 and 4 spread IFS points along the top of the square, the 3–4 line .

But we also see so much more. Faint echoes of the 2–3 diagonal in squares 1 and 4 are obtained by applying T_1 and T_4 to points on the 2–3 diagonal; that is, a string of data points in bins 2 and 3 is followed by a point in bin 1 or 4. Also, the diagonal points in square 4 give rise to the diagonal points in square 44, which in turn give rise to the diagonal points in square 444: many combinations of 2 and 3 can be followed by one, two, or three 4s. On the other hand, despite the abundance of points in square 111, we see only a very weak echo of the square 1 diagonal in squares 11 and 111. From this we deduce that except for points very near corner 2, points along the 2–3 diagonal are only rarely followed by two or three consecutive points in address 1.

And of course there's still more. What about the points along the horizontal line through the center of the square? How did they get there? Do the center line points to the left of the 2–3 diagonal come from applying T_3 to points along the line at the bottom of the square or from applying T_1 to points along the 3–4 line at the top of the square? How could we tell? We'll let you think a bit about that.

Further, by counting the number of points in any given address square, we can estimate the probability of the corresponding combination of bins. For instance, this time series has $1,969$ data points, and 204 of these have address 222, so the probability of finding three consecutive data points in bin 2 is about $204/1969 \approx 0.103$. If three consecutive points in bin 2 correspnd to something noxious – heartbeats slower than normal for example – knowing this probability could help your cardiologist adjust the dosage of your medcine.

Let's look at this more closely. Suppose we have two consecutive data points in bin 2. What are the probabilities the next data point will lie in bin 1, in bin 2, in bin 3, or in bin 4? Writing $N(jk)$ for the number of driven IFS points with address jk, and $N(ijk)$ for the number with address ijk – we call these the *occupancies* of the addresses – we count

$$N(22) = 308, \ N(122) = 51, \ N(222) = 204, \ N(322) = 44, \ N(422) = 9$$

Then we estimate

$$\text{Prob}(2 \to 2 \to 1) = \frac{N(122)}{N(22)} = \frac{51}{308} \approx 0.166$$

and similarly

$$\text{Prob}(2 \to 2 \to 2) \approx 0.662, \ \text{Prob}(2 \to 2 \to 3) \approx 0.143,$$
$$\text{and } \text{Prob}(2 \to 2 \to 4) \approx 0.029$$

We can therefore say that if we observe two consecutive data points in bin 2, about 2/3 of time the next data point will also lie in bin 2. Many more quantitative deductions can be made, but to help develop your intuition it is useful to look at the driven IFS picture. Let it speak to you.

The third and fourth driven IFS of Fig. 7.9 use a binning centered on the data median, with the other bin boundaries placed symmetrically above and below the median. Unlike the equal-size and equal-weight bins, both of which give just one way to sort

Figure 7.10: A driven IFS and attempts to produce this image with forbidden pairs and triples.

the data, the median-centered approach yields a whole collection of sorting strategies, a different strategy for each placement of the boundaries between the bottom two and between the top two bins. Placing the two outer boundaries near the median pushes most of the data points into bins 1 and 4, giving a more detailed view of the points in bins 2 and 3. Moving the outer boundaries out pushes most of the data points into bins 2 and 3, giving a more detailed view of the points in bins 1 and 4. In the third image of Fig. 7.9 the outer boundaries are farther from the median, in the

7.1. DRIVEN IFS

fourth, closer. Studying how the bin occupancy changes as the bin boundaries are moved can give a detailed picture of the dynamical relations that generated the data.

The older author has been using driven IFS in his classes since the early 1990s. In the late 1990s, some students ran our driven IFS software using ice core data – annual records of atmospheric CO_2 levels – and got a picture much like the first image of Fig. 7.10. Though in retrospect the source of this pattern should have been immediately clear, for a short while it was puzzling. The empty length-2 addresses are 14, 24, 41, and 42. Forbidding these pairs produces the second image of the figure, not a good match with the driven IFS. Taking into account the additional forbidden triples – 114, 124, 134, 214, 224, 234, 314, 324, 334, 341, 342, 441, and 442 – gives the third image of the figure, also not a good match, in part because much of the top edge of the square is empty.

The solution lay in looking at the time series, shown in Fig. 7.11. We see two distinct regimes: the points of the left 90% of the time series lie in bins 1, 2, and 3; those of the right 10% in bins 3 and 4. Regime changes signal an alteration of the dynamics, or perhaps the presence of several time scales: over the short term, the time series is generated by a process whose parameters may change very slowly.

Figure 7.11: A time series revealing two distinct regimes.

In the first image of Fig. 7.12 we see another cardiac time series (top) and the driven IFS with equal-weight bins (bottom). The driven IFS suggests two gaskets face-to-face. The lower left gasket is generated by the transformations T_1, T_2, and T_3, the upper right by T_2, T_3, and T_4. So in the second image we have a two-regime time series (top), the first with points in bins 1, 2, and 3, the second with points in bins 2, 3, and 4. Sure enough, the driven IFS is two gaskets face-to-face. Now certainly the cardiac time series does not fall into two neat regimes, but look more closely. Some parts of the cardiac time series lie entirely in bins 1, 2, and 3, some parts entirely in bins 2, 3, and 4. Recognizing the pattern of this driven IFS helped us see two distinct time scales, the rapid measurement-by-measurement fluctuation, and the slower drift between the 1, 2, 3 clusters and the 2, 3, 4 clusters.

To be sure, a driven IFS does not generate new information: every bit of the driven IFS comes from the time series. But what we see, and how easily we see it, can depend delicately on how the data are presented. Benoit often commented on the importance of looking, of seeing. The powerful pattern recognition software of our brains is a resource to treasure. Driven IFS is a tool that sometimes can help that software.

Part of this is learning to recognize regime change, that is, places in a

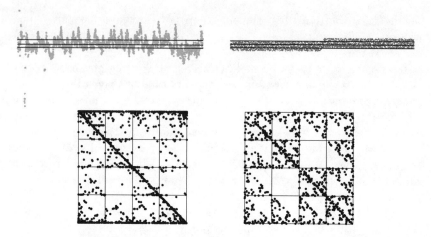

Figure 7.12: First: A cardiac time series with equal-weight bins. Second: A time series with two regimes.

time series where the data seem to shift abruptly. For practice, we'll do one more example. In Fig. 7.13, we see a time series that can be divided into five regimes. In the first regime (labeled A), points are scattered randomly among bins 1, 3, and 4. In B, all points lie in bin 3. In C, points are scattered randomly among bins 2 and 3; in D, all points lie in bin 2; and in E, points are scattered randomly among bins 1, 2, and 4.

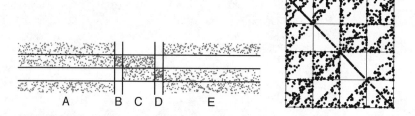

Figure 7.13: First: Labeling the regimes. Second: The driven IFS.

Points in regime A visit bins 1, 3, and 4 in many combinations and so produce driven IFS points on the gasket with corners 1 (the point $(0,0)$), 3 (point $(0,1)$), and 4 (point $(1,1)$), the fractal generated by T_1, T_3, and T_4. Regime B produces driven IFS points converging to corner 3. Regime C produces driven IFS points on the line between corners 3 and 2. Regime D produces driven IFS points converging to corner 2. And finally, regime E produces driven IFS points on the gasket with corners 1, 2, and 4. The result can be see in the second image of Fig. 7.13.

Without recognizing these regimes, we run into some difficulty when we try to interpret the driven IFS. No single set of forbidden combinations

7.1. DRIVEN IFS

of bins can produce this particular pattern. We can see that no length-2 address is empty, so we would have to start by forbidding some length 3-addresses. But even this wouldn't suffice. In fact, we'd have to forbid arbitrarily long addresses (e.g., forbid 141, 1441, 14441, etc.), an approach that only rarely can give a simple description of the shape. For an indication of this trouble, note that the 2–3 diagonal is not repeated in any other square. These complications may signal a change in the nature of the process being measured. Though regime change is pretty obvious after the fact (Well, ... the intermittent regime change in the first time series of Fig. 7.12 wasn't so obvious.), when we began studying driven IFS we did not expect to be able to see the shifts of the very processes generating the data.

This is a young technique, largely untried except in simple examples. What's lacking to turn this into a good scientific tool is proper statistics to interpret driven IFS. But this situation may change soon. If it does, we will see a new approach to data analysis, sensitive simultaneously across many scales. This application of fractals is just beginning.

Exercises

In problems 1, 2, and 3 sketch or describe the driven IFS for these time series.

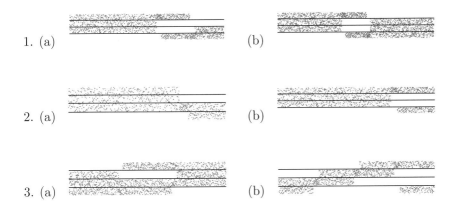

4. Sketch a time series that could generate each of these driven IFS.

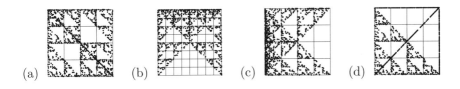

7.2 Driven IFS and synchronization

Driven IFS can reveal surprising patterns in many types of data. In this section we'll see IFS driven by chaotic signals and use what we learn to recognize when some networks of chaotic processes synchronize, when they begin moving in step even though they remain chaotic.

First, let's be more precise about what we mean by "chaotic." Though this notion has been rediscovered sporadically since it was recognized first in the 1890s by the great French mathematician Henri Poincaré, chaos entered popular scientific culture in the 1980s. Before this, the common wisdom was that simple behavior such as the steady swing of a pendulum has simple causes, while complicated behavior such as the growth of a population or spread of an epidemic, has complicated causes. Contradicting this dichotomy, some chaotic systems that exhibit very complicated behavior have simple causes. The most familiar aspect of chaos is *sensitivity to initial conditions*: tiny changes in the starting point of a system can lead to later behavior that is worlds apart from the later behavior of the original starting point. Or, quoting Grandpa Simpson from the "Time and Punishment" segment of *Treehouse of Horror, V*, "Even the tiniest change can alter the future in ways you can't imagine."

Our tool for generating chaotic signals is the *logistic map*, $L_r(x) = rx(1-x)$, where r is a constant, related to birth and death rates and to competition for resources. To guarantee the iterates do not run off to infinity (in fact, to negative infinity, very bad behavior for a population) r must lie in the range $0 \leq r \leq 4$. In 1976, the mathematical biologist Robert May used the logistic map as the distillation of the growth of a single species with non-overlapping generations in an environment with limited resources. Iterating this map from some starting point x_0 (the population size when we start measuring it) will generate a time series x_0, $x_1, x_2, x_3, \ldots$ by

$$x_1 = L_r(x_0), x_2 = L_r(x_1), x_3 = L_r(x_2), \ldots$$

The time series of Figs. 7.3 and 7.5, models of population dynamics, were generated by iterating the logistic map for $r = 4$ and $r = 3.97$. Not surprisingly, the pattern of points depends on the number r. In Fig. 7.14 we see the first 200 points of logistic time series for $r = 3.5, 3.75, 3.8275$, and 4.0. Of these four, the only repeating pattern we see is in the $r = 3.5$ time series. Here we see a 4-cycle: the points repeat the same pattern of four different values. Mathematically we write this as $x_{n+4} = x_n$ for all n. This is about as far as possible from any definition of chaos.

The other three time series in Fig. 7.14 are chaotic. Note that the $r = 3.8275$ time series appears to differ from the other chaotic series in that it exhibits a 3-cycle, or something close to a 3-cycle, for some iterates. Then the cycle dissolves into chaos, only for the iterates to return to dance close to the points of the 3-cycle. This combination of behaviors repeats

7.2. DRIVEN IFS AND SYNCHRONIZATION

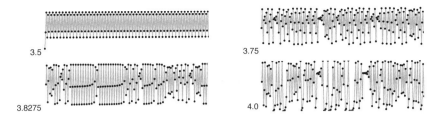

Figure 7.14: Time series for the logistic map and the associated r values.

again and again. Called an *intermittent* 3-cycle, this is a crystal clear example of interleaving predictability and novelty in a chaotic system.

Looking more closely at the other two chaotic time series, we see that both have intermittent fixed points, iterates oscillating about a horizontal line, and at least for the $r = 3.75$ series we see an intermittent 4-cycle in two places. Mathematical (and physical and biological and chemical) chaos is filled with short-lived order. Despite any dictionary-inspired expectations, chaos is not at all a synonym for randomness.

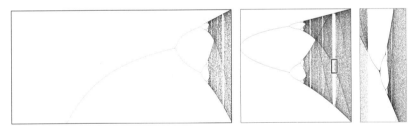

Figure 7.15: The logistic bifurcation diagram and two magnifications.

The time series of Fig. 7.14 show that the simple logistic map can produce a variety of behaviors. An iconic image of the "chaos revolution" of the 1980s, the logistic *bifurcation diagram*, Fig. 7.15, is one way to encode some aspects of these behaviors and to see just how thoroughly chaos and order are mixed together. Along the horizontal axis we plot the r value, $0 \leq r \leq 4$; along the vertical axis we plot the x-values. In more detail, to allow the iterates to settle down to their eventual behavior, we select some number d of iterates to compute without plotting and some number p of additional iterates to compute and plot. Then for each value of r, we compute $x_1 = L_r(x_0), \ldots, x_{d+p} = L_r(x_{d+p-1})$ and plot the points $(r, x_{d+1}), \ldots, (r, x_{d+p})$.

The first quarter of the first image of Fig. 7.15, $0 \leq r \leq 1$, shows iterates converging to 0: the population dies out. For $1 < r < 3$ the iterates converge to a fixed point, $x = 1 - 1/r$. At $r = 3$ we see that the fixed point splits into a 2-cycle, a literal bifurcation. (In fact, the fixed point still is present, but now is unstable. The 2-cycle appears at

$r = 3$.) With increasing r, the 2-cycle bifurcates into a 4-cycle, that to an 8, that to a 16, and so on. After an infinite sequence of bifurcations through 2^n-cycles for all $n \geq 0$, the diagram exhibits bands of chaos, themselves filled with windows containing new cycles, each of which goes through its own sequence of period-doubling bifurcations. The largest of these windows exhibits a 3-cycle, visible in the second image of Fig. 7.15. The third image is a magnification of the box in the second image. Here we see the middle branch of the 3-cycle gives rise to its own small copy of the bifurcation diagram. Yes, the logistic bifurcation diagram contains infinitely many small copies of itself, each of which contains infinitely many still smaller copies of the diagram, and so on. Do you detect a familiar theme? Pretty complicated for iterates of a humble parabola, but we're thinking of another theme.

In A.93 we show the bifurcation diagrams of two other functions. One is non-differentiable at a single point; the other has a discontinuity at a single point. The bifurcation diagrams of these functions look nothing like that of the logistic map.

Before leaving this digression, we consider one more issue. How can we be sure these dark bands are chaos and not, say, million-cycles? One answer is that at least some are real chaos, because a theorem of M. Jakobson shows that the logistic map exhibits real chaos for positive length of r-values. (More precisely, the collection of r-values exhibiting chaos form a set of positive 1-dimensional Lebesgue measure, but thinking of this as length is close enough.)

Another way to establish chaos for particular r-values is given in A.94, where we show that the sensitivity to initial conditions that characterizes chaos can be measured by the average rate at which the iterates of nearby points diverge. This is called the *Liapunov exponent*; a positive value indicates chaos, and a negative value indicates repeating behavior. For the four values of r in Fig. 7.14, we'll see that the Liapunov exponent is about -0.89, 0.36, 0.33, and 0.69.

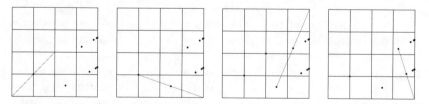

Figure 7.16: Lines indicating the first four applications of the driven IFS to the $r = 3.5$ time series.

In Fig. 7.16 we see the driven IFS for the $r = 3.5$ time series, using equal-size bins. One puzzle is the first IFS, which appears to be a simple collection of points converging to a 2-cycle on the right edge of the square. The time series is a 4-cycle, so shouldn't the driven IFS converge to four

7.2. DRIVEN IFS AND SYNCHRONIZATION

points? Let's look more closely. The first point is the lowest point of the time series and the only point in bin 1. Consequently, the first driven IFS point is $(1/4, 1/4) = T_1(1/2, 1/2)$. Then the other time series points alternate between bins 2 and 4, and the driven IFS is a sequence of points moving halfway between corners 2 and 4. The detail – two different points of the 4-cycle in bin 2 and two different points of the 4-cycle in bin 4 – is lost with equal-size bins. With equal-weight bins, the driven IFS would reveal the 4-cycle.

The driven IFS for the other three time series are shown in Fig. 7.17. The $r = 4.0$ driven IFS is particularly interesting for its complex yet regular structure.

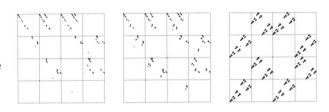

Figure 7.17: IFS driven by the $r = 3.75$ (first), 3.8275 (second), and 4.0 (third) time series.

We've seen this before in Figs. 7.4 and 7.7. In the first image of Fig. 7.18 we see the forbidden pairs and triples for this driven IFS. The second image of Fig. 7.18 is the IFS with memory determined by these eight forbidden pairs. These are the same fractals. Unlike most values of r, the $r = 4.0$ logistic map gives a surprisingly simple driven IFS. See A.90. Even from the apparent chaos of the time series, order emerges.

 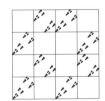  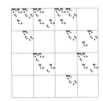

Figure 7.18: Forbidden pairs and triples for the $r = 4.0$ (first image) and $r = 3.8275$ (third image) driven IFS, together with the IFS with memory determined by these forbidden pairs (second and fourth images).

The third and fourth images of Fig. 7.18 are more complicated. In the third we see the forbidden pairs and triples for the $r = 3.8275$ logistic map, using equal-size bins. In this case, some of the forbidden triples (141, 342, and 334) do not contain forbidden pairs, so memory longer than pairs is needed to specify this fractal. We can easily see this by comparing the IFS determined by these forbidden pairs (fourth image) with the $r = 3.8275$ logistic driven IFS (third image). Unlike the example seen in the first two images of Fig. 7.18, these fractals clearly differ.

Though useful for these kinds of distillations, a single logistic map can model only a system whose agents all behave the same way. But many systems admit groups of agents acting according to different rules. For these, we take several logistic maps and couple them together, so the behavior of each map has a definite influence on the neighboring maps. These are called *coupled map lattices (CML)*. For two logistic maps with variables x_n and y_n, a CML is defined by these relationships:

$$x_{n+1} = (1-c)L_r(x_n) + cL_s(y_n)$$
$$y_{n+1} = cL_r(x_n) + (1-c)L_s(y_n)$$

Note that we allow different parameters, r and s, for the x and y logistic maps. The number c, $0 \leq c \leq 1$, is the *coupling constant*. The $1-c$ and c factors are placed as they are so that $c = 0$ gives a completely uncoupled system, where x_{n+1} depends on just x_n, and y_{n+1} on just y_n. Non-zero values of c give maps coupled in nontrivial ways.

By extension, the simplest three logistic map CML is

$$x_{n+1} = (1-c)L_r(x_n) + (c/2)L_s(y_n) + (c/2)L_t(z_n)$$
$$y_{n+1} = (c/2)L_r(x_n) + (1-c)L_s(y_n) + (c/2)L_t(z_n)$$
$$z_{n+1} = (c/2)L_s(x_n) + (c/2)L_r(y_n) + (1-c)L_t(z_n)$$

Here the c factor has been replaced by $c/2$ to guarantee that if x_n, y_n, and z_n all lie between 0 and 1, then so will x_{n+1}, y_{n+1}, and z_{n+1}. More general networks allow different coupling values, for instance,

$$x_{n+1} = c_{11}L_r(x_n) + c_{12}L_s(y_n) + c_{13}L_t(z_n)$$

provided $c_{11} + c_{12} + c_{13} = 1$.

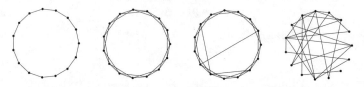

Figure 7.19: Nearest neighbor coupling, next nearest neighbor coupling, next nearest neighbor with some random coupling, all random coupling.

With more maps, more coupling patterns are available. Fig. 7.19 shows four ways to couple 15 maps together, with each map represented by a point. There are many, many other combinations. Especially interesting is hierarchical clustering, in which strongly coupled clumps of logistic maps are grouped together, and the coupling between clumps is weaker than the coupling within the clumps. These compose still larger clusters of clusters with even weaker coupling, and so on. Think of logistic maps as vertices of a Sierpinski gasket of connections.

7.2. DRIVEN IFS AND SYNCHRONIZATION

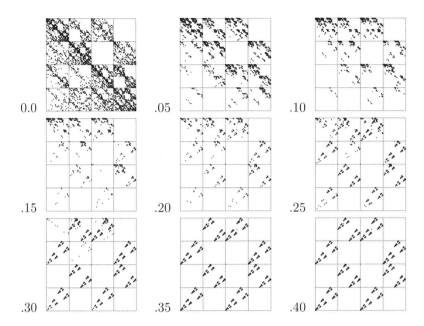

Figure 7.20: IFS driven by the average of three $r = 4.0$ logistic maps, with coupling constant c indicated to the lower left of each driven IFS.

To explain why we are interested in coupled map lattices, think about how we measure the air temperature with a thermometer, not analogous to how we measure air temperature with an iPhone. Air molecules collide with the thermometer, transferring some energy with each collision. The thermometer measures the average energy of the air molecules that collide with it, which have in turn received energy from other air molecules, and so on. Like temperature, some physical measurements are averages of many smaller-scale events. Motivated by this averaging effect, we build networks of coupled logistic maps and at each iteration record the average of the values produced by all these maps. Then we drive the IFS with these averages.

In Fig. 7.20 we see IFS driven by the average of three coupled $r = 4.0$ logistic maps, with different starting points x_0, y_0, and z_0, and with the coupling constant c ranging from 0 to 0.4 in steps of .05. In some of these images, the coupling can be seen by wisps of several familiar fractal patterns in the same grid. The middle row shows some traces of the $r = 4.0$ single logistic map driven IFS, as well as a clustering of points along the top similar to that of the first row images. Note the $c = 0.35$ and $c = 0.40$ driven IFS resemble very closely the $r = 4.0$ single logistic map driven IFS, the third image of Fig. 7.17. At least as far as the driven IFS is concerned, for these values of c, the average of the coupled logistic maps acts like a single logistic map. This is the driven IFS signature of the synchronizing of the three chaotic logistic maps, with of all three doing the same chaotic dance.

Maybe the driven IFS, or the averaging process itself, has lost some information. To investigate more fully, for the moment let's ignore the averages and plot the individual points x_n, y_n, and z_n. In Fig. 7.21 we do this for $c = 0.2$, 0.3, and 0.4. For $c = 0.2$ we see that the three series do not appear particularly related; for $c = 0.3$ they follow similar, but not identical, patterns; for $c = 0.4$ after a few iterates these points coincide almost exactly. But look carefully at what has happened. This is not the synchronization of pendulum clocks discovered in 1665 by the Dutch physicist Christiaan Huygens, or the synchronized flashing of fireflies in Southeast Asia, both examples of *periodic* processes synchronizing. The $c = 0.4$ example shows three *chaotic* processes synchronizing. A single chaotic process exhibits sensitivity to initial conditions; tiny variations in the starting value can lead to entirely different behaviors. Yet here we see chaotic processes that start at different values and rapidly converge to follow *the same* chaotic process.

The logistic map bifurcation diagram presents an organized picture of how the pattern of iterates varies with the map parameter r. While driven IFS can teach us much about iterates for some r-values, a stack of driven IFS pictures, each with its own r-value, would be difficult to unpack visually. (An animation – a logistic driven IFS CAT scan – is much easier to interpret, but except as a sequence of stills, it cannot be put into the pages of a physical book. However, we can put animations on the webpages that accompany this book.)

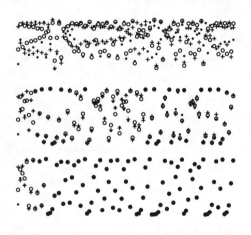

Figure 7.21: Plots of x_n (o), y_n (+), and z_n (·) for (top to bottom) $c = 0.2$, $c = 0.3$, and $c = 0.4$.

Analogous to the logistic map bifurcation diagram, is a CML bifurcation diagram with c-values replacing the r-values of the logistic bifurcation diagram. We call this a *coupling-bifurcation diagram*. Above each c-value we plot the eventual values of the averages of the coupled maps. We see six examples in Fig. 7.22, with the r-values of the maps given below the diagram. In A.95 we give a bit more information about these diagrams.

What can we see from these? Do the coupling-bifurcation diagrams

7.2. DRIVEN IFS AND SYNCHRONIZATION

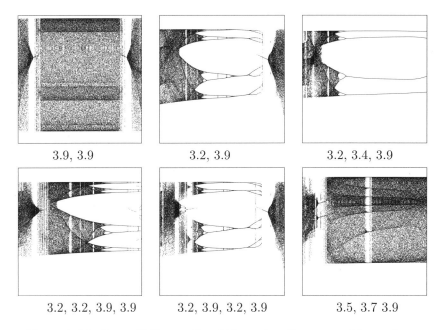

Figure 7.22: Some CML coupling bifurcation diagrams with r-values.

suggest additional tests? Top first is the diagram for two coupled logistic maps, both with $r = 3.9$. For two maps with the same r-value, we'll see in A.95 that finding a range of c-values guaranteeing synchronization is straightforward. And indeed the large middle band of the top first diagram corresponds to synchronized chaotic logistic maps. But what about the regions at either end? In Fig. A.67 of Sect. A.95 we'll see that something completely different is going on there.

The top second diagram also is generated by two coupled maps, one in the periodic range (a single $r = 3.2$ logistic map exhibits a 2-cycle), the other ($r = 3.9$) chaotic. Surely, mixing equal portions of chaos and order should give chaos on a smaller range. But this is not at all what we see. The left third of this diagram looks a bit like part of the single logistic map bifurcation diagram, but left-right reversed. Again near the middle of the diagram we find that the logistic maps synchronize, but now to a 4-cycle. And the diagram contains other peculiar features.

The top third diagram adds a third logistic map, with $r = 3.4$. Here we are coupling one chaotic map and two that produce 2-cycles. Over more than half of the range of c-values, all three synchronize to a 4-cycle. At the moment we have no idea why this happens. If we do figure it out, the answer will go on the webpage.

The bottom first and second diagrams are for four coupled maps, two with $r = 3.2$, two with $r = 3.9$. Because each map is coupled to only its immediate left and right neighbors (with wrap-around: the right neighbor of the right-most map is the left-most map, for example, illustrated in the first diagram of Fig. 7.19), we are not surprised that changing the order of the maps alters the coupling-bifurcation diagram.

The bottom third diagram is a puzzle. The 7-cycle window near the middle c-range appears to follow seven darker curves crossing the wide band that occupies about 3/4 of the c-range. Again, at the moment we have no idea why.

We've included these diagrams to show that as interesting as synchronization is, it isn't the only story in CML town. In fact, the younger author exhibited considerable patience while the older author was distracted for a while, late in the preparation of the text, by this shiny toy. But eventually she put on the brakes. Nevertheless, both authors think these diagrams are fascinating.

The natural world has a wealth of coupled processes, though we may not know the number of oscillators, how closely they are coupled, or which processes are coupled. Examples abound: heart muscle cells pulsing in synchrony, sections of tectonic plates releasing stresses through earthquakes, neural reinforcement when we learn, the response of our immune systems to pathogens, ecosystems and metabolic networks adapting to variations in resource abundance, gene regulatory networks during development of an embryo, and on and on. If we lack a clear grasp of a process because we don't understand some of the fundamental mechanisms, are unclear about the relations between agents, or do not know how many agents make up the system, how can we guess the coupling patterns or make any kind of model? Without a model, predicting any behavior might seem impossible.

One approach is to build a dictionary of coupled maps. One map, two maps, three maps, and larger collections; different r values; different coupling patterns and strengths. Then take the driven IFS for a data series, and compare it with those for the dictionary entries.

What do we mean by compare? One way to find the closest match is to look at the data-driven IFS and seek the dictionary entry that looks most like it. A less taxing and more objective approach, based on the project by Simo Kalla and Nader Sobhan in the older author's fall 2000 fractal geometry course, is to compare the fraction of the total number of points in each length-2, -3, -4 and -5 addresses. The optimal address length depends on the size of the data set and on how evenly the points are spread among the bins. Complete details have not been worked out yet.

Other aspects were worked out, and a preliminary version of the dictionary was produced by C. Noelle Thew in her 2014 applied math senior thesis at Yale. The choices of coupling constants and logistic map parameters had to be managed carefully, to keep the dictionary from becoming too large. Each dictionary entry consists of average occupancies for addresses of lengths 1 through 5, for a given collection of logistic maps with specified r- and c-values. These occupancies are computed from the average values

7.2. DRIVEN IFS AND SYNCHRONIZATION

of the logistic maps, using equal-size bins.

To compare a data series with a dictionary entry, the data bin boundaries are adjusted so the length-1 address occupancies match, as closely as possible, those of the dictionary entry. This is the approach we adopt to deal with the difficulty of choosing optimal bin boundaries for the data series. Denoting by X_{ij} the number of data-driven IFS entries with address ij, and by D_{ij} the number of dictionary entries with this address, the 2-*address correlation* is

$$\kappa_2(X, D) = \frac{N - \frac{1}{2}\sum_{i,j=1}^{4}|X_{ij} - D_{ij}|}{N}$$

with the condition that we ignore any address with fewer than five points. We still are working on rigorous bounds for the sparseness of an address that should be ignored, but ignoring the sparse addresses is a reasonable first step because we seek to describe the bulk behavior of the system and not rare transitions that may be artifacts of the bin boundary placement.

The correlations $\kappa_3(X, D)$, $\kappa_4(X, D)$, and $\kappa_5(X, D)$ are defined similarly.

For a given time series X, the comparison tool calculates $\kappa_2(X, D)$ through $\kappa_5(X, D)$ for each dictionary entry D, and the best matches are determined by these address correlations.

As a test, take the time series generated by three asymmetrically coupled logistic maps,

$$x_{n+1} = 0.3 \cdot L_{r_1}(x_n) + 0.4 \cdot L_{r_2}(y_n) + 0.3 \cdot L_{r_3}(z_n)$$
$$y_{n+1} = 0.4 \cdot L_{r_1}(x_n) + 0.4 \cdot L_{r_2}(y_n) + 0.2 \cdot L_{r_3}(z_n)$$
$$z_{n+1} = 0.3 \cdot L_{r_1}(x_n) + 0.2 \cdot L_{r_2}(y_n) + 0.5 \cdot L_{r_3}(z_n)$$

with $r_1 = 3.754$, $r_2 = 3.985$, and $r_3 = 3.971$. Then drive the IFS by the average value, $(x_n + y_n + z_n)/3$. This driven IFS is the top first image of Fig. 7.23. Because all the dictionary entries are generated with symmetric coupling, for example,

$$x_{n+1} = (1 - c) \cdot L_{r_1}(x_n) + (c/2) \cdot L_{r_3}(z_n) + (c/2) \cdot L_{r_2}(y_n)$$

none of the dictionary entries will be this test system. The comparison tool found several close matches, for example, a single $r = 3.9$ logistic map (top second picture of Fig. 7.23), and two coupled logistic maps with $r_1 = 3.94$, $r_2 = 3.86$, and $c = 0.5$ (top third).

To illustrate how these were selected, the bottom graphs of Fig. 7.23 plot κ_2, κ_3, and κ_4, shown with the black, dark gray, and light gray curves, for dictionary entries 1 through 100 and 2901 through 3000. The top second picture was generated by dictionary entry 33, the top third by dictionary entry 2908.

That the test-series driven IFS matches that of a single logistic map suggests that the test logistic maps have synchronized. And the dictionary

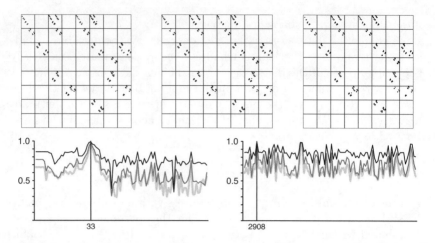

Figure 7.23: Top first: The test-driven IFS. Top second and third: Two matches from the dictionary. Bottom: graphs of κ_2, κ_3, and κ_4 for the portions of the dictionary giving these matches.

entry 2908 pair of logistic maps have synchronized to the same $r = 3.9$ logistic map. This suggests that the driven IFS statistics stratify the dictionary parameter space. If this coarse-graining of behavior holds more generally, we have some reason to hope that this method will be effective in studying necessarily noisy experimental data.

Our first test against cardiac data revealed a good match, with $\kappa \approx 0.73$, for four uncoupled chaotic maps, suggesting that the heart cells act in four autonomous groups, all synchronized within their own group. We're just beginning to analyze these results. More to follow, in the obvious place

Another approach to studying networks is to build scaling into the network, to couple the maps in a fractal lattice. We describe some investigations of these networks in A.96. While these results are preliminary, these lattices appear to support a new type of synchronized behavior. What else will we find?

So we see that driven IFS may be useful not only for data analysis but also for model building. Even if the process generating the data has no obvious fractal characteristics, viewing it through a fractal lens may reveal some structural aspects that are difficult to detect by other means.

7.3 Multifractals from IFS

If we think of fractal dimension as a measure of roughness, assigning a single dimension to a shape implies homogeneity, the same roughness everywhere. This is a decent-enough description for some natural objects,

7.3. MULTIFRACTALS FROM IFS

but more often we find some parts smoother and other parts rougher. These shapes are not well described by a single scaling exponent, so we divide the object into regions of the same roughness – similar to the way a topographical map divides a landscape into regions of the same altitude – and find the dimension of each of those regions.

In broad strokes (we resist the temptation to say "Roughly"), to proceed pick a small patch of the object and measure its roughness. Call the value of the roughness α. Now find all the parts of the object having roughness $= \alpha$ and compute the dimension of those parts. Call the dimension $f(\alpha)$. Pick an unmeasured patch of the object and repeat these steps. Continue until all of the object has been measured. Assemble all the $f(\alpha)$ values into a graph, one $f(\alpha)$ for each measured α, and you have a compact representation of how the different roughnesses are spread across the object.

Because such objects are made up of many fractals, they are called *multifractals*. Direct computation of α and $f(\alpha)$ can be challenging. We'll take a different, but related, approach based on the random IFS algorithm. Suppose we build a fractal using transformations $T_1, \ldots, T_n$, where r_i is the contraction factor of T_i, and p_i is the probability that T_i is applied. We know the probabilities don't influence the shape, but as illustrated in Fig. 7.1, they can have a substantial effect on the rate at which parts of the shape fill. We'll investigate the probability of landing in a region rather than the roughness of the region, though it will turn out that these concepts are related.

If A is the shape generated by this IFS, then the probability of winding up in the region $A_{i_1 i_2 \ldots i_m} = T_{i_1} \circ T_{i_2} \circ \cdots \circ T_{i_m}(A)$ is

$$\text{Prob}(A_{i_1 i_2 \cdots i_m}) = p_{i_1} \cdot p_{i_2} \cdot \cdots \cdot p_{i_m}$$

Following the idea of varying roughness, we'll look at the probability of winding up in each region, grouping together the addresses of those regions which have the same probability. But we will not exactly use the probability: roughness is measured by dimension, which is the exponent in a power law scaling. So we'll guess that the probability of an address scales with the length of the region with that address. Applying T_i scales by a factor of r_i, applying $T_i \circ T_j$ scales by a factor of $r_i r_j$, and so $A_{i_1 \cdots i_m}$ has size $r_{i_1} \cdots r_{i_m}$. Then our scaling hypothesis for probabilities is

$$\text{Prob}(A_{i_1 \cdots i_m}) = (r_{i_1} \cdots r_{i_m})^{\alpha(i_1 \cdots i_m)}$$

where we've used the shorthand notation $\alpha(i_1 \cdots i_m) = \alpha(A_{i_1 \cdots i_m})$.

This exponent already was familiar in mathematics. It's called the *Hölder exponent* and was developed to study the roughness of graphs that wiggle so much they have no tangent lines. (We've encountered the notion of a curve without tangents before: the Koch curve of Sect. 2.2.) For shapes determined by a random IFS we can combine the last two displayed equations, after taking the logs of both sides, to find a formula for the Hölder exponent:

$$\alpha(i_1 \cdots i_m) = \frac{\log(p_{i_1} \cdots p_{i_m})}{\log(r_{i_1} \cdots r_{i_m})}$$

Recall that for each value of α, $f(\alpha)$ is the dimension where that α occurs. Because it's a dimension, we might expect that we can compute $f(\alpha)$ by the Moran equation. In fact, that will turn out to be true, though not surprisingly, we'll have to modify the Moran equation to include the probabilities.

First, however, to help develop our intuition we'll compute the minumum and maximum α values, and their $f(\alpha)$ values, for three examples. That is, for these examples we'll find the left and right endpoints of the $f(\alpha)$ curves. All use the square IFS rules we've seen before

$$T_i(x, y) = (x/2, y/2) + (e_i, f_i)$$

with

$(e_i, f_i) = (0,0), (1/2, 0) (0, 1/2),$ and $(1/2, 1/2)$ for $i = 1, 2, 3, 4$.

The difference between the examples involves varying the probabilities.

Example 1. Suppose $p_1 = 0.1$ and $p_2 = p_3 = p_4 = 0.3$. The random IFS with these probabilities will eventually fill in the unit square, but certainly not in a uniform fashion. The first image of Fig. 7.24 shows 10,000 points generated this way.

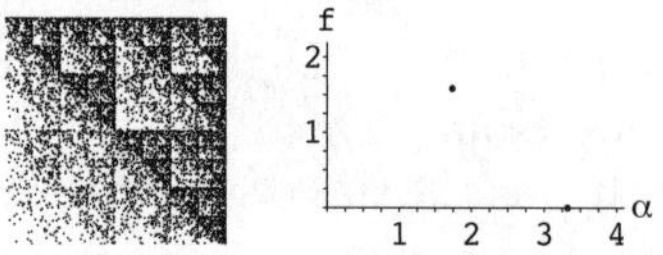

Figure 7.24: Example 1. Driven IFS and $f(\alpha)$ endpoints.

To begin to understand the endpoints of the $f(\alpha)$ curve of this example, we find the Hölder exponents of the addresses $1, 11, 111, \ldots$, the regions with the lowest probabilities.

$$\alpha(1) = \frac{\log(p_1)}{\log(.5)} = \frac{\log(.1)}{\log(.5)} \approx 3.32$$

$$\alpha(11) = \frac{\log(p_1^2)}{\log(.5^2)} = \frac{2\log(.1)}{2\log(.5)} \approx 3.32$$

$$\alpha(111) = \frac{\log(p_1^3)}{\log(.5^3)} = \frac{3\log(.1)}{3\log(.5)} \approx 3.32$$

$$\cdots$$

And for all n, $\alpha(1^n) = \log(.1)/\log(.5)$.

The highest probabilities are $p_2, p_3,$ and p_4. A similar calculation gives

$$\alpha(2^n) = \alpha(3^n) = \alpha(4^n) = \frac{\log(.3)}{\log(.5)} \approx 1.74$$

7.3. MULTIFRACTALS FROM IFS

In fact, for any address $i_1 \ldots i_n$ consisting just of 2, 3, or 4,

$$\alpha(i_1 \ldots i_n) = \frac{\log(.3)}{\log(.5)}$$

These are the minimum and maximum α values,

$$\max(\alpha) = \frac{\log(\min(p_i))}{\log(.5)} \qquad \min(\alpha) = \frac{\log(\max(p_i))}{\log(.5)}$$

This claim should be plausible, because the calculations are based on the minimum and maximum probabilities. But note that the minimum α corresponds to the maximum probability. We'll fill in the details in A.97.

Now that we have found the minimum and maximum values of α, we find the dimensions of the portion of the square where these values occur. That is, we calculate $f(\min(\alpha))$ and $f(\max(\alpha))$. One way to do this is to find familiar geometric descriptions, if there are such descriptions, of the set of points where the minimum and maximum values of α occur.

For example, the minimum value of α occurs at the region of maximum probability of being filled. Since T_2, T_3, and T_4 all are applied with the maximum probablity $p_2 = p_3 = p_4 = 0.3$, the minimum value of α occurs on the attractor of these three transformations, that is, the gasket with corners $(1,0)$, $(0,1)$, and $(1,1)$. From this we see that the value of f at this minimum α is equal to the dimension of the gasket:

$$f(\min(\alpha)) = \dim(\text{gasket}) = \frac{\log(3)}{\log(2)}$$

By contrast, the maximum value of α occurs at the region with the minimum probability, and T_1 is applied with the minimum probability 0.1. The single transformation T_1 produces the point $(0,0)$. (If this is unclear, sketch the regions A_1, A_{11}, and A_{111} to see that the region with address $111\ldots$ is a point.) Then,

$$f(\max(\alpha)) = \dim(\text{point}) = 0$$

These calculations account for the left and right points of the $f(\alpha)$ curve. These points are shown on the second image of Fig. 7.24.

The next two examples will show us how changing the probabilities can produce different geometric shapes where the minimum and the maximum values of α occur.

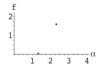

Figure 7.25: Example 2. Driven IFS and $f(\alpha)$ endpoints.

Example 2. Now suppose $p_1 = 0.4$ and $p_2 = p_3 = p_4 = 0.2$. The first image of Fig. 7.25 shows 10,000 points generated with these probabilities.

Applying the reasoning of Example 1 to these probabilities, we can find the minimum and maximum values of the Hölder exponent α.

$$\min(\alpha) = \frac{\log(\max(p_i))}{\log(.5)} = \frac{\log(.4)}{\log(.5)} \approx 1.32$$

$$\max(\alpha) = \frac{\log(\min(p_i))}{\log(.5)} = \frac{\log(.2)}{\log(.5)} \approx 2.32$$

Since $\max(p_i)$ occurs on the attractor of T_1, the point $(0,0)$, we see

$$f(\min(\alpha)) = \dim(\text{point}) = 0$$

And of course $\min(p_i)$ occurs on the attractor of T_2, T_3, T_4, the gasket with corners $(1,0)$, $(0,1)$, and $(1,1)$. Then

$$f(\max(\alpha)) = \dim(\text{gasket}) = \frac{\log(3)}{\log(2)}$$

This accounts for the left and right points of the second image of Fig. 7.25. Note that the range of α values is reduced compared to those in Example 1. A smaller range of α values signals a more uniform distribution of points across the fractal and so less variation in the roughness.

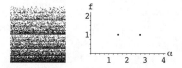

Figure 7.26: Example 3. Driven IFS and $f(\alpha)$ endpoints.

Example 3. Finally, suppose $p_1 = p_2 = 0.35$ and $p_3 = p_4 = 0.15$. The first image of Fig. 7.26 shows $10,000$ points generated with these probabilities.

We begin with the minimum and maximum α values, as usual.

$$\min(\alpha) = \frac{\log(\max(p_i))}{\log(.5)} = \frac{\log(.35)}{\log(.5)} \approx 1.51$$

$$\max(\alpha) = \frac{\log(\min(p_i))}{\log(.5)} = \frac{\log(.15)}{\log(.5)} \approx 2.74$$

Here, $\max(p_i)$ occurs on the attractor of T_1 and T_2, just the 1–2 line, so

$$f(\min(\alpha)) = \dim(\text{line segment}) = 1$$

Similarly, $\min(p_i)$ occurs on the attractor of T_3 and T_4, the 3–4 line, so

$$f(\max(\alpha)) = \dim(\text{line segment}) = 1$$

This accounts for the left and right points of the second image of Fig. 7.26. We give some more details in A.97.

7.3. MULTIFRACTALS FROM IFS

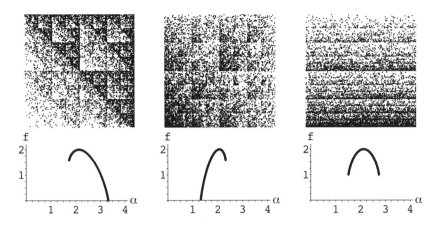

Figure 7.27: Top: 10000 points of the square IFS with the probabilities of Examples 1, 2, and 3. Bottom: the $f(\alpha)$ curves of these examples.

If, unlike in these examples, we are unable to find a familiar set on which the maximum and minimum α values occur, we can resort to another generalization of the ever-generalizable Moran equation. This time, we must take into account the probablities of applying the transformations.

In order to incorporate probabilities in the standard Moran equation, $r_1^d + \cdots + r_n^d = 1$, we first multiply each r_i^d by p_i. Now the exponent of r_i no longer is the dimension but is related to the Hölder exponent α. In fact, just multiplying by the p_i is no longer enough. We'd like to be able to focus on different values of α, finding their dimensions separately. So instead of multiplying by p_i, we multiply by p_i^q, where q can range from $-\infty$ to ∞. A large negative q emphasizes the smaller probabilities, a large positive q the larger probabilities. Different probabilities correspond to different α, so emphasizing different probabilities lets us investigate various αs. In order to fit these terms into the Moran equation, the products of p_i^q and $r_i^{\text{some power}}$ must sum to 1. But for this to happen, the exponent of the r_i may need to depend on q. So we call it $\beta(q)$ With these changes, the *generalized Moran equation* is

$$p_1^q r_1^{\beta(q)} + \cdots + p_n^q r_n^{\beta(q)} = 1$$

In the special case that $r_1 = \cdots = r_n = r$, the generalized Moran equation can be solved explicitly for $\beta(q)$:

$$\beta(q) = -\frac{\log(p_1^q + \cdots + p_n^q)}{\log(r)}$$

Since the square IFS rules have $r_i = 1/2$ across the board, this is good news for us.

From this we can find the Hölder exponent α as a function of q instead of address. It is the negative of the slope of the tangent line of the $\beta(q)$

curve. Using the language of calculus, this is $\alpha = -d\beta/dq$. We'll give a hint of why this is true in A.98. Because we can write β in terms of q, we also can write α in terms of q.

Finally, $f(\alpha)$ is given by

$$f(\alpha) = \alpha \cdot q + \beta(q)$$

That this formula gives the dimension of the part of the attractor with that α is not straightforward or simple. We'll give a signpost in A.99.

Although most often we cannot write $f(\alpha)$ as an explicit function of α, now we have both α and $f(\alpha)$ as functions of q, and the points on the $f(\alpha)$ curves of Fig. 7.27 are obtained by plotting $(\alpha(q), f(\alpha(q)))$ for q values in the range -20 to 20, in steps of size 0.1.

We've mentioned that the $f(\alpha)$ curve gives us a picture of how the dimension of the points having a specific value of α changes with α. That is, it measures the roughness of the distribution of points determined by the random IFS algorithm. A narrow curve indicates a small range of probabilities, while an extended curve suggests a large range. The concavity of the curve shows the acceleration of dimension with α, and there is some preliminary evidence that the curvature of the $f(\alpha)$ curve gives information about how the amount of memory of past behaviors is responsible for roughness.

Another feature of the $f(\alpha)$ curve is hinted at by the observations that all the $f(\alpha)$ curves of Fig. 7.27 have maximum value of 2 and that the IFS generates a filled-in square, of dimension 2. In fact, the highest point of the $f(\alpha)$ curve is the dimension of the whole shape generated by the IFS. We'll see why in A.97.

Multifractals probably are the most active part of fractal geometry now. Their ability to accommodate a range of roughnesses makes multifractals suitable for more nuanced analyses of complex phenomena, from heartbeats to traffic on the Internet and on to the large-scale distribution of galaxy clusters. The $f(\alpha)$ curve can be calculated in many ways in addition to IFS with probabilities, but we believe this is the most straightforward way to introduce multifractals.

In the next section we'll discuss some applications; we end this section with some exercises.

Exercises

1. Which random IFS picture of Fig. 7.28 corresponds to which $f(\alpha)$ curve? Explain your choices by estimating the relative values of p_1, p_2, p_3, and p_4 from the random IFS picture
2. Suppose a multifractal consists of 11 pieces each scaled by $r = 1/4$, generated with these probabilities: $p_1 = \cdots = p_6 = 0.05$, $p_7 = p_8 = p_9 = 0.1$, and $p_{10} = p_{11} = 0.2$. Find $\min(\alpha)$, $\max(\alpha)$, $f(\min(\alpha))$, $f(\max(\alpha))$, and the maximum value of $f(\alpha)$.

7.4. APPLICATIONS OF MULTIFRACTALS

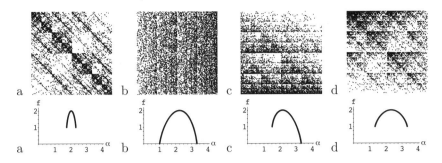

Figure 7.28: Random IFS plots and $f(\alpha)$ curves. Which goes with which, and why?

3. Suppose a multifractal consists of 7 pieces with $r_1 = r_2 = 1/2$, $p_1 = p_2 = 0.25$ and with $r_3 = \cdots = r_7 = 1/4$, $p_3 = \cdots = p_7 = 0.1$. Find $\min(\alpha)$, $\max(\alpha)$, $f(\min(\alpha))$, $f(\max(\alpha))$, and the maximum value of $f(\alpha)$.

4. Suppose a multifractal consists of 8 pieces with $r_1 = \cdots = r_5 = 1/3$, $p_1 = p_2 = 0.3$, $p_3 = 0.17$, $p_4 = p_5 = 0.1$ and with $r_6 = r_7 = r_8 = 1/9$, $p_7 = p_8 = p_9 = 0.01$. Find $\min(\alpha)$, $\max(\alpha)$, $f(\min(\alpha))$, $f(\max(\alpha))$, and the maximum value of $f(\alpha)$.

7.4 Applications of multifractals

Now that we have a grasp of the basic mechanics of multifractality, we can turn to PhysicalWorld for new examples and applications. With experimental data, we don't have an IFS and probabilities, but, for example, we have measurements of the distribution of a natural resource or of earthquakes. Can we find approximate α and $f(\alpha)$ values from this information?

The answer, predictably, is yes, but it will require some work on our part. Recall that when we learned box-counting dimension, we covered the shape by boxes of side length ϵ for a sequence of $\epsilon \to 0$, found $N(\epsilon)$, the minimum number of boxes of side length ϵ needed to cover the object, and tested the existence of a power law written as

$$N(\epsilon) \approx (1/\epsilon)^d = \epsilon^{-d}$$

A more nuanced question to ask is not "Does this box contain any of the shape?" but rather "How much of the shape is in this box?" For every box B we denote by $\mu(B)$ the amount of the shape in box B. Mathematicians call this the *measure* of B. The Hölder exponent of a box B of side length ϵ is

$$\alpha(B) = \frac{\log(\mu(B))}{\log(\epsilon)}$$

That is, we've replaced the probability of visiting a given address with the measure of the amount of the shape in box B. The connection between these two approaches should be clear: the probability of visiting a box is about the fraction of all the points that lie in the box.

Now for each value of α, you can find the number $N_\alpha(\epsilon)$ of boxes of side length ϵ needed to cover the parts of the shape having Hölder exponent α. Then $f(\alpha)$ is given by the power law

$$N_\alpha(\epsilon) \approx \epsilon^{-f(\alpha)}$$

This rough observation can be refined to approximate the $f(\alpha)$ curve by several approaches, including histograms, moments, and wavelets, among others. Though we won't go into these here, it is worth pointing out that the mathematical approach we have been describing can be adapted to specific kinds of experimental data.

7.4.1 Turbulence

Originally proposed in 1974 by Benoit to model turbulence, multifractals have indeed proved an invaluable tool for measuring properties of turbulent fluid flow. Andrey Kolmogorov's cascade model describes turbulence as dissipation of energy by an eddy flowing into several smaller eddies, and from each of those into more still smaller eddies, and on and on. Fig. 7.29 shows eddies of several sizes. Richardson (familiar to us from his careful observation that coastline lengths exhibit a power law scaling), described Kolmogorov's model by this parody of Jonathan Swift's "big-fleas have little fleas" verse

> Big whirls have little whirls that feed on their velocity
> and little whirls have lesser whirls and so on to viscosity.

Multifractality arises when the energy of one eddy is distributed in different fractions to smaller eddies. Detailed calculations and experiments have been carried out by many, including Charles Meneveau and Katepalli Sreenivasan.

Figure 7.29: Turbulent water flow.

7.4. APPLICATIONS OF MULTIFRACTALS

Understanding at least some aspects of turbulence is an essential step toward controlling its effects, with applications ranging from chemical mixers to airplane wings and hulls of ships. In addition, random multifractals associated with turbulence provided an important step in Benoit's work on negative dimension, a project begun but so far left underdeveloped.

7.4.2 Earthquake distributions

Patterns of earthquakes across the land tend to group into clusters that are divided into smaller subclusters, themselves divided into still smaller subsubclusters, and so on. Because the level of seismic activity is not necessarily evenly divided among subclusters, multifractals provide an appropriate language to describe the distribution of earthquakes. This is not a surprise: earthquakes occur along fault lines, tectonic plate boundaries on the largest scale, but then smaller faults branch off these, still smaller off those, and so on, creating a network of faults, self-similar over some range of scales. The uneven hierarchical clustering of seismic activity in space and time suggests a multifractal distribution of earthquakes.

Geologists Paul Okubo and Keiiti Aki found a suggestive correlation between the fractal dimension of regions of the San Andreas fault and earthquake propagation: higher-dimensional faults impede or stop the progression of a rupture, while lower-dimensional faults allow the rupture to progress.

Seismic activity appears to be clustered hierarchically in space and in time. Uneven energy distribution from one fault to several smaller faults suggests a multifractal distribution. Multifractals are a tool, embraced by many geologists in their efforts to understand and eventually (we hope) predict location, time, and magnitude of earthquakes. Here we'll survey a few instances of multifractals in earthquake distributions.

Geophysicists Ali Öncel, Ian Main, Ömer Alptekin, and Patience Cowie study variations in the fractal distribution of earthquakes in Anatolia. Their data set was all earthquakes in that region that had a magnitude greater than 4.5 and that occurred between 1900 and 1992. They found a negative correlation between the fractal dimension of the distribution of earthquake epicenters and the exponent b in the *Gutenberg-Richter law*,

$$\log(N) = -bM + a$$

where M is the earthquake magnitude and N is the number of earthquakes of magnitude $\geq M$. The authors discussed the complications induced by the multifractality of the various scaling behaviors in seismic data.

Geologists Tadashi Hirabayashi, Keisuke Ito, and Toshikatsu Yoshi studied the distributions of locations and of energy for earthquakes in California (1971 – 1985, 7,467 earthquakes), eastern Japan (1983 – 1987, 6,878 earthquakes), and Greece (1971 – 1985, 14,603 earthquakes). In all three studies, the distribution of earthquake locations was heterogeneous, clumped in ways more complex than fractal. The degree of heterogeneity (multifractality) of the Japan earthquake data was higher than that of the California data, and earthquakes in Greece were more heterogeneous than those of Japan. The energy distributions were even more heterogeneous (had a wider $f(\alpha)$ curve) than the location distributions. This is hardly a surprise because the energy distributions follow a power law, the Gutenberg-Richter law, superimposed on the multifractal distribution of locations. In this paper we find the wonderful quotation

"Earthquakes can be considered as turbulence in solid."

Seismologists Li Dongsheng, Zheng Zhaobi, and Wang Binghong studied 1,400 earthquakes of magnitude $M \geq 1.8$ occurring between January 1, 1970, and December 31, 1981, in the Tangshan region of China. They, too, found a heterogeneous distribution of earthquakes, but went a bit further. The multifractal dimensions dropped abruptly in the months before the July 29, 1976, magnitude 7.8 earthquake in Tangshan, leading Li and his coworkers to speculate that multifractals may have some utility in predicting earthquakes. So far as we know, work in this direction has not advanced significantly ... yet.

Finally, geophysicists Luciano Telesca and Vincenzo Lapenna and physicist Maria Macchiato analyzed the times between consecutive earthquakes of magnitude ≥ 2.4 from 1986 to 2001 in three seismic zones of Italy: Irpina (south), Friuli (north), and Marche (middle). The $f(\alpha)$ curves for Irpina and Friuli are similar: they have a wide range of α values (about 0.75 to about 3.5), high values of $f(\alpha_{\min})$ and low values of $f(\alpha_{\max})$, and are asymmetric, with the maximum of the curves near their left endpoints. By contrast, the Marche $f(\alpha)$ curve has a smaller range of α values (about 0.6 to about 1.6), a low value of $f(\alpha_{\min})$ and a high value of $f(\alpha_{\max})$, and is symmetric. The middle region has a more homogeneous distribution of inter-event times, the other two, a more heterogeneous distribution. That the northern and southern $f(\alpha)$ curves are similar to one another, and substantially different from the $f(\alpha)$ curve of the middle region, is trying to tell us something. We know a few specifics, but so far the big picture remains elusive.

Even if the possibility of making predictions about earthquakes is slight, the horrific fact that the 1976 Tangshan earthquake killed 240,000 people is adequate motivation to study multifractals and seismology.

7.4.3 Multifractals and the Internet

Unlike voice traffic, data traffic is widely variable, both in connection duration and in information rate. Consequently, incoming traffic at hubs can exceed the rate at which the routers can direct it downstream. Excess packets go to buffers, but the observed power law scaling of packet volume guarantees that eventually the buffer capacity will be exceeded and some packets lost. This is called data congestion. Congestion control protocols

7.4. APPLICATIONS OF MULTIFRACTALS

decrease packet transmission rates as needed, making packet traffic depend on past traffic. This and power law scaling are good indicators that data traffic follows a fractal or a multifractal distribution.

Bursts of data traffic occur on many time scales, and careful analysis shows that the longer we look, the wider the spread of the data. Statisticians call this unbounded variance. If Internet traffic continues to grow as it has so far, traffic fluctuations, as well as the temporary demand on buffers, will continue to grow. While some kinds of Internet traffic (local area networks, for example) are known to be self-similar, described by a single scaling exponent, other types (wide area networks and transmission control protocols) are multifractal. After seeing that multifractals arise in fluid turbulence and in earthquakes because energy is distributed unevenly when going from big bits to smaller bits, the ever-growing variability of data traffic makes its multifractality no surprise.

The Internet is the most complicated structure humans have created, with a data capacity now of several thousand human minds. This number will be larger, maybe much larger, when you read this sentence. In under two decades the Internet has become central to almost every aspect of modern life. We need to understand something about this thing which amounts to the central nervous system of the human world.

7.4.4 Multifractal finance cartoons

First off, to say neither author is very familiar with finance is a kindness. We're quite clueless, so don't expect to learn anything useful about stock trading from us. Benoit was interested in finance because the data are so complicated, offering the possibility for fractal analysis. When fractals proved inadequate, finance became one of his motivations for multifractal cartoons, a way for unraveling some of the complex behavior of markets. Taking our cue from him, we present this discussion of finance cartoons as a geometry problem. Geometry is a subject we do understand, a bit.

Why did Benoit name these "cartoons" instead of "models"? Benoit regretted giving the name "model" to his early efforts to reproduce the statistics of galaxy distributions after some astronomers criticized his work because it did not account for known physical processes. Cartoons, on the other hand, may replicate some aspects of the appearance of an object without reference to the actual mechanisms it. No one thinks the simple IFS of this section describe how stocks are traded in the real world, but they are an interesting way to produce a similar graph by a very different route.

To begin, we use the IFS with the transformations of Table 7.1. We start with the line segment between $(0,0)$ and $(1,1)$, then show the first, second, and sixth iterates in Fig. 7.30. The effect of each iterate is to replace the straight line segments of the current iteration with a scaled

r	s	θ	φ	e	f
4/9	2/3	0	0	0	0
1/9	-1/3	0	0	4/9	2/3
4/9	2/3	0	0	5/9	1/3

Table 7.1: IFS for a non-random cartoon.

Figure 7.30: The initial segment, and the first, second, and sixth iterate of our first IFS finance cartoon.

copy of the first iterate (the second image of Fig. 7.30), flipped upside down if the line segment has negative slope.

The result is a graph which approximates the *kind* of roughness observed in graphs of real-world data. Of course, these cartoon graphs are far too regular to represent the apparent randomness of the complex interactions that govern trading. To compensate for that, the cartoons can be randomized by shuffling the order in which the three transformations are applied to each scaled version. To do this, we replace each upward-directed segment with one of the three broken lines of Fig. 7.31. Downward-directed segments are replaced with one of these, flipped upside down.

Brownian motion satisfies the *square root* scaling, that is, the (magnitude of the) change in y is the square root of the change in t. In the IFS table this is revealed by the observation that $|s| = \sqrt{r}$ for each transformation. We give some more details in A.100.

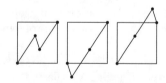

Figure 7.31: Shuffling the order of the generator sequence for Fig. 7.30.

Fig. 7.32 shows three examples of this shuffling, for the sixth stage of the construction, sampled at equal time intervals. The images differ only in the random sequences of choices of the three shuffled generators.

Much closer, but we have a way to go. Even randomized, this graph still appears to have about the same degree of roughness everywhere. This is because all the IFS r and s values have the same power law relation, $|s| = r^{1/2}$. Now suppose we keep the same s values, and also r_3, but decrease r_1 and increase r_2 while still preserving $r_1 + r_2 + r_3 = 1$. With

7.4. APPLICATIONS OF MULTIFRACTALS

this simple change, the r and s values of the IFS no longer have the same power law exponent; with multiple exponents the cartoon becomes multifractal.

In Fig. 7.33 we start with $r_1 = 4/9$, and then go through $r_1 = 1/3, 2/9$, and $1/9$, with the corresponding changes compensating in r_2. To the right of each graph is the graph of the *differences*, $y(t_{i+1}) - y(t_i)$, sampled at equal time intervals, hence representing the change in the

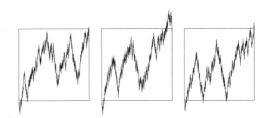

Figure 7.32: Three instances of the right image of Fig. 7.30 after shuffling.

daily (or hourly) price of the stock. Except for the top ($r = 4/9$) row, these simple cartoons exhibit large jumps seen as spikes much larger than the surrounding values, suggesting that the distribution of jumps does not follow a bell curve. (In fact, it doesn't.) Jumps following a power law instead of a bell curve is a characteristic of Lévy flights, random processes studied by Benoit's teacher Paul Lévy. In addition, we see memory effects, the clumping together of large values and of small values. This is a property of fractional Bronwnian motion.

Certainly, the graphs look more like what we expect from finance data as soon as r_1 is changed to less than $4/9$. But creating convincing forgeries of real financial data is not the point here, though watching experts try to select – incorrectly, but with considerable confidence – the real data from the forgeries has been quite entertaining. One motivation is to develop an understanding of how simple processes can generate convincing forgeries of real, complicated data. Again, though no one suggests that these cartoons respect real market forces, they may suggest that a small number of dominant components account for much of the observed variability of the markets.

Fairly often, Benoit was asked how to solve the "inverse problem:" given a financial time series, find a set of generators that will reproduce – approximately, of course – the time series. Several people have worked on this, but every time Benoit mentioned the inverse problem, he said he did not know even how to approach the question. Of those approaches we have seen, the closest match (and that match was not so very close) was to by eye identify the *turning points* in the time series. These are the points that correspond to the corners in the cartoon generator graph. Certainly, someone may apply a more sophisticated technique with some success, but the inherent complexity, the high-dimensionality, of financial markets make this seem unlikely. The best we likely will be able to do is learn some of the language of financial time series. Cartoons may help

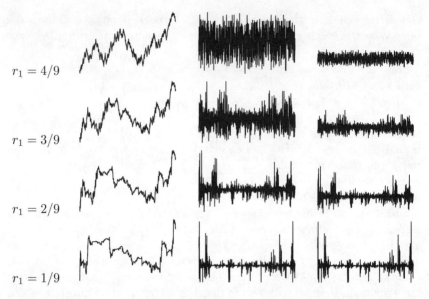

Figure 7.33: First: Price graphs. Second: Difference graphs, all adjusted to have the same height. Third: Difference graphs on the vertical scale of the 1/9 graph. All graphs are plotted from 1,000 points sampled at equal time intervals from ninth iterate cartoons.

with the grammar.

In other words, with these cartoons we are trying to train our eyes to find the stories of the data.

7.4.5 Fractals, multifractals, and climate

Early work on scaling relations in climate was done by British hydrologist Harold Hurst when he studied the annual flooding of the Nile in order to estimate the size of dam needed to contain the flood waters. Hurst found a considerable variability in the flood magnitude, a hint of the wild swings caused by the complex interactions of water, atmosphere, land, Earth's orbit, the Sun, and lately, human activity.

In a sense, the study of fractals in climate predates fractals. In the late 1960s, Benoit and James Wallis applied fractional Brownian motion to study the variability of rainfall patterns. The same forces acting over a range of space or time scales would yield self-similar distributions, so we should expect to see fractals – and sometimes we do.

But studies of temperature variations, especially over times spanning some glaciations, often reveal multiple scaling factors. This is how multifractals appear in climate records. *The Weather and Climate: Emergent Laws and Multifractal Cascades*, a recent book by Shaun Lovejoy and

7.4. APPLICATIONS OF MULTIFRACTALS

Daniel Schertzer, tells this story with great clarity and considerable detail. They show that techniques developed to study turbulence can be applied to find emergent laws of atmospheric dynamics, and that multifractals provide an effective measure of the time and space complexity of the atmosphere.

Lovejoy and Schertzer divide atmospheric dynamics into three time regimes: weather (up to 10 days), macroweather (10 days to 10 to 30 years), and climate (10 to 30 years up to several hundred thousand years). By analyzing large data sets – the Greenland and Vostock (Antarctica) ice cores are over 3 km long and go back in time 240,000 years and 420,000 years, respectively – they observe that while weather and climate are unstable (fluctuation ranges increase with time scale), macroweather is stable. This important observation provides guidelines for interpreting climate data.

But scaling, both in space and in time, is their main result. They show multifractals are in the air. You knew this was coming, didn't you?

Despite these successes, Lovejoy and Schertzer bemoan the fact that significant resources are directed to traditional climate modeling while multifractal analysis is underappreciated. This is particularly unfortunate, because more fully developed models that reflect scaling cascades of the atmosphere could provide more rigorous statistical tests of the levels of anthropogenic climate change. The more empirically based our study of what we're doing to our world, the more difficult it will be for people to fault either analysis or conclusion.

Although multifractals are not yet universally embraced in climate studies, they are a tool used by many. We'll mention one example, a recent paper by Zhi-Gang Shao and Peter Ditlevsen. They study temperature variations using recent instrument recordings, ice core data going back several hundred thousand years, and stacked deep ocean sediment records going back about 5 million years. They found that in recent times (back about 12,000 years) the tempertaure variations are unifractal. In contrast to a multifractal, a unifractal is characterized by a single Hölder exponent, in this case, $\alpha \approx 0.7$. (Shao and Ditlevsen use the term "monofractals" for what we call "unifractals," and our α they call H, the Hurst exponent.)

Looking further back into the past, back into glaciations, they find the temperature variations are multifractally distributed. The α values range from about 0.7 to a bit above 1.3, $f(\min(\alpha)) \approx 0.6,$, $f(\max(\alpha)) \approx 0.8$, and the maximum point on the $f(\alpha)$ curve has $f = 1$. By contrast, the $f(\alpha)$ curve for the recent temperature variations collapses to a single point, $(1,1)$, on the $f(\alpha)$ graph. The range of scaling factors suggests a richer variety of forces influencing the climate. They present evidence that Dansgaard-Osecger events (rapid rises in temperature, followed by gradual declines) are part of the natural variability of climate.

Further back than 20,000 years the Hölder exponent drops of 0.5, that of Brownian motion. Here the future is decoupled from the past, suggesting on that time scale climate is dominated by glacial cycles, which are driven by outside influences, variations in the Earth's orbit, for example.

Shao and Ditlevsen show that multifractal analysis can shed light on several complex features of climate. To the extent that some aspects of climate are characterized by cascades of scaling processes, this is not a surprise.

One final comment: the older author has never understood how or why the Trading Time Theorem, discussed in A.100, could apply to finance. The key step of the theorem is to rescale time (compress it during quiescent periods, expand it during active periods), converting a complex multifractal signal in clock time into a simple fractional Brownian motion signal in Trading Time. Biological data looked more proimising, but climate data may be more promsing still. In particular, if the conversion of clock time to Trading Time could estimate the relative probabilities of active and quiescent periods, this could be useful. So far, no one knows how to extract Trading Time from time series, but the older author knows one very bright person working on this problem, so our current understanding of these matters may be about to improve substantially.

By unpacking complex scaling relations in weather data, fractals and multifractals are useful tools for studying climate. Let's get to work. We may not have long to think about this problem.

7.5 Fractals and stories, again

Just as the Sierpinski gasket is determined by the rules and repetitions that create it, the branching geometry of a snowflake is determined by the local conditions – temperature, pressure, humidity – in which it grows. The shape of a snowflake therefore is the story of how it traveled through the snow cloud before it drifted to the ground. Similarly, the profile of a mountain is a story of erosion and geological forces, as are the course of a river, the dunes of a desert, and the roughness of a coastline. The branching of a tree is a story about light, wind, water, soil; about insects, birds, small mammals. In all these cases, the object contains hints of its own story.

We are no exception. The branching of our lungs is a story about our genetics and the environmental conditions in which our lungs developed. Even the things we make can bear the mark of fractal stories. For example, the Internet was built, though not designed, by us. Existing in the world of ideas and the world of things, at both conceptual and structural levels, the Internet is exactly the story of how it grew from a few networked servers to a whole planet of connections.

As we move into the realm of ideas, the possibilities expand. Perhaps stories themselves can be fractal. If the Euclidean version of a story is beginning-middle-end, a fractal story has its ends tangled up in beginnings,

7.5. FRACTALS AND STORIES, AGAIN

its middle pocketing out into other stories, branching and folding again and again. Or maybe the fractal patterns are more subtle than that – maybe a story can exhibit scaling symmetry in the comparison of large, historic events to small, personal ones (with steps in between, the middle-personal, the near-historical), or perhaps with plot patterns that repeat in an iterative, looping way, so a single event is multiplied and modified to make a complicated, intricate curve. Or the language of a story could be fractal, exhibiting a fractal or multifractal distribution of key words, or perhaps of sentence lengths.

There are certain challenges to embedding fractal structures in narratives. Of necessity, stories must be followed linearly, even stories such as Julio Cortázar's *Hopscotch*, in which the chapters can be read in two orders. (Cortázar's two chapter orderings are a bit of a cheat. The first order is to read chapters 1 through 56 sequentially, stopping at the end of chapter 56. The other order inserts later chapters between sequential chapters of the first 56, maintaining the order of principal narrative presented in these 56 chapters, and using the later chapters to insert additional information between these.) Even though the large-scale order is perturbed, still each chapter flows in only one direction. Can even this order be disrupted, so that a narrative jumps like a particle obeying rules of fractional Brownian motion? We assume that a story's reader must be able to divine some relationship between one event and another, but perhaps there are other ways that a story's signal can emerge from its text's noise.

Unlike mathematical fractals, which show their self-similarity as we zoom in, words do not have (much) substructure, so fractality cannot be revealed through symmetry under magnification below the scale of words. Any fractals that might be found in stories must be seen by zooming out from the level of words.

Because fractal geometry can give simple descriptions of some objects that appear to be complicated, in order for fractality to be relevant, the object must exhibit complications. So a story must be fairly long in order to own enough complexity, allow enough scales, to be plausibly fractal. We do not expect to find fractal haikus.

Given these restrictions – adequate length, complication, and patterns seen when zooming out – how can fractality be sensibly embedded in the structure of a story, if indeed it can be?

Keep in mind that we can look for patterns of ideas, not just of words or of sounds. This flexibility gives a counterweight to the restrictions just mentioned. We'll sketch four approaches that might reveal, or allow us to build, fractality in stories. We expect there are many other ways.

We'll start with a simple, visual approach which we'll call *mountain range fractals*, suggested by Prokofiev's method of composing music for *Alexander Nevsky* mentioned in Sect. 4.7.3. Think of some aspect of a story, emotional state for instance, that can take on a large range of values. As the story progresses, the emotional state of one character rises and falls, with the overall trends embellished by rises and falls on ever-smaller scales. The figure illustrates this process. Of course, without seeing the graph, a reader might never consciously acknowledge the changing of scales. In fact, it may be that many stories do show a mathematically sound expression of fractality without ever our – or the writers – knowing it.

There may even be a naturalness to a fractal narrative like this. If eye and mind can implicitly recognize the likeness of these scaling patterns to the way some parts of nature change with time, then the story may seem 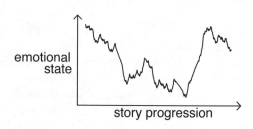 more real, more relevant to the world it occupies. We are accustomed to having small disappointments next to large ones, or little pleasures compared to great joys. From Homer to Leo Tolstoy, Virginia Woolf to Toni Morrison, writers present us with a version of reality, familiar though changed, with a rich texture of lives and events. Should we be surprised if one day we find fractals among them? By now, you must know the answer.

Another path is to build the story around a *branching structure*, our own garden of forking paths. A challenge is how to reveal a branching structure, at home in 2 dimensions, in the necessarily 1-dimensional flow of a story. A greater challenge is for the author to do this in a natural way, not a contrived, awkward way, with a smug awareness of being clever.

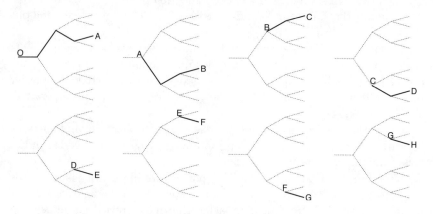

Figure 7.34: A sequence of paths through a story tree.

Fig. 7.34 illustrates one way to take a sequential path through a story. Suppose the horizontal direction represents the flow of time, and different branches signal an emphasis on different characters. Each branching point corresponds to a clearly identifiable event. The story begins at O

7.5. FRACTALS AND STORIES, AGAIN

and continues to A in the upper left tree, following the viewpoint of one character. Then the narrative loops back to A in the next tree and tracks a different character through to B. And so on. Without the graphs, readers could recognize this pattern by the scaling relation between the number of story segments – passages bookended by jumps in time or locale, segment changes signaled by "meanwhile, back at the ranch" and its ilk – and the duration of the segments.

The order of the paths in this sequence can be altered to add variety to the durations. This general construction is common: in many stories the narrative jumps across space or time. So far as we know, no one has designed a scaling distribution of jumps in a story. If these have occurred, we expect they are accidents, results of rewriting until the rhythm sounded right. We haven't read stories looking for such scalings, but this is on our list of explorations to do.

Scale ambiguity is another approach. For instance, the time scale can be ambiguous. Here is an illustration, doggerel of the older author. The younger author bears no responsibility for this.

> dark and cold,
>   just awake
> first light,
>   warming,
>   a few ideas clear
> bright middle,
>   surrounded by many,
>   talking and thinking
> lights dim,
>   mature thoughts,
>   results of long efforts, shared notions
> dark and cold,
>   all the questions that can be answered
>   have been
> curtain coming down,
>   a good X

If X is *day*, the sections refer to predawn, morning, afternoon, evening, night, midnight. If X is *year*, to winter, spring, summer, autumn, winter, year's end. If *life*, to childhood, adolescence, early adulthood, late adulthood, old age, death. If *field of study*, to problems unapproachable in an old field, the stumbling development of new tools, the beginning perception of autonomy, mature and efficient applications, the declining population of young practicioners, falling out of fashion.

We wonder if stories can involve ambiguous scales in aspects other than time. Perhaps geographical scale can be ambiguous. Or the actions of a person can be mistaken for those of a crowd, or of a nation. More abstractly, how the mind models part of the world may have features common across many scales.

Scale ambiguity offers the possibility for one story to be many. Put another way, a single story can open our eyes to rhythms of life that recur on several levels.

Here we note a hint of an exception to one of our list of requirements for fractality. The verse above is not very long, so where's the complexity we need? It's implied by scale ambiguity. For example, if X is *life*, then each year of that life has its own pattern, and each day of each year shows a similar pattern, too. Fractals can be found in the world of things and in the world of ideas, and sometimes straddling the border between these worlds.

Projections of higher-dimensional fractals is the last method we'll sketch. Although this notion has inhabited our conversations almost from the start, and we have worked out a few crude example stories (about cats – no surprise there), this is more abstract and less familiar than the others mentioned in this section.

Stories can be subtle and may extend in dimensions beyond those of the obvious narrative. This gives the possibility of several stories being projections, in different dimensions, of a single higher-dimensional fractal story. For instance, in the space of character and time, a story may be encoded as a Sierpinski gasket. How? As time passes, each character is in a given place (filled-in portion of the gasket) or is not in that place (empty part of the gasket). Along the mood axis, we might find a Cantor set: a range of moods accessed by a character (Cantor set point) or not (Cantor set gap).

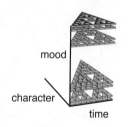

Figure 7.35: Part of story space.

The whole story is high-dimensional. Think of the story's characters, locations, moods, actions, and interactions seen geometrically by representing each along a different axis. A sketch of a part of this is shown in Fig. 7.35. Does this sound far-fetched? Think of the concurrent lives of all the people you know. Any story we can tell is a way to make sense of a small portion of the whole, a shadow that shows only part of the whole, though the shadow is cast by it. Must all these shadows be fractal? Certainly not, but this notion of each story's being a projection of some high-dimensional ur-story came to us when we were pinging back and forth ideas about how stories can be fractal. By itself, this is an interesting, if indirect, application of fractals.

One of the implications of deducing the high-dimensional story from several projections is the ability to predict some broad features of other projections. For example, focus on one character, ignore place and time,

7.5. FRACTALS AND STORIES, AGAIN

and tell the story of how that character's emotional state progresses from depression to happiness. Keeping the time jumps from becoming dizzying would require some thought. Then focus on one place and tell the stories of the characters who visit that location, in the order of occurrence, or perhaps in another order. Then freeze time: describe a snapshot of all the characters and the conditions surrounding them. What happens in one story informs what can happen in others.

If we notice a pattern at one level, we can look for a similar pattern on other levels. Finding a few suggests scaling, so we could seek the source of this behavior, or predict the pattern on still other levels. Scaling in one projection implies scaling in some other projections: a Cantor set shadow in one direction suggests a Cantor set shadow in some others. Don't think of these as constraints to writing. They are a superstructure for building natural rhythms into our writing.

And just perhaps, writing or reading stories that are projections of a higher-dimensional story will help us recognize some aspects of the whole story of which our lives are shadows. Even if there are mysteries too big for us to see, still we should, we must, keep trying to understand.

We'll end with another quote from Lightman's *Mr g* (a story that is not fractal so far as we know, but we are ready to be surprised).

> Perhaps just at the moment of death, they could feel a little piece of the Void.
> Some of them have that feeling now, I said. At least, they feel a mystery. They sense that there are big things they do not know, even though they have no way to know them.

This is pretty much how we feel all the time.

Chapter 8

Valediction

This is the younger author writing. The phrase "thinking outside the box" comes to mind when trying to explain the importance of fractals. It is a familiar cliché, but in this case it becomes an apt metaphor in its own right. In the world we are used to, a box is simply a box. Straight lines, a finite blank space corralled within, and everything else left outside. This is the standard geometric formulation of a box and, by extension, the world, in its simplest form.

What is amazing about fractals is that they have the power to redefine the box itself. Fractal geometry proves that simplicity is not merely the straightest set of lines enclosing a geometrically uniform space. Simplicity can be the seed that encodes the whole tree. Simplicity can be an if-then proposition which builds the universe, the planets, the mountains and rivers and forests that make up the landscape, the organisms which inhabit these landscapes, and the smaller organisms which inhabit their landscape. The box, when it comes to fractals, is made up of other boxes, made up of even smaller boxes. What is inside and what is outside is no longer entirely clear. Instead, witness a chain-linked set of ever-widening possibilities, shades and shades of distinction held within a complex net revealed by one simple rule: each part looks like, or acts like, the whole.

While at work on this book, we had fun finding instances of this rule. We kept turning up more authors who wrote about fractals, even if they never knew that's what they did. Every week we found more examples from science. Some were topics new to us – fractal capacitors and DNA globules – while others were familiar features seen afresh through the light of fractals. We both had seen pictures of the Crab Nebula – my parents are astrophysicists; Michael's undergraduate minor was astronomy – but recognizing the fractal nature of its networked filaments made us feel as though we were discovering something brand-new. This sense of hidden aspects revealed has not diminished yet. I imagine this same excitement marked so many of the free-ranging conversations Michael and Benoit had.

Bringing the idea of fractals to people has been one of the most rewarding aspects of our work on this book. What is important about fractal geometry is not just IFS tables and nice pictures of the Mandelbrot set; it is also the knowledge that the world is stranger than you know. Learning about fractals is like learning that the language you have been speaking your whole life is one possibility out of thousands, any of which would provide a valid method for describing the world.

When Mandelbrot and colleagues started finding fractals everywhere they looked both in the most straightforward and in the most unlikely places, in the mathematics of Newton and Einstein and in the physics of atoms and stars. It may have seemed like a gimmick: the magician revealing that the man in the audience held the dollar bill all along. We had become used to certain patterns. After all, Pythagoras and Euclid are the oldest names in the book when it comes to geometry.

But mathematics is not a book; it is a world. That, I think, is the true lesson of fractals. The world can be read in several different ways, in several directions. There may still be fractals underlying the math we know as fundamental, all predictable, figured out – waiting for someone to notice they've been there the whole time. Maybe a reader of this book.

This is the older author writing. Since the early 1980s, fractal geometry has grown and branched, again and again. First a collection of ideas related in ways only Benoit noticed, now the field is seen to be an intricate tapestry with threads from every part of science and most parts of the humanities and arts. Still, the field is young and continues to develop, uncaptained after Benoit's death on October 14, 2010. Driven by the momentum of its ever-growing group of practitioners, fractal geometry continues to mature. Techniques developed to study fractals now belong in the toolboxes of most young scientists and many writers and artists. This was Benoit's dream; it is wonderful that he lived to see so much of his dream come to pass.

How many times in our lives have we, within the span of a few hours, learned a genuinely new way to look at shapes in the world? Not just to look at shapes, but to understand the mechanics of how they are assembled. Seldom, maybe only once or twice. Did Benoit really give us a truly new way to view the world? Oh, yes. Yes, indeed.

From early in his life, Benoit wanted his own Keplerian revolution. He wanted to find a new way to understand some aspects of the world. When asked why Keplerian and not Copernican – surely "Copernican revolution" is the more familiar phrase – Benoit replied that Copernicus had a brave idea, but Kepler quantified it, gave us three laws of planetary motion. Measurement and quantification were important themes for Benoit . Although he often spoke of the role fractals play in making pictures important in scientific investigation, Benoit was careful with the many difficult calculations he performed.

But he did not calculate the way I do, the way I believe most scientists do. In a national mathematics examination he took as a student, Benoit amazed his teacher by correctly evaluating a complicated triple integral. "How did you do that? I couldn't have done that integral in the allotted time," his teacher said. Benoit replied that he changed x this way, changed y that way, and changed z that way. "Then the problem turns into finding the volume of a sphere, and I know the volume of a sphere." Shapes moved effortlessly, fluidly in Benoit's mind. He soared through geometry, where most of us swim or plod.

As we've mentioned earlier, some mathematical examples of fractals were known decades before Benoit developed the concept, and some fractals were used as artistic and architectural decorations centuries before that. Benoit's extraordinary geometric intuition brought together these few examples and made a coherent field of study that has touched almost every aspect of our world. Watching, and to a very minor level participating in, the development of a new field from so close to its center has been a dizzying experience. Even now, every day brings fresh understanding of lessons learned from Benoit. I expect this to continue as long as I live.

This brings us to my last point. Years ago, one of Benoit's many visitors remarked that he'd read papers by Benoit on math, physics, finance, hydrology, probability, meteorology, ... What did Benoit consider himself? Was he a mathematician, a physicist, an economist, what? Without a moment's hesitation, Benoit replied, "I'm a storyteller."

Benoit loved to tell stories. Because he knew so much, and saw such surprising connections, his stories were quite entertaining, but eventually I understood another meaning of his statement. When we met, in Albany in 1988, part of my introductory talk about fractals included characterizing them as dynamical geometry, shapes described not as static objects but as processes that grow the shapes. Benoit spoke of pictures, of the importance of the eye in science. But finding a fractal description of an object is finding a story about how the object grows. The delicate arms of a snowflake are a story about the humidity, temperature, and pressure along its path through a cloud. A coastline is a story of rocks, waves, wind, and tides. A river basin is a story of plate tectonics, rain patterns, gravity, and erosion. So much of the world is filled with stories if only we knew how to listen, how to read them. Fractal geometry is another way to listen to, another way to see, another way to read, much of the world around us.

For most of the thirty years I've studied the math of fractals (though as my first-grade drawings shown on the right suggest, the idea of self-similarity was percolating  in my mind almost sixty years ago), I've known they are a way to read stories about nature. And I've read lots of stories, so I thought writing a story about fractals would be straightforward. It was not. Reading has a subtlety I had missed entirely, and this blindness hobbled my writing, forcing it to resemble the noise of a cow clomping through the bushes to drink at a stream. Amelia opened up the language of this book; she added a dimension to how we tell stories about fractals. Benoit described fractals as a house of many rooms – a common-enough description of most fields – but he added that from each room we can view all the others with a light unique to that room. Not only can math tell us some things about literature, literature can tell us some things about math. Fractals are bridges for so many fields to talk with one another. Benoit knew this, but I didn't get it. Working with Amelia, learning how she thinks about this subject, finally I do get it. And I hope you do, too. She and I have explored a few of these bridges. We've left most of them for you.

When Benoit knew his life was ending, he was unhappy about the work that remained unfinished, the stories untold. I know this frustrated him, and I understand why. Rather than thinking only of the tasks remaining, I tried to get him to focus on what he had done, on the tens of thousands, maybe hundreds of thousands, of lives he had touched. But he would have none of this; he wanted only to be as sure as he could be that work begun by him would be completed by others. Benoit made the world better by giving all of us another way to exercise our curiosity, by giving geometry a beautiful new language for telling stories about Nature.

Asking, and trying to answer, questions about the natural world is the very finest thing we do. It draws us, as a species, back to the optimism of childhood, when every kid is an explorer, every morning an opportunity to wander through the woods, looking up into trees and under rocks. Most of these days of my chronological childhood were accompanied by a gentle breeze, rippling the smooth surface of a pond, turning leaf shadows into hypnotic dances of beautiful hieroglyphs. My untrained mind didn't notice the long time scale of growth and decay, but the breeze brought change, dynamics, stories, to a scale I could understand; it began my training in reading the text of Nature; it paved the way for me to understand fractals.

As I wrote these words, a cat asleep on my lap, workroom window open to the nighttime concert of crickets and frogs, a breeze came in through the window, touseled my hair, and said Hi, remember me? And I'm a little kid again, walking through the woods, looking at everything. For a moment, concerns about ill health and the ache of loss of loved ones drop away. For a moment, I become curiosity embodied.

Our curiosity has carried us from caves toward the stars. Benoit gave us new tools to speed this trip. But for me, personally, he opened another path to return to the wonder of childhood, deepest of all joys.

Appendix A

A look under the hood: Some technical notes

In this appendix we give supplementary notes, additional examples, bits of mathematical background not included in the main body of the text. Assuming a familiarity with basic calculus, we develop some of the mathematics needed to understand how and why fractals work.

We introduce the other techniques as needed. Notes are divided into the sections to which they refer. Some sections have several appendix entries. Of course, as with the rest of this book, feel free to skip around as you please.

Sources for material quoted or described in the Appendix are listed in the References. The sources in the References are grouped by chapter and section; Appendix sources are cited in the relevant sections.

Because this appendix contains many sections, here's a roadmap.

A.1	IFS table	256
A.2	Three properties of the Cantor set	256
A.3	Points that are not limit points	260
A.4	Tree and fern IFS tables	261
A.5	IFS rules from point images	262
A.6	Spiral IFS tables	271
A.7	IFS topological types	272
A.8	Expected time to visit an address	274
A.9	Benoit's circle inversion method	280
A.10	Three random fractal construction	281
A.11	Strange types of self-similarity	288
A.12	Coastlines, mountains, rivers	289
A.13	Estimating lung area	290
A.14	Power spectra and correlation functions	292
A.15	Metabolic scaling models	294
A.16	Distribution of galaxies	297

A.17	Fractals and perception	301
A.18	Ammonite sutures	302
A.19	Dielectric breakdown and relatives	302
A.20	Bacterial growth	303
A.21	Earthquakes	304
A.22	EEG, again	307
A.23	Sunspots	307
A.24	Solar flares	308
A.25	Solar prominences	309
A.26	Supernova remnants	309
A.27	Saturn's rings	311
A.28	Fractal cracks and fractures	312
A.29	Soil pores	314
A.30	Distribution of resources	315
A.31	Clouds	316
A.32	Turbulence	317
A.33	Snowflakes	317
A.34	Foraging paths	319
A.35	Distribution of epidemics	319
A.36	Ion channels	322
A.37	Fractal fitness landscapes	324
A.38	Nature-inspired chemical engineering	325
A.39	Lateral capacitors and parasitic capacitance	326
A.40	Scaling solutions of Maxwell's equations	326
A.41	Scaling in invisibility cloaks	327
A.42	Fractal models of Internet traffic and infrastructure	327
A.43	Fractal music	328
A.44	Structural scaling in *Arcadia* and other comments.	329
A.45	Mandelbrot set picture coordinates	330
A.46	The escape criterion	331
A.47	Julia set c-values	333
A.48	Julia set connectivity and critical points	333
A.49	The fixed point at infinity	334
A.50	The technical definition of chaos	334
A.51	Polar representation	335
A.52	Chaotic dynamics on a Julia set	337
A.53	Starting with $z_0 = 0$	340
A.54	Stability of fixed points and cycles	340
A.55	Stable fixed point	342
A.56	Stable 2-cycle	343
A.57	Counting discs and cardioids	344
A.58	Stable fixed point at infinity	347
A.59	The angle-doubling map	347
A.60	A proof of Fermat's Little Theorem	348
A.61	Misiurewicz points and Julia sets	350

A.62	The magnification factor in Tan Lei's theorem	350
A.63	Locations and magnifications of Figs. 5.15 – 5.17	351
A.64	The general escape criterion	351
A.65	Critical points of other functions	352
A.66	The formula for Newton's method	353
A.67	Newton's method for $f(z) = z^2 - 1$	353
A.68	The roots of $z^3 - 1$	355
A.69	Lakes of Wada	355
A.70	Stable cycles and Newton's method	356
A.71	Removing the next-to-highest power	357
A.72	Cubes in the hypercube boundary	358
A.73	The Sierpinski hypertetrahedron	359
A.74	Parameters for Fig. 5.26	360
A.75	Some variations in the $1/n^2$ rule	360
A.76	Moran equation: unique solution	361
A.77	The infinite Moran equation	362
A.78	More infinite Moran examples	366
A.79	Box-counting dimension: limit form	368
A.80	Box-counting dimension: sequential form	369
A.81	Some other choices for $N(\epsilon)$	370
A.82	Random Moran equation	371
A.83	Computing eigenvalues	372
A.84	Finding eigenvalues with *Mathematica*	374
A.85	Moran: memory and infinite	375
A.86	The Open Set condition	376
A.87	Preserving fractal dimension	377
A.88	Careful statements of dimension rules	378
A.89	Dimensions of self-affine fractals	378
A.90	Markov partitions and forbidden pairs	380
A.91	Driven IFS and graph shapes	382
A.92	Markov chains and forbidden pairs	383
A.93	Other bifurcation diagrams	387
A.94	Liapunov exponents	389
A.95	Networks of logistic maps	390
A.96	Fuzzy synchronization	393
A.97	Drawing $f(\alpha)$ curves	394
A.98	The slope of the $\beta(q)$ curve is $-\alpha$	397
A.99	Why $f(\alpha) = \alpha \cdot q + \beta(q)$	399
A.100	Trading time	402

Of course, we could include many more topics than those we selected. By now, fractal geometry touches every part of mathematics. Topology, geometry, and analysis are obvious, but algebra and number theory have fractal aspects, too. For example, look up recent elaborations of Ramanujan's work on partition numbers. Fractal geometry is not just a new kind of geometry, although that already makes it an interesting subject. Rather, it is a new way of looking at math (and physics and chemistry and biology and astronomy and ...), new questions, new perspectives. Many lifetimes of fascinating work. Part of our goal for this appendix is to give you a glimpse of that breadth.

Chapter 2. Self-similarity in geometry

This note refers to Sect. 2.1, **A simple way to grow fractals**.

A.1 IFS table

r	s	θ	φ	e	f
-1/2	1/2	90	90	1/2	1/2
1/2	1/2	-90	-90	1/2	1/2
1/2	1/2	0	0	0	1/2

Table A.1: The IFS table for the third fractal of Fig. 2.1.

Table A.1 is the IFS rules to generate the third fractal of Fig. 2.1. The placement of the outlines on the right gives some guidance in determining the orientations of the pieces. Can you find which piece corresponds to which row of the IFS table?

The next two notes refer to Sect. 2.2, **Some classical fractals**

A.2 Three properties of the Cantor set

We'll show the Cantor middle-thirds set has length 0, is uncountably infinite, and each of its points is a limit point. These three properties are valid for many other Cantor sets. We chose the Cantor middle-thirds set because it illustrates these properties cleanly, without most of the bookkeeping complications needed for other Cantor sets.

In preparation for showing that the Cantor set has length 0, we derive the formula for summing a geometric series. Recall that a geometric series has this form:
$$1 + r + r^2 + r^3 + \cdots$$
That is, successive terms of the series have a common ratio, r.

Write S_n for the sum of the terms 1 through r^n:
$$S_n = 1 + r + r^2 + r^3 + \cdots + r^n$$

A.2. CANTOR SET PROPERTIES

Multiplying both sides by r gives

$$rS_n = r + r^2 + r^3 + \cdots + r^n + r^{n+1}$$

Adding 1 to both sides:

$$\begin{aligned} 1 + rS_n &= 1 + r + r^2 + r^3 + \cdots + r^n + r^{n+1} \\ &= (1 + r + r^2 + r^3 + \cdots + r^n) + r^{n+1} \\ &= S_n + r^{n+1} \end{aligned}$$

Solving the resulting equation, $1 + rS_n = S_n + r^{n+1}$, for S_n:

$$S_n = \frac{1 - r^{n+1}}{1 - r} = \frac{1}{1-r} - \frac{r^{n+1}}{1-r}$$

If $|r| < 1$, then higher powers of r get ever smaller, and in fact $\lim_{n \to \infty} r^{n+1} = 0$. The sum of the whole infinite geometric series is $\lim_{n \to \infty} S_n$, so taking the $n \to \infty$ limit of the displayed expression for S_n we obtain

$$1 + r + r^2 + r^3 + \cdots = \frac{1}{1-r}$$

so long as $|r| < 1$.

A.2.1 The Cantor middle-thirds set has length 0

To construct the Cantor middle-thirds set, we start with the interval $[0,1]$ and remove one interval of length $1/3$, two intervals of length $1/9$, four intervals of length $1/27$, and so on. We can sum the lengths of the intervals removed from $[0,1]$:

$$\begin{aligned} \frac{1}{3} + \frac{2}{9} + \frac{4}{27} + \frac{8}{81} + \cdots &= \frac{1}{3}\left(1 + \frac{2}{3} + \frac{4}{9} + \frac{8}{27} + \cdots\right) \\ &= \frac{1}{3} \cdot \frac{1}{1 - 2/3} = 1 \end{aligned}$$

That is, the entire length of the interval $[0,1]$ has been removed incrementally, leading us to say the leftovers – that is, the Cantor set – has length 0.

A.2.2 The Cantor middle-thirds set is uncountably infinite

The German mathematician Georg Cantor began much of our understanding of the different sizes of infinities. An infinite set is *countable* if it can be matched, member by member, with the positive integers. That is, there is a first element, a second, a third, and so on: the set can be counted. Taken together, the endpoints 0, 1, 1/3, 2/3, 1/9,... of the initial interval and of the intervals removed when we build the Cantor set are countable. Perhaps more surprisingly, the (positive) rational numbers can be counted by placing them in a table with denominators increasing along the side and numerators increasing along the top, and then by threading a clever path, illustrated in Fig. A.1, through all the fractions. You might notice that Cantor's table lists some of the fractions more than once. For example, every fraction along the diagonal – 1/1, 2/2, 3/3, etc. – equals 1. (In fact, every fraction appears infinitely many times in Cantor's table.) But as far as countability is concerned, this doesn't matter. We can count this whole table, and every rational number is in this table, so we can count all the rationals.

By another clever construction, *Cantor's diagonal argument*, Cantor showed that it is impossible to count the real numbers, not even the real numbers between 0 and 1. For suppose we could count the real numbers. Then say r_1 is the first real number, r_2 the second, and so on. Now write these, one below the other, and to the right of each number r_i write its decimal expansion, $r_i = 0.r_{i1}r_{i2}r_{i3}\ldots$. We get this table:

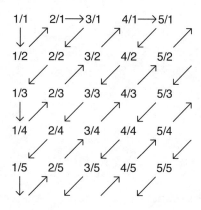

Figure A.1: How to count the rationals.

$$r_1 = 0.r_{11}r_{12}r_{13}r_{14}r_{15}\ldots$$
$$r_2 = 0.r_{21}r_{22}r_{23}r_{24}r_{25}\ldots$$
$$r_3 = 0.r_{31}r_{32}r_{33}r_{34}r_{35}\ldots$$
$$r_4 = 0.r_{41}r_{42}r_{43}r_{44}r_{45}\ldots$$
$$r_5 = 0.r_{51}r_{52}r_{53}r_{54}r_{55}\ldots$$
$$\ldots$$

Remember, if the real numbers are countable, then every real number between 0 and 1 is somewhere in this table. Now make the decimal expansion $t = 0.t_1t_2t_3t_4t_5\ldots$, where

$$t_1 \neq r_{11}, \ t_2 \neq r_{22}, \ t_3 \neq r_{33}, \ t_4 \neq r_{44}, \ t_5 \neq r_{55}, \ \ldots$$

What real number is t? Because $t_1 \neq r_{11}$, t cannot be r_1. Because $t_2 \neq r_{22}$, t cannot be r_2. Because $t_3 \neq r_{33}$, t cannot be r_3. Continuing in this way, we see that t cannot be any of the real numbers $r_1, r_2, r_3, \ldots$. That is, t is

A.2. CANTOR SET PROPERTIES

not any of the numbers in our list, but our list was supposed to contain *all* the real numbers between 0 and 1. Is this really a problem? Can't we just add t to the list? Sure, but then the same argument can be applied to the new list. Always. And think of this: for each decimal digit of t, we have nine choices, so for every list we can make many, many, many numbers not on the list. A few more details need to be filled in, but these are the main steps in Cantor's argument that the real numbers are uncountable.

To show the Cantor set also is uncountable, we'll show that it can be mapped point for point to all the numbers in the interval $[0, 1]$, which we've just seen are uncountable. We need one more ingredient: base-3 expansions. The familiar decimal expansion is just the base-10 expansion. For example,

$$3.14159\cdots = 3 + \frac{1}{10} + \frac{4}{10^2} + \frac{1}{10^3} + \frac{5}{10^4} + \frac{9}{10^5} + \cdots$$

For a base-3 expansion, replace the powers of 10 by powers of 3, and the numerators can be $0, 1$, or 2 only. For instance, you can use the formula for summing geometric series to show that

$$\frac{1}{3} + \frac{2}{3^2} + \frac{1}{3^3} + \frac{2}{3^4} + \frac{1}{3^5} + \frac{2}{3^6} + \cdots = \frac{5}{8}$$

Hint: sum all the terms with numerator 1, then sum all the terms with numerator 2, and add these sums.

To map Cantor set points to the interval $[0, 1]$, we observe that every number x in the Cantor set has a base-3 expansion

$$x = \frac{a_1}{3} + \frac{a_2}{3^2} + \frac{a_3}{3^3} + \cdots$$

where the a_i are 0s or 2s, with no 1s allowed. For example, $a_1 \neq 1$ means x does not lie between $1/3$ and $2/3$. Then $a_2 \neq 1$ means x does not lie between $1/9$ and $2/9$ or between $7/9$ and $8/9$, and so on. Once again, a few details remain, but the main step to show that the Cantor set is uncountable is this correspondence, matching the base-3 expansions of the Cantor set points with the base-2 expansions of the points in the interval $[0, 1]$.

$$\frac{a_1}{3} + \frac{a_2}{3^2} + \frac{a_3}{3^3} + \cdots \to \frac{a_1/2}{2} + \frac{a_2/2}{2^2} + \frac{a_3/2}{2^3} + \cdots$$

For example, the point $2/3 + 0/9 + 2/27$ in the Cantor set is sent to the point $1/2 + 0/4 + 1/8$ in the unit interval. As the a_i run through all combinations of 0 and 2, the $a_i/2$ run through all combinations of 0 and 1, and so the right side runs through the base-2 expansions of all real numbers between 0 and 1, uncountable according to Cantor.

A.2.3 Limit points

A point x is a *limit point* of a set A of points on the x-axis if every interval centered about x contains a point of A distinct from x. For example, the point $x = 0$ is a limit point of $A = \{1/3, 1/9, 1/27, \ldots\}$, because no matter how small, every interval centered about $x = 0$ must contain some positive numbers, and consequently some numbers in A, because A contains numbers as close to 0 as we wish.

To see why every point of the Cantor set is a limit point, recall that every point x of the Cantor set has a base-3 expansion,

$$x = \frac{a_1}{3} + \frac{a_2}{3^2} + \frac{a_3}{3^3} + \cdots$$

where the a_i are 0s or 2s. Then for every positive distance ϵ we must show how to find a point y of the Cantor set, $y \neq x$, with $|x - y| < \epsilon$. Here's how. Given any positive ϵ, no matter how tiny, $2/3^n < \epsilon$ as long as n is large enough. Then if x has this base-3 expansion,

$$x = \frac{a_1}{3} + \frac{a_2}{3^2} + \frac{a_3}{3^3} + \cdots + \frac{a_{n-1}}{3^{n-1}} + \frac{a_n}{3^n} + \frac{a_{n+1}}{3^{n+1}} + \cdots$$

take y to be the point with this expansion,

$$y = \frac{a_1}{3} + \frac{a_2}{3^2} + \frac{a_3}{3^3} + \cdots + \frac{a_{n-1}}{3^{n-1}} + \frac{2 - a_n}{3^n} + \frac{a_{n+1}}{3^{n+1}} + \cdots$$

Then y belongs to the Cantor set because its base-3 expansion contains only 0s and 2s, and $|x - y| = 2/3^n$. This last fact shows both that $y \neq x$ and that $|x - y| < \epsilon$. Consequently, every point of the Cantor set is a limit point of the set.

A.3 Points that are not limit points

Here we show how to construct an uncountable length-0 set in which not every point is a limit point.

This is easy if we start with a Cantor set. For example, to the Cantor middle-thirds set add the point $x = 1/2$, the midpoint of the largest interval removed. Call this new set X. Adding a point to an uncountable set gives an uncountable set; adding a point to a set of length 0 doesn't change the length. The point $1/2$ is not a limit point of this set, because, for example, the interval $(1/2 - 1/6, 1/2 + 1/6)$ intersects X only in the point $1/2$. Because this interval contains no point of X other than $1/2$, the point $1/2$ is not a limit point of X.

Higher-dimensional sets also can have limit points. For example, for subsets of the plane, replace intervals about x with discs centered at a point, and proceed as usual.

The next two notes refer to Sect. 2.3, **Fractal trees and ferns**

A.4 Tree and fern IFS tables

For relatively simple mathematical fractals, shapes that look a bit like the gasket, with some practice we can find the IFS rules by visual inspection. This is because we know the scaling factors will be simple fractions and any rotations will be by simple fractions of 90°. Natural-looking fractals present a different kind of challenge. In this section we'll describe a process, based on measurements, by which we can find the IFS rules for these fractals. In A.5 we develop a method that works from measurements of the coordinates of points in the fractal and of the corresponding points in each piece of the fractal. For now, the tree will be our guide.

Here we see the fractal tree rendered in light gray, so our choice of measurements will be more visible. We need to identify roughly vertical and horizontal lines in the whole fractal, as well as the corresponding lines in each piece of the fractal. In general, the longer the lines, the smaller the error in computing the scaling factors. On the other hand, we must choose lines easy to identify in the fractal and its pieces.

We'll illustrate the method by finding the r, s, θ, φ, e and f values for the lower left branch of the tree. Take the line between $O = (0,0)$ and $A = (0, 1.15)$ to be the vertical reference line in the whole tree. This is a reasonable choice: base to the top of

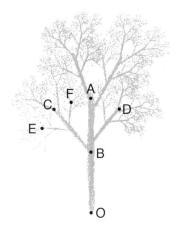

Figure A.2: Measuring tree IFS parameters.

the split of the trunk. In the lower left branch, the corresponding line lies between $B = (0, 0.61)$ and $C = (-0.375, 1.04)$. For this branch, the vertical scaling factor is

$$s = \frac{BC}{OA} = \frac{0.571}{1.15} \approx 0.496$$

We'll say $s = 0.5$.

For the horizontal reference line we'll take the line between the points $C = (-0.375, 1.04)$ and $D = (0.3, 1.04)$, the line between the top splitting in the lower left and lower right branches. In the lower left branch, the corresponding line is between $E = (-0.5, 0.85)$ and $F = (-0.2, 1.12)$. For this branch, the horizontal scaling factor is

$$r = \frac{EF}{CD} = \frac{0.404}{0.657} \approx 0.598$$

We'll say $r = 0.6$.

The rotation θ is the angle between CD and EF; the rotation φ is the angle between OA and BC. Both of these measure about $40°$.

The e-value for this branch is easy: it's 0. After we shrink and rotate the tree to form that branch, we lift it straight up to its position in the tree. Before commenting of the f-value, we must make one more observation about the magnitudes of the translation factors in the IFS table. Once again, the gasket is our guide.

A good way to find out what happens when we multiply all the e and f of an IFS table by the same factor is to start with the familiar Sierpinski gasket rules, multiply all the e and f values by $1/2$ and try to guess what will happen. Many people guess that the pieces of the gasket will overlap, but they won't. That the r and s values still are $1/2$ means that the gasket still will be made of three pieces each scaled by $1/2$. The e and f values determine the location of the pieces, and consequently the size of the gasket, so multiplying the e and f values by $1/2$ produces a gasket half the size. The first image of Fig. A.3 has non-zero e and f values equal to $1/2$; for the second image, both these values are $1/4$. Both gaskets are contained in the unit square. Put another way, we have the *scaling rule*:

Figure A.3: Rescaling e and f.

Multiplying all the e and f values of an IFS table by the same number multiplies the size of the IFS attractor by that same factor.

Consequently, we have some latitude in the choice of e and f values. So long as we use the same measuring scale, any scale will do, though of course measuring translations in centimeters instead of inches will give different numerical values for e and f, and these will alter the size if the image rendered on the computer screen.

Keeping this in mind, measurements give us the vertical translations of the branches of the tree. Similar measurements and calculations determine the IFS table for the fern.

In the next section we'll discuss another approach that does not depend on recognizing horizontal and vertical line segments.

A.5 IFS rules from point images

The method of A.4 relies on finding lines in the whole fractal and the corresponding lines in each piece of a decomposition of the fractal. While this has a visual appeal, there's a more direct way. If we select three points, not all lying on the same line, and find their images in a piece of the fractal, from the coordinates of these points we can find the IFS parameters for the rule that takes the whole fractal to that piece.

A.5. IFS RULES FROM POINT IMAGES

We'll approach this in three steps. First, we introduce a bit of linear algebra, just multiplying matrices, computing determinants, and inverting matrices. If these topics already are familiar, skip ahead to the second step, our discussion of the matrix formulation of IFS in A.5.4. The third step, A.5.5, is the real goal of this section: finding IFS rules from the coordinates of points and their images.

First, we introduce a bit of linear algebra.

A.5.1 Multiplying matrices

An example illustrates the general principle, "across on the left, down on the right."

$$\begin{bmatrix} a & b \\ c & d \end{bmatrix} \begin{bmatrix} e & f \\ g & h \end{bmatrix} = \begin{bmatrix} ae+bg & af+bh \\ ce+dg & cf+dh \end{bmatrix}$$

This extends in the obvious way to larger matrices, so long as the number of columns of the left matrix equals the number of rows of the right matrix.

Unlike simple arithmetic multiplication, the order can matter when multiplying matrices:

$$\begin{bmatrix} 1 & 2 \\ 3 & 4 \end{bmatrix} \cdot \begin{bmatrix} 1 & 4 \\ 2 & 0 \end{bmatrix} = \begin{bmatrix} 5 & 4 \\ 11 & 12 \end{bmatrix} \text{ but } \begin{bmatrix} 1 & 4 \\ 2 & 0 \end{bmatrix} \cdot \begin{bmatrix} 1 & 2 \\ 3 & 4 \end{bmatrix} = \begin{bmatrix} 13 & 18 \\ 2 & 4 \end{bmatrix}$$

Here is a more extreme example of how the order of multiplication matters: changing the order of multiplication can change the number of rows and columns of the product matrix.

$$\begin{bmatrix} 1 & 2 & 3 \\ 4 & 5 & 6 \end{bmatrix} \cdot \begin{bmatrix} 1 & 2 \\ 3 & 4 \\ 5 & 6 \end{bmatrix} = \begin{bmatrix} 22 & 28 \\ 49 & 64 \end{bmatrix} \text{ but } \begin{bmatrix} 1 & 2 \\ 3 & 4 \\ 5 & 6 \end{bmatrix} \cdot \begin{bmatrix} 1 & 2 & 3 \\ 4 & 5 & 6 \end{bmatrix} = \begin{bmatrix} 9 & 12 & 15 \\ 19 & 26 & 33 \\ 29 & 40 & 51 \end{bmatrix}$$

For some matrices the order of multiplication does not matter. An important example of this is the *identity matrix*, a square matrix consisting of 1s along the diagonal and 0s everywhere else.

$$\begin{bmatrix} a & b \\ c & d \end{bmatrix} \cdot \begin{bmatrix} 1 & 0 \\ 0 & 1 \end{bmatrix} = \begin{bmatrix} a & b \\ c & d \end{bmatrix} = \begin{bmatrix} 1 & 0 \\ 0 & 1 \end{bmatrix} \cdot \begin{bmatrix} a & b \\ c & d \end{bmatrix}$$

A.5.2 Computing determinants

The rule for finding the determinant of a 2 × 2 (number of rows × number of columns) matrix is simple.

$$\det \begin{bmatrix} a & b \\ c & d \end{bmatrix} = ad - bc$$

This can be generalized to 3 × 3 matrices.

$$\det \begin{bmatrix} a & b & c \\ d & e & f \\ g & h & i \end{bmatrix} = a \cdot \det \begin{bmatrix} e & f \\ h & i \end{bmatrix} - b \cdot \det \begin{bmatrix} d & f \\ g & i \end{bmatrix} + c \cdot \det \begin{bmatrix} d & e \\ g & h \end{bmatrix}$$

$$= -d \cdot \det \begin{bmatrix} b & c \\ h & i \end{bmatrix} + e \cdot \det \begin{bmatrix} a & c \\ g & i \end{bmatrix} - f \cdot \det \begin{bmatrix} a & b \\ g & h \end{bmatrix}$$

and four other expressions, altogether one for each row and each column of the matrix. The first expression is called *expanding along the first row*; the second, *expanding along the second row*. We can evaluate a determinant by expanding along any row or column, but note that the signs of the coefficients of the determinants of the 2 × 2 matrices alternate, starting with a positive sign for the upper left matrix element.

A.5.3 Inverting matrices

The *inverse* of an $n \times n$ matrix A is an $n \times n$ matrix A^{-1}, which is for matrix multiplication what the reciprocal is for real numbers:

$$A \cdot A^{-1} = A^{-1} \cdot A = I$$

where I is the $n \times n$ identity matrix. A real number r has an inverse (reciprocal) if and only if $r \neq 0$. The corresponding condition for square matrices is that $\det(A) \neq 0$. For example, for 2 × 2 matrices

$$A^{-1} = \begin{bmatrix} a & b \\ c & d \end{bmatrix}^{-1} = \frac{1}{\det(A)} \begin{bmatrix} d & -b \\ -c & a \end{bmatrix}$$

where multiplying a matrix by a number, here $1/\det(A)$, means multiplying each entry of the matrix by that number. That is, as long as $\det(A) \neq 0$,

$$A^{-1} = \begin{bmatrix} a & b \\ c & d \end{bmatrix}^{-1} = \begin{bmatrix} d/\det(A) & -b/\det(A) \\ -c/\det(A) & a/\det(A) \end{bmatrix}$$

For $n > 2$, finding the inverse of an $n \times n$ is more work. But if we only need to know when a matrix is invertible, the determinant condition holds regardless of the size of the matrix. That is, an $n \times n$ matrix is invertible if and only if $\det(A) \neq 0$.

Now that we know how to invert and multiply matrices, we're ready to calculate IFS parameters from the images of three non-collinear points.

A.5.4 IFS matrix formulation

We know what each one of the IFS parameters (r, s, θ, φ, e, and f) does to an image, but the software needs a more mathematical expression of the transformation. Here it is:

$$\begin{bmatrix} x_{i+1} \\ y_{i+1} \end{bmatrix} = T \begin{bmatrix} x_i \\ y_i \end{bmatrix} = \begin{bmatrix} r\cos(\theta) & -s\sin(\varphi) \\ r\sin(\theta) & s\cos(\varphi) \end{bmatrix} \begin{bmatrix} x_i \\ y_i \end{bmatrix} + \begin{bmatrix} e \\ f \end{bmatrix}$$

Pick some IFS parameters and a few points (x_i, y_i). A bit of experimentation with how the images of these points move as the IFS parameters change can help develop intuition about these transformations.

Now we're ready for the final step.

A.5.5 IFS rules from point images

Suppose we have three non-collinear points, the *initial points*

$$p_1 = (x_1, y_1), \ p_2 = (x_2, y_2), \text{ and } p_3 = (x_3, y_3)$$

and three other points, the *image points*

$$q_1 = (u_1, v_1), \ q_2 = (u_2, v_2), \text{ and } q_3 = (u_3, v_3)$$

Think of the initial points as three points we select on the fractal. Because we choose the points, we can guarantee they are non-collinear. The image points are the images of the initial points in a piece of the fractal. We'll find a transformation T that does this

$$T(p_1) = q_1, \ T(p_2) = q_2, \text{ and } T(p_3) = q_3$$

and we'll show that only one transformation T does this. We'll do this in two steps, first, writing

$$T \begin{bmatrix} x \\ y \end{bmatrix} = \begin{bmatrix} a & b \\ c & d \end{bmatrix} \begin{bmatrix} x \\ y \end{bmatrix} + \begin{bmatrix} e \\ f \end{bmatrix}$$

we'll show how to find a, b, c, d, e, and f from the coordinates of p_1, p_2, p_3, q_1, q_2, and q_3. Then we'll show how to find r, s, θ, φ, e, and f from a, b, c, d, e, and f. The labels show how to find two of these: the e and f of the first formulation are the e and f of the second.

Using the matrix multiplication rule, the equation $T(x, y) = (u, v)$ means

$$ax + by + e = u$$
$$cx + dy + f = v$$

Then the equations

$$T(p_1) = q_1, \ T(p_2) = q_2, \text{ and } T(p_3) = q_3$$

mean

$$ax_1 + by_1 + e = u_1 \quad ax_2 + by_2 + e = u_2 \quad ax_3 + by_3 + e = u_3$$
$$cx_1 + dy_1 + f = v_1 \quad cx_2 + dy_2 + f = v_2 \quad cx_3 + dy_3 + f = v_3$$

Now group together the equations containing a, b, and e, and those containing c, d, and f:

$$\begin{aligned} ax_1 + by_1 + e &= u_1 & cx_1 + dy_1 + f &= v_1 \\ ax_2 + by_2 + e &= u_2 & cx_2 + dy_2 + f &= v_2 \\ ax_3 + by_3 + e &= u_3 & cx_3 + dy_3 + f &= v_3 \end{aligned}$$

In most of our experience with math, we see letters from the end of the alphabet – x, y, u, and v for example – as variables, and letters from the beginning of the alphabet as constants. But now it's the other way around. The xs, ys, us, and vs are coordinates of points, so are constants known to us. We're trying to find a, b, c, d, e, and f. To do this, we'll rewrite these last six equations as two matrix equations

$$\begin{bmatrix} x_1 & y_1 & 1 \\ x_2 & y_2 & 1 \\ x_3 & y_3 & 1 \end{bmatrix} \begin{bmatrix} a \\ b \\ e \end{bmatrix} = \begin{bmatrix} u_1 \\ u_2 \\ u_3 \end{bmatrix} \quad \begin{bmatrix} x_1 & y_1 & 1 \\ x_2 & y_2 & 1 \\ x_3 & y_3 & 1 \end{bmatrix} \begin{bmatrix} c \\ d \\ f \end{bmatrix} = \begin{bmatrix} v_1 \\ v_2 \\ v_3 \end{bmatrix}$$

Now both of these equations have the same coefficient matrix, so if that matrix is invertible, the solutions are easy:

$$\begin{bmatrix} a \\ b \\ e \end{bmatrix} = \begin{bmatrix} x_1 & y_1 & 1 \\ x_2 & y_2 & 1 \\ x_3 & y_3 & 1 \end{bmatrix}^{-1} \begin{bmatrix} u_1 \\ u_2 \\ u_3 \end{bmatrix} \quad \begin{bmatrix} c \\ d \\ f \end{bmatrix} = \begin{bmatrix} x_1 & y_1 & 1 \\ x_2 & y_2 & 1 \\ x_3 & y_3 & 1 \end{bmatrix}^{-1} \begin{bmatrix} v_1 \\ v_2 \\ v_3 \end{bmatrix}$$

To show the coefficient matrix is invertible, remember the coordinates of the points $p_1 = (x_1, y_1)$, $p_2 = (x_2, y_2)$, and $p_3 = (x_3, y_3)$. Then form the vectors $\vec{A} = p_2 - p_1$ and $\vec{B} = p_3 - p_1$, illustrated in Fig. A.4.

If the points p_1, p_2, and p_3 are not collinear, the angle between $\vec{A}$ and $\vec{B}$ is not $0°$ or $180°$. This is easy to see from the picture.

The cross-product of vectors measures the angles between the vectors, but the cross-product is defined only for vectors in 3-dimensional space. (Well, it works in 7-dimensional space, too, but we won't get into this now, or at all.) Making 3-dimensional vectors out of $\vec{A}$ and $\vec{B}$ is easy. Call the 3-dimensional vectors $\vec{A}'$ and $\vec{B}'$:

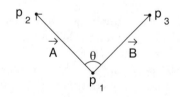

Figure A.4: The angle θ between the vectors $\vec{A}$ and $\vec{B}$.

$$\vec{A}' = \langle x_2 - x_1, y_2 - y_1, 0 \rangle \quad \vec{B}' = \langle x_3 - x_1, y_3 - y_1, 0 \rangle$$

A.5. IFS RULES FROM POINT IMAGES

Now we know the length of the cross-product $|\vec{A}' \times \vec{B}'|$ satisfies

$$|\vec{A}' \times \vec{B}'| = |\vec{A}'||\vec{B}'|\sin(\theta)$$

If p_1, p_2, and p_3 are distinct points, then $|\vec{B}'| \neq 0$ and $|\vec{A}'| \neq 0$, so the only way $|\vec{A}' \times \vec{B}'| = 0$ is if $\sin(\theta) = 0$, that is, if $\theta = 0°$ or $\theta = 180°$. But this isn't possible, because p_1, p_2, and p_3 are not collinear. All this is just a careful way to show that so long as p_1, p_2, and p_3 are distinct, non-collinear points, then $|\vec{A}' \times \vec{B}'| \neq 0$.

Recalling the matrix definition of the cross-product

$$\vec{A}' \times \vec{B}' = \det \begin{bmatrix} \vec{i} & \vec{j} & \vec{k} \\ x_2 - x_1 & y_2 - y_1 & 0 \\ x_3 - x_1 & y_3 - y_1 & 0 \end{bmatrix}$$
$$= 0\vec{i} - 0\vec{j} + ((x_2 - x_1)(y_3 - y_1) - (x_3 - x_1)(y_2 - y_1))\vec{k}$$

Consequently,

$$|\vec{A}' \times \vec{B}'| = x_2 y_3 - x_2 y_1 - x_1 y_3 + x_1 y_1 - x_3 y_2 + x_3 y_1 + x_1 y_2 - x_1 y_1$$
$$= x_2 y_3 - x_2 y_1 - x_1 y_3 - x_3 y_2 + x_3 y_1 + x_1 y_2$$
$$= \det \begin{bmatrix} x_1 & y_1 & 1 \\ x_2 & y_2 & 1 \\ x_3 & y_3 & 1 \end{bmatrix}$$

That is, $|\vec{A}' \times \vec{B}'|$ is the determinant of the coefficient matrix above. Recall that a matrix is invertible if and only if its determinant is nonzero, we see that the coefficient matrix is invertible if the points p_1, p_2 and p_3 are non-collinear.

The inverse of the coefficient matrix is

$$\frac{1}{\Delta} \begin{bmatrix} y_2 - y_3 & y_3 - y_1 & y_1 - y_2 \\ x_3 - x_2 & x_1 - x_3 & x_2 - x_1 \\ x_2 y_3 - y_2 x_3 & x_3 y_1 - y_3 x_1 & x_1 y_2 - y_1 x_2 \end{bmatrix}$$

where $\Delta = x_2 y_3 - x_3 y_2 - x_1 y_3 + x_3 y_1 + x_1 y_2 - x_2 y_1$ is the determinant of the coefficient matrix. So we see that the coordinates of the three (non-collinear) initial points and the three image points (These can be collinear. That's not a problem at all.) determine the matrix parameters a, b, c, d, e, and f.

Finally, we convert the matrix coefficients a, b, c, and d into r, s, θ and φ. The translations e and f are the same for both matrix representations.

We can make this a bit simpler by noting that reflection across the x-axis is equivalent to reflection across the y-axis, followed by a 180° rotation. Draw a picture on a square of paper and trace it on the back of the square, to allow for reflections. Then verify this statement by appropriate motions of the paper square. We see that both reflections can be achieved by reflection across the y-axis and possibly a rotation.

The equivalence of the matrices

$$\begin{bmatrix} r\cos(\theta) & -s\sin(\varphi) \\ r\sin(\theta) & s\cos(\varphi) \end{bmatrix} \quad \text{and} \quad \begin{bmatrix} a & b \\ c & d \end{bmatrix}$$

means

$$r\cos(\theta) = a, \ -s\sin(\varphi) = b, \ r\sin(\theta) = c, \text{ and } s\cos(\varphi) = d$$

The magnitudes of r and s are easy to find:

$$a^2 + c^2 = r^2\cos^2(\theta) + r^2\sin^2(\theta) = r^2, \quad \text{so } r = \pm\sqrt{a^2 + c^2}$$
$$b^2 + d^2 = s^2\sin^2(\varphi) + s^2\cos^2(\varphi) = s^2, \quad \text{so } s = \sqrt{b^2 + d^2}$$

The sign of r is $+$ if the transformation does not involve a reflection across the y-axis and $-$ if it does. The sign of s we always can take to be $+$ because as we have seen, a reflection across the x-axis can be achieved by a reflection across the y-axis followed by a 180° rotation.

To determine if the transformation T involves a reflection, look at the initial points

$$p_1 = (x_1, y_1), \quad p_2 = (x_2, y_2), \text{ and } p_3 = (x_3, y_3)$$

and their images

$$T(p_1) = q_1 = (u_1, v_1), \quad T(p_2) = q_2 = (u_2, v_2), \text{ and } T(p_3) = q_3 = (u_3, v_3)$$

Viewing these as points in the xy-plane in 3-dimensional space, we can form the cross-products

$$(p_2 - p_1) \times (p_3 - p_1)$$
$$= 0\vec{i} + 0\vec{j} + ((x_2 - x_1)(y_3 - y_1) - (x_3 - x_1)(y_2 - y_1))\vec{k}$$
$$(q_2 - q_1) \times (q_3 - q_1)$$
$$= 0\vec{i} + 0\vec{j} + ((u_2 - u_1)(v_3 - v_1) - (u_3 - u_1)(v_2 - v_1))\vec{k}$$

If both vectors point in the same direction, the orientation of the triple of image points is the same as that of the triple of initial points, so T does not involve a reflection. If these vectors point in opposite directions, then T involves a reflection and in this case we reverse the sign of both a and c before finding the value of θ. This is because $a = r\cos(\theta)$ and $c = r\sin(\theta)$, so accounting for the reflection reverses the sign of r, hence of both a and c.

A.5. IFS RULES FROM POINT IMAGES

Now to find the angles, from the first circle of Fig. A.5 we see

$$a = r\cos(\theta)$$
$$c = r\sin(\theta)$$

and so

$$\frac{c}{a} = \frac{r\sin(\theta)}{r\cos(\theta)} = \tan(\theta)$$

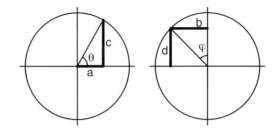

Figure A.5: The angles θ and φ.

From the second circle we see

$$b = -s\cos(\varphi)$$
$$d = s\sin(\varphi)$$

and so

$$\frac{b}{d} = \frac{-s\sin(\varphi)}{s\cos(\varphi)} = -\tan(\varphi)$$

Inverting each relation gives $\theta = \arctan(c/a)$ and $\varphi = \arctan(-b/d)$.

There is one complication: different angles can have the same tangent. But between $-90°$ and $90°$, each value of the tangent corresponds to only one angle. Taking some care with the signs of a, b, c, and d, we can determine θ and φ in the range $-180°$ to $180°$. By looking at the relations between the signs of a and c and the range of θ-values, using Fig. A.5 we can work out these cases

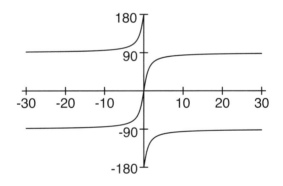

Figure A.6: The arctan graph.

- If $a > 0$ then $\theta = \arctan(c/a)$.

- If $a = 0$ and $c > 0$, then $\theta = 90°$.

- If $a < 0$ and $c > 0$, then because $\tan(c/a) = \tan(-c/-a)$ and $(-a, -c)$ lies in the range where we have computed arctan, we have $\theta = \arctan(c/a) + 180°$.

- If $a = 0$ and $c < 0$, then $\theta = -90°$.

- If $a < 0$ and $c < 0$, then because $(-a, -c)$ lies in the range where we have computed arctan, we have $\theta = \arctan(c/a) - 180°$.

The same kind of analysis works for φ. Some care is needed with signs, so we'll spell out the cases.

- If $d > 0$ then $\varphi = \arctan(-b/d)$.
- If $d = 0$ and $b < 0$, then $\varphi = 90°$.
- If $d < 0$ and $b < 0$, then $\varphi = \arctan(-b/d) + 180°$.
- If $d = 0$ and $b > 0$, then $\varphi = -90°$.
- If $d < 0$ and $b > 0$, then $\varphi = \arctan(-b/d) - 180°$.

This may seem complicated, so we'll run through an example.

Suppose the initial points are $p_1 = (0,0)$, $p_2 = (1,0)$, and $p_3 = (0,1)$, and the image points are $q_1 = (0.5, 0.75)$, $q_2 = (-0.15, 0.375)$, and $q_3 = (0.25, 1.18)$. Then the coefficient matrix

$$\begin{bmatrix} x_1 & y_1 & 1 \\ x_2 & y_2 & 1 \\ x_3 & y_3 & 1 \end{bmatrix} = \begin{bmatrix} 0 & 0 & 1 \\ 1 & 0 & 1 \\ 0 & 1 & 1 \end{bmatrix}$$

has inverse

$$\begin{bmatrix} 0 & 0 & 1 \\ 1 & 0 & 1 \\ 0 & 1 & 1 \end{bmatrix}^{-1} = \begin{bmatrix} -1 & 1 & 0 \\ -1 & 0 & 1 \\ 1 & 0 & 0 \end{bmatrix}$$

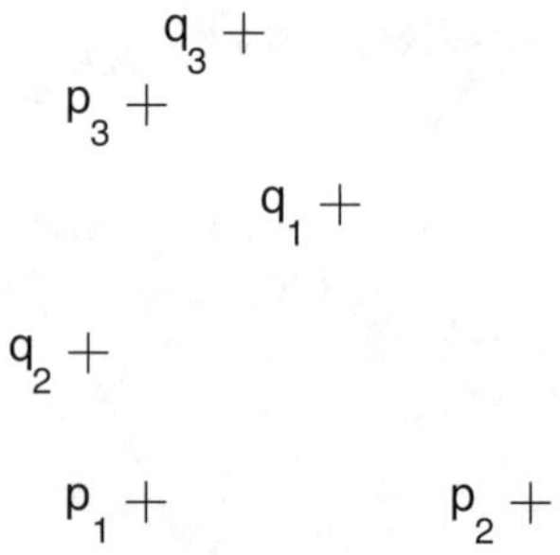

Figure A.7: Three points and their images.

We find a, b, c, d, e, and f by

$$\begin{bmatrix} a \\ b \\ e \end{bmatrix} = \begin{bmatrix} -1 & 1 & 0 \\ -1 & 0 & 1 \\ 1 & 0 & 0 \end{bmatrix} \begin{bmatrix} u_1 \\ u_2 \\ u_3 \end{bmatrix} = \begin{bmatrix} -1 & 1 & 0 \\ -1 & 0 & 1 \\ 1 & 0 & 0 \end{bmatrix} \begin{bmatrix} 0.5 \\ -0.15 \\ 0.25 \end{bmatrix} = \begin{bmatrix} -0.65 \\ -0.25 \\ 0.5 \end{bmatrix}$$

and by

$$\begin{bmatrix} c \\ d \\ f \end{bmatrix} = \begin{bmatrix} -1 & 1 & 0 \\ -1 & 0 & 1 \\ 1 & 0 & 0 \end{bmatrix} \begin{bmatrix} v_1 \\ v_2 \\ v_3 \end{bmatrix} = \begin{bmatrix} -1 & 1 & 0 \\ -1 & 0 & 1 \\ 1 & 0 & 0 \end{bmatrix} \begin{bmatrix} 0.75 \\ 0.375 \\ 1.18 \end{bmatrix} = \begin{bmatrix} -0.375 \\ 0.43 \\ 0.75 \end{bmatrix}$$

Then

$$r = \pm\sqrt{a^2 + c^2} = \pm\sqrt{(-0.65)^2 + (-0.375)^2} \approx \pm 0.75$$

and

$$s = \sqrt{b^2 + d^2} = \sqrt{(-0.25)^2 + 0.43^2} \approx 0.50$$

The sign of r is determined by the absence or presence of a reflection. To test that, we compute

$$(x_2 - x_1)(y_3 - y_1) - (x_3 - x_1)(y_2 - y_1) = 1$$

and

$$(u_2 - u_1)(v_3 - v_1) - (u_3 - u_1)(v_2 - v_1) \approx -0.373$$

because the vectors $(p_2 - p_1) \times (p_3 - p_1)$ and $(q_2 - q_1) \times (q_3 - q_1)$ point in the same direction if their $\vec{k}$ components have the same sign. Here the $\vec{k}$ components have opposite signs so the vectors point in opposite directions and that means we take r to be negative.

So far, we have $r \approx -0.75$, $s \approx 0.50$, $e = 0.5$ and $f = 0.75$. Because there is a reflection, to find θ we reverse the sign of a and c before consulting the table. Then

$$\theta = \arctan(0.375/0.65) \approx 30° \text{ and } \varphi = \arctan(-(-0.25)/0.43) \approx 30°$$

This is a bit tedious, but it can be coded pretty easily. Then measuring the coordinates of three non-collinear points in a fractal and of their images in each piece is all we need do to be able to calculate the IFS rules.

The next note refers to Sect. 2.4, **Spirals**.

A.6 Spiral IFS tables

To find IFS rules for these spirals, first decompose them into smaller copies of themselves. The first spiral consists of three copies: a small copy at the end of the left arm, a small copy at the end of the right arm, and everything else. Adapting this approach to the second spiral, we see five copies: one at the end of each arm, and everything else. Recalling the scaling rule, only the relative values of the translations are are important.

Figure A.8: Top: Two fractal spirals and their pieces.

r	s	θ	φ	e	f
0.30	0.30	0	0	0.7	0
0.30	0.30	0	0	-0.7	0
0.85	0.85	20°	20°	0	0

Table A.2: IFS for the first spiral of Fig. A.8.

r	s	θ	φ	e	f
0.20	0.20	0	0	0.7	0
0.20	0.20	0	0	-0.7	0
0.20	0.20	0	0	0	0.7
0.20	0.20	0	0	0	-0.7
0.85	0.85	20°	20°	0	0

Table A.3: IFS for the second spiral of Fig. A.8.

The IFS for these spirals are given in Table A.2 and in Table A.3

The next note refers to Sect. 2.5, **IFS with memory**

A.7 IFS topological types

Here our use of the word "topology" refers to the arrangement of pieces of a fractal. Specifically, we're interested in these properties:

- *connected* means all one piece

- *disconnected* means at least two pieces

- *totally disconnected* means single points are the only connected pieces

- *simply-connected* means no loops

The *Sierpinski gasket relatives* are a collection of examples effective for becoming familiar with these topological notions. Recall the first instance of the Sierpinski gasket we saw consisted of three pieces, each scaled by 1/2, one placed in the lower left corner, one in the lower right corner, and one in the upper left corner. By a gasket relative we mean a fractal consisting of three pieces, each scaled by 1/2, with each piece possibly reflected across the x- or y-axes, rotated 90°, 180°, or 270°, and translated so one piece occupies the lower left corner, one the lower right corner, and one the upper left corner. Some combinations of rotations and reflections have identical effects. For example, reflecting across the y-axis and rotating 270° isthe same transformation as reflecting across the x-axis and

A.7. IFS TOPOLOGICAL TYPES

rotating 90°. Taking all these agreements into account, there are 224 gasket relatives. All these are shown (as small images, to be sure) on pages 246–248 of the large, beautiful "fractal encyclopedia," *Chaos and Fractals: New Frontiers in Science* by Heinz-Otto Peitgen, Hartmut Jürgens, and Dietmar Saupe. In Fig. A.9 we see gasket relatives that are
(a) connected but not simply-connected, (In fact, we see infinitely many loops.)
(b) connected and simply-connected,
(c) disconnected, and
(d) totally disconnected.

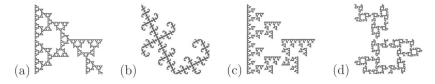

Figure A.9: Examples of gasket relatives.

Tara Taylor has studied gasket relatives and found conditions on the low-order iterates of the deterministic IFS algorithm that give rise to each type (a), (b), (c), and (d).

Inspired by Taylor's work, Kacie Saxer-Taulbee and the older author have studied the topological type of IFS generated by 1-step memory.

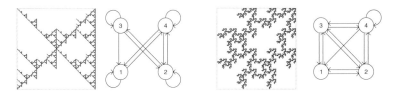

Figure A.10: Connected and not simply-connected examples.

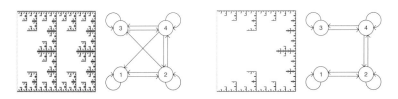

Figure A.11: Connected examples with one loop (first) and with no loops (second).

The two fractals of Fig. A.10 are connected and not simply-connected. In fact, both have infinitely many loops. Both fractals of Fig. A.11 are

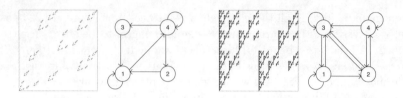

Figure A.12: Totally disconnected (first) and disconnected but not totally disconnected (second).

connected. The first has a single loop, the second has no loops; it is simply-connected. Both fractals of Fig. A.12 are disconnected. The first is totally disconnected, the second is not.

A surprise of our investigations is the variety of fractals generated. While those of Figs. A.11 and A.12 are not so unusual, both of Fig. A.10 exhibit patterns we had not seen. And we have found others just as peculiar. Take a look. See what you can find.

As of this writing, just before sending the final version of this book to the publisher, our understanding of the conditions responsible for these topological types is mechanical, at what level of iteration are connections preserved or broken. Yet there must be a more holistic approach, some way to organize properties of the transition graph , that will reveal the topological type of the fractal. Perhaps you'll find it before we do.

The next note refers to Sect. 2.6, **Random rendering of fractal images**

A.8 Expected time to visit an address

Here we study how to calculate the expected time needed to visit a given address with the random IFS algorithm. Also, we see how to select the order in which to apply the IFS rules to fill in the picture of the attractor as quickly as possible. For both problems we use the bit of linear algebra introduced in A.5.

First, if we are interested in the mathematical object instead of an image on a computer screen, we use the limit set of the iterates of (x_0, y_0) instead of the iterates themselves. The *limit set* of a set A is the collection of all limit points of A. (Limit points are defined in the Sect. A.2.) This is the answer to the question, occasionally posed to the older author, of how the random IFS algorithm can generate a Cantor set, because we know the Cantor set is uncountable, but the random IFS iterates of a point are countable. This is an invisible distinction on a computer screen, but it is glaring in MathWorld.

The second issue is a bit more involved. We have seen that for any length n we care to specify, the random IFS algorithm will eventually

A.8. EXPECTED TIME TO VISIT AN ADDRESS

generate points in all length-n address regions. But so far, we have not raised the issue of how much time is needed for "eventually." The answer can be discouraging. For example, if the image is contained in 1024×1024 pixels and is generated by four transformations with scaling factor $1/2$, then a single pixel has address length 10. If, say, the probability of applying transformation T_1 is 0.1, on average how long must the random algorithm run in order to put a point in the pixel with address 1^{10}? The rough guess is that on average we need 10 iterates to see one application of T_1, $10^2 = 100$ iterates to see two consecutive applications of T_1, and on till we need about 10^{10}, ten billion, iterates to find ten consecutive applications of T_1. The careful calculation is a bit more complicated, but we'll do it now because it's a very pretty construction that we'll also use later.

To answer the question of how long we should expect to wait to see ten consecutive applications of T_1, we'll build a mathematical construction called a *Markov chain*. In fact, we'll build a simpler chain to estimate how long we must wait to see four consecutive applications of T_1. The answer for ten requires no new ideas, but is a bit more involved. Rather than go through abstract definitions, we'll get right to it and describe how this example is put together.

First is a list of *states*, stages along the way to getting four consecutive 1s. We'll use five states. Recall that in driven IFS we append the digit of the current transformation on the left of the address. We'll use the symbol ¬1 to mean "not 1," that is, 2, 3, or 4. The states are

A: the string 1111 has not occurred, and the left-most digit is ¬1.
B: the string 1111 has not occurred, and the left-most pair is 1¬1.
C: the string 1111 has not occurred, and the left-most triple is 11¬1.
D: the string 1111 has not occurred, and the left-most quadruple is 111¬1.
E: the string 1111 has occurred.

States A, B, C, and D describe every string that does not contain 1111. As a string grows, it travels along a path through these states. If 1111 occurs, the string enters state E and we're done.

The *transition graph*, a map of the allowed paths between states, is shown in Fig. A.13. To calculate the expected number of steps needed to reach state E, we must estimate the probability of moving from one state to another.

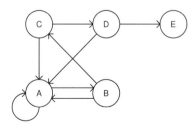

Figure A.13: The transition graph.

To go from state A to state B, we must apply T_1, and so $\Pr(A \to B) = 0.1$. Applying any other transformation to an element of state A gives an element of state A, so $\Pr(A \to A) = 0.9$. Arranging the state going *from* A through E along the top, and the state going *to* A through E along the side, we can rerpresent the probabilities of all these transitions by this *transition matrix*:

$$M = \begin{bmatrix} .9 & .9 & .9 & .9 & 0 \\ .1 & 0 & 0 & 0 & 0 \\ 0 & .1 & 0 & 0 & 0 \\ 0 & 0 & .1 & 0 & 0 \\ 0 & 0 & 0 & .1 & 1 \end{bmatrix}$$

For example, suppose we start in state A, represented by the column vector $\vec{1}$ with 1 in the first entry and with all the rest 0s. One iteration gives $M\vec{1}$,

$$M\vec{1} = \begin{bmatrix} .9 & .9 & .9 & .9 & 0 \\ .1 & 0 & 0 & 0 & 0 \\ 0 & .1 & 0 & 0 & 0 \\ 0 & 0 & .1 & 0 & 0 \\ 0 & 0 & 0 & .1 & 1 \end{bmatrix} \begin{bmatrix} 1 \\ 0 \\ 0 \\ 0 \\ 0 \end{bmatrix} = \begin{bmatrix} .9 \\ .1 \\ 0 \\ 0 \\ 0 \end{bmatrix}$$

This shows that if we start in state A, we'll stay in A 90% of the time and move to B 10% of the time. After two iterations, the likelihood of being in each state, starting from A, is $M^2\vec{1}$. After calculating M^2, we see that we stay in state A 90% of the time, move to state B 9% of the time, and move to state C 1% of the time, and so on.

State E is called an *absorbing* state, because once we enter state E, we cannot leave: as soon as 1111 has occurred, it always will have occurred. (Older author writing: Some of my students have suggested that absorbing states be renamed "Hotel California states." We'll stick with absorbing states.) To see how we can use this to find the expected number of iterates before reaching state E, we subdivide the transition matrix M into pieces,

$$M = \begin{bmatrix} Q & O \\ R & I \end{bmatrix}$$

where Q is the transition matrix for going from states $A, B, C,$ and D into states $A, B, C,$ and D, and R is the transition matrix for going from states $A, B, C,$ and D into state E.

$$Q = \begin{bmatrix} .9 & .9 & .9 & .9 \\ .1 & 0 & 0 & 0 \\ 0 & .1 & 0 & 0 \\ 0 & 0 & .1 & 0 \end{bmatrix}, O = \begin{bmatrix} 0 \\ 0 \\ 0 \\ 0 \end{bmatrix}, R = \begin{bmatrix} 0 & 0 & 0 & .1 \end{bmatrix}, I = \begin{bmatrix} 1 \end{bmatrix}$$

Then it's simple, if a bit tedious, to verify that

$$M^2 = \begin{bmatrix} Q^2 & O \\ R(I+Q) & I \end{bmatrix} \quad \text{and, in general,} \quad M^n = \begin{bmatrix} Q^n & O \\ R(I+Q+\cdots+Q^{n-1}) & I \end{bmatrix}$$

A.8. EXPECTED TIME TO VISIT AN ADDRESS

Taking the $n \to \infty$ limit,

$$M^\infty = \begin{bmatrix} Q^\infty & O \\ R(I + Q + Q^2 \cdots) & I \end{bmatrix}$$

Let's figure out what Q^∞ and $R(I + Q + Q^2 \cdots)$ are. First, every entry of Q is smaller than 1, the columns sum to ≤ 1, and one column (column 4) sums to < 1. With a bit of work, we can see that these conditions guarantee $\lim_{n \to \infty} Q^n = 0$.

Next, to understand $R(I + Q + Q^2 \cdots)$, recall that the geometric series $1 + r + r^2 + \cdots$ sums to $1/(1-r)$ if $|r| < 1$. The sum $I + Q + Q^2 \cdots$ looks a lot like a geometric series, so we might expect that under some condition on Q analogous to $|r| < 1$, $I + Q + Q^2 \cdots = (I - Q)^{-1}$. To show this is true, multiply

$$(I + Q + Q^2 + \cdots + Q^n)(I - Q)$$
$$= I - Q + Q - Q^2 + Q^2 - Q^3 + \cdots - Q^n + Q^n - Q^{n+1}$$
$$= I - Q^{n+1}$$

Because $\lim_{n \to \infty} Q^n = 0$, we see

$$I + Q + Q^2 + Q^3 + \cdots = (I - Q)^{-1}$$

Combining these results, we can rewrite M^∞ as

$$M^\infty = \begin{bmatrix} 0 & O \\ R(I - Q)^{-1} & I \end{bmatrix}$$

The expected number of iterates to go from state A to state E is the sum of the entries of the first column of $(I - Q)^{-1}$. (Seeing why this is true requires some work. Simplest is to Google "absorbing Markov chain.") For the matrix Q of the example we're discussing,

$$(I - Q)^{-1} = \begin{bmatrix} .1 & -.9 & -.9 & -.9 \\ -.1 & 1 & 0 & 0 \\ 0 & -.1 & 1 & 0 \\ 0 & 0 & -.1 & 1 \end{bmatrix}^{-1} = \begin{bmatrix} 10000 & 9990 & 99990 & 9000 \\ 1000 & 1000 & 990 & 900 \\ 100 & 100 & 100 & 90 \\ 10 & 10 & 10 & 10 \end{bmatrix}$$

Then the expected number of iterations to reach state E – that is, to apply T_1 four times consecutively – is $10000 + 1000 + 100 + 10 = 11110$.

The analogous construction, with substantially larger matrices, shows that with the random algorithm we should expect to take 11111111110 iterates to apply T_1 10 times consecutively. Our rough guess, 10^{10}, was pretty close. Depending on your computer speed, this can be a long time to wait. Can we do better?

Suppose we have an IFS with k transformations and we want to visit all addresses of length n. Rather than applying the transformations in random order, the fastest way to visit all length-n addresses is to apply the transformations in the order specified by a *de Bruijn sequence*, a sequence in which every length-n sequence of k numbers occurs exactly once. For the $n = 10$ and $k = 4$ example mentioned above, a de Bruijn sequence has length about $4^{10} = 10,488,576$, so will take about one thousandth of the time the random algorithm will need to visit every pixel.

To illustrate de Bruijn sequences, we'll use binary sequences, but with some more work these results hold for any IFS with at least two transformations.

Here are two binary de Bruijn sequences:

$$00110 \text{ for } n = 2 \qquad 0001011100 \text{ for } n = 3$$

For example, reading the first sequence from left to right, we find 00, 01, 11, and 10, all the $n = 2$ binary sequences.

But how do we know there are de Bruijn sequences for larger n? This isn't obvious. The existence of binary de Bruijn sequences for arbitrary n was proved in 1894 by the French mathematician Camille Flye Sainte-Marie and generalized for all k by the Dutch mathematicians Nicolaas de Bruijn and Tatyana van Aardenne-Ehrenfest. An elegant way to generate de Bruijn sequences is given by Franklin Mendivil.

We'll describe a simpler approach, first showing that a length-n de Bruijn sequence is an Eulerian cycle in an order $n - 1$ de Bruijn graph; both are defined in a moment. Next, we present the *prefer* 1 *algorithm* for constructing de Bruijn sequences.

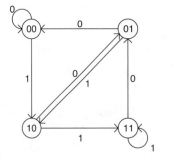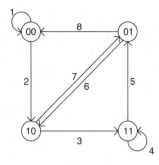

Figure A.14: First: An order-2 de Bruijn graph. Second: Using this graph to construct an $n = 3$ de Bruijn sequence.

For example, to construct an $n = 3$ de Bruijn sequence, we begin with the order-2 de Bruijn graph. This is the graph whose vertices are the length-2 binary sequences, and an edge goes from vertex ab to vertex cd if and only if $a = d$. This edge is labeled c. We interpret this graph as showing the effect of applying transformations on length-2 addresses. For example, applying T_1 to a point in the region with address 00 gives a point in the region with address 10. This graph is the first image of Fig. A.14. If

A.8. EXPECTED TIME TO VISIT AN ADDRESS

we think of these vertices as the vertices of a square, those of the order-3 de Bruijn graph can be viewed as vertices of a cube, those of the order-4 de Bruijn graph as the vertices of a hypercube, and so on.

For each binary de Bruijn graph, each vertex is the terminal point of two edges, and each vertex is the initial point of two edges. Specifically, the vertex ab is the terminal point of the edges

$$b0 \xrightarrow{a} ab \quad \text{and} \quad b1 \xrightarrow{a} ab$$

and the initial point of the edges

$$ab \xrightarrow{0} 0a \quad \text{and} \quad ab \xrightarrow{1} 1a$$

Some authors append the next entry on the right instead of on the left. We append the new entries on the left so the order of the strings is the order of the corresponding composition of functions, and in turn, the address of the region which is the result of applying the transformations in that order.

An Eulerian cycle in a graph is a path that traverses each edge of the graph exactly once, and begins and ends at the same vertex. Then an Eulerian cycle in an order-$(n-1)$ de Bruijn graph is a length-n de Bruijn sequence. We illustrate this correspondence with an example built with the prefer 1 algorithm for binary length-n de Bruijn sequences. By an n-string we mean a string of length n.

1. Begin with an empty list $\mathcal{L}$ of n-strings visited so far, and with the n-string $00\ldots0$, called the *current n-string*.

2. Add the current n-string to $\mathcal{L}$.

3. Append 1 to the left of, and remove the right-most entry of, the current string. If this gives a string not already in $\mathcal{L}$, then loop to step 2.

4. If the operation of step 3 gives an n-string already in $\mathcal{L}$, append 0 to the left of, and remove the right-most entry of, the current string. If this gives a string not already in $\mathcal{L}$, then loop to step 2.

5. If neither steps 3 nor 4 gives rise to an n-string not already in $\mathcal{L}$, we are finished.

To see how the prefer 1 algorithm generates an Eulerian cycle in a de Bruijn graph, consider the second image of Fig. A.14. The 3-string 000 is generated by by the edge $00 \xrightarrow{0} 00$ on the left graph and indicated by the edge labeled 1 on the right graph. Next apply 1 by traversing the edge $00 \xrightarrow{1} 10$ on the left graph, labeled 2 on the right graph. In Table A.4 we show the Eulerian path generated by the prefer 1 algorithm. "Edge label" refers to the labels of the edges on the second graph of Fig. A.14.

edge label	current 3 string	growing string
1	000	000
2	100	1000
3	110	11000
4	111	111000
5	011	0111000
6	101	10111000
7	010	010111000
8	001	0010111000

Table A.4: Growing an Eulerian path.

The entries of the middle column are all the binary 3-strings. Reading the last entry of the right column from right to left, we find all these 3-strings.

For even modest lengths n there are a *lot* of de Bruijn sequences. Applying transformations in the order determined by a de Bruijn sequence can render fractal images much more rapidly than the random IFS algorithm, an important issue for high-resolution images, especially if the fractal consists of pieces with substantially different sizes. But generating the sequences does take some work.

The next note refers to Sect. 2.7, **Circle inversion fractals**.

A.9 Benoit's circle inversion method

Figure A.15: The first few stages of constructing the limit set by removing discs disjoint from the limit set.

As mentioned in Sect. 2.7, because circle inversions contract distances by factors that vary with location, adapting the random IFS algorithm to circle inversions fills in some parts of a fractal with maddening sloth. Benoit solved this problem for some families of circles in his 1983 paper in the *Mathematical Intelligencer*. His method is this: find the smallest disc containing the whole limit set. Within that disc, find a set of smaller discs whose interiors contain nothing of the limit set. Then inverses of these smaller discs, and inverses of inverses, and inverses of inverses of inverses, will also contain nothing of the limit set. In the cases we consider, removing the inverses of these smaller discs eats away at the big disc, eventually leaving only the limit set. This method is illustrated in Fig. A.15, based on the inverting circles shown in the first image of Fig. A.16. The fourth picture is a later stage of the construction, with the removed discs painted white instead of grey.

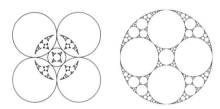

The IFS algorithm is guaranteed to generate a limit set if all the transformations are contractions. Circle inversion is a contraction only when points outside the circle are inverted to points inside the circle. If each inverting circle lies outside all the others, this is easy to do. If the circles overlap, guaranteeing that each inversion acts as a contraction is more complicated. But help with this and other situations was given in 1987 by John Elton, who proved that the random IFS algorithm almost always produces the limit set if *on average* the transformations act as contractions.

Figure A.16: First: Five inverting circles and points generated by adapting the random IFS algorithm. Note the uneven fill. Second: The limit set, magnified, generated by Benoit's method.

The next note refers to Sect. 2.8, **Random fractals**.

A.10 Three random fractal constructions

Here we sketch three random fractal constructions that require a bit more mathematics than do most of the randomizations of IFS.

A.10.1 Brownian motion

The oldest is Brownian motion, first reported around 1827 by Scottish botanist Robert Brown. In his microscope, Brown observed that tiny particles emitted by pollen grains he had placed in a drop of water, were zigging and zagging in an intricate way. The particles appeared to be swimming. Could this be true? Similar swimming was seen by lampblack (this is carbon soot, alive only in Hayao Miyazaki's soot sprites) and by granite dust. What could animate these inanimate particles?

Although it was a mystery to Brown, now we know that this dance of particles is the signature of the particles being buffeted by the thermal motion of water molecules. A mathematical study of Brownian motion was done in 1905 by Albert Einstein. (This calculation, not special relativity, was one of the reasons cited for Einstein's Nobel Prize.) An earlier mathematical exploration of Brownian motion is contained in Louis Bachelier's 1900 doctoral dissertation *Théorie de la spéculation* (translated as *Theory of Speculation* in Paul Cootner's *The Random Character of Stock Market Prices*) as a model of the stock market. Einstein's work is based on well-substantiated physics, while Bachelier's work, foundational to the Black-Scholes-Merton options pricing model, is not supported by abundant evidence. In 1910, physicist Jean Perrin used Einstein's work to derive the value of Avogadro's number; in 1923 mathematician and cybernetics founder Norbert Wiener gave a rigorous mathematical development of Brownian motion, providing an abstract backbone for Einstein's heuristics.

Here we'll restrict our attention to 1-dimensional Brownian motion $X(t)$. But note that Brownian motion in any dimension

$$(X_1(t), \ldots, X_n(t))$$

consists of n independent copies $X_i(t)$ of 1-dimensional Brownian motion. What are the properties of 1-dimensional Brownian motion?

1. With probability 1, $X(t)$ is continuous.

2. The probability distribution of *increments* or *jumps* $X(t+h) - X(t)$ is independent of t. That is, for all t_1, t_2, and x,

$$Pr(X(t_1 + h) - X(t_1) \leq x) = Pr(X(t_2 + h) - X(t_2) \leq x)$$

We say the increments are *stationary*.

This next property involves the distribution of the jumps $X(t+h) - X(t)$ for all t. In light of property (2), we can phrase this in terms of $t = 0$.

3. For all $h > 0$, the increment $X(h) - X(0)$ is normally distributed with mean 0 and standard deviation $\sqrt{h}$ – i.e.,

$$Pr(X(h) - X(0) \leq x) = \frac{1}{\sqrt{2\pi h}} \int_{-\infty}^{x} \exp\left(\frac{-u^2}{2h}\right) du$$

4. For $t_1 \leq t_2 \leq t_3 \leq t_4$, the increments $X(t_2) - X(t_1)$ and $X(t_4) - X(t_3)$ are independent.

The scaling property of Brownian motion is a consequence of property (3). This is true for all t, but again using property (2), we can express it for $t = 0$.

$$Pr(X(s \cdot h) - X(0) \leq s^{1/2}x) = Pr(X(h) - X(0) \leq x)$$

A.10. THREE RANDOM FRACTAL CONSTRUCTIONS

That is, time and space scale differently. Multiplying time by s and multiplying space by $s^{1/2}$ leaves the statistics unchanged. Here's why.

$$Pr(X(s \cdot h) - X(0) \leq s^{1/2}x) = \frac{1}{\sqrt{2\pi hs}} \int_{-\infty}^{s^{1/2}x} \exp\left(\frac{-u^2}{2hs}\right) du$$

$$= \frac{1}{\sqrt{2\pi h}} \int_{-\infty}^{x} \exp\left(\frac{-v^2}{2h}\right) dv, \text{ with } v = us^{-1/2}$$

$$= Pr(X(h) - X(0) \leq x)$$

That is, by a change of variables we see that the integrals defining

$$Pr(X(s \cdot h) - X(0) \leq s^{1/2}x) \text{ and } Pr(X(h) - X(0) \leq x)$$

are the same.

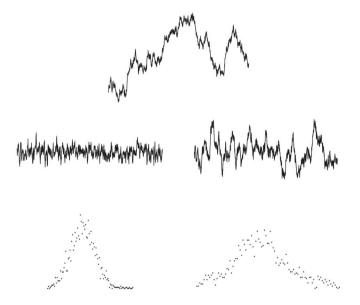

Figure A.17: Top: A Brownian path of 2000 points. Middle: Brownian increments, $h = 10$ (first), $h = 50$ (second). Bottom: Histogram of jump sizes.

Fig. A.17 illustrates this scaling. The top image is a Brownian motion path; the middle images show increments with a delay $h = 10$ (first) and $h = 50$ (second). By a delay of h we mean plotting the differences $X(t+h) - X(t)$ to study how the past influences the present. The bottom plots are histograms of the differences, using 100 bins, plotted with the same horizontal and vertical scales. The standard deviation of the second histogram is about $2.2 \approx \sqrt{5}$ times that of the first histogram.

One of the simplest random fractals, Brownian motion is a good model of the collective behavior of independent agents with identical distributions. But not everything we see behaves that way. For example, some processes remember their past actions. Fractional Brownian motion adds memory to Brownian motion.

A.10.2 Fractional Brownian motion

In the late 1960s, Benoit and statistician J. W. Van Ness developed fractional Brownian motion (fBm), a variation on Brownian motion, to understand how current rainfall depends on previous amounts. Like the increments of Brownian motion, those of fBm are normally distributed. Where fBm differs from Brownian motion is that the increments of fBm are not independent, as they are for Brownian motion (property (4)).

In fact, fBm is a family of random processes, indexed by a number α, $0 < \alpha < 1$. An *index α fBm* is a random process $X(t)$ with stationary increments $X(t+h) - X(t)$ that are normally distributed and that have mean 0 and standard deviation h^α. That is,

$$Pr\bigg(X(t+h) - X(t) \leq x\bigg) = Pr\bigg(X(h) - X(0) \leq x\bigg)$$
$$= \frac{1}{\sqrt{2\pi}} h^{-\alpha} \int_{-\infty}^{x} \exp\left(\frac{-u^2}{2h^{2\alpha}}\right) du$$

Note $\alpha = 1/2$ fBm is standard Brownian motion.

A slight modification of the proof of standard Brownian motion's scaling shows that the same is true for fBm:

$$Pr\bigg(X(s \cdot h) - X(0) \leq s^\alpha \cdot x\bigg) = Pr\bigg(X(h) - X(0) \leq x\bigg)$$

To show the increments for fBm are not independent, we need a preliminary result. For all t and for all $h > 0$, the expected value $\mathbb{E}$ of index α fBm satisfies

$$\mathbb{E}((X(t) - X(0)) \cdot (X(t+h) - X(t))) = \frac{1}{2} \cdot ((t+h)^{2\alpha} - t^{2\alpha} - h^{2\alpha})$$

To see this, recall that for a normally distributed random variable X with mean 0 and standard deviation h^α, the expected value of X^2 is

$$\mathbb{E}(X^2) = \int_{-\infty}^{\infty} u^2 \cdot \exp\left(\frac{-u^2}{2h^{2\alpha}}\right) du$$

Applying this to the delay $X(a+b) - X(a)$, we obtain

$$\mathbb{E}((X(a+b) - X(a))^2) = \frac{1}{\sqrt{2\pi}} b^{-\alpha} \int_{-\infty}^{\infty} u^2 \cdot \exp\left(\frac{-u^2}{2b^{2\alpha}}\right) du = b^{2\alpha}$$

A.10. THREE RANDOM FRACTAL CONSTRUCTIONS

Then for the delays $X(t) - X(0)$ and $X(t+h) - X(t)$,

$$\mathbb{E}((X(t) - X(0))^2) = t^{2\alpha} \text{ and } \mathbb{E}((X(t+h) - X(0))^2) = (t+h)^{2\alpha}$$

Consequently, we see that

$$\begin{aligned}
(t+h)^{2\alpha} &- t^{2\alpha} - h^{2\alpha} \\
&= \mathbb{E}((X(t+h) - X(0))^2) - \mathbb{E}((X(t) - X(0))^2) - \mathbb{E}((X(t+h) - X(t))^2) \\
&= \mathbb{E}((X(t+h) - X(0))^2 - (X(t) - X(0))^2 - (X(t+h) - X(t))^2)) \\
&= 2\mathbb{E}((X(t) - X(0)) \cdot (X(t+h) - X(t))).
\end{aligned}$$

The first equality is a substitution from the line above, the second follows from $\mathbb{E}(X + Y) = \mathbb{E}(X) + \mathbb{E}(Y)$ (the average of a sum is the sum of the averages), and the third is just algebra.

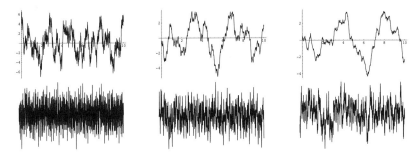

Figure A.18: Top: Fractional Brownian motion simulations with $\alpha = 0.25$ (anti-persistent), $\alpha = 0.5$ (standard, non-fractional, Brownian motion), and $\alpha = 0.75$ (persistent). Bottom: difference plots $X(t+1) - X(t)$ of the graphs above.

From this we see that if $\alpha = 1/2$, then

$$\mathbb{E}((X(t) - X(0)) \cdot (X(t+h) - X(t))) = \frac{1}{2} \cdot ((t+h)^1 - t^1 - h^1) = 0$$

and again we obtain the independence of increments of Brownian motion. If $\alpha > 1/2$, then

$$(t+h)^{2\alpha} - t^{2\alpha} - h^{2\alpha} > 0$$

and because we expect that $X(t) - X(0)$ and $X(t+h) - X(t)$ to have the same sign, this is called *persistent* fBm. Similarly, if $\alpha < 1/2$, then we expect that $X(t) - X(0)$ and $X(t+h) - X(t)$ will have opposite signs, and this is called *anti-persistent* fBm.

For standard Brownian motion ($\alpha = 1/2$) the present is unaffected by the past. For persistent ($\alpha > 1/2$) and anti-persistent ($\alpha < 1/2$) fBm the past does influence the present, but in opposite ways. For persistent we remember what we've just done and we like it, so let's do it again. Another fresh strawberry? Don't mind if I do. For anti-persistent we remember what we've just done and we don't like it, so let's do the opposite. I'm tired of running up and down these stairs. Time for a nap.

Looking at the top graphs of Fig. A.18 we notice that the graph of anti-persistent fBm is rougher than the graph of standard Brownian motion, which is itself rougher than the graph of persistent fBm. We can quantify this roughness of fBm graphs by relating the fBm Hölder exponent α to the box-counting dimension of the graph. To understand this part of the argument, you'll need to be familiar with the ideas of Sect. 6.2. In Fig. A.19 we see a graph of index α fBm. Suppose t ranges between 0 and 1. Divide the t axis into intervals of length dt. We'll need $1/dt$ intervals to cover the range of t-values.

Now look at the portion of the graph lying above an interval of width dt. Because this is the graph of an index α fBm X, we expect that the range dX of X values lying above dt is $dX = (dt)^\alpha$. We'll need $(dt)^\alpha/dt = (dt)^{\alpha-1}$ boxes of side length dt to cover the part of the graph lying above this interval dt. The number of these intervals is $1/dt$, so the total number of boxes of side length dt that we need to cover the graph is

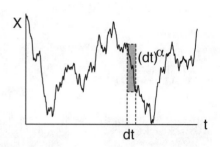

Figure A.19: Counting boxes of an fBm graph.

$$N(1/dt) = (\text{number of boxes per column}) \cdot (\text{number of columns})$$
$$= ((dt)^{\alpha-1}) \cdot (1/dt)$$
$$= (dt)^{\alpha-2}$$

From this we see that the box-counting dimension of the graph of index α fBm is

$$d = \lim_{dt \to 0} \frac{\log(N(dt))}{\log(1/dt)} = \lim_{dt \to 0} \frac{\log((dt)^{\alpha-2})}{-\log(dt)}$$
$$= \lim_{dt \to 0} \frac{(\alpha-2)\log((dt))}{-\log(dt)} = 2 - \alpha$$

This validates what our eyes told us. With its higher α, persistent fBm has a lower dimension than does anti-persistent fBm.

A.10.3 Lévy flights

While fBm introduces correlations between past and future, the increments remain normal and so large jumps are extremely rare. The normal

A.10. THREE RANDOM FRACTAL CONSTRUCTIONS

distribution is characterized by its mean and variance, but these are not well-defined concepts for some natural phenomena. For example, in some phase transitions (water to ice under high pressure, for instance), the mean ice cluster size of a sample keeps growing as the sample size increases.

Perhaps motivated by these considerations, Benoit's teacher Paul Lévy studied random processes for which the jump distributions have infinite mean and variance, and the distribution after N jumps is a rescaling of the distribution after one jump. The results of this study are called *Lévy stable processes* or *motions*. The large outliers result from their power law distribution: $Pr(X > x) \approx x^{-\delta}$. Lévy processes usually are defined by their Fourier transforms, but we'll take a more geometrical approach.

The *unit step function* $\xi(t)$ is defined by

$$\xi(t) = \begin{cases} 0 & \text{for } t < 0 \\ 1 & \text{for } t \geq 0 \end{cases}$$

That is, the graph takes a step of height 1 at $t = 0$. A (1-dimensional) Lévy stable process is a sum,

$$f(t) = \sum_{i=1}^{\infty} \lambda_i \xi(t - t_i)$$

where the step times t_i and amplitudes λ_i are chosen according to the Lévy distribution: given t and λ, the probability of choosing (t_i, λ_i) in the rectangle $t < t_i < t + dt$, $\lambda < \lambda_i < \lambda + d\lambda$ is $C\lambda^{-\delta-1}d\lambda dt$, where δ is the exponent of the power law above. Fig. A.20 shows an example of such a construction.

Figure A.20: Lévy flight. First: The path. Second: The $h = 1$ increment.

In 1963 Mandelbrot modeled the stock market with Lévy processes to emphasize the presence of large excursions on all time scales. Because they have independent increments, Lévy processes alone do not capture richness of market behavior. We need a single model amalgamating the dependent jumps of fBm and the large jumps of Lévy flights. Fractal sums of pulses is one such approach. Another, much simpler, is given in A.100.

The next note refers to Sect. 2.9, **And flavors stranger still**.

A.11 Strange types of self-similarity

The transformations used to generate the fractal of Fig. A.21 are combinations of scalings, translations, reflections and circle inversions, particular combinations that can be represented as complex functions $f(z) = (az+b)/(cz+d)$ called linear fractional transformations. The image is generated by a variation of the deterministic IFS algorithm with linear fractional transformations replacing our familiar scaling transformations. This is an immense subject and could fill another book. Oh, wait, it already has: *Indra's Pearls: The Vision of Felix Klein*.

Figure A.21: A Kleinian group limit set.

Semems and Guy define BPI spaces by requiring that for any pair of discs in the space, a substantial portion of the part of the space in one disc should look nearly the same geometrically as a portion of the space in the other disc, after rescaling by the ratio of the radii of the discs.

Here are the first two steps in constructing a BPI space that is not a self-similar shape. Start with the product of a Cantor middle-thirds set K and the unit interval I. Denote by K_0 the part of the Cantor set in $[0, 1/3]$, by K_1 the part of the Cantor set in $[2/3, 1]$, by K_{00} the part of the Cantor set in $[0, 1/9]$, and so on. Glue each point of $K_0 \times 0$ to the corresponding point of $K_1 \times 0$,

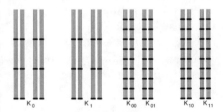

Figure A.22: Constructing a BPI spce.

glue each point of $K_0 \times 1/3$ to the corresponding point of $K_1 \times 1/3$, and glue each point of $K_0 \times 2/3$ to the corresponding point of $K_1 \times 2/3$. These points are indicated by the dark dots in the first image of Fig. A.22. The second image of that figure shows the next step: for example, glue each point of $K_{00} \times 0$ to the corresponding point of $K_{01} \times 0$, glue each point of $K_{10} \times 0$ to the corresponding point of $K_{11} \times 0$, and so on. Do the analogous glueings for all parts of the Cantor set. If you think this looks complicated, but maybe not too strange, consider this: this BPI space cannot be embedded with limited distortion in any Euclidean space of any finite dimension. Every picture will stretch some parts enormously, and some other parts even more.

Chapter 3. Self-similarity in the wild

This note refers to Sect. 3.3, **Coastlines, Mountains, Rivers**.

A.12 Coastlines, mountains, rivers

Here we present a few details about fractal aspects of coastlines, mountains, and rivers.

A.12.1 Coastlines

Benoit's 1967 paper on coastlines was, for the general scientific world, the beginning of fractals as a field of study. Benoit's analysis was based on Richardson's 1961 observations, derived from measurements on detailed maps, that the measured length of a coastline increases, apparently without bound, as the coastline is measured at smaller distance scales.

In his papers on fractal drums, Sapoval shows that drums with fractal perimeters can damp vibrations so effectively that the vibrations do not spread across the whole drum membrane. Because a fractal perimeter has many length scales, it can efficiently absorb energy over many wavelengths. This is why fractal drums are so quiet, and, Sapoval thinks, why coastlines are fractal.

A.12.2 Mountains

Handelman made some of the earliest fractal mountain constructions. Here we see something similar to Handelman's images: triangles subdivided into smaller triangles, with vertices pushed up or down by amounts following a power law. The artificial sea level of the second image makes this an island simulation.

Voss is well known for fractal landscapes. Musgrave's fractal mountains and landscapes have appeared on many fractal calendars, in some motion pictures, and in his chapter, "Fractal Forgeries of Nature," of *Fractal Geometry and Applications*.

Benoit called Handelman's work "heroic," because of the substantial effort needed to produce these images with the relatively primitive computing equipment available then. Voss's work, expressing a mature artistry, Benoit called "classical." Musgrave's work Benoit called "romantic," because these landscapes were parts of stories. We wonder what will come next.

A.12.3 Rivers

Google Maps reveals the fractality of river networks for all to inspect. The source of this fractality in the geometry of river networks and basins has been studied from several viewpoints, by Banavar, Maritan, Rinaldo, and their coworkers, among others. At least some aspect of the fractality of these networks, and of the boundaries of river drainage basins, is a consequence of the fractality of the landscape. Lest we get too confident about our understanding of nature, look at the rivers of Fig. A.23.

The first image exhibits the properties of two drainage patterns: *trellis* and *rectangular*. Both involve approximately 90° angles. Trellis patterns usually occur in folded mountains, smaller tributaries joining the main branch at right angles giving the appearance of a vine growing on a trellis. Rectangular patterns occur on rectangular bulks of rock, along the joints, less erosion-resistant than the rectangular blocks. Given than part of the Quinnipiac basin is approximately horizontal, we expect this is a rectangular pattern.

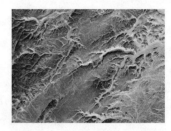

Figure A.23: The Quinnipiac river in Connecticut, rivers of sand in the Sahara.

As for the second image of Fig. A.23 we'll say only this: the dessert is home to many mysteries.

The next note refers to Sect. 3.4, **Fractal lungs**.

A.13 Estimating lung area

Here we give some detail about the lungs and, in an illustrative special case, derive the formula $A/V = 2I/L$ for computing lung area.

Weibel has written extensively about the structure of the lungs. Hess and Murray studied the mother-daughter pulmonary branch diameter relation and obtained a ratio of 0.79, the optimal value for minimizing turbulent airflow. Weibel, Sapoval, and their coworkers B. Mauroy and M. Filoche measured the mother-daughter airway diameter ratio as about 0.85, rather than 0.79. To guard against the effects of diseases such as asthma, some inefficiency may be beneficial.

Another piece of the picture, estimating the number of alveoli, has proven challenging. An early approach, far too crude, consisted of dividing the volume of the lungs by that of a single alveolus. Weibel and Domingo

A.13. ESTIMATING LUNG AREA

Gomez developed a much better method by counting the number of alveoli transections per unit area by random cut surfaces. This gave an estimate of about 300,000,000 alveoli. Over forty years later, Matthias Ochs and coworkers increased the estimate to about half a billion.

The transection method for counting alveoli is inspired by the geometry of the Buffon needle problem, whose solution is straightforward. Denote by θ the angle between the needle and a horizontal line. Observing, for example, that the picture for $\theta = 10°$ is identical to the picture for $\theta = 190°$, we can take θ to lie between $0°$ and $180° = \pi$ radians. Next, as illustrated in Fig. refFi:Buffon, the vertical distance spanned by the needle is $L\sin(\theta)$. For a given θ the probability that the needle crosses one of the lines is $P(\theta) = L\sin(\theta)/d$. To find the probability for all θ, integrate $P(\theta)$ over the range $0 \leq \theta \leq \pi$ and divide by this range, π. That is, the probability that a needle of length $L < d$ randomly dropped on an array of parallel lines separated by a distance d is

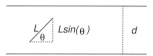

Figure A.24: Solving the Buffon needle problem.

$$\frac{1}{\pi}\int_0^\pi \frac{L\sin(\theta)}{d}d\theta = -\frac{L\cos(\theta)}{\pi d}\bigg|_0^\pi = \frac{2L}{d\pi}$$

Here we'll sketch a derivation of the formula $A/V = 2I/L$ in the special case of a sphere in a cube – specifically, for a sphere of radius r situated inside a cube of side length ℓ, and for sample lines parallel to a cube edge and uniformly spread across the cube faces perpendicular to that edge. First, the easy part:

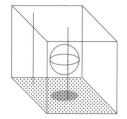

$$\frac{A}{V} = \frac{4\pi r^2}{\ell^3}$$

Next, take a grid of n points uniformly spaced on the bottom face of the cube. The total length of these lines is $n\ell$. The lines that intersect the sphere are those from grid points that lie in the shadow of the sphere. The number of those points is very close to the total number of points, n, multiplied by the ratio of the shadow area to that of the base of the cube,

$$n\pi r^2/\ell^2$$

Unless a line intersects the sphere exactly along its equator, which is very unlikely, each line that intersects the sphere does so in two points, one in the northern hemisphere, one on the southern. Then

$$\frac{I}{L} = \frac{2n\pi r^2/\ell^2}{n\ell} = \frac{2\pi r^2}{\ell^3} = \frac{1}{2}\frac{A}{V}$$

The general argument is more sophisticated, and more involved, because it must treat irregular surface shapes and the random nature of slicing a cross-section. Two proofs can be found in Sect. 3.2.2 of Volume 2 of Weibel's *Stereological Methods*. But the result remains the same. This is the principal tool used to estmate the surface area of the lungs.

The next note refers to Sect. 3.5, **Power laws**.

A.14 Power spectra and correlation functions

Here we fill in some details about the relation between power spectra and correlation functions.

First, we recall, or introduce if you haven't seen Fourier series, a method for finding the periodic components of a signal. Most sounds we know – voices of friends, chirps of birds, rumbles of thunder, purrs of cats, songs of Itzhak Perlman's violin – are combinations of simple oscillations. Generally, these are sums of sines and cosines, but we'll describe sums of cosines only. Including sines adds some bookkeeping complications but no new ideas.

Suppose we represent the signal by a function $f(t)$ for $0 \leq t \leq L$, where L is the duration of the signal, the length of time we hear it. If the signal is reasonably well behaved, we can write

$$f(t) = \sum_{n=0}^{\infty} a_n \cos(n\pi t/L)$$

This is called the *Fourier cosine series* for f. Part of its appeal is that this method works for many signals; another part is that the coefficients a_n can be found by simple integration:

$$a_0 = \frac{1}{L} \int_0^L f(t) dt$$

$$a_n = \frac{2}{L} \int_0^L f(t) \cos(n\pi t/L) dt \quad \text{for } n > 0$$

Deriving these formulas for the coefficients a_n is straightforward. Multiply both sides of the equation for $f(t)$ by $\cos(m\pi t/L)$ and integrate both sides between $t = 0$ and $t = L$:

$$\int_0^L f(t) \cos(m\pi t/L) dt = \int_0^L \cos(m\pi t/L) \sum_{n=0}^{\infty} a_n \cos(n\pi t/L) dt$$

$$= \sum_{n=0}^{\infty} a_n \int_0^L \cos(m\pi t/L) \cos(n\pi t/L) dt$$

A.14. SPECTRA AND CORRELATIONS

The formulas for a_0 and a_n follow because

$$\int_0^L \cos(m\pi t/L)\cos(n\pi t/L)\,dt = \begin{cases} 0 & \text{if } n \neq m \\ L/2 & \text{if } n = m \neq 0 \\ L & \text{if } n = m = 0 \end{cases}$$

The coefficients a_n are the amplitudes of the signal at frequencies that get higher with increasing n. The *power spectrum* of the signal is the list of the squared amplitudes a_n^2 present in the original signal.

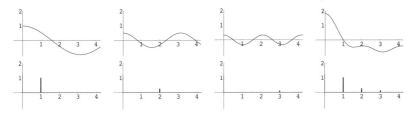

Figure A.25: Top: Three cosines and their sum. Bottom: The corresponding power spectra.

Fig. A.25 shows several signals and their (very simple) power spectra. The first graph of the top row is that of $\cos(t)$ for $0 \leq t \leq 4$. All the signal graphs use the same t range, $0 \leq t \leq 4$. The first graph of the bottom row is the power spectrum of this signal, a single spike, of height 1 and frequency 1. The next two signals are $(1/2)\cos(2t)$ and $(1/3)\cos(3t)$ and their power spectra, single spikes of height $1/4$ and $1/9$, at frequencies 2 and 3. The last signal is the sum of the others, that is, $\cos(t) + (1/2)\cos(2t) + (1/3)\cos(3t)$; its power spectrum has all three spikes.

Now we'll illustrate the relation between power spectra and the correlation function. Recall that the correlation function $C(\delta)$ of a time series $z_1, \ldots, z_N$, which we can view as $z_j = f(jL/N)$ (remember that the function $f(t)$ has domain $0 \leq t \leq L$), is defined by

$$C(\delta) = \frac{1}{N-\delta} \sum_{j=\delta+1}^{N} z_j z_{j-\delta}$$

The number $C(\delta)$ represents the average relation between a time series entry and the entry δ steps in the past.

Then the power spectrum $P(\omega)$ is related to the correlation function $C(\delta)$ by

$$P(\omega) = \sum_{\delta=1}^{N-1} C(\delta) \cos(\delta \cdot \omega)$$

Rather than sketch the proof, we'll show how to use this method to obtain the power spectrum of the last signal of Fig. A.25, the part for $0 \leq t \leq 1$.

In the first image of Fig. A.26 we plot the signal $f(t) = \cos(t) + (1/2)\cos(2t) + (1/3)\cos(3t)$, $0 \leq t \leq 1$, sampled at 100 evenly spaced points. Next, we plot the correlation function. For example,

$$C(1) = \frac{1}{99}\left(f\left(\frac{2}{100}\right)f\left(\frac{1}{100}\right) + f\left(\frac{3}{100}\right)f\left(\frac{2}{100}\right) + \cdots + f\left(\frac{98}{100}\right)f\left(\frac{97}{100}\right) + f\left(\frac{99}{100}\right)f\left(\frac{98}{100}\right)\right)$$

The third graph is the power spectrum, obtained from the formula giving the power spectrum in terms of the correlation function. Note the presence of three spikes, the second and third occurring at frequencies 2 and 3 times that of the first.

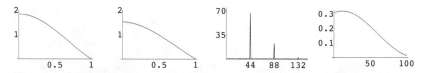

Figure A.26: First the signal, then the correlation function, the power spectrum, and the graph of the difference between signal and correlation function.

Just in case you think the signal and its correlation function look very similar, the last graph of Fig. A.26 is a plot of their differences.

Finding power spectra for real data is much trickier than this simple example suggests. The finiteness of real data sets can introduce artifacts, false positives in the power spectrum. Noise in the data will have an effect, as can a choice of sampling time close to the period of one of the cosines making up the signal. To get reliable results from real data can require considerable effort. Careers have been built around this problem. Still, even from this simple example we can get a sense of the basic relation between correlation functions and power spectra.

The next note refers to Sect. 3.8, **Metabolic rates**

A.15 Metabolic scaling models

Here we present some details of two approaches to explain the observed $M^{-1/4}$ metabolic rate per unit mass scaling.

The analysis of West, Brown, and Enquist is based on the hypothesis that metabolic processes use fractal networks over only a finite range of length scales, all terminating in a smallest scale ℓ_0, the size of capillaries for animals, of chloroplasts for plants, of mitochondria for cells.

Very roughly, their argument goes this way. Recall that the biological area is the total area across which nutrients and energy are exchanged with

A.15. METABOLIC SCALING MODELS

the organism's environment. Think area of the lungs more than area of the skin. Biological volume is the volume of the biologically active material of the organism. For example, from the volume of water displaced by an animal, subtract the volume of the gut and the airways of the lungs, among others.

In addition to ℓ_0, the biological area a and biological volume v may depend on other lengths, $\ell_1, \ldots, \ell_n$, the length of the aorta or the distance between mitochondria, for example. Now scale the organism by a length factor of λ, by which we do not mean make its cells larger, but rather add same-size cells until the organism is larger by λ. Alternatively, think that the larger organism is of a different species with linear size λ times that of the smaller organism. Recalling that the smallest length scale ℓ_0 remains unchanged, with a bit of work West, Brown, and Enquist show the biological area and volume scale not as Rubner suggested,

$$a(\lambda \ell) = \lambda^2 a(\ell) \quad \text{and} \quad v(\lambda \ell) = \lambda^3 v(\ell)$$

but rather as

$$a(\ell_0, \lambda \ell_1, \ldots, \lambda \ell_n) = \lambda^{2+\epsilon_a} a(\ell_0, \ell_1, \ldots, \ell_n)$$

and

$$v(\ell_0, \lambda \ell_1, \ldots, \lambda \ell_n) = \lambda^{3+\epsilon_v} v(\ell_0, \ell_1, \ldots, \ell_n)$$

where the exponent parameters, ϵ_a and ϵ_v, are related to the geometrical complexity of the metabolic network. Recalling the notation $p \propto q$ for "p is proportional to q," we have the relations

$$v \propto \lambda^{3+\epsilon_v} \quad \text{and} \quad a \propto \lambda^{2+\epsilon_a}$$

Eliminating λ from these relations, because mass $M \propto v$ we see

$$a \propto v^{(2+\epsilon_a)/(3+\epsilon_v)} \propto M^{(2+\epsilon_a)/(3+\epsilon_v)}$$

Aguing that metabolic rate is proportional to biological area, we find

$$\text{metabolic rate} \propto M^{(2+\epsilon_a)/(3+\epsilon_v)}$$

To express this scaling in terms of area and length, instead of area and volume, West, Brown, and Enquist use the relation $v \propto a \cdot l$, where the biological length l, a length characteristic of the metabolic mechanisms of the organism, scales as

$$l(\ell_0, \lambda \ell_1, \ldots, \lambda \ell_n) = \lambda^{1+\epsilon_l} l(\ell_0, \ell_1, \ldots, \ell_n) \propto \lambda^{1+\epsilon_l}$$

Then noting that $v \propto a \cdot l$ gives $\lambda^{3+\epsilon_v} \propto \lambda^{2+\epsilon_a} \cdot \lambda^{1+\epsilon_l}$, we see we can replace ϵ_v in the exponent by $\epsilon_a + \epsilon_l$, obtaining

$$\text{metabolic rate} \propto M^{(2+\epsilon_a)/(3+\epsilon_a+\epsilon_l)}$$

Conjecturing that organisms have evolved to maximize the scaling of biological area a, hence of metabolic rate, we see this is achieved by maximizing the exponent b,

$$b = \frac{2 + \epsilon_a}{3 + \epsilon_a + \epsilon_l}$$

subject to the constraints $0 \leq \epsilon_l \leq 1$ and $0 \leq \epsilon_a \leq 1$ imposed by the limiting dimensions of biological length and area. The maximum of b is $b = 3/4$, obtained with $\epsilon_l = 0$ and $\epsilon_a = 1$. Then the metabolic rate per unit mass scales as $M^{3/4}/M = M^{-1/4}$. That is, West, Brown, and Enquist recover Kleiber's observation.

Banavar, Maritan, Rinaldo, and coworkers take a different approach to explaining the metabolic rate per unit mass scaling. Their assumption is that on microscopic scales nutrient transfer is independent of organism size and occurs at L^D sites, where L is the linear size of the organism and D is the organism's dimension. The metabolic rate, B, is proportional to the total amount of nutrients delivered per unit time, so

$$B \propto L^D$$

They argue that evolution favors efficiency and therefore minimizes C, the total blood volume, which is proportional to the product of the number of transfer sites and the mean distance from the source to the transfer sites. Because it must reach all transfer sites, the blood distribution network is a *spanning network*. The crux of Banavar, Maritan, and Rinaldo's argument is their proof that the total blood volume C is bounded between powers of L,

$$L^{D+1} \leq C \leq L^{2D}$$

This calculation does not assume hierarchical aspects of the distribution network, though a hierarchical network has shorter total length and so would be selected for efficiency of resource use.

The left bound and right bound give

$$L \leq C^{1/(D+1)} \quad \text{and} \quad C^{1/(2D)} \leq L$$

From these bounds, we see that L^D, and so the metabolic rate B, lies in a limited range:

$$C^{1/2} \leq L^D \leq C^{D/(D+1)}$$

Also from $L^{D+1} \leq C \leq L^{2D}$ we see that C is minimized if $C = L^{D+1}$. The mass M scales as the blood volume C, and the metabolic rate scales as L^D, so

$$\text{metabolic rate} \propto L^D = (M^{1/(D+1)})^D = M^{D/(D+1)}$$

For $D = 3$, this gives metabolic rate $\propto M^{3/4}$ and so the metabolic rate per unit mass scales as $M^{-1/4}$, retrieving Kleiber's scaling once again.

This argument has its own hypotheses, some different from those of West, Brown, and Enquist. A new prediction of this model is that the speed of blood flow scales as $M^{1/12}$.

Of course, neither explanation may be correct. But this is how science progresses: we build models that reproduce observed phenomena. With new observations, the models stand or fall. When different models account for the same observations, we're going to learn something interesting.

The next note refers to Sect. 3.12, **The distribution of galaxies**.

A.16 Distribution of galaxies

Here we present some additional background information on the distribution of galaxies.

For current information about the number of stars, the sizes of the smallest and largest galaxies, the number of galaxies in the observable universe, the percentage of dark matter in a galaxy, and so on, Google is the first place to check. A wonderful source, especially for historical information, is Norriss Hetherington's *The Encyclopedia of Cosmology*.

Newtonian cosmology was presented in Newton's *PhilosophiæNaturalis Principia Mathematica* (1687). Newton's universal laws of motion and gravitation imply Kepler's three laws of planetary motion. Here the word "universal" refers to the fact that these laws describe the orbits of the planets around the sun, the orbits of Jupiter's moons about Jupiter, and the path of projectiles launched on the surface of the Earth. From this, Newton derived a model of the cosmos, an infinite collection of stars spread throughout infinite space, held in their positions by mutual gravitational attraction. Newton's cosmology was homogeneous (looks about the same in every location) and isotropic (looks about the same in every direction). Many details are given in the article "Newtonian Cosmology," in Hetherington's *Encyclopedia*.

A brief detour

Homogeneity and isotropy have interesting physical consequences, uncovered by the German mathematician Emmy Noether, one of the most original 20th-century mathematicians. She worked with David Hilbert, and Albert Einstein praised the depth of her mathematical thinking. Noether's theorem is this:

> For every continuous symmetry there is a conservation law.
> For every conservation law there is a continuous symmetry.

A good exposition of Noether's theorem is *Symmetry and the Beautiful Universe* by Nobel laureate Leon Lederman and Christopher Hill.

Homogeneity is equivalent to symmetry under translation (motion). Wherever we look, however we translate our viewpoint, the universe looks about the same. The corresponding physical law is the conservation of momentum.

Isotropy is equivalent to symmetry under rotation. No matter in what direction we look, the universe appears about the same. The corresponding physical law is conservation of angular momentum.

Another principle is symmetry under time translation. The laws of physics tomorrow will be the same laws as those today. The corresponding physical law is conservation of energy.

The older author heard these correspondences in Paul Yergin's Physics 1 class at RPI. This was so mysterious, so profound. It convinced me that I must learn a lot more physics. One question: can Noether's theorem be applied to scaling symmetry? At least as we've stated it here, the symmetry must be continuous, Noether's theorem wouldn't apply to the gasket, because the gasket's symmetries are discrete: magnify by 2, 4, 8, and so on. So maybe Noether's theorem doesn't apply to fractals. But what about random fractals, the kind we see in PhysicalWorld? For these, scaling symmetry may well be continuous. So do physical fractals point to an as-yet-undiscovered conservation law? Benoit and I discussed this several times, but got no further than to agree it is a wonderful question.

End of the detour, back to the distribution of galaxies

Modern cosmology really started with Einstein's general theory of relativity, his beautiful marriage of gravity and geometry. Gravity is not a force; instead, it is the curvature of space and time. In the absence of forces – and remember, gravity is not a force – particles travel on paths called *time-like geodesics*, paths that maximize time measured by anyone riding along on the particle. This is a remarkable construct of mathematical physics, holding surprises that amply reward the effort needed to understand it.

Einstein applied the *cosmological principle*, that at large scales the universe is homogeneous and isotropic, to general relativity and found that there is no static solution to his gravity equations for the whole universe.

At that time, around 1917, the standard interpretation of the cosmology had the universe neither expanding nor contracting, so the absence of a static solution to his gravity equations induced Einstein to modify them, adding the *cosmological constant*, a large-scale repulsive force that balances the large-scale attractive effects of gravity. With the addition of the cosmological constant, a static universe is a solution of Einstein's gravity equations. But this story does not end here. Between 1914 and 1922, Vesto Slipher, an American astronomer working at Lowell Observatory in Flagstaff, Arizona, measured the motion of the 25 spiral nebulae – now we know these are galaxies outside our own – and found 21 of the 25 are moving away from us. This suggested that the universe is not static but dynamic, and likely expanding.

Then, in the 1920s, the Russian cosmologist Alexander Friedmann and

A.16. DISTRIBUTION OF GALAXIES

the Belgian astronomer and priest Georges Lemaître showed that Einstein's original gravity equations, without the cosmological constant, has a solution in which the universe expands. The general expansion of the universe was established in 1929 by American astronomers Edwin Hubble and Milton Humason.

Currently, something like the cosmological constant is thought to be responsible for the observed acceleration of the expansion of the universe. What Einstein considered his "biggest blunder" (at least, as reported by the Russian-American physicist George Gamow) appears to be an important part of cosmology, though not for the reasons Einstein originally thought.

Hierarchical cosmologies are an alternative to homogeneous. Swedenborg published his cosmological views in 1734, in his three-volume *Principia: Opera Philosophica et Mineralia*. Johann Lambert and Immanuel Kant corresponded about this, and Lambert recorded his ideas in his *Cosmological Letters* (1761). Carl Charlier's map of nebulae is reproduced as Fig. 7.1 of *Discovery of Cosmic Fractals* by Baryshev and Teerikorpi. A consequence of the hierarchical distribution is that density decreases with distance.

As we mentioned in Sect. 3.12, hierarchical cosmology could offer a solution to Olbers' paradox, also called the *dark night sky paradox*. It was named for the German astronomer and physician Heinrich Olbers, though Olbers was not the first to recognize this problem. A thorough account of the history of the paradox is given in Edward Harrison's *Darkness at Night*.

Poe's *Eureka* was his last major work, an elaboration of his February 3, 1848, lecture, "On the Cosmography of the Universe," at the Society Library of New York. Both lecture and essay were not well received, but Poe's intuition found many notions that can be interpreted as related to themes of modern cosmology. In particular, he addressed a resolution of Olbers' paradox.

Interpreting the observed recession of the galaxies as a consequence of the expansion of the universe in 1927, Lemaître imagined running the film in reverse: if the universe is larger now, then at some time in the distant past it must have been much smaller. In 1932 he called the initial state of universe the "primeval atom."

Gamow called the primordial atom "yelm." Ralph Alpher, Gamow's graduate student, worked out detailed calculations of the relative abundances of the light elements (mostly hydrogen and helium) cooked in the vast heat – billions of degrees – immediately after the big bang. Famous for his sense of humor, Gamow added physicist Hans Bethe's name to the paper, making its authors Alpher, Bethe, and Gamow.

The expansion of the universe cooled the radiation from the big bang. With Robert Herman, also Gamow's student, Alpher calculated the current temperature of this echo of the explosion that created the universe.

In 1964 at Holmdel, New Jersey, this *cosmic background radiation* was accidentally detected by radio astronomers Arno Penzias and Robert Wilson while they were identifying the sources of all the radio signals received by the Holmdel antenna. The antenna was designed to receive signals bounced off the Echo satellite. Because these signals were very faint, all other radio noise had to be filtered out. After accounting for all known sources, including pigeon droppings in the antenna, a faint signal remained. It turned out that this was the echo of the Big Bang, over 13 billion years ago. For this discovery, Penzias and Wilson received the Nobel Prize. Data from the COBE and WMAP satellites, and more recently the European Space Agency's Planck satellite, have confirmed and refined the microwave background radiation measurements. The Planck data reveal some subtle anomalies in the distribution of the background radiation. We have not seen the end of this story.

Details about these calculations can be found in Alpher's article "Origins of Primordial Nucleosynthesis and Prediction of Cosmic Background Radiation," in Hetherington's *Encyclopedia*. There are also two good popular accounts of primordial nucleosynthesis and cosmic background radiation, written by Nobel Prize winners: Steven Weinberg's *The First Three Minutes* and George Smoot's *Wrinkles in Time*, coauthored by Keay Davidson.

A.16.1 Peebles' analysis of distances galaxy distances

Peebles analyzed the distribution of galaxies by the method of correlation functions in his books *The Large-Scale Structure of the Universe* and *Principles of Physical Cosmology*. The actual distribution of galaxies is compared with a model of uniformly randomly distributed galaxies. A correlation between galaxies occurs if the actual number of galaxies in a region exceeds that expected from the random model. The correlation function represents how this excess depends on distance. Typically, the correlation function decreases with distance, and the *correlation length* is the distance beyond which the excess drops below twice the value expected by the uniform random model. The distribution is approximately uniform for distances longer than the correlation length. Discussions of galaxy distribution fractality often are expressed in terms of the correlation length. Peebles argues the galaxy distribution is fractal below the correlation length, homogeneous above.

A.16.2 Pietronero's analysis of galaxy distances

Pietronero developed another approach, *conditional density*, a measure of how the average density of galaxies varies with the distance over which the average is computed. For each galaxy in a sample, count the number of galaxies in the spheres of radius 1 Mpc, 2 Mpc, 3 Mpc, and so on, centered

A.17. FRACTALS AND PERCEPTION   301

on that galaxy. Perform the analogous counts for the other galaxies in the sample. The average of the number of galaxies in each of these 1 Mpc spheres is the average 1 Mpc galaxy count; the average 2 Mpc, 3 Mpc, etc. counts are obtained similarly. The conditional density is the average count divided by the volume of the sphere. A fractal (hierarchical) distribution is implied if the conditional density decreases with distance; constant conditional density implies homogeneity.

A.16.3 Peacock's fractal metric

Physicist John Peacock argues that the *metric* of the universe – the way time and space are measured by an observer – exhibits fluctuations that are scale-invariant. Fractality may be part of the very fabric of the cosmos. He writes:

> We therefore argue that the inflationary process produces a universe that is fractal-like in the sense that scale-invariant fluctuations correspond to a metric that has the same 'wrinkliness' per log-length scale.

He notes further:

> Since potential perturbations govern the flatness of specetime, this says that the scale-invariant spectrum corresponds to a metric that is *fractal*: spacetime has the same degree of 'wrinkliness' on each resolution scale. ... [S]uch a universe is self-similar in the sense of always appearing the same under the magnification of cosmological expansion.

The next 21 notes refer to Sect. 3.13, **Is this all?**

A.17 Fractals and perception

Quite a lot has been written about this general topic, and certainly some of it is nonsense, much deriving from misunderstandings of what Benoit intended by the notion "fractal." Among the interesting and plausible applications, we mention two: our ability to perceive fractal shapes, and fractal aspects of our EEG patterns.

James Cutting and Jeffrey Garvin showed eight subjects fractal curves having dimensions 1.125 to 1.750 and asked them to rate the complexity of the curves. Dimension was reasonably strongly related to perceived complexity, though some other factors also were important predictors. Caroline Hagerhall and her colleagues measured the EEG signals of 31 subjects shown fractal curves. They found that the largest changes in EEG response were produced by curves of dimension about 1.31, and speculated that shapes of this dimension might dominate the competition for processing resources in a visually mixed environment.

Neurobiologist Walter Freeman and neuroscientist Christine Skarda's work on olfaction focused on finding evidence of chaos in the way scents are processed and recognized. Freeman's reconstruction of EEG signals by delay embedding (We give an illustration of delay embedding in Sect. A.35) suggests fractal patterns in the dynamical relations between activated regions, but not necessarily in the physical location of activated regions of the olfactory bulb. This offers a possible explanation for preattentive processing: that familiar sensations are stored as fractals may allow them to be recognized from only partial sensory information.

A.18 Ammonite sutures

A good source of amazing pictures of ammonite sutures is Moore's *Treatise on Invertebrate Paleontology*. This book has images of hundreds and hundreds of sutures, exhibiting a wide range of complexities. The intricacy of these curves dominates our initial impressions, but look a while longer and you'll see they are remarkably beautiful. The original papers reporting the calculations of suture box-counting dimensions were written by Boyajian and Lutz. Gould's wonderful book describing the left wall effect is *Full House*. In addition to presenting an alternative to the standard evolutionary narrative of ammonite suture complexity, Gould makes a bold application of the left wall effect to re-evaluate the evolution of biological complexity in general. Gould suggests that the steps that led to humans was not the likely outcome of a single unlikely evolutionary step, a radical contingency. Rather, he posits the path that leads to humans is filled with such steps. Gould writes

> Radical contingency is a fractal principle, prevailing at all scales with great force.

– an interesting way for fractals to appear in evolution.

A.19 Dielectric breakdown and relatives

Dielectric breakdown, electrodeposition, and viscous fingering are described nicely in Vicsek's book *Fractal Growth Phenomena*. All three processes can generate striking fractal images.

Of the three, viscous fingering is easiest to see for yourself. Take two hard surfaces with some viscous substance between them and find a way to inject air near the center of the goo; this is done most often through a hole drilled in the top plate. This arrangement of plates is called a *Hele-Shaw cell*. The distance between the plates, the viscosity of the goo, and the air pressure are the principal experimental parameters. Knut Måløy, Feder, and Torstein Jøssang report much more convincing fractals if the fluid is in a *random porous medium*. For example, the bottom plate is coated with epoxy on which is spread a layer of glass beads 1 mm in diameter. After the epoxy hardens, the loose beads are removed, the top plate is replaced, and the region between the plates is filled with fluid. Then the introduction of air forms much more complex fractals.

Electrodeposition requires a petri dish filed with a 2M zinc sulfate aqueous solution, a zinc strip anode placed along the inside wall of the petri dish, and a thin wire cathode lying on the top surface of the zinc sulfate solution. A DC current of several volts will grow a branched crystal, several cm in diameter, in a few minutes. Voltage, current, and temperature are the main variables. The paper of physicist Mitsugu Matsushita et al. shows some examples.

Dielectric breakdown is the most challenging for the kitchen science enthusiast. Place a grounded conductor under a glass plate 2 mm thick and apply a very short (about 1 μsec) high voltage (about 30 kV) electrical current to an electrode placed on top of the glass near the center. Examples are shown in the paper of L. Niemeyer et al. The opportunity for – let's say mistakes – is significant. Maybe a safer approach is to watch the sky during a thunderstorm, but not standing in the middle of an open field, please..

A.20 Bacterial growth

In the late 1980s and early 1990s, Matsushita and biologist Hiroshi Fujikawa grew bacteria on nutrient-depleted agar. They experimented on colonies of *Bacillus subtilis*, in part because it is hydrophobic and so growth occurs only on the surface of the agar, simplifying analysis of the growth pattern. Typically, colonies were grown for about three weeks, on agar with 1g/L of nutrient (peptone) instead of the common 15g/L. In this food-scarce environment, the bacteria grow in a pattern reminiscent of DLA (diffusion-limited aggregation). Quantitatively, too: box-counting of digitized images gives $d_b \approx 1.7$, close to the theoretical dimension of planar DLA clusters.

Of course, just matching dimensions does not establish DLA as the mechanism responsible for growth. Careful observations reveal that growth of the outer branches screens that of the inner branches, and two colonies from seeds about 1 cm apart do not grow together but maintain a clear channel between them. The branches of each colony appear to be repelled by those of the other. Both are behaviors exhibited by DLA, but this still isn't proof. To test if diffusion really contributes to the observed growth patterns, nutrient was placed only in one small portion of the agar, near the edge of the dish. The bacteria colony grew, again in a branching pattern, but this time toward the nutrient. This supports the notion that the bacteria growth is directed by the diffusion of the nutrients. Under these conditions, the bacteria exhibit *diffusion-limited growth*.

Complex patterns, some fractal, some not, are produced by other bacteria, for example *Salmonella typhimurium* and *Serratia marcescens*. By comparing the growth patterns of wild types and some mutants, Tohey Matsuyama, Rasika Harshey, and Mitsugu Matsushita identified some membrane lipids responsible for fractal growth.

In Fig. A.27 we see examples of electrodeposition, dielectric breakdown, and bacterial growth. Similar shapes certainly; similar causes, maybe.

  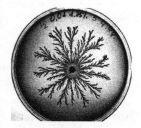

Figure A.27: First: Electrodeposition. Second: Dielectric breakdown. Third: Bacteria on a nutrient-depleted medium.

Some scientists think DLA alone is not sufficient to explain the complex patterns of bacterial growth. Rather, they propose a hybrid of DLA and biological factors, the *communicating walkers model* of Eshel Ben-Jacob and colleagues, in which bacteria communicate by exchanging chemical and genetic material. They write

> Using the model we demonstrate how communication enables the colony to develop complex patterns in response to adverse growth conditions. Efficient response of the colony requires self-organization on all levels, which can be achieved only via cooperative behavior of the bacteria.

Finally, among biological systems, DLA patterns are not the sole province of bacteria. Under some conditions, fungi also exhibit fractal growth.

A.21 Earthquakes

The most obvious occurrence of scaling in earthquakes is the *Gutenberg-Richter law* of Beno Gutenberg and Charles Richter:

$$\log(N) = -bM + a$$

where N is the number of earthquakes of magnitude $\geq M$. The magnitude scale is logarithmic in the amplitude S of the earthquake waves; that is,

A.21. EARTHQUAKES

the violence of the shaking: $M = \log(S)$. A magnitude increase of 1 corresponds to a shaking amplitude increase by a factor of 10 and, it turns out, to an energy release increase by a factor of 32. In the Wikipedia entry on the Richter magnitude scale, we find amounts of TNT that will release the approximate energy equivalent to that of the earthquake. See Table A.5. Recall that the prefix "kilo" means thousand, so 15KT is 15,000 tons, and "mega" means million. Geophysicist Donald Turcotte related the scaling exponent b in the Gutenberg-Richter law to the dimension d of the earthquake fault break, through the relation $d = 2b$.

Magnitude M	TNT equivalent
5	480 tons
6	15 kilotons
7	480 kilotons
8	15 megatons
9	480 megatons

Table A.5: Approximate energy equivalents for some earthquake magnitudes.

Another example of scaling is *Omori's law*, the decay of the number of aftershocks with the passage of time. In 1894, Fusakichi Omori proposed that the rate of aftershocks a time t after the main shock is proportional to $1/(c+t)$. In 1961 Tokuji Utsu modified this to $1/(c+t)^p$, where p lies in the range 0.7 to 1.5. Since we can estimate how quickly the probability of the aftershocks drops off after the main shock, we can estimate the likelihood of aftershocks occurring as a function of time. Hence the interest in Omori's law to people living in active earthquake regions.

Okubo and Aki studied the complexity of a portion of the San Andreas fault by approximating the fault fractal dimension from maps of the region. We must mention two possible complications. First, quoting the authors:

> It is possible that the mapped complexity is strictly an effect of the properties of the materials which constitute the uppermost crust and that faults appear as smoother surfaces at depth.

However, the authors point out that seismic studies have indicated that fault complexity increases with depth, so this possible complication does not seem to be an actual complication. The other possible complication is that scale range that can be investigated is limited by the resolution of the map, not much over one order of magnitude for even the largest map. Still, their results are interesting. Between Parkfield and the Salton Sea, the dimension of the San Andreas fault ranges from $1.12 \pm .02$ to $1.42 \pm .02$. Particularly intriguing is the observation that the 1857 Fort Tejon earthquake with magnitude 7.9 occurred near Parkfield, where the measured dimension of the fault is 1.1, and stopped near San Bernardino, where the fault dimension is 1.4. For the 1966 magnitude 6.0 Parkfield earthquake, the rupture stopped at a region where the fault dimension is 1.2. This suggests that high fractal dimension of a fault may decrease the ease with which one plate slips past another, hence stops the progression of an earthquake rupture. Lower fault dimension does less to impede rupture propagation. Can analyses of this sort predict the locations and extents of earthquakes? For now, only in our dreams, but they may still be a component of predictive strategies. Or maybe protective strategies, if there were a way to increase fault dimension.

A study of time series of real and model earthquakes, and an explanation of an earthquake model based on the sandpile (self-organized criticality or SOC) of Per Bak, Chao Tang, and Kurt Wiesenfeld, is given by Harold Hastings and George Sughara.

Applications of modern network theory to earthquakes are presented by Sumiyoshi Abe and Norikazu Suzuki and by Raúl Madariaga. Here seismic data are mapped to a graph with vertices corresponding to land regions, and an edge connects two vertices if consecutive earthquakes occur in the regions represented by those vertices. For example, consider the earthquake network constructed from Southern California data for 1984–2003. First, we define some terms we'll use to study this graph. The *connectivity distribution* $P(k)$ is the probability of finding a vertex with k edges in a random graph. A graph is *scale-free* if its connectivity distribution $P(k)$ satisfies a power law relation with k. Using 10 km ×10 km regions, the earthquake graph connectivity distribution satisfies $P(k) = k^{-1.2}$, supporting the notion that the earthquake graph is scale-free.

A network is a *small worlds network* if, after removing all loops and multiple edges between vertices, the average path length is small and the clustering coefficient is large. Here the clustering coefficient represents the likelihood that two vertices connected directly to a third vertex are connected directly to one another. Let k_i denote the number of vertices sharing an edge with the ith vertex. The maximum number of these edges is $k_i(k_i - 1)/2$. Then the clustering coefficient of the ith vertex is

$$c_i = \frac{\text{number of edges between the neighbors of the } i^{\text{th}} \text{ vertex}}{k_i(k_i - 1)/2}$$

The clustering coefficient C of the graph is the average of the c_i. The earthquake graph has a low average path length (under 3) and a clustering coefficient more than 10 times higher than that of a random graph. This combination supports the claim that the earthquake graph is a small worlds network. Relations between small worlds networks and fractals may account for some of the fractal characteristics observed in earthquake graphs.

In addition, the California earthquake graph exhibits hierarchical clustering, quantified by the scaling of the clustering coefficient with the con-

A.22 EEG, again

nectivity. Recent years have seen an explosive development of mathematical and computational tools for analyzing network dynamics. We expect these first steps in applying network tools to earthquakes will lead to very interesting developments.

EEG data are used clinically to distinguish epileptic from other seizures, and to locate and characterize seizures. Although MRI and CT scans offer higher spatial resolution than EEGs, EEGs have higher temporal resolution, so give a more detailed picture of brain dynamics.

An excellent reference for interpreting EEG data is *Imaging Brain Function with EEG* by Walter Freeman and bioengineer Rodrigo Quiroga. Physicist Agnes Babloyantz and her coworkers were early investigators of the fractal and chaotic aspects of EEG signals. In particular, they found evidence that during sleep, brain dynamics are dominated by low-dimensional chaos. The complexity of chaos may help us understand the meanders of dreams, what our brains are doing when our eyes are shut and no one is home.

A.23 Sunspots

Sunspots are relatively dark regions on the sun's photosphere, its visible surface. In fact, sunspots are dark only in contrast to the brightness of the photosphere. Isolated from the rest of the sun's image, sunspots are still very bright. Although their study began with Galileo, the exact mechanism responsible for sunspots remains an active topic of research. We do know they are caused by intense magnetic fields and often occur in pairs, one correspoding to the north magnetic pole, the other to the south. In size, sunspots vary from about 16 km to about 160,000 km. Careful counts of sunspot numbers indicate an approximate 11-year cycle, though this may be superimposed on a much slower cycle. Individual sunspots persist for days or weeks. Russian astrophysicists Alexander Milovanov and Lev Zelenyi explain sunspots as a consequence of fractal clusters of magnetic field lines. The formation and lifetime of sunspots simulated with this model agree well with observations.

Physicist Yusuf Shaikh and his colleagues studied monthly and yearly averages of sunspot numbers from 1818 to 2002 and deduced that the sunspot time series data exhibit persistent fractional Brownian motion behavior. Using R/S analysis (We'll describe this in a moment.) to estimate the time series dimension from several sets of monthly data, Shaikh obtained dimension values ranging from 1.1 to 1.3. Using Takens' delay embedding method (in A.35 we give a simple example of Takens' method), physicist Zhang Qin got higher values of the dimension, between about 3 and 4. Rather than measuring the roughness of the sunspot time series – Shaikh's result – Qin sought the dimension of the attractor of the time series, a possibly high-dimensional space encapsulating all time series points and how each is related to the others. Little surprise that Shaikh and Qin got different results: they looked at different aspects of the sunspot data.

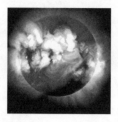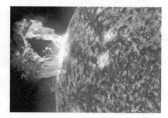

Figure A.28: Sunspots, solar flares, and a solar prominence.

R/S analysis

Developed by Hurst to study the variations in the annual flooding of the Nile, R/S analysis is relatively simple. Applied to a time series $X_1, \ldots, X_M$, for increasing values of n, compute the range of $X_1, \ldots, X_n$ (the difference between the maximum and minimum values) and divide by the standard deviation of this sample. This is the rescaled range; call it $R/S(n)$. Typically, R/S exhibits a power law scaling with n. When this is the case, a plot of $\log(R/S(n))$ vs. $\log(n)$ produces points falling near a straight line. The slope of this line is called the *Hurst exponent H*, and often equals the Hölder exponent α. At the end of A.10.2 we sketched why the dimension of a graph with Hölder exponent α is $2 - \alpha$. This is how Shaikh calculated the dimension of the sunspot time series.

A.24 Solar flares

Solar flares are intense storms on the sun. They are loosely associated with clusters of sunspots, though so far they have eluded prediction. Prediction is an important goal because energetic flares release a tremendous amount of energy, the equivalent of several tens of billions of megatons of TNT. (Compare this number with the earthquake energy chart in A.21 for a dizzying exercise in perception.) Periods of high solar activity may see several flares per day, while the periods of least activity see fewer than one per week. In addition to electromagnetic radiation, solar flares generate particle radiation, which is potentially dangerous to spacewalking astronauts. Observations of EM radiation from a flare can give some advance warning, usually an hour or so, but more advance notice of flares is very desirable.

On Earth, intense flares can have considerable effects despite the shielding benefits of Earth's atmosphere and magnetic field. For example, a solar flare on March 13, 1989, disrupted Québec Hydro's transmission lines, leaving six million without power for nine hours. The same storm caused power transformers in New Jersey to melt and disrupted radio signals worldwide.

Astrophysicists Marcus and Pascal Aschwanden analyzed the fractal scaling of areas and volumes of 20 solar flares. These fractal scaling relations have implications for the conductive radiance and cooling time of flares and for the frequency distribution of flare energies. Numerical results are important – indeed, central to the scientific application of fractals. But Benoit emphasized the importance of pictures for recognizing fractals. Do look at the remarkable images in the Aschwandens' first paper on flare geometries. Who would see those and not think of fractals?

In addition, physicists E. Mogilevsky and N. Shilova analyzed the fractal structure of solar flares and proposed a model of flare formation and development based on Bak's self-organized criticality. Mathematician Z. Yu and coworkers have found evidence of multifractality in the daily fluctuations in solar flare X-ray brightness.

A.25 Solar prominences

Solar prominences are immense ejections of plasma, often in the shape of a loop, extending outward from the sun's surface. Prominences can extend several hundred thousand km above the sun. For comparison, the diameter of the Earth is about 12,750 km. Prominences last for a few days to a few months. As suggested by the complex patterns seen in the third image of Fig. A.28, the plasma flow in solar prominences is turbulent. Astrophysicist Ersilia Leonardis and others have found multifractal scalings in the intensity fluctuations of some prominences, both in *Hinode* Solar Optical Telescope, observations spanning three orders of magnitude in space and in time, as well as in 3-dimensional simulations. They found that the conversion of magnetic energy of the plasma to kinetic energy exhibits multifractal characteristics. This provides another piece in our understanding of solar prominences.

The sun is not just a bright spot in the sky. It is a complex, highly nonlinear system, a superheated plasma interacting with magnetic fields and moving charges that generate their own magnetic fields. The same forces act over many distance scales: little wonder that fractals and multifractals have been found. Have we found them all? Probably not.

A.26 Supernova remnants

A supernova is the immense explosion of a star. For a brief time, many supernovae outshine an entire galaxy, throwing much of the star's mass outward at speeds up to 10% that of light. The resulting shock wave sweeps up the dust and gas surrounding the star and forms the supernova remnant.

If you think that supernova remnants are impossibly far removed from daily life, consider this. Elements heavier than iron 56, including the gold atoms of your ring, come from supernovae. Shortly after their formation, these gold atoms were traveling at a noticeable fraction of the speed of light. Now they are wrapped comfortably around your finger.

About 3200 light years from Earth is the Crab Nebula, the remnant of a supernova observed in 1054. Almost a thousand years later, the diameter of the Crab Nebula is now about 6.5 light years, and is still increasing at about 2200 km/sec. At the center of the nebula is a pulsar, a neutron star rotating 30 times per second. This beautiful, complex nest of filaments made of filaments made of filaments, similar structures over so many levels, looks wonderfully fractal.

 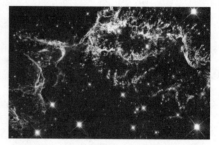

Figure A.29: First: The Crab Nebula. Second: Cassiopeia A.

Cassiopeia A is another supernova remnant, with a diameter of about 10 light years and located about 11,000 light years from Earth. Light from the explosion would have reached earth about 300 years ago, though in that time period unambiguous historical records of a supernova sighting are absent. The best candidate is a dim star observed in 1680 by British Astronomer Royal John Flamsteed. Because it is one of the brightest radio sources in the sky, Cas A is frequently investigated by radio astronomers. Astronomer Yeunjin Kim and coworkers found power law scalings in the distributions of filaments and sheets of the supernova remnant. They investigated the structure of the remnant by using *IR echoes*, infrared energy produced when the dust surrounding Cas A is heated by light from the supernova remnant. The sensitivity of this method enables the construction of 3-dimensional images of the dust cloud. Earlier studies were only able to deduce fractality from 2-dimensional images, while these 3-dimensional reconstructions provide more direct evidence supporting the commonly held belief that significant turbulence in supernova remnants gives rise to fractal structures.

Radio telescope observations reveal a similar a power law in the 21 cm (hydrogen) line, providing additional support for the hypothesis that the

A.27 Saturn's rings

The photo of Saturn's rings certainly looks similar to a simple fractal shape, the product of a Cantor set and a circle. Mathematicians Joseph Avron and Barry Simon investigated whether gravitational resonances with Saturn's moons and the sun could cause the complex, nuanced ring structure. Using a simplified solution to a differential equation called Hill's equation, they deduced that the ring fractality may well

Figure A.30: First: Saturn's rings, real Second: Saturn's rings, simulated: a product of a Cantor set and a circle.

be caused by resonances. But the Cantor set they find is a *fat Cantor set*, that is, a Cantor set with positive length. Recall that the familiar Cantor middle-thirds set starts with the unit interval, from which we remove one open interval of length $1/3$, 2 open intervals of length $1/9$, 4 open intervals of length $1/27$, and so on. The total length removed is obtained by summing a geometric series,

$$\frac{1}{3} + \frac{2}{3^2} + \frac{2^2}{3^3} + \cdots = \frac{1}{3}\left(1 + \frac{2}{3} + \frac{2^2}{3^2} + \cdots\right) = \frac{1}{3}\frac{1}{1 - 2/3} = 1$$

and so as we saw in A.2, the length of the Cantor middle-thirds set is 0. This is the general rule for Cantor sets with a constant scaling factor. For the middle-thirds set, each piece, no matter how tiny, is made of two copies scaled by $1/3$. For the middle-halves set, each piece is made of two copies scaled by $1/4$. By allowing the scaling factor to change between one level and the next, we can construct Cantor sets with positive lengths. Technically, with positive 1-dimensional Hausdorff (or Lebesgue: same thing in integer dimensions) measure, but this is just a generalization of length.

For any number q, $0 < q < 1$, we build a fat Cantor set by removing one open interval of length $q/3$, 2 intervals of length $q/3^2$, 4 intervals of length $q/3^3$, and so on. The the total length removed is

$$\frac{q}{3} + \frac{2q}{3^2} + \frac{q2^2}{3^3} + \cdots = \frac{q}{3}\left(1 + \frac{2}{3} + \frac{2^2}{3^2} + \cdots\right) = q$$

and so this Cantor set has length $1 - q$. Because it has a positive length, this fat Cantor set has dimension 1.

What are the scaling factors for this fat Cantor set? Because we begin by removing an interval of length $q/3$, the Cantor set consists of two pieces scaled by $(1-q/3)/2 = (3-q)/6$. From each of these, we remove an interval of length $q/9$, so the Cantor set consists of four pieces scaled by $(9-5q)/18$. This is not $((3-q)/6)^2$, so the fat Cantor set does not have a constant scaling factor.

Because this fat Cantor set has dimension 1, the product of this fat Cantor set and a circle has dimension 2. This calculation is reassuring, because if Avron and Simon's analysis had shown the dimension of Saturn's rings is less than 2, they would have predicted that the rings are pretty much invisible. A shape having dimension less than 2 would be so sparse it would reflect very little light.

A.28 Fractal cracks and fractures

Benoit began studying fractal dimension as a measure of roughness in 1984, though at the time he did not recognize the wide applicability of that interpretation. Written with Dann Passoja and Alvin Paullay, this study reports the results of a clever experiment to measure the roughness of a break. Fractured steel was coated with nickel, which then was ground down until just the highest parts of the steel were exposed, forming steel islands in a sea of nickel. In the next two paragraphs we describe a method by which the *area-perimeter relation* is calculated. Applied to finding the roughness of fractured metal surfaces, this is called the *slit island method*. . These measurements revealed that the surface dimension is about 2.28, using a respectable data range of $10^{-2}m$ to $10^{-5}m$.

The area-perimeter relation is introduced in chapter 12 of Benoit's *The Fractal Geometry of Nature* and also in chapter 12 (coincidentally, we believe) of Feder's book *Fractals*. Briefly, the area-perimeter relation works this way. Suppose A_1 and A_2 are the areas of two similar geometric shapes, squares, for instance, and P_1 and P_2 are their perimeters. Then

$$\frac{P_2}{P_1} = \left(\frac{A_2}{A_1}\right)^{1/2}$$

For example, suppose square 1 has side length 1 and square 2 has side length 2. Then $P_1 = 4$, $P_2 = 8$, $A_1 = 1$, $A_2 = 4$, and we see the area-perimeter relation holds. The numerator of the exponent is the dimension

A.28. FRACTAL CRACKS AND FRACTURES

of the perimeter, 1 in the case of simple Euclidean shapes. For shapes with fractal perimeters of dimension d, the relation is

$$\frac{P_2}{P_1} = \left(\frac{A_2}{A_1}\right)^{d/2}$$

Interpreting this relation requires some care. For example, if $d > 1$, then the perimeter is infinite and the left side of the equation becomes challenging. For shapes with fractal perimeters, we need to compare areas and perimeters *measured at the same scale*.

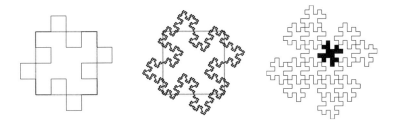

Figure A.31: The area-perimeter relation for shapes with fractal boundaries.

For instance, the first image of Fig. A.31 we see the beginning step of turning a square into a fractal island. Looking at each side of the process, we see that every straight line segment is replaced by $N = 8$ segments, each of length $r = 1/4$. Iterating this process generates the fractal island shown in the second image. Applied to any one of the four island sides, the basic similarity dimension formula shows that the island perimeter has dimension d given by

$$d = \log(N)/\log(1/r) = \log(8)/\log(4) = 3/2$$

In the third image, we see that resolved to the same scale as a small island, a large island consists of 16 copies of the small island, so we see that $A_2 = 16 A_1$, and by counting the number of straight line segments in the perimeters of both islands, we see $P_2 = 8 P_1$. Sure enough,

$$8 = \frac{P_2}{P_1} = \left(\frac{A_2}{A_1}\right)^{d/2} = \left(\frac{A_2}{A_1}\right)^{3/4} = 16^{3/4}$$

Several computer simulations have produced convincing digital versions of cracks. Hans Herrmann reports on 2-dimensional simulations of the effects of external strains on the *displacement field*, the change of position of bits of the substance from their at-rest positions. The material is modeled by points on a lattice – the corners of the squares of a checkerboard – with bonds between adjacent corners. The distributions of stresses in the material is governed by a relation called the *Lamé equation*. First one bond is broken. Solving the Lamé equation gives the stresses on the adjacent bonds, and the bond with highest stress is broken. This break changes the forces, so the Lamé equation must be solved again, and so on. A crack or fracture grows across the lattice. The fractality of the crack is a consequence of the competition between the global direction of the stresses and any direction of local growth imposed by lattice structure.

Simulations on a triangular lattice gave analogous results. C. Lung used the fractality observed in simulations to suggest that the critical crack extension force rises more rapidly than expected for metals composed of very small grains, because for such metals the fracture surface, being fractal over a larger range of scales, has a larger area than non-fractal models predict, and the crack extension force depends on the fracture surface area.

In elementary school we learned that every non-living thing around us fits into one of three boxes: solid, liquid, and gas. But curious kids know this is an oversimplification. What's a sandpile? It's made of solid bits, but as a whole, it doesn't act solid. Kids wonder why a rock dropped straight down will sink in a pond, but thrown at a low angle, the rock bounces off of the water. And what about the right mixture of cornstarch and water? Most kids don't know the term "non-Newtonian fluid," but they are fascinated by its life-like behavior, not at all how a solid, a liquid, or a gas should act.

The world is much more complicated, much more interesting, than solids, liquids, and gases. And cracks are not just straight lines across the pavement. Why do so many look like trees, like lightnoing? If only they are encouraged to look, kids see a wonderful world filled with lifetimes and lifetimes of puzzles to explore. Many of these puzzles involve fractals.

A.29 Soil pores

Mathematician R. Uthayakumar and coworkers analyzed the porosity of two soil samples using box-counting. The object of their study was to assess whether treating the soil with pyroligneous acid increases soil porosity, which they viewed as correlating inversely with dimension. They showed that both untreated samples have dimensions of about 1.9, and both treated samples have dimensions of about 1.7. But we must note that their work is based on 256×256 pixel images, suggesting that we use some care in interpreting their results.

Agricultural and environmental scientist Apostolos Papadopoulos and colleagues investigated applications of the slit island method (described in A.28) to measure the roughness of soil pore perimeters. Pore perimeters influence, among other things, the ability of soil to hold water and the amount of available habitat space for microarthropods. Application of the slit island method to one sample, measuring pores with areas spanning a range of about a factor of 100, gives a fairly tight scattering of points in a log-log plot about a line with slope 1.4, suggesting that this is the dimension of the pore perimeters. But these authors observe that the slit island method requires that the pores have about the same shape, which is not necessarily the case in this sample. Selecting pores of about the same shape does not solve all the problems with interpreting this slope as the perimeter dimension. Papadopoulos concludes that rather than perimeter dimension, their study measures the increase of perimeter complexity with pore size.

In a survey article, Earth scientist Edmund Perfect and land resource scientist B. Kay sample applications of fractal geometry to soil science. They point out that traditional mathematical tools focus on average properties, but many important aspects of soil depend on heterogeneity, which can be approached using fractal geometry. The authors review dimension computations to measure bulk density, the distribution of pore sizes, pore surface area, and size distributions for both particles and aggregates, as well as for adsorption, diffusion, transport, fracture, and fragmentation. Fractal analysis sheds light on the movement of microbes in the soil, on the architecture of root systems, and on the relation between root branching and soil structure. The apt use of fractals to model soil behavior allows observations at one scale to give predictions at another. In addition, the authors point out that soil types may be better distinguished by multifractals than by fractal dimension alone.

Finally, to illustrate the extent to which fractals have become involved in soil science, we point out that in November 2013, Amazon listed two different books titled *Fractals in Soil Science.*

A.30 Distribution of resources

A good reference is the book *Wildlife and Landscape Ecology*, and especially the chapters by Milne and Peak. Biologist Bruce Milne surveys applications of fractal geometry to wildlife biology, including how the distribution of resources influences population dynamics. Because many habitats are fractal, treating them as Euclidean regions can lead to a substantial underestimation of the area and resources needed to support a species. Encounters between predators and prey usually are modeled by the principle of mass action: the probability of an encounter is proportional to the product of the populations. But if the prey respond to a fractal resource distribution, or if fractal landscape aspects constrain the foraging routes of either, then the encounter probability is more complex and may depend on the distance scales involved. Fractal landscapes imply that the areas of available cover or habitat depend on the scale at which a species operates. Such regions can support a wide range of body sizes, because animals of different sizes perceive the environment at different scales. On an evolutionary timescale, rough terrain fosters biodiversity, so highly fractal landscapes like some mountains and rain-forests feature lots of different species.

An important tool for estimating biodiversity is the *species-area relationship*, often seen as a power law,

$$\text{number of species} = k \cdot (\text{habitat area})^d$$

One proposed explanation: this is a response to fractal resource distributions.

Peak studied the dynamics of interacting propulations, elaborating on the usual approach by following the growth of portions of species in geographically distinct locations, allowing the incorporation of fractal landforms. He applied to wildlife population management the method of Edward Ott, Celso Grebogi, and James Yorke for temporarily stabilizing unstable periodic behavior in a chaotic system. A great strength of the OGY method is that it does not require any model of the underlying process. All we need can be deduced from the time series of measurements. Peak interpreted the results of the OGY method in terms of relocating some of the population into or out of the study area.

Real populations may only rarely be described by low-dimensional chaotic dynamics. Peak demonstrated how to implement the OGY protocol to systems with random elements, which are more realistic models of populations. Among other things, he found that stocking (importing animals) always is a more stable control mechanism than is harvesting (removing animals).

Here the fractals are hidden in the dynamics of the interacting populations. Complicated shapes and complicated behaviors often, but not always, are the result of fractals, either in physical space or in the abstract spaces of pathways and interactions.

A.31 Clouds

The fractal geometry of clouds has been investigated in many places. Good starts are Benoit's *The Fractal Geometry of Nature* and Feder's book *Fractals*. The original source is Shaun Lovejoy's paper on the area-perimeter relationship for clouds.

Artists have long know about the the scaling nature of clouds. Hokusai as we have seen, but also Leonardo da Vinci, Cluade Monet, John Constable, and Jonah Dahl,, among many others. As careful observers of the world, artists noticed the scaling of clouds, though they did not always depict it convincingly. Still, most got that big clouds are made of smaller clouds, in turn made of even smaller clouds, and so one.

A simple way to show the fractality of clouds is through scale ambiguity: a small cloud nearby reveals about the same structure as a large cloud seen far away. The boxes in Fig. A.32 illustrate this.

And the variety of clouds, names learned by grade school kids, sketches memorized for quizzes. That there are different types is important, remembering their names is less so. Why don't we teach scaling of clouds? Use them as a gateway to scaling in nature. A tedious exercise could be an enlightening event.

Figure A.32: The scale ambiguity of clouds. Oh, and a rocket.

A.32 Turbulence

Some authors believe the fractal aspect of clouds is due, at least in part, to turbulence in clouds. This brings us to references for fractal (and multifractal) aspects of turbulence. A brief summary of the main ideas is in Chapters 10, 11, and 30 of Benoit's *The Fractal Geometry of Nature*.

Very briefly, turbulence occurs when energy of a flowing fluid gives rise to vortices, which in turn produce smaller vortices, and the process continues in an approximately self-similar cascade. When the energy is divided unevenly between daughters of a parent vortex (and this always happens), turbulence generates a multifractal distribution of energies. In fact, Benoit's studies of turbulence in the 1970s were where he began to form the notion of multiifractals.

A great deal of work, experimental as well as theoretical, on multifractals and turbulence was done by Katepalli Sreenivasan, physicist, mathematician, engineer, and long-time friend of Benoit. . Sreenivasan and his coworkers wrote a lot of papers on this subject, including reports of some of the first experimental work on turbulence, multifractals, and negative dimensions.

A.33 Snowflakes

During the late 19th and early 20th centuries, in Jericho, Vermont, Wilson Bentley developed a method for catching and photographing snowflakes, over 5000 in all. Half these are published in the remarkable book *Snow Crystals*, the source of the images in Fig. A.33. Bentley focused on the more symmertical snowflakes, but he did include a small collection of anomalies, snowflakes in intersecting planes, parallel snowflakes attached by hexagonal columns, and snowflakes with other peculiarities.

Figure A.33: Snowflake photographs by Wilson Bentley.

Every kid in climates with snowy winters knows two things about snowflakes: they have six (usually) approximately identical sides, and no two are exactly alike. Snowflakes form when diffusing water molecules aggregate on a nucleus of soot or an ice crystal. So they grow by diffusion-limited aggregation, which gives complex, delicate branches. The thickness and multiplicity of branches is determined by local atmospheric conditions – temperature, humidity, pressure – approximately constant over a snowflake, so the six branches are approximately identical. The branching pattern of a snowflake is a story about how these conditions varied along the snowflake's path through the cloud. Turbulent airflow in snow clouds guarantees that no two flakes follow the same paths, so all are at least slightly different. Irregular snowflakes tell a different set of stories.

This description of how snowflakes grow is itself just a story. Why should you believe it? The story sounds plausible, but that's not enough to make it true. The next question we should ask is could the story's process work? For this, we turn to computer simulations. Johann Nittmann and H. Eugene Stanley programmed DLA simulations on hexagonal lattices and obtained convincing forgeries of some of Bentley's images. This is another step in understanding the mechanism responsible for the fractality of snowflakes.

A.34 Foraging paths

Many animals, on land and in the sea, follow foraging patterns that exhibit a power law distribution of search lengths. The most familiar way to understand this is through the kids' game hide-and-seek. Run to the back yard, then look around the tree, behind the shrubs, past the corner of the shed, under the picnic table. Then run to the front yard. Look around each tree, behind the flower boxes, past the rose bushes. Then run to the neighbor's yard. The few long treks are punctuated by many smaller excursions.

As reported in the papers of M. Etzenhauser, J. Bascompte, and L. Seuront, and others, animals move in similar patterns. The fractality of the paths likely is influenced by the fractality of resource distribution, but perhaps also by efficiency. After exhausting the food in one area, they travel a bit before hunting again, because finding food is easier in an environment that is not already resource-depleted.

A.35 Disribution of epidemics

One of the earliest, and most successful, applications of mathematics to problems in medicine was to model the spread of an epidemic in a population. Comparing a model's predictions to data from real epidemics establishes the model's general effectiveness. Once this is done, we can ask more subtle questions. A natural question for us is, do the model solutions exhibit fractal properties? Studies of this sort are reported by evolutionary biologists William Schaffer and Mark Kot. In particular, they discuss the SEIR (Susceptible, Exposed, Infected, Recovered) model of disease spread. Upon encountering an infected person (or an *infected*), some fraction of susceptible people (*susceptibles*) acquire the pathogen and enter a period of latency, *exposed* but not yet infectious. Eventually, exposeds become infecteds, and, in this model, then become *recovereds* and are immune from additional infection. The SEIR model is a collection of differential equations, specifying the rate of change of each of the four populations, S, E, I, and R, in terms of the current values of those populations.

$$dS/dt = a - aS - bIS$$
$$dE/dt = bIS - aE - cE$$
$$dI/dt = cE - aI - dI$$
$$dR/dt = dI - aR$$

Here a, b, c, and d are constants we'll interpret in a moment. Built into this model are some unrealistic assumptions made to make the equations more tractable. If this is the first mathematical model of a complex biological system that you've seen, these simplifications may make you question the utility of the model. We'll address this point at the end of this section.

The first simplification is to let S, E, I, and R represent the fraction of the total population in each of these four categories, so $S+E+I+R = 1$. Next, the total population size is taken to be constant, so the per capita birth rate equals the per capita death rate. Call this rate a. In the dS/dt equation, the $-aS$ term corresponds to the decrease in the S population due to mortality, the per capita death rate times the fraction of the population in state S. The a term is the birth rate – in more detail, it's $a(S + E + I + R)$ – and reflects the assumption that the infection is not passed from mother to child. Next, b is the probability that an encounter between an infected and a susceptible results in the transmission of the pathogen to the susceptible, moving the susceptible into the exposed population. This term in the equation is based on the principle of mass action, mentioned in A.30: the probability of an encounter between members of two populations is proportional to the products of the populations. This is sensible: if either population is small, encounters will be rare; if both are small, encounters will be very rare; if both are large, members of each populations will be tripping over members of the other.

In the dE/dt equation, the $+bIS$ term balances the $-bIS$ term in the dS/dt. Mortality of the exposed population is represented by $-aE$. The $-cE$ term represents movement from the exposed population to the infected. The coefficient c is the rate at which exposed people become infectious. In fact, c is the reciprocal of the average time spent exposed before becoming infectious. To understand this, suppose that on average, 10 days are spent exposed but latent. Then every day we expect $1/10$ of the exposed population to become infectious. Finally, d is the rate at which infecteds recover, that is, the reciprocal of the average time spent infected before recovering.

The solutions of these differential equations spiral toward a fixed point, a stable equilibrium mix of the four populations. Yet many epidemics exhibit repeated outbreaks: for instance, in the United States measles outbreaks occur every two to five years. The SEIR model is simple and sensible, so what's wrong? One of the simplifying assumptions is that b, the probability of disease transmission in any encounter, is constant. By simply replacing b with a seasonally varying function $b = b_0(1 + b_1 \cos(2\pi t))$, the model can be made to yield periodic outbreaks and even more complex behaviors for some parameter values.

OK, but where are the fractals? For one thing, some solution curves have fractal characteristics, representing fractal patterns in time, revealed through dimension computations. A solution curve of the SEIR model is a curve

$$(S(t), E(t), I(t), R(t))$$

in 4-dimensional space, whose tangent vector

$$(S'(t), E'(t), I'(t), R'(t))$$

satisfies the SEIR equations for all times t. Because this curve is in 4 dimensions, visualizing its fractality is tricky. We would like to see how a solution curve of differential equations can be fractal, so we'll look at a 3-dimensional system with similar solution curves. Fig. A.34 shows solution curves for a system of one predator and two prey, studied by conservation biologist Michael Gilpin. Here are the equations:

$$dV_1/dt = r_1 V_1 - (a_{11} V_1^2 + a_{12} V_1 V_2 + a_{13} V_1 P)$$
$$dV_2/dt = r_2 V_2 - (a_{21} V_2 V_1 + a_{22} V_2^2 + a_{23} V_2 P)$$
$$dP/dt = r_3 P - (a_{31} P V_1 + a_{32} P V_2 + a_{33} P^2)$$

where V_1 and V_2 are the populations of the two prey species (V for "victim," we suppose), P is the predator population, and the constants are the growth and competition factors. Once again, we have used the mass action principle: for example, $-a_{12} V_1 V_2$ is the negative impact of the competition of V_1 and V_2 on the growth rate of V_1. The solution curve appears to be wrapped around a cone; a line from base to apex intersects the curve in a complicated collection of points – in fact, points on a Cantor set.

Another appearance of fractals: for some parameter values of the SEIR model we find several coexisting, stable, constant solutions, fixed points of the differential equations. To which of these a given solution curve converges depends, sometimes very delicately, on the starting point. The collection of all starting points from which the solution curve converges to a given fixed point constitutes the *basin of attraction* of the fixed point. (We've encountered this concept in our study of Newton's method.) Some of these basins of attraction have fractal boundaries.

Schaffer and Kot report similar curves for field data of measles epidemics in New York City and Baltimore, based on monthly physicians' reports from 1928 to 1963, after which the disease was largely eradicated by the wide-spread use of vaccines. This gives about 430 data points; call them $x_1, x_2, \ldots, x_{430}$.

Figure A.34: First: A solution curve of Gilpin's two-prey, one-predator system. Second: A Cantor set of circles wrapped on a cone.

How can these monthly measurements be turned into curves like those in the first image of Fig. A.34? Schaffer and Kot used a method called delay embedding, developed by the Dutch mathematician

Floris Takens and elaborated by others. As applied by Schaffer and Kot, the delay embedding consisted of plotting the points

$$(x_1, x_{1+3}, x_{1+6}), (x_2, x_{2+3}, x_{2+6}), \ldots, (x_{424}, x_{427}, x_{430})$$

The choice of 3 corresponds to tracking changes over a 3-month period. Connecting successive points with line segments gives a curve similar to that in the second image of Fig. A.34.

Other examples can be found in papers by these authors. Radiation oncologist Wayne Kendal studied weekly data on measles epidemics in England and Wales between 1944 and 1966 and found evidence for random fractal processes underlying the spread of these epidemics. Computer scientist Richard Crandall studied the space and time dependence of the spread of an epidemic. He found fractals not only in the propagating front of the epidemic but also in the distribution of the islands of survivors within the spreading epidemic.

In landscape ecology studies (A.34) we saw that fractality of foraging paths resulted at least in part from the fractality of the terrain or distribution of vegetation. AIDS researcher R. Wallace suggests that the HIV epidemic spread may be more effectively modeled as occurring on a fractal, but not a physical fractal, rather a fractal sociogeographic network.

Now, a final comment about models and simplifying assumptions, which are of some use in fractal modeling of natural phenomena. Often simplifications are justified with the assertion that more detailed models are mathematically intractable. This may be true, but it feels cheap, an unnecessary surrender. If only we knew more math, we could do better. But we think this misses the main point of mathematical modeling. Although it is true that the detailed behavior of a system depends on the detailed interactions of all its pieces, perhaps some of its large-scale behavior results from cruder relations between big chunks of the system. Simplifications allow us to look for those aspects of behavior that do not depend on system details. These are the parts that most likely generalize to other systems. With simplified models, we are looking for the forest, not the trees.

The cleanest expression of this idea is Jorge Luis Borges' parable "On exactitude in science." Borges describes an empire where cartography was pursued with such enthusiasm that a map of the empire was produced that coincided, point for point, with the empire itself. Later generations regarded this map as too cumbersome and abandoned it to the elements. Eventually wild animals used bits of the map for shelter.

A.36 Ion channels

Ion channels are proteins that span a cell membrane and can configure their structure to open a pore in the membrane for the flow of specific ions, primarily potassium, sodium, and calcium. Ion channels are always

A.36. ION CHANNELS

in one of two states, open or closed, allowing or preventing the flow of their specific ions across the cell membrane. Quantitative study of ion channel kinetics was accelerated by the development of the *patch clamp*, a microscopic technology for measuring the electrical current flowing through a single channel. Patch clamp measurements produce detailed records of the time intervals during which a channel is open and the time intervals during which it is closed. From these measurements, biologists deduce the number of molecular configurations that correspond to each channel state. In the standard model, transitions between these configurations are caused by random impacts of surrounding molecules with the channel molecules, so the probability of making the transition from one configuration to another is proportional to the energy difference between the configurations and does not depend on the history of configurations preceding the current one. Because of this lack of memory beyond the immediate past, these are examples of *Markov models*, forgetful systems. Markov models have been used successfully in many settings, including the pioneering study of Colleen Clancy and Yoram Rudy, and a multitude of other analyses. Full disclosure: Colleen was an undergraduate student of the older author and has remained a dear friend. With these models, Colleen and her coworkers have made significant contributions to our understanding of the precise genetic basis of cardiac diseases, including identifying a mutation in the cardiac sodium channel that can lead to long QT-syndrome and sudden cardiac death..

However, some have criticized the validity of these exponential fits, and even the Markov hypothesis of no memory beyond the immediate past. Perhaps configuration memory is important in the kinetics of some membranes. Biologist and complex systems scientist Larry Liebovitch and others, devised a different approach, fractal ion channel kinetics. In this model, state transitions are governed by a nonlinear deterministic process, giving rise to a power law scaling of residence times in each state. As evidence for fractality, Liebovitch noted that a higher frequency sampling of patch clamp data shows that the statistics of open and closed states is similar to the statistics of the original data. That is, patch clamp data appear to exhibit scaling, a sign of fractality.

Some variants on Liebovitch's fractal ion channel model include one proposing mostly fractal kinetics for closed states and another proposing a combination of fractal and Markov kinetics for open states.

Why do we care about ion channels? Clancy and Rudy's work already has identified the genetic basis for some cardiac arrythmias. Applying similar techniques to neurons may lead to a better understanding of neurological diseases. And then, there's the bigger picture: ion channels are responsible for the propagation of signals along neurons, and these produce all your thoughts, hopes, fears, loves. Yes, ion channels are important to everyone.

A.37 Fractal fitness landscapes

Mathematical biologist Sewall Wright developed the notion of a fitness landscape in the 1930s. Species are arranged on a horizontal plane with neighboring species determined by genetic similarity; the landscape is made by placing above each species a mountain whose height represents the fitness of the species. How is fitness measured? Typically, biologists use *reproductive success* as a surrogate for overall fitness. The landscape is a good way to visualize how evolution follows a path which always leads to higher ground. There's no mysterious evolutionary force driving toward greater fitness; rather, more fit species have more offspring, increasing the fraction of the population with those traits that improve fitness.

Biophysical chemist and Nobel laureate Manfred Eigen and theoretical chemist Peter Schuster extended the fitness landscape model by replacing the horizontal plane of species with the space of genotypes, that is, specific DNA sequences. Nearby sequences differ at a small number of positions. This is the natural context to describe their notion of *quasispecies*, collections of nearby sequences generated from wild types by mutation and selection. This approach has been very useful for understanding the evolution of viruses, for example.

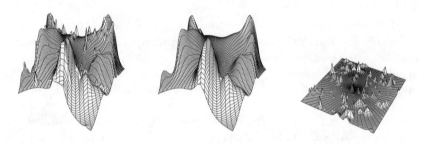

Figure A.35: First: An approximation of a fractal fitness landscape. Second: The largest-scale landscape features only. Third: The difference between the first and second images.

Our interest in the topic, other than its being a beautiful application of geometry, is that at least some fitness landscapes seem to be fractal. For smooth landscapes, nearby sequences have similar levels of fitness; for rugged landscapes, nearby sequences can have substantially different levels of fitness. So we are interested in the decay of the similarity of fitness as a function of the distance between the sequences, measured by yet another application of the correlation function. Essentially, for self-similar fitness landscapes, we expect that the more closely we look, the more uphill paths we will find. Gregory Sorkin defines a fitness landscape as fractal of type

A.38. NATURE-INSPIRED CHEMICAL ENGINEERING

H if the average

$$\langle (f(\vec{x}) - f(\vec{y}))^2 \rangle \sim d^{2H}$$

where the average, indicated by the pointy brackets, is taken over all sequences $\vec{x}$ and $\vec{y}$ a distance d apart.

Fitness landscapes for many models are fractal in Sorkin's sense. For example, in Christopher Adami's book *Artificial Life*, Adami gives evidence of the fractality of RNA landscapes. The first image of Fig. A.35 shows an approximation of a fractal fitness landscape. Note that the wider peaks are augmented by a collection of thinner peaks. In real landscapes, this cascade of smaller peaks continues over even more levels, but this illustration captures the idea.

This property of fitness landscapes is an important part of our understanding of evolution, but this is just a first step. In fact, evolution does not occur on a fixed landscape, because all the other species also are evolving simultaneously. We really need a coevolutionary landscape – a more interesting, and more challenging, problem.

Chapter 4. Manufactured fractals

The next note refers to Sect. 4.1, **Chemical mixers**

A.38 Nature-inspired chemical engineering

In addition to using the fractal structure of the lungs to guide the design of chemical mixers in 2 and 3 dimensions, Coppens and his team have found other applications of fractal structures – for example, that a hierarchical distribution of pore sizes and numbers improves the performance of fuel cell catalysts, and mimicking the growth of bacteria colonies to produce materials exhibiting bacteria colony dynamics, including fractal growth.

Also, Coppens has taken an engineering approach to studying the efficiency of air flow through the lungs. In the first 15 levels of lung branching (the conducting airways), transport is mostly by convection; in the remaining 8 or so levels (the acinar airways or exchange zone), transport is mostly by diffusion. Coppens' group found that air reaches a large part, but not all, of the acinar airways while the body is at rest. However, air reaches all the acinar airways during exercise. This is sensible: evolution selected toward a design with some flexibility to account for different demands of air usage. Coppens' approach, an understanding of nature's geometry to inform engineering designs, here has a clever inversion: using his understanding of engineering design principles to guide his calculations of nature's efficiency. Most roads – even those passing through fractals – can be traveled in two directions.

This note refers to Sect. 4.2, **Fractal capacitors**

A.39 Lateral capacitors and parasitic capacitance

One additional way lateral capacitors increase capacitance is by reducing the *parasitic capacitance* effect of the substrate. In lateral capacitors, electric field lines between the metal films reduce the field line intensity to the substrate, thus reducing the capacitor's current loss caused by field lines to the substrate.

Looking for some additional sources, we Googled "fractal capacitor" and found about a quarter million results (many of which dealt with the on-line role-playing game "Guild Wars 2," which is not related to science). Some did deal with constructions similar to those we've shown. For example, Amro Elshurafa and coworkers have investigated applications of fractal capacitors to MEMS circuitry (Microelectromechanicalsystems – Google again for some very interesting pictures, among other things). The original application of fractal capacitors was increased capacitance density, which reduced the fraction of etched circuit space needed for capacitors. MEMS applications are different, including very high frequency of operation.

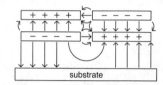

Figure A.36: Reduction of substrate parasitic capacitance.

This note refers to Sect. 4.3, **Fractal antennas**

A.40 Scaling solutions of Maxwell's equations

The dynamical behavior of electric and magnetic fields, and consequently also of light and radio waves, is described by Maxwell's equations. Robert Hohlfeld and Nathan Cohen show that Maxwell's equations are unchanged if space coordinates are scaled by a factor of s and frequency by $1/s$. They go on to show that origin symmetric fractals (for example, two equilateral Sierpinski gaskets attached at their apexes) are frequency-independent, at least for frequencies that are multiples of the size scaling of the antenna's scaling factor.

Cohen points out the advantages of fractal antennas, including their smaller size, sensitivity to multiple frequencies, and higher gain. All this through the antenna geometry alone, without adding electronic components to the antenna, is pretty impressive. An interesting application of fractal design is to smart antennas, necessary for SDR (software-defined radio – ask Google).

This note refers to Sect. 4.4, **Invisibility cloaks**

A.41 Scaling in invisibility cloaks

Cohen adapted fractal antenna design to build a wide band, person-sized microwave invisibility cloak. Because microwaves and light both are governed by Maxwell's equations, if the fractal elements of a microwave cloak were rendered on a much smaller scale, they could be used to build an optical invisibility cloak. Microwave wavelengths range from 10mm to 200mm; visible light from 400 to 700nm (nanometers, billionths of a meter), so optical invisibility cloak elements will have to be *much* smaller than those of a microwave invisibility cloak. Certainly, some significant engineering problems remain. We imagine Harry Potter's invisibility cloak is based on different principles, but still, the result appears worth some effort.

Applications, besides the obvious, may include shielding camera supports in telescopes, thus reducing the diffraction artifacts in deep sky photographs.

Cohen suspects this is just the beginning. Recently, his group applied the same fractal cloaking design to infrared radiation, moving heat from one location to another rapidly and without any power consumption. The technology is completely passive. Interesting times may well be ahead of us.

This note refers to Sect. 4.6, **Web infrastructure**

A.42 Fractal models of internet traffic and infrastructure

Network traffic scientist Walter Willinger and computer scientist Vern Paxson, reviewing the history of voice traffic infrastructure, have shown that voice traffic architecture will not work for Internet traffic because of the self-similarity of Internet traffic. Instead, the Internet must be designed for a large range of traffic volume fluctuations. A problem, *congestion*, occurs when the volume of packets that arrive at a hub exceeds the hub's capacity, the rate at which packets can be sent on, plus the number than can be stored temporarily in hub's buffer space waiting for the downstream traffic to thin. To avoid this congestion, Internet traffic is rerouted around a hub when current traffic approaches the hub's capacity. This rerouting introduces power law time correlations in data traffic, violating the exponential time correlations of voice traffic. For typical data networks, power law scaling spans 3 to 5 orders of magnitude, plenty of evidence that fractality. While fractal modeling is a vast improvement over Poisson modeling (voice traffic), patterns in data fluctuations suggest that at least some Internet traffic is more properly modeled by multifractals.

One of the first suggestions that Internet traffic can be modeled with fractals is the paper "On the self-similar nature of ethernet traffic," by

Will Leland, Murad Taqqu, Willinger, and Daniel Wilson. Lacking an explanation for fractality of data traffic, the authors suggested seeking heavy tails (the probabilities of rare events decay following a power law, not an exponential, as with the normal distribution) in Internet statistics. Once people knew to look for heavy tails, they appeared everywhere, and the fractal nature of the Internet became mainstream.

Internet researchers Gary Flake and David Pennock think of the web as an ecosystem. One of the fascinating results they report is that the local organization of the web resembles the shape of a bow tie. That is, many pages (the left side of the bow tie) link to a small number of pages (the knot of the bow tie), which in turn link to another large number of pages (the right side of the tie). Pages not part of this big bow tie form other, smaller bow ties – but really, this can't be a surprise.

Physicists Romualdo Pastor-Satorras and Vespignani support the fractality of the Internet by pointing out that the geographical density of routers exhibits a power law scaling with exponent 1.5. Because this exponent is less than 2, routers are arranged in clumps, each of which breaks into smaller clumps, and so on. In fact, this exponent is the fractal dimension we discuss in Chapter 6.

In addition, Pastor-Satorras and Vespignani analyze the self-similarity of many aspects of Internet traffic: queueing times, inter-arrival times, and round-trip times, among others. The authors survey several mechanisms that might generate these self-similar behaviors.

Finally, recall that one consequence of self-similarity is that fractals have no characteristic length scale. A circle has its radius, a square its side, but what would be the characteristic length of the gasket? The side of the whole gasket? But the gasket is made of three smaller exact replicas of the gasket, so why not the side length of one of these replicas? Or the side length of one of the three still smaller gaskets making up each of these? You see the problem with trying to assign a characteristic length scale to the gasket. The Internet is complex in a much more sophisticated way. In fact, Pastor-Satorras and Vespignani show that the Internet does not even have a characteristic hierarchical level.

Together with physicists Alain Barrat and Marc Barthélemy, Vespignani studies the dynamics of complex networks, including the Internet. In particular, they present a class of model systems whose large-scale properties can be predicted from local interactions of the model constituents. Because the Internet is self-organizing, such models are particularly appropriate.

This note refers to Sect. 4.7, **Fractal music**

A.43 Fractal music

In commenting on Voss's work, Martin Gardner wrote

Good music, like a person's life or the pageant of history, is a wonderous mixture of explanation and unanticipated turns. There is nothing new about this insight, but what Voss has done is to suggest a mathematical measure of the mixture.

Some of Harlan Brothers' ideas on fractals and music were field-tested by him at the summer workshops he, Nial Neger, Benoit, and the older author ran. His applications of fractals to music were convincing to, and very popular with, the teachers who attended.

This note refers to Sect. 4.8, **Fractal literature**

A.44 Structural scaling in Stoppard's *Arcadia*, and other comments

Here we have a few more comments, plus calculations of the time period scaling in scene 7 of Stoppard's *Arcadia*.

Fracals as subject

William Bloch points out that in Borges' review of *Mathematics and the Imagination* by Edward Kasner and James Newman, he mentions many of the geometrical and logical constructions described by Kasner and Newman and that we expect would interest him. But Bloch notes that Borges makes no mention of the Koch curve, which occupies a dozen pages of Kasner and Newman's book. Bloch suggests that this omission speaks against the view of some that Borges would have been interested in fractals. On the other hand, Borges' review is hardly one page long and focuses on logical paradoxes, especially the Liar's Paradox. Borges' copy of Kasner and Newman was not among his books that were donated to the National Library, so we do not know if he annotated the Koch curve section. Incomplete information, several possible conclusions, a mystery. Perfectly appropriate for Borges.

Structural fracals

Recall that in her analysis of Scene 7 of Stoppard's *Arcadia*, Rodberg counted the durations, the number of lines by individuals from one time period, bracketed by lines from the other time period, obtaining twenty-two durations of length 1, three of length 2, four of length 3, one each of lengths 4, 5, and 7, two of length 9, one each of lengths 10, 15, and 16, two each of lengths 26 and 43, and one each of lengths 73 and 86.

To test the presence of a power law scaling we must bin the data. But how? We could use powers of 2 for the right end of each bin

$$[1], [2], [3,4], [5,8], [9,16], [17,32], [33,64], [65,128]$$

or powers of 3

$$[1], [2,3], [4,9], [10,27], [28,81], [82,243]$$

Powers of larger integers wouldn't give enough data points, so we'll compare these two. In Fig. A.37 we plot $(i, \log_p(B_p(i)))$, where i is the bin number, so the bin size is $B_p(i) = p^i$ and $i = \log_p(B_p(i))$. Circles are the $p = 2$ bins, solid dots are the $p = 3$ bins. The best-fitting lines have $r^2 = 0.52$ for $p = 2$, not so impressive, and $r^2 = 0.90$ for $p = 3$, quite a bit better and a reasonably good indication of some fractality in the time period changes.

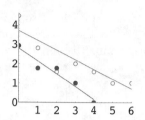

Figure A.37: Log-log plots of the duration between time period changes.

Metastructural fractals

The Mitchell quote is from the Wednesday, April 9, 2014, issue of the *Yale Daily News*. The text surrounding the quote elaborates Mitchell's point:

> All of Mitchell's works are in conversation with each other, he claimed, adding that all his novels are "individual chapters in a mega-novel."

The Dickens quote was from his February 9, 1856, letter to Angela Bassett Coutts. Dickens had purchased a house, Gad's Hill Place, near where he lived as a child. A paragraph about the house concludes with this sentence:

> To crown all, the sign of Sir John Falstaff is over the way, and I used to look at it as a wonderful Mansion (which God knows it is not), when I was a very odd little child with the first faint shadows of all my books, in my head – I suppose.

Chapter 5. The Mandelbrot set: infinite complexity from a simple formula

This note refers to Sect. 5.1, **Some pictures**

A.45 Mandelbrot set picture coordinates

For the Mandelbrot set pictures of Fig. 5.1 we'll give the c value of the lower left corner, and the width, δ, of the image. For each row 1 through 5, here's a key to which

row	first	second
	third	fourth

	c	δ	c	δ
1	$-2-2i$	3	$-0.65+0.5$	0.2
	$-0.535+0.663i$	0.01	$-0.235+0.685i$	0.025
2	$0.42778+0.22287i$	0.01027	$-0.8419+0.0819i$	0.1779
	$-0.69426+0.29761i$	0.01846	$-0.685405+0.307944i$	0.002269
3	$-1.710806+0.002179i$	0.000917	$-1.7318489+0.0005205i$	0.0002416
	$-1.746506+0.0132736i$	0.00015	$-1.7464763+0.0133562i$	0.00004
4	$0.25808+0.0008i$	0.0008	$0.2585+0.0012i$	0.0004
	$0.321+0.034i$	0.003	$0.323+0.0365i$	0.0005
5	$0.38151-0.10308i$	0.00715	$-1.25442+0.04218i$	0.008
	$-1.7477921+0.001482i$	0.000001	$-0.0577-0.668i$	0.001

Table A.6: Parameters for the images of Fig. 5.1.

entry in Table A.6 refers to which image of Fig. 5.1.

This note refers to Sect. 5.2, **The algorithm**

Here we show why the escape criterion works and mention an algebraic trick that speeds up the test of whether iterates of a point escape to infinity.

A.46 Proof of the escape criterion and a faster algorithm

Recall the escape criterion: if some $|z_n| > 2$, then later iterates run away to infinity.

First, suppose $|z_n| > \max\{2, |c|\}$. Because $|z_n| > 2$, we can write $|z_n| = 2 + \epsilon$, for some $\epsilon > 0$. We want to find a relation between $|z_{n+1}|$ and $|z_n|$. The first step is easy. Because $z_{n+1} = z_n^2 + c$,

$$|z_{n+1}| = |z_n^2 + c|$$

Next, we need a relation between $|z_n^2|$ and $|z_n^2 + c|$. We'll use the triangle inequality, $|a+b| \le |a| + |b|$, so called because the complex numbers a, b, and $a+b$ can be viewed as the three vertices of a triangle, and the length of any side of a triangle is less than or equal to the sum of the lengths of the other two sides.

$$\begin{aligned} |z_n^2| &= |z_n^2 + c + (-c)| & \text{by adding and subtracting } c \text{ to } z_n^2 \\ &\le |z_n^2 + c| + |-c| & \text{by the triangle inequality} \\ &= |z_n^2 + c| + |c| & \text{because } |-c| = |c| \end{aligned}$$

Rearranging this inequality, we obtain

$$|z_n^2 + c| \ge |z_n^2| - |c|$$

Now back to $|z_{n+1}| = |z_n^2 + c|$.

$$|z_{n+1}| = |z_n^2 + c|$$
$$\geq |z_n^2| - |c| \qquad \text{by the inequality above}$$
$$= |z_n|^2 - |c| \qquad \text{because } |z_n^2| = |z_n|^2$$
$$> |z_n|^2 - |z_n| \qquad \text{because } |z_n| > |c|$$
$$= (|z_n| - 1)|z_n| \qquad \text{factoring out } |z_n|$$
$$= (1 + \epsilon)|z_n| \qquad \text{because } |z_n| = 2 + \epsilon$$

Iterating $|z_{n+1}| > (1 + \epsilon)|z_n|$, we obtain

$$|z_{n+k}| > (1 + \epsilon)^k |z_n|$$

As $k \to \infty$, $(1 + \epsilon)^k \to \infty$, this last inequality shows that $|z_{n+k}| \to \infty$.

To complete the proof that $|z_n| > 2$ implies the $|z_{n+k}|$ run away to infinity, we consider two cases: $|c| \leq 2$ and $|c| > 2$.

If $|c| \leq 2$, then $|z_n| > 2$ implies $|z_n| > \max\{2, |c|\}$ and the argument above shows the $|z_{n+k}|$ run away to infinity.

If, on the other hand, $|c| > 2$, then look at the first few iterates of z_n.

$$z_0 = 0$$
$$z_1 = c, \qquad \text{so } |z_1| = |c| > 2$$
$$z_2 = c^2 + c = c(c + 1)$$

and so $|z_2| = |c| \cdot |c + 1| > |c|$. That is, $|z_2| > \max\{2, |c|\}$ and again the argument above shows the $|z_{2+k}|$ run away to infinity.

We have shown that if any iterate z_n is farther than 2 from the origin, then later iterates must necessarily run away to infinity. The only iterates that do not run away to infinity stray no farther than 2 from the origin.

Saving computational time

Generating these images requires many iterations, so any economy is welcome. Computing $\sqrt{x_k^2 + y_k^2}$ at each iterate takes time, not much, but if we do a hundred iterates for each of 400,000 pixels, even tiny amounts of time add up. To avoid this unnecessary computational expense, programs test if $x_k^2 + y_k^2 > 4$ instead of $\sqrt{x_k^2 + y_k^2} > 2$. These are algebraically equivalent, and the former saves the unnecessary computation of the square root.

The next six notes refer to Sect. 5.3, **Julia sets**.

Here we cite the theorem of Fatou and Julia on the role of the iterates of critical points, show the point at infinity is a fixed point of $f(z) = z^2 + c$, develop the polar representation of complex multiplication, establish the technical conditions for chaos, and show that the dynamics of $z^2 + c$ are chaotic for one Julia set. This last result generalizes: for every c, the dynamics of $z^2 + c$ are chaotic on the Julia set J_c.

A.47 The c-values for the Julia sets in Figs. 5.3 and 5.5

The Julia sets of Fig. 5.3 are for the c-values shown in Table A.7, the correspondence between Julia set and parameter is shown in this map.

row	first	second
	third	fourth

Those of Fig. 5.5 are for $c = -0.12 + 0.75i$, $c = 0.284 + 0.535i$, $c = 0.3863 + 0.3163i$, and $c = -0.0244 + 0.7402i$, that is, they are the Julia sets of the top row of Fig. 5.3.

1	$-0.12 + 0.75i$	$0.284 + 0.535i$
	$0.3863 + 0.3163i$	$-.0244 + .7402i$
2	$-0.533 + 0.602i$	$-0.52771 + 0.61936i$
	$0.3869 + 0.1401i$	$0.37344 + 0.13602i$
3	$0.290941 + 0.016428i$	$-0.7909 + 0.13669i$
	$-0.78541 + 0.13713i$	$0.2427 + 0.5127i$
4	$0.291031 + 0.015i$	$-0.79037 + 0.13743i$
	$-0.78517 + 0.14i$	$0.241 + 0.51i$
5	$-0.238 + 0.694i$	$-0.239 + 0.961i$
	$0.218 + 0.594i$	$0.35454 + 0.34478i$

Table A.7: Parameters for the images of Fig. 5.3.

A.48 Julia set connectivity and critical points

Early in the twentieth century, Fatou and Julia studied the long-term behavior of iterates of complex functions $f(z)$. One of their results is this theorem:

> The Julia set is connected (all one piece) if the iterates of all the critical points of f do not run away to infinity, and the Julia set is a Cantor set if the iterates of all the critical points of f do run away to infinity.

By critical points we mean those points z for which the derivative $f'(z) = 0$. For the function $f(z) = z^2 + c$, the function iterated in the Mandelbrot formula, the derivative $f'(z) = 2z$ and so the only critical point is $z = 0$. Interpreted for $f(z) = z^2 + c$, the theorem of Fatou and Julia says this:

> For $f(z) = z^2 + c$, the Julia set is connected if the iterates of $z_0 = 0$ remain bounded, and the Julia set for c is a Cantor set if the iterates of $z_0 = 0$ run off to infinity.

That is, because $f(z) = z^2 + c$ has only one critical point, the only possibilities for Julia sets are that they are connected or Cantor sets.

A.49 The point at infinity is a fixed point of $f(z) = z^2 + c$.

We can't show this by simply plugging in $z = \infty$, because ∞ is not a number. Instead, we'll change variables to $w = h(z) = 1/z$. Note that $z = \infty$ when $w = 0$, so we must show that $w = 0$ is a fixed point, but for what function? We can use the substitution $w = h(z)$ to change variables:

$$w_n = \frac{1}{z_n} \to \frac{1}{z_n^2 + c} = \frac{1}{(1/w_n)^2 + c} = \frac{w_n^2}{1 + cw_n^2}$$

So the function $z_{n+1} = z_n^2 + c$ translates to $w_{n+1} = w_n^2/(1 + cw_n^2)$. We'll define a function g by $g(w) = w^2/(1 + cw^2)$, and we see that $w = 0$ is indeed a fixed point for the function g. If you're uncomfortable about using the function h to interchange 0 and ∞, all this can be made precise using a construction called the Riemann sphere. This is a bit out of our way, but Google will satisfy your curiosity.

In A.58 we'll show that the fixed point at infinity has a property called stability. This means all points far enough from $c = 0$ iterate to infinity.

A.50 The technical definition of chaos

What, precisely, do we mean by chaotic? In his important text *An Introduction to Chaotic Dynamical Systems*, mathematician Robert Devaney characterizes chaos by three properties. We're going to give a complete proof that the iteration of the Mandelbrot function is chaotic on the Julia set of $c = 0$, so we need to be careful with the definitions. This is why you'll see more precise – fussier – statements than we usually make. We'll give the definition in terms of a general function $f : [0, 1) \to [0, 1)$. The domain and range of this function is $[0, 1) = \{x : 0 \leq x < 1\}$, and not $[0, 1] = \{x : 0 \leq x \leq 1\}$, because the $c = 0$ Julia set is the unit circle, and we can represent every point on the circle by a number τ representing what fraction of the way around the circle the point lies. Both $\tau = 0$ and $\tau = 1$ represent the same point, $\theta = 0$, on the unit circle, and we needn't define a function twice at this, or any, point.

1. *Sensitivity to initial conditions* Associated with the function f there is a positive constant S, called the *sensitivity constant*. For every point $x \in [0, 1)$ and for every number $\delta > 0$, there is a point $y \in [0, 1)$ with the distance $d(x, y) < \delta$ and the distance $d(f^n(x), f^n(y)) \geq S$ for some number n of iterations. This is the bad news of chaotic

physical systems. Because we can't know the starting point with infinite precision, the system might be at y instead of x, so we can predict the behavior of the system only for a limited time. How often does your local weather report get the eight-day forecast right?

2. *Periodic points are dense* For every point $x \in [0,1)$ and every number $\delta > 0$, there is a point $y \in [0,1)$ with distance $d(x,y) < \delta$ and y is periodic. Here periodic means that for some n, $f^n(y) = y$. So chaos is filled with order, though always it is unstable order. If some of these periodic points were stable, then eventually the system would get stuck in one of these or cycle between them and not be chaotic at all. These unstable periodic points are the basis of the technique called controlling chaos, the OGY method mentioned in A.30. Occasional small changes to the system can stabilize, temporarily, any one of these periodic points.

3. *Mixing* Given any two intervals I and J in $[0,1)$, the iterates of I eventually will intersect J. That is, for some positive number n, $f^n(I) \cap J \neq \emptyset$. The name "mixing" is apt: each interval gets mixed around until it touches every other interval. Put another way, $[0,1]$ cannot be split into pieces that are forever kept separate under repeated iterations of f.

A.51 Polar representation of complex multiplication

In order to demonstrate that iterating the Mandelbrot function on the Julia set exhibits chaotic dynamics, we need the polar coordinate representation of complex numbers and, in particular, of complex number multiplication. In the first image of Fig. A.38 we see that the complex number $a + ib$ also can be represented by a distance $r = \sqrt{a^2 + b^2}$ from the origin and an angle $\theta = \arctan(b/a)$ from the positive x-axis. The distance from the origin is called the *modulus* of the complex number and is written $|a+ib|$; the angle is called the *argument*, written $\arg(a+ib)$. As illustrated in the second image of the figure, complex number multiplication has a simple polar formulation: multiply the moduli and add the arguments.

Starting with the standard rules for distributing products over sums, we can see why the modulus of the product of two complex numbers is the product of the moduli of the numbers.

$$(a+ib) \cdot (c+id) = (ac - bd) + i(bc + ad)$$

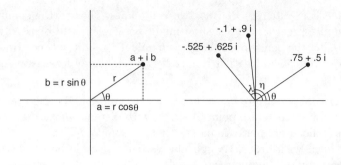

Figure A.38: First: The polar representation of complex numbers. Second: An example of complex multiplication, $(.75+.5i)\cdot(-.1+.9i)n = -.525+.625i$. Note $\lambda = \eta + \theta$, the arguments add.

Then

$$|(a+ib)\cdot(c+id)| = \sqrt{(ac-bd)^2 + (bc+ad)^2}$$
$$= \sqrt{a^2c^2 + b^2d^2 + b^2c^2 + a^2d^2}$$
$$= \sqrt{(a^2+b^2)(c^2+d^2)}$$
$$= |a+ib||c+id|$$

Here's why the argument of the product is the sum of the arguments. The argument of $a+ib$ is $\arctan(b/a)$, the argument of $c+id$ is $\arctan(d/c)$, and the argument of $(a+ib)\cdot(c+id) = (ac-bd)+i(bc+ad)$ is $\arctan((bc+ad)/(ac-bd))$. We want to show

$$\arctan(b/a) + \arctan(d/c) = \arctan((bc+ad)/(ac-bd))$$

Take tan of both sides. Recall that $\tan(\arctan(z)) = z$. (This way is fine; it's the other way around, $\arctan(\tan(\theta)) = \theta$ that requires some care because many angles have the same tangent.) To show that arguments add we need

$$\tan(\arctan(b/a) + \arctan(d/c)) = \frac{bc+ad}{ac-bd}$$

Applying the tangent angle-sum formula to the left-hand side, our goal becomes

$$\frac{\tan(\arctan(b/a)) + \tan(\arctan(d/c))}{1 - \tan(\arctan(b/a))\tan(\arctan(d/c))} = \frac{(b/a) + (d/c)}{1 - (b/a)(d/c)}$$

Then noting

$$\frac{(b/a) + (d/c)}{1 - (b/a)(d/c)} = \frac{bc+ad}{ac-bd}$$

we've shown that the argument of the product of two complex numbers is the sum of their arguments.

A.52 Chaotic dynamics on a Julia set

Now using the polar representation of complex multiplication, we'll see that the Julia set for $c = 0$, that is, for $f(z) = z^2$, is the unit circle, and that on the unit circle, $f(z) = z^2$ exhibits chaotic dynamics. For a proof that $f(z) = z^2 + c$ exhibits chaotic dynamics of J_c, see Theorem 5.12 of Chapter 3 of Devaney's *An Introduction to Chaootic Dynamical Systems*.

The first part is relatively easy, once we note that in the polar representation, $f(z) = z^2$ squares the modulus of z and doubles the argument of z. We consider three cases: $|z_0| < 1$, $|z_0| = 1$, and $|z_0| > 1$. The effect of successive squarings of the modulus is

$$|z_0| < 1 \quad \text{implies } |z_n| \to 0 \text{ as } n \to \infty,$$
$$|z_0| = 1 \quad \text{implies } |z_n| = 1 \text{ as } n \to \infty, \text{ and}$$
$$|z_0| > 1 \quad \text{implies } |z_n| \to \infty \text{ as } n \to \infty$$

That is, starting from a point z_0 inside the unit disc, future iterates converge to 0; future iterates of z_0 on the unit circle remain on the unit circle; and future iterates of z_0 outside the unit disc run away to infinity. Then K_0, the starting points z_0 whose iterates do not run away to infinity, is the closed unit disc. Then J_0, the boundary of K_0, is the unit circle.

That part was easy. To show $f(z) = z^2$ is chaotic on the unit circle, we'll express angles, not in radians, θ, but in fractions τ of turns around the circle. That is, $\theta = \pi/2$ corresponds to $\tau = 1/4$, $\theta = \pi$ corresponds to $\tau = 1/2$, and so on. The variable τ measures the position on the unit circle relative to the point $z = 1$; that is, $\tau = 0$. If $0 \leq \tau < 1/2$, then our argument-doubling result shows $f(\tau) = 2\tau$. On the other hand, if $\tau \geq 1/2$, then $2\tau \geq 1$. For example, if $\tau = 2/3$, then $2\tau = 4/3$, but this is the same point on the unit circle as $4/3 - 1 = 1/3$. So the angle-doubling map, $f(z) = z^2$ on the unit circle, can be expressed as

$$f(\tau) = \begin{cases} 2\tau & \text{if } 0 \leq \tau < 1/2 \\ 2\tau - 1 & \text{if } 1/2 \leq \tau < 1 \end{cases}$$

To establish the three defining properties of chaos for our angle-doubling map, we'll be helped by a thoughtful choice of coordinates for points of $[0, 1)$. Because the angle-doubling map involves multiplying by 2, we'll use powers of $1/2$. Specifically, we can write $x \in [0, 1)$ as

$$x = \frac{x_1}{2} + \frac{x_2}{2^2} + \frac{x_3}{2^3} + \frac{x_4}{2^4} + \cdots$$

with each x_i either 0 or 1. Now, what does the angle-doubling map do to the numerators x_i of x? If $x_1 = 0$, then $x < 1/2$ and $f(x) = 2x$. That is,

$$f(x) = 2\left(\frac{x_1}{2} + \frac{x_2}{2^2} + \frac{x_3}{2^3} + \frac{x_4}{2^4} + \cdots\right)$$

$$= 2\left(\frac{0}{2} + \frac{x_2}{2^2} + \frac{x_3}{2^3} + \frac{x_4}{2^4} + \cdots\right)$$

$$= \frac{x_2}{2} + \frac{x_3}{2^2} + \frac{x_4}{2^3} + \frac{x_5}{2^4} + \cdots$$

What if $x_1 = 1$? The $x \geq 1/2$, so $f(x) = 2x - 1$ and

$$f(x) = 2\left(\frac{x_1}{2} + \frac{x_2}{2^2} + \frac{x_3}{2^3} + \frac{x_4}{2^4} + \cdots\right) - 1$$

$$= 2\left(\frac{1}{2} + \frac{x_2}{2^2} + \frac{x_3}{2^3} + \frac{x_4}{2^4} + \cdots\right) - 1$$

$$= 1 + \frac{x_2}{2} + \frac{x_3}{2^2} + \frac{x_4}{2^3} + \frac{x_5}{2^4} + \cdots - 1$$

$$= \frac{x_2}{2} + \frac{x_3}{2^2} + \frac{x_4}{2^3} + \frac{x_5}{2^4} + \cdots$$

Adopting the *sequence notation*, $x = (x_1 x_2 x_3 x_4 \ldots)$, we see in both cases, $x_1 = 0$ and $x_1 = 1$, that $f(x)$ has the sequence expression $(x_2 x_3 x_4 x_5 \ldots)$. That is, for all sequences $(x_1 x_2 x_3 x_4 \ldots)$,

$$f(x_1 x_2 x_3 x_4 \ldots) = (x_2 x_3 x_4 x_5 \ldots)$$

Not surprisingly, when thinking of its action on sequences, the angle-doubling map sometimes is called the *shift map*.

Now we're ready to prove the angle-doubling map is chaotic.

To prove sensitivity to initial conditions, we can take the sensitivity constant $S = 1/2$. We're given any point $x = (x_1 x_2 x_3 x_4 \ldots)$ and any $\delta > 0$. To find the point y, take any n large enough that $\delta > 1/2^n$. Now take any point $y = (y_1 y_2 y_3 y_4 \ldots)$ that agrees with x on the first n terms and disagrees on the $(n+1)^{\text{st}}$, that is,

$$y_1 = x_1, \; y_2 = x_2, \; \ldots, \; y_n = x_n, \; y_{n+1} \neq x_{n+1}$$

First, we show that x and all these y are close together:

$$d(x,y) = \frac{|x_1 - y_1|}{2} + \frac{|x_2 - y_2|}{2^2} + \cdots + \frac{|x_n - y_n|}{2^n} + \frac{|x_{n+1} - y_{n+1}|}{2^{n+1}} + \cdots$$

$$= \frac{1}{2^{n+1}} + \frac{|x_{n+2} - y_{n+2}|}{2^{n+2}} + \frac{|x_{n+3} - y_{n+3}|}{2^{n+3}} + \cdots$$

$$\leq \frac{1}{2^{n+1}} + \frac{1}{2^{n+2}} + \frac{1}{2^{n+3}} + \cdots$$

$$= \frac{1}{2^n} < \delta$$

A.52. CHAOTIC DYNAMICS ON A JULIA SET

where the last equality comes from summing the geometric series. So all these points y lie within a distance of δ from x.

Next, we show that the nth iterates of x and y are at least $1/2$ apart.

$$\begin{aligned} d(f^n(x), f^n(y)) &= d(f^n(x_1x_2x_3x_4\ldots), f^n(y_1y_2y_3y_4\ldots)) \\ &= d((x_{n+1}x_{n+2}x_{n+3}x_{n+4}\ldots), (y_{n+1}y_{n+2}y_{n+3}y_{n+4}\ldots)) \\ &= \frac{|x_{n+1} - y_{n+1}|}{2} + \frac{|x_{n+2} - y_{n+2}|}{2^2} + \frac{|x_{n+3} - y_{n+3}|}{2^3} \\ &\quad + \frac{|x_{n+4} - y_{n+4}|}{2^4} + \cdots \\ &= \frac{1}{2} + \frac{|x_{n+2} - y_{n+2}|}{2^2} + \frac{|x_{n+3} - y_{n+3}|}{2^3} + \frac{|x_{n+4} - y_{n+4}|}{2^4} \\ &\quad + \cdots \\ &\geq \frac{1}{2} \end{aligned}$$

That is, the points x and y eventually iterate at least a distance S apart. This proves the shift map exhibits sensitivity to initial conditions.

To prove that periodic points are dense, we'll start with any point $x = (x_1x_2x_3x_4\ldots)$ and any $\delta > 0$. Here's one way to build a periodic point y within a distance δ of x. Again, we take any n large enough that $\delta > 1/2^n$. Now define y this way:

$$y_1 = x_1, \; y_2 = x_2, \; \ldots \; y_n = x_n, \; y_{n+1} = y_1, \; y_{n+2} = y_2, \; y_{n+3} = y_3, \; \ldots$$

The computation of $d(x, y)$ in the sensitivity proof can be adapted here to show $d(x, y) < \delta$. And by looking at their sequence representations, $f^n(y) = y$, so y is a periodic point. This shows periodic points are dense.

To prove mixing, given intervals I and J, take any n large enough that the length of I is greater than $1/2^{n-1}$. Take $x = (x_1x_2x_3x_4\ldots)$ to be the midpoint of I, and $y = (y_1y_2y_3y_4\ldots)$ to be any point of J. Now look at the point z having this sequence:

$$z = (x_1x_2x_3x_4\ldots x_ny_1y_2y_3y_4\ldots)$$

One more time, the familiar calculation shows $d(z, x) \leq 1/2^n$. Because x is the midpoint of I, and the length of I is greater than $1/2^{n-1}$, the point z lies in the interval I. From their sequence representations, we see $f^n(z) = y$. That is, a point in I iterates into J.

When we stated the three conditions for chaos, many people would expect proving them – they do seem pretty involved – would be long and tricky. But after the conditions are translated into sequence manipulations, they become straightforward. This general method falls under the heading of *symbolic dynamics*, a very powerful suite of tools for studying iteration. In fact, our binning of data for driven IFS (Sect. 7.1) gives a way to visualize symbolic dynamics.

The next 10 notes refer to Sect. 5.4. **the Mandelbrot set** Here we'll give reasons for some of the properties of the Mandelbrot set. Others require far more math than we use in this book. For those, we'll provide references.

A.53 Why the Mandelbrot set iteration starts with $z_0 = 0$

In A.48 we mentioned the version for quadratic polynomials of a theorem that Fatou and Julia proved for any complex polynomial $f(z)$:

- the Julia set of f is connected if and only if the iterates of all critical points of f remain bounded, and

- the Julia set is a Cantor set if and only if the iterates of all critical points run away to ∞.

Recall a critical point of f is a point where $f'(z) = 0$. For $f(z) = z^2 + c$, $f'(z) = 2z$, a function that has only one zero $z = 0$. Therefore, for every complex number c, the Julia set of $f(z) = z^2 + c$ is either connected or is a Cantor set. There is no middle ground; for example, no Julia set of the Mandelbrot equation consists of exactly two pieces, or three, or a million. One piece or dust, those are the only possibilities. Moreover, a Julia set is connected if and only if the iterates of $z_0 = 0$ remain bounded, and it is a Cantor set if and only if these iterates run away to infinity. Then the Mandelbrot set, the set of all c for which the Julia set of $f(z) = z^2 + c$ is connected, can be drawn by plotting those c for which the iterates of $z_0 = 0$ remain bounded.

A.54 Stability of fixed points and cycles

We'll introduce this idea with real functions in order to use familiar tools from calculus, then extend these results to complex functions, the functions of interest to us for studying Julia sets and the Mandelbrot set.

First, we'll talk about fixed points of a function f. These are points not moved by f; that is, the fixed points are the solutions of the *fixed point equation*

$$f(x) = x$$

Because we're going to be working with them for a while, we need a notation for fixed points. We can't use x, because that's a variable; we can't use $x_0, x_1, x_2, \ldots$ because these denote successive iterates. We'll use x_* to denote a fixed point of a function f. For now, we'll consider only functions with continuous derivatives. If $|f'(x_*)| < 1$, then for all points x near enough to x_*, we'll show that $f^n(x) \to x_*$ as $n \to \infty$.

A.54. STABILITY OF FIXED POINTS AND CYCLES

To see this, first note that $|f'(x_*)| < 1$ means $|f'(x_*)| = \epsilon$ for some $\epsilon < 1$. Because f' is continuous, there is some positive δ for which $|x - x_*| < \delta$ implies $|f'(x)| < \epsilon_1 = (\epsilon + 1)/2 < 1$. Then for all x with $|x - x_*| < \delta$

$$\begin{aligned}|f(x) - x_*| &= |f(x) - f(x_*)| && \text{because } x_* = f(x_*) \\ &= |f'(x')||x - x_*| && \text{by the mean value theorem} \\ &< \epsilon_1 |x - x_*| && \text{because } x' \text{ lies between } x \text{ and } x_*\end{aligned}$$

That is, $f(x)$ is closer to the fixed point x_* than x is.

Next, writing $f^2(x)$ for the composition $f(f(x))$, we see

$$\begin{aligned}|f^2(x) - x_*| &= |f(f(x)) - f(x_*)| && \text{because } f^2(x) = f(f(x)) \\ &&& \text{and } x_* = f(x_*) \\ &= |f'(x'')||f(x) - x_*| && \text{mean value theorem, again} \\ &< \epsilon_1 |f(x) - x_*| && \text{because } x'' \text{ lies between } f(x) \text{ and } x_*\end{aligned}$$

Combining these last two results, we see that

$$|f^2(x) - x_*| < \epsilon_1 |f(x) - x_*| < \epsilon_1^2 |x - x_*|$$

That is, $f^2(x)$ is closer to the fixed point x_* than $f(x)$ is, and so closer still than x is.

Continuing in this way, we find $f^n(x) \to x_*$ as $n \to \infty$. For this reason we say a fixed point x_* is *stable* if $|f'(x_*)| < 1$.

Derivatives can also be used to test the stability of cycles. Suppose $x_1, x_2, \ldots, x_n$ is an n-cycle. That is,

$$x_2 = f(x_1), \ x_3 = f(x_2), \ \ldots, \ x_n = f(x_{n-1}), \ x_1 = f(x_n).$$

Then it's easy to see each of these x_i are fixed points of the function f^n, the composition of f with itself n times. We can test any of these points for stability using the method just described. For example, the point x_1 is stable if

$$\begin{aligned}1 > |(f^n)'(x_1)| && \text{the derivative condition} \\ = |f'(f^{n-1}(x_1))||f'(f^{n-2}(x_1))| \cdots |f'(x_1)| && \text{by the chain rule} \\ = |f'(x_n)||f'(x_{n-1})| \cdots |f'(x_1)| && \text{because } x_n = f^{n-1}(x_1), \text{ etc.}\end{aligned}$$

and from this we see $|(f^n)'(x_i)| = |(f^n)'(x_1)|$ for all x_i of the cycle. This condition,

$$|(f^n)'(x_i)| < 1$$

for any x_i of an n-cycle, guarantees the stability of the cycle.

Translating this to complex variables z, the *cycle stability condition* for an n-cycle $z_1, z_2, \ldots, z_n$ is

$$|(f^n)'(z_i)| < 1$$

for any z_i in the cycle. Let's use this to find where the Mandelbrot set has a stable fixed point and where it has a stable 2-cycle.

A.55 Where the Mandelbrot function has a stable fixed point

The fixed points of $f(z) = z^2 + c$ are the solutions of $f(z) = z$, that is, $z^2 - z + c = 0$. This quadratic equation has two solutions, so we'll call them $z_\pm$, one for the positive square root, one for the negative:

$$z_\pm = \frac{1 \pm \sqrt{1 - 4c}}{2}$$

The derivative of $f(z) = z^2 + c$ is $2z$, so the derivative of f at these fixed points is

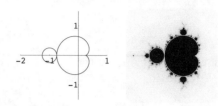

Figure A.39: First: Boundary curves for the stable fixed point and 2-cycle regions. Second: These curves superimposed on the Mandelbrot set.

$$f'(z_\pm) = 1 \pm \sqrt{1 - 4c}$$

To find the c for which $|1 \pm \sqrt{1 - 4c}| < 1$, we must take the square root of the complex number $1 - 4c$. This is not difficult if we rewrite the complex number in polar form. Begin by writing

$$1 - 4c = re^{i\theta} = r\cos(\theta) + ir\sin(\theta)$$

In A.51 we've seen that viewed in its polar form squaring a complex number amounts to squaring the modulus and doubling the argument. Then taking the square root of a complex number is just taking the square root of the modulus and halving the argument. Of course, there are two square roots, positive and negative.

For the polar expression for $1 - 4c$, we take θ to lie in the range $-\pi \leq \theta < \pi$, because then

$$\sqrt{1 - 4c} = \sqrt{r}(\cos(\theta/2) + i\sin(\theta/2))$$

has positive real part ($-\pi \leq \theta < \pi$ implies $\cos(\theta/2) \geq 0$ and is 0 only at $\theta/2 = -\pi/2$), and $-\sqrt{1 - 4c}$ has negative real part.

Then

$$\begin{aligned} |f'(z_+)| &= |1 + \sqrt{1 - 4c}| \\ &= |1 + \sqrt{r}\cos(\theta/2) + i\sqrt{r}\sin(\theta/2)| \\ &= \sqrt{(1 + \sqrt{r}\cos(\theta/2))^2 + (\sqrt{r}\sin(\theta/2))^2} \\ &= \sqrt{1 + 2\sqrt{r}\cos(\theta/2) + r} \end{aligned}$$

A.56. STABLE 2-CYCLE

With this expression, we'll show that z_+ is not a stable fixed point. Because $-\pi/2 \leq \theta/2 \leq \pi/2$, we see that $\cos(\theta/2) \geq 0$ and so

$$|f'(z_+)| = \sqrt{1 + 2\sqrt{r}\cos(\theta/2) + r} \geq \sqrt{1+r} > 1$$

That is, the fixed point z_+ is not stable for any c. For the other fixed point, z_-, the analogous calculation gives

$$|f'(z_-)| = \sqrt{1 - 2\sqrt{r}\cos(\theta/2) + r}$$

This fixed point is stable for those c for which $|f'(z_-)| < 1$. If $|f'(z_-)| > 1$, the fixed point is unstable. The boundary of the parameter plane region where this fixed point is stable occurs when $|f'(z_-)| = 1$. Squaring both sides gives

$$1 - 2\sqrt{r}\cos(\theta/2) + r = 1$$

Solving for r gives

$$r = 4\cos^2(\theta/2) = 2(1 + \cos(\theta))$$

This is the polar coordinate description of a cardioid, the shape of a valentine heart, not the heart in your chest. But this is not quite the cardioid we want. Remember, we want the c values, and so far we have found a polar description for $1 - 4c$. Subtracting 1 and multiplying by $-1/4$ gives the cardioid shown in the first image of Fig. A.39. For all points c inside this cardioid, the Mandelbrot function has a stable fixed point z_-. In the second image we see the cardioid superimposed on the Mandelbrot set.

A.56 Where the Mandelbrot set has a stable 2-cycle

The 2-cycle equation is $f^2(z) = f(f(z)) = z$; that is,

$$(z^2 + c)^2 + c = z$$

This simplifies to

$$z^4 + 2cz^2 - z + c^2 + c = 0$$

You needn't worry that we're going to haul out the quartic formula, because there is a clever trick to reduce this to a quadratic. We know the fixed points are the solutions of $f(z) = z$; that is, $z^2 - z + c = 0$. And we know that a fixed point necessarily belongs to a 2-cycle: because $f(z_\pm) = z_\pm$, we certainly have

$$f^2(z_\pm) = f(f(z_\pm)) = f(z_\pm) = z_\pm$$

Knowing this, we can find the factorization of $z^4 + 2cz^2 - z + c^2 + c$ by dividing $z^2 - z + c$ into $z^4 + 2cz^2 - z + c^2 + c$, obtaining

$$z^4 + 2cz^2 - z + c^2 + c = (z^2 - z + c)(z^2 + z + (c+1))$$

This shows the 2-cycle points are the solutions of $z^2 + z + (c+1) = 0$; that is,

$$z_\pm = \frac{-1 \pm \sqrt{-3 - 4c}}{2}$$

Finding the c for which the 2-cycle is stable turns out to be algebraically easier than finding where the fixed point is stable. Here we won't need to take square roots of complex numbers. Using the cycle stability condition, we see that the boundary of the 2-cycle region is given by

$$\begin{aligned} 1 = |(f^2)'(z_+)| &= |f'(f(z_+))f'(z_+)| = |2z_- 2z_+| \\ &= |(-1 - \sqrt{-3-4c})(-1 + \sqrt{-3-4c})| \\ &= |4 + 4c| \\ &= 4|1 + c| \end{aligned}$$

So the 2-cycle is stable for all c inside the disc of radius $1/4$ centered at $c = -1$, which you can see in Fig. A.39.

A.57 Counting discs and cardioids

For every disc and cardioid component of the Mandelbrot set there is a corresponding stable cycle. For each point c in each component, the cycle stability condition shows that the derivative evaluated on the cycle has modulus < 1. The derivative is 0 at exactly one c, called the *center* of the component. For the main cardioid, the center is the solution of

$$0 = f'(z_-) = 1 - \sqrt{1 - 4c}$$

that is, $c = 0$. For the 2-cycle disc, the center is the solution of

$$0 = f'(z_+)f'(z_-) = 4 + 4c$$

so $c = -1$. We can find the centers one by one in this way, but if we look a bit differently at the problem of locating them, we find a simpler way. Recalling that $f'(z) = 2z$, we see that for any n-cycle $z_1, z_2, \ldots, z_n$,

$$(f^n)'(z_1) = 2z_n 2z_{n-1} \cdots 2z_2 2z_1$$

Consequently, in order for the derivative to be 0, one of the cycle points z_i must itself be 0, and so 0 must be a fixed point of $f^n(z)$. In other words, the centers of the n-cycle components are exactly the solutions of

$$0 = f^n(0) = f^{n-1}(c)$$

A.57. COUNTING DISCS AND CARDIOIDS

For the last equality we have used $f^n(0) = f^{n-1}(f(0))$ and $f(0) = c$. Each solutiuon corresponds to the center of its own n-cycle component, so we call $0 = f^{n-1}(c)$ the *center-locating equation*. Are there any other components, areas of the Mandelbrot set with positive areas, that do not correspond to a stable cycle? That, it turns out, is a very hard question. The *hyperbolicity conjecture* is that there are none, and as of this writing, no one knows the answer.

Applying the center-locating equation carefully, we can find the centers of all the disc and cardioid components of the Mandelbrot set. To illustrate this, we'll find the centers of the 3-cycle components. These are the solutions of $0 = f^2(c) = c^4 + 2c^3 + c^2 + c$. While exact solutions can be found by factoring out c and applying the cubic formula, three of the four solutions are a mess. We'll settle for numerical approximations:

$$0, -1.75488, -.122561 \pm .744862i$$

The point $c = 0$ is the center of the Main cardioid, the 1-cycle component. Because a 1-cycle is necessarily a 3-cycle, it is no surprise that when applied to 3-cycle components, the center-locating equation also detects the center of the 1-cycle component. The points $c = -.122561 \pm .744862i$ are the centers of the 3-cycle discs of the principal sequence. We'll see these are the only 3-cycle discs, so $c = -1.75488$ is the center of the 3-cycle cardioid.

Locating the centers involves solving polynomial equations. Except for the first few, centers can be found only approximately, by numerical, not algebraic, methods. Much easier is to count the centers, thus counting the number of cycle components of the Mandelbrot set. Two items require our attention before we can start counting.

First, we see that if n is a multiple of m, this method of counting n-cycle components will count m-cycle components as well. A 2-cycle is automatically a 4-cycle, a 6-cycle, an 8-cycle, and an n-cycle for all even numbers n. Two paragraphs earlier we saw an illustration of this for 3-cycles and 1-cycles.

To see the second issue, which is a bit more subtle, let's look at the first few functions that locate the centers.

$f^0(c) = c$
$f^1(c) = c^2 + c$
$f^2(c) = (c^2 + c)^2 + c = c^4 + 2c^3 + c^2 + c$
$f^3(c) = ((c^2 + c)^2 + c)^2 + c = c^8 + 4c^7 + 6c^6 + 6c^5 + 5c^4 + 2c^3 + c^2 + c$
$\ldots$

In general, the polynomial $f^{n-1}(c)$ has order 2^n, that is, the highest power of c in $f^{n-1}(c)$ is c^{2^n}. Counting Mandelbrot set components requires counting the roots of these polynomials, so how do we know that each polynomial has distinct roots? After all, some polynomials, $c^2 + 2c + 1 = (c+1)^2$ for example, have repeated roots. The answer involves a useful fact from algebra: for any polynomial $p(z)$, $z = a$ is a repeated root if and only if $p'(a) = 0$. (This is pretty easy to prove, using the fact that if $z = a$ is a root of $p(z)$, then for some positive integer m, $p(z) = (z-a)^m q(z)$, where $z = a$ is not a root of the polynomial $q(z)$.) Using this, and a bit more work than we want to do here, we can show that the polynomials $f^n(c)$ have no repeated roots.

Now we can count the cycle components in the Mandelbrot set. We'll go up through 7-cycle components, but this process can continue through any cycle number we wish.

- Fixed point: $f^0(c) = c$ has only one root, so the Mandelbrot set has only one fixed point component.

- 2-cycle: $f^1(c) = c^2 + c$ has two roots, one for the fixed point component, leaving one 2-cycle component.

- 3-cycle: $f^2(c)$ has four roots, one for the fixed point component, leaving three 3-cycle components.

- 4-cycle: $f^3(c)$ has eight roots, one for the fixed point component and one for the 2-cycle component, leaving six 4-cycle components.

- 5-cycle: $f^4(c)$ has 16 roots, one for the fixed point component, and because 5 is prime, each of the other 15 roots correspond to distinct 5-cycle components.

- 6-cycle: $f^5(c)$ has 32 roots, one for the fixed point component, one for the 2-cycle component, and three for the 3-cycle components, leaving 27 6-cycle components.

- 7-cycle: $f^6(c)$ has 64 roots, one for the fixed point component, and because 7 is prime, each of the other 63 roots correspond to 7-cycle components.

Let's think a bit more about the 7-cycle components. By the multiplier rule, another consequence of 7 being prime is that the only 7-cycle disc components are those attached to the Main cardioid, because the period of a disc component attached to any other component must be a multiple of the period of that component. There is one 7-cycle component in each of the upper and lower principal sequence, one in the upper and lower Farey sequences between the principal 3 and 4, and one in the upper and lower Farey sequence between the principal 2 and the Farey 5. Altogether, there are six 7-cycle discs attached to the main cardioid. These are the only 7-cycle discs, leaving 57 other distinct 7-cycle components, which must be cardioids.

A.58 Why the fixed point at infinity is stable

Recall that in A.49 in order to show that the point at infinity is fixed for $z_{n+1} = f(z_n) = z_n^2 + c$, we changed variables to $w_n = h(z_n) = 1/z_n$. Then the iteration $z_{n+1} = f(z_n)$ becomes $w_{n+1} = w_n^2/(1 + cw_n^2) = g(w_n)$, and $z = \infty$ corresponds to $w = 0$. To test the stability of this fixed point, we evaluate the derivative

$$\frac{dg}{dw} = \frac{2w}{(1 + cw^2)^2}$$

Plugging in $w = 0$, we obtain $dg/dw|_{w=0} = 0$. This shows the fixed point at infinity is stable.

A.59 The angle-doubling map in Lavaurs' algorithm

In Lavaurs' algorithm, recall that we identify points on the unit circle with numbers x, $0 \leq x < 1$, with each x representing the fraction of the circumference a point on the circle is from $(1, 0)$. For example, the point $(0, 1)$ corresponds to $x = 1/4$. Recall the angle-doubling map from Sect. 5.3, reformulated here as

$$f_2(x) = \begin{cases} 2x & \text{if } x < 1/2 \\ 2x - 1 & \text{if } x \geq 1/2 \end{cases}$$

The reason for the subscript 2 will appear in A.60. Another formulation is

$$f_2(x) = 2x \pmod 1$$

where for any real number y, by $y \pmod 1$ we mean y minus its integer part. For example,

$$1/2 \pmod 1 = 1/2 - 0 = 1/2$$
$$5/3 \pmod 1 = 5/3 - 1 = 2/3$$

and so on.

As promised, we'll show that fractions with denominators $2^k - 1$ belong to k-cycles for the angle-doubling map f_2. In A.57 we saw that points belonging to a k-cycle for a function f are fixed points of f^k, the composition of f with itself k times. So every point x of a k-cycle of f_2 is a solution of

$$f_2^k(x) = x$$

To understand $f_2^2(x) = f_2(f_2(x))$, consider the cases

$$f_2(f_2(x)) = \begin{cases} 2f_2(x) & 0 \leq x < 1/2 \\ 2f_2(x) - 1 & 1/2 \leq x < 1 \end{cases}$$

$$= \begin{cases} 2(2x) & 0 \leq x < 1/4 \\ 2(2x) - 1 & 1/4 \leq x < 1/2 \\ 2(2x - 1) & 1/2 \leq x < 3/4 \\ 2(2x - 1) - 1 & 3/4 \leq x < 1 \end{cases}$$

That is, $f_2^2(x) = 2^2 x - b$, where $b = 0, 1, 2$, or 3. In other words, $f_2^2(x) = 2^2 x \pmod 1$, and in general,

$$f_2^k(x) = 2^k x \pmod 1$$

Then the k-cycle equation $f_2^k(x) = x$ is just

$$x = 2^k x \pmod 1, \text{ or } x = 2^k x - b$$

where $b = 0, 1, 2, \ldots, 2^k - 1$. Solving this last equation for x, we see x is a fraction with denominator $2^k - 1$, as promised.

Comparing the component counts by Lavaurs' method and the center-locating polynomials of A.57, we see they differ by 1, because Lavaurs' method does not count the fixed point component.

A.60 A proof of Fermat's Little Theorem

We can use some relatives of the angle-doubling map to give a very simple proof of a theorem of Fermat. Not the big one, Fermat's Last Theorem, but instead Fermat's Little Theorem.

We use these relatives of the angle-doubling map:

$$f_n(x) = nx \pmod 1$$

These functions all have range $[0, 1)$. Later bookkeeping is simpler if we extend all these functions by defining

$$f_n(1) = 1$$

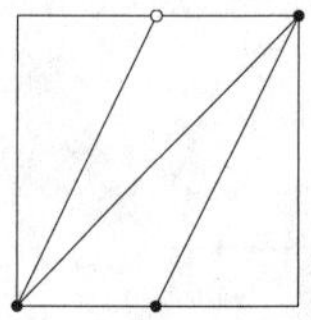

Figure A.40: The graph of f_2.

for all n.

The graph of f_n consists of n straight line segments, each of slope n. For example, in Fig. A.40 we see the graph of $y = f_2(x)$. Also in this graph we have drawn the diagonal line $y = x$. The points of intersection of these graphs are those points where $y = f_2(x)$ and $y = x$. Consequently these points satisfy $f_2(x) = x$: they are the fixed points of f_2. That is,

A.60. A PROOF OF FERMAT'S LITTLE THEOREM

the fixed points of f are the points of intersection of the graphs $y = f(x)$ and $y = x$.

From near the end of A.59 recall that points of an k-cycle for a function f are fixed points for f^k, the composition of f with itself k times. So we recognize k-cycle points visually as the intersections of the graphs $y = f^k(x)$ and $y = x$.

We recall a relation between cycles and fixed points, mentioned earlier. Note that a fixed point x_0 for f also is a fixed point of f^2:

$$f^2(x_0) = f(f(x_0)) = f(x_0) = x_0$$

so the fixed points of f^2 are either the fixed points of f or the 2-cycle points of f. Put another way, to count the number of 2-cycles of f, first count the number of fixed points of f^2, then subtract the number of fixed points of f, and finally divide by 2, because each 2-cycle contains two points.

To count the number of 4-cycles of f, count the fixed points of f^4, subtract the number of fixed points of f, then subtract the number of 2-cycle points of f (because every point of a 2-cycle is a fixed point for f^4), and then divide by 4.

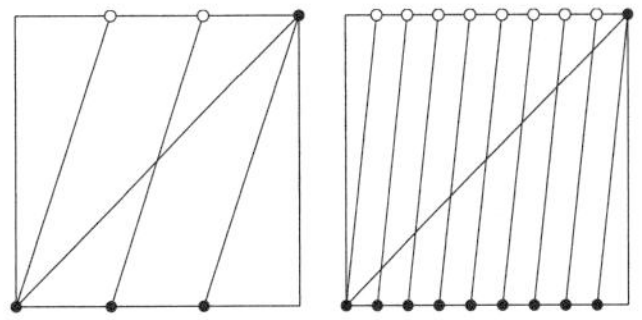

In general, enumerating the k-cycles of a function requires subtracting the number of points in every j-cycle, where j divides k without remainder. The special case of this that we'll use to prove Fermat's Little Theorem is this: *if k is a prime number, we need to subtract only the number of fixed points of f.*

Figure A.41: Graphs of f_3 and f_3^2.

In Fig. A.41 we see the graphs of f_3 and f_3^2. These illustrate the two facts we need to prove Fermat's Little Theorem:

1. f_n has exactly n fixed points, and

2. f_n^k has exactly n^k fixed points.

The second item follows from the fact that the graph of f_n^k consists of n^k straight line segments, each of slope n^k.

Now suppose $k = p$, a prime number. Then f_n^p has n^p fixed points, and because p is prime, these fixed points of f_n^p must either be fixed points of f_n or belong to a p-cycle of f. Because f_n has n fixed points, the remaining $n^p - n$ points all must belong to p-cycles. Each p-cycle consists of p points, and we can't have fractions of cycles, nor can cycles overlap: either they have no points in common or they are identical.

Consequently, for every prime number p and for every positive integer n,

$$p \text{ divides } n^p - n$$

This is Fermat's Little Theorem, a result from number theory, proved here just by counting the cycles of relatives of the angle-doubling map. Were you expecting to see a proof of a result from number theory in a book about fractals? Probably not. This is a surprising field.

A.61 Why every Misiurewicz point c belongs to the Julia set J_c

In Sect. 5.4 we defined a Misiurewicz point as a point c for which $z_0 = 0$ iterates under $z_{n+1} = z_n^2 + c$ to a cycle that does not include $z_0 = 0$. In terms of the family of polynomials introduced in A.57, c is a Misiurewicz point if for some $m > 0$ and some $n \geq 0$,

$$f^m(0) = f^{m+n}(0)$$

Because $m > 0$ and $f(0) = c$, this gives $f^{m-1}(c) = f^{m-1+n}(c)$. Consequently, the iterates of $z_0 = c$ do not escape to infinity, and so we know that c belongs to K_c. We'll need to do some more work to show that the point c belongs to J_c.

For each Misiurewicz point c, the cycle to which the iterates of 0 converge is unstable. The general proof is more involved than we want to go into here, but here are some examples of unstable cycles to which the iterates of 0 converge.

- $c = -2$: the first few iterates are 0, -2, 2, 2, 2; all later iterates are stuck at 2, a fixed point. The derivative at $z = 2$ is 4, so this fixed point is unstable.

- $c = i$: the first few iterates are 0, i, $-1+i$, $-i$, $-1+i$, $-i$, $-1+i$; all later iterates alternate between $-i$ and $-1+i$, the points of a 2-cycle. The derivative at this 2-cycle is $2(-1+i)2(-i) = 4(1+i)$, so the 2-cycle is unstable.

If there were a stable cycle for a Misiurewicz point c, the iterates of 0 would converge to that cycle, but since the iterates of 0 converge to an unstable cycle, this value of c has no stable cycle. The solid parts of the filled-in Julia set are the z_0 that converge to a stable cycle, so if there is no stable cycle, K_c has no solid parts. Without solid parts, K_c is all boundary, and so $K_c = J_c$.

A.62 The magnification factor in Tan Lei's theorem

The Julia set of Fig. 5.13 is for $c = -0.67 + 0.458i$. The Julia sets for Fig. 5.14 are for $c = 0.2552444 + 0.00061i$ and $c = 0.2714 + 0.00513i$. These illustrate Tan Lei's theorem.

Suppose c is a Misiurewicz point with the iterates of $z = 0$ converging to a k cycle $z_m, z_{m+1}, \ldots, z_{m+k-1}$. Then each point of this k-cycle is a fixed point of f^k, and the derivative

$$(f^k)'(z_m) = f'(f^{k-1}(z_m)) \cdot f'(f^{k-2}(z_m)) \cdots f'(z_m)$$
$$= f'(z_{m+k-1}) \cdot f'(z_{m+k-2}) \cdots f'(z_m)$$
$$= 2z_{m+k-1} \cdot 2z_{m+k-2} \cdots 2z_m$$

Call this derivative ρ. Because this cycle is unstable, $\rho > 1$. Tan Lei proved there is a number λ_c for which magnifying J_c about c by ρ^n, and magnifying the Mandelbrot set about c by $\lambda_c \rho^n$, makes the parts of both sets in a small circle centered at c look more and more similar with increasing n.

Tan Lei's theorem is why little bits of so many Julia sets look like little bits of the Mandelbrot set. It's also an important ingredient in Shishikura's proof that the Mandelbrot set boundary has dimension 2.

The next three notes refer to Sect. 5.5, **Other Mandelbrot sets**. First, we give the locations of the magnifications of Fig. 5.15. Then we address two technical issues about the Mandelbrot set for functions other than $f(z) = z^2 + c$. For $f(z) = z^n + c$, we establish the analog of the familiar escape criterion for $z^2 + c$. And we find the critical points for other complex functions.

A.63 Locations of the magnifications of Fig. 5.15 – Fig. 5.17

The magnifications of Fig. 5.15 have lower left corner c and image width δ, for $c = 0.351 + 0.633i$ and $\delta = 0.023$ ($n = 2$), $c = -0.293 + 1.228i$ and $\delta = 0.06$ ($n = 3$), $c = 0.305 + 1.036i$ and $\delta = 0.08$ ($n = 4$), and $c = -0.9 + 0.73i$ and $\delta = 0.06$ ($n = 5$). The magnification constituting the fourth image of Fig. 5.16 has lower left corner $c = -0.95 + 1.74i$ and width $\delta = 0.04$. The Julia set on the left of Fig. 5.17 has $c = -0.236 + 1.734i$. The magnification on the right has lower left corner $c = -0.45 - 0.65i$ and width $\delta = 1$.

A.64 The escape criterion for $z_{k+1} = z_k^n + c$

Much of the argument is similar to our derivation of the escape criterion for $z^2 + c$ given in A.46. There we began with $|z_k| > \max\{2, |c|\}$.

Here, we begin by supposing $|z_k| > \max\{2^{1/(n-1)}, |c|\}$. Then

$$|z_k| > 2^{1/(n-1)}$$

and by raising both sides to the $(n-1)$st power,

$$|z_k|^{n-1} > (2^{1/(n-1)})^{n-1} = 2$$

so $|z_k|^{n-1} = 2 + \epsilon$ for some $\epsilon > 0$.

As with the escape criterion for $z^2 + c$, we find a lower bound for $|z_k^n + c|$ by using the triangle inequality.

$$\begin{aligned} |z_k^n| &= |z_k^n + c - c| \\ &\leq |z_k^n + c| + |-c| \qquad \text{by the triangle inequality} \\ &= |z_k^n + c| + |c| \end{aligned}$$

Subtracting $|c|$ from both sides and rearranging gives

$$\begin{aligned} |z_k^n + c| &\geq |z_k^n| - |c| \\ &\geq |z_k^n| - |z_k| \qquad \text{because } |z_k| \geq |c| \\ &= |z_k|^n - |z_k| \\ &= (|z_k|^{n-1} - 1)|z_k| \\ &= (1 + \epsilon)|z_k| \qquad \text{because } |z_k|^{n-1} = 2 + \epsilon \end{aligned}$$

Since $z_{k+1} = z_k^n + c$, we have shown

$$|z_{k+1}| \geq (1 + \epsilon)|z_k|$$

Iterating, we obtain

$$|z_{k+m}| \geq (1 + \epsilon)^m |z_k|$$

Consequently, if $|z_k| > \max\{2^{1/(n-1)}, |c|\}$, then $|z_{k+m}| \to \infty$ as $m \to \infty$.

To complete the proof of the general escape criterion, that $|z_k| > 2^{1/(n-1)}$ implies that later iterates run to infinity, we must consider two cases: $|c| \leq 2^{1/(n-1)}$ and $|c| > 2^{1/(n-1)}$. Again, the derivation of the escape criterion for $z^2 + c$ is our guide.

In the first case, $|z_k| > 2^{1/(n-1)}$ implies $|z_k| > \max\{2^{1/(n-1)}, |c|\}$, so the argument above shows that later iterates run to infinity.

In the second case, $|c| > 2^{1/(n-1)}$ implies $|c|^{n-1} > 2$. Then, $z_1 = c$ and $z_2 = c^n + c$, so

$$|z_2| = |c^n + c| = |c||c^{n-1} - 1| \geq |c|(|c^{n-1}| - 1) > |c| = \max\{2^{1/(n-1)}, |c|\}$$

so again, the argument above shows that later iterates run to infinity.

A.65 Critical points of other functions

For $f(z) = z^n + c$, the derivative is $df/dz = nz^{n-1}$, so the only critical point is $z_0 = 0$.

For $g(z) = z^3/3 + z^2/2 + c$, the derivative is $dg/dz = z^2 + z$ and so the critical points are the solutions of

$$0 = z^2 - z = z(z - 1)$$

That is, $z_a = 0$ and $z_b = -1$.

The next five notes refer to Sect. 5.6, **Universality of the Mandelbrot set**. Here we derive the formula for Newton's method and present Cayley's analysis of the basins of attraction for the roots of $f(z) = z^2 - 1$. Next, we cite the solution of the corresponding problem for $z^3 - 1$ and illustrate the *Lakes of Wada* construction, the original example of the. We end the notes on this section with a derivation of the method for finding stable cycles for Newton's method.

A.66 The formula for Newton's method

The geometry of Newton's method is straightforward. Because he had invented calculus, Newton was able to replace the geometry with an algebraic formula, easy to iterate. Here's how.

The slope of the tangent line to the graph of $y = f(x)$ at the point $(x_0, f(x_0))$ is the derivative $f'(x_0)$; the slope of the line through the points $(x_0, f(x_0))$ and $(x_1, 0)$ is $(f(x_0) - 0)/(x_0 - x_1)$. Equating these two expressions for the slope and solving for x_1 gives

$$x_1 = x_0 - \frac{f(x_0)}{f'(x_0)}$$

So, given an initial guess x_0, the sequence $x_1, x_2, x_3, \ldots$ is generated by iterating the *Newton's method formula*

$$x_{n+1} = N_f(x_n) = x_n - \frac{f(x_n)}{f'(x_n)}$$

So long as no $f'(x_n) = 0$ – that is, so long as the tangent line to the graph of f at the point $(x_n, f(x_n))$ is not horizontal – we can continue generating more points, until they converge to a root, ... or do something different and more interesting.

A.67 Newton's method for $f(z) = z^2 - 1$

British mathematician Arthur Cayley studied the *basins of attraction* for Newton's method, that is, the areas from which the iterates converge to each of the roots. Cayley started with the simple function $f(z) = z^2 - 1$, for which the Newton function is

$$N_f(z) = z - \frac{f(z)}{f'(z)} = z - \frac{z^2 - 1}{2z} = \frac{z^2 + 1}{2z}$$

The roots of $f(z) = 0$ are $z = 1$ and $z = -1$, but figuring out what $N_f(z)$ does is not straightforward. To make the behavior of iterates easy to understand, Cayley's clever idea was to change the variables from z to $u = T(z)$, where

$$u = T(z) = \frac{z-1}{z+1}, \quad \text{so} \quad z = T^{-1}(u) = \frac{1+u}{1-u}.$$

Then

- $z = 1$ corresponds to $u = 0$,
- $z = -1$ corresponds to $u = \infty$, and
- the imaginary axis $z = iy$ corresponds to the unit circle $|u| = 1$.

To see this last point, recall that for any complex number $u = v + iw$, the *conjugate* is $\bar{u} = v - iw$ and the modulus $|u| = \sqrt{v^2 + w^2} = \sqrt{u\bar{u}}$. Then $1 = |u|$ is equivalent to

$$1 = |u|^2 = u\bar{u} = \frac{z-1}{z+1}\frac{\bar{z}-1}{\bar{z}+1}$$

That is,

$$(z+1)(\bar{z}+1) = (z-1)(\bar{z}-1)$$

This simplifies to $\bar{z} = -z$. Writing $z = x + iy$ so $\bar{z} = x - iy$, we see that the condition $\bar{z} = -z$ gives $x = -x$ and $-y = -y$, so $x = 0$. In other words, the unit circle in u coordinates corresponds to the imaginary axis in z coordinates.

The final step of Cayley's argument is to find an expression for $N_f(z)$ in terms of the variable u. To do this, we send u to z using T^{-1}, then apply N_f to z, and finally send z back into u using T. We'll call this function $R(u)$. Then we can find an expression for $R(u)$:

$$R(u) = T N_f T^{-1}(u)$$

$$= T N_f \left(\frac{1+u}{1-u}\right) \quad \text{using the formula for } T^{-1}(u)$$

$$= T \left(\frac{1+u^2}{1-u^2}\right) \quad \text{using the formula for } N_f, \text{ and more algebra}$$

$$= u^2 \quad \text{using the formula for } T, \text{ and another bit of algebra}$$

Since we know that points z with positive real parts translate to points u *inside* the unit circle, iterating R sends the points, through successive squaring, to 0. This corresponds to z going to 1. Similarly, any points z with negative real parts translate to u *outside* the unit circle. Iterating R sends these points to ∞, which corresponds to z going to -1.

Furthermore, points z on the imaginary axis correspond to points u on the unit circle. Iterating R on the unit circle keeps points on the unit circle, so iterating $N_f(z)$ on the imaginary axis keeps points on the

A.68 The roots of $z^3 - 1$

The three lower images of Fig. 5.20 are the basins of attraction for the roots $1, -0.4 \pm 0.866i$ (first), $1, -0.6 \pm 0.866i$ (second), and $1, -0.8 \pm 0.866i$ (third).

A.69 Lakes of Wada

In a dark ocean we find a white island with two lakes, one lighter and one darker. The first goal is to dig a canal so every location on the island is within 1 mile of the ocean, or at least within 1 mile of the canal leading to the ocean (first picture of Fig. A.42). Next, we dig a canal so every location on the island is within 1/2 mile of the lighter lake or the canal leading to the lighter lake (second picture). Next, we dig a canal so every location on the island is within 1/4 mile of the darker lake or the canal leading to the darker lake. At this point, the people on (what's left of) the island who want to get to the ocean complain, and dig another canal putting every location on the rapidly disappearing island within 1/8 of a mile of water leading to the ocean. Continue. What remains of the island – mathematical points only, nothing as substantial as atoms – is the common boundary of the ocean and the two lakes.

This old topological construction was published in 1917 by Kunizo Yoneyama, who attributed it to, and named it for, Takeo Wada, his teacher.

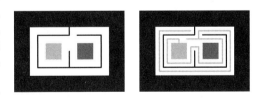

Figure A.42: Two steps in constructing the Lakes of Wada.

Here's why this is a surprising, maybe seemingly impossible, shape. Suppose you want to paint the ceiling of your bedroom blue, three walls tan, and one wall green. Except for answering some questions about your color sense, this won't be a problem. Three edges of the ceiling will separate tan from blue, one edge will separate green from blue, and the edges of two walls will separate tan from green. Only at two corners will all three colors meet. But suppose you want every point of every edge that shares two colors to share all three. How to do this isn't so clear. In fact, it might seem to be impossible. In the Wada construction, take the ocean to be the three tan walls, one lake to be the green wall, and the other lake to be the blue ceiling. Do you feel compelled to get a collection of ever-finer paint brushes and try to paint your bedroom this way?

A.70 Stable cycles and Newton's method

The paper of Curry, Garnett, and Sullivan is the first example we know where the Mandelbrot set appeared in a context other than the iteration of polynomials. Curry, Garnett, and Sullivan were looking for significant ways that Newton's method can fail, and found a constellation of Mandelbrot sets.

First note that the roots of f are fixed points of N_f. If z_* is a root of f, then $f(z_*) = 0$ and so

$$N_f(z_*) = z_* - \frac{f(z_*)}{f'(z_*)} = z_* - \frac{0}{f'(z_*)} = z_*$$

This is provided, of course, that $f'(z_*) \neq 0$, a case which must be treated with a bit more care than we pursue here.

If Newton's method exhibits a stable cycle, then the theorem of Fatou and Julia implies that some critical point of the Newton function N_f must converge to that cycle. To find the critical points of N_f, compute the derivative of $N_f(z)$.

$$\frac{d}{dz} N_f(z) = 1 - \frac{f'f' - ff''}{(f')^2}$$
$$= \frac{(f')^2}{(f')^2} - \frac{(f')^2 - ff''}{(f')^2}$$
$$= \frac{ff''}{(f')^2}$$

Then the critical points of N_f are the roots of f, of f', and of f''. However, we can rule out two options. We know that the Newton function N_f is not defined at the points where $f' = 0$, and the roots of f are fixed points of N_f. So if there is a stable cycle, the only critical points it could attract are the roots of f''.

For $f(z) = z^3 + (c-1)z - c$, the second derivative $f''(z) = 6z$. Therefore, the only critical point of f'' is $z = 0$. So to find attracting cycles for these polynomials, scan across the c plane and check if the iterates of N_f starting from $z_0 = 0$ converge to some cycle. This is how Curry, Garnett, and Sullivan constructed their remarkable picture, a portion of which we see on the right.

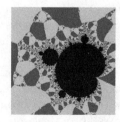

The three images of Fig. 5.23 have lower left corners c and widths δ, with $c = -2 - 2i$, $\delta = 4$ (first), $c = -0.08 + 1.46i$, $\delta = 0.48$ (second), and $c = 0.228 + 1.58i$, $\delta = 0.072$ (third).

The next four notes refer to Sect. 5.7, **The Mandelbrot set in 4 dimensions**. First, we show how a simple linear change of variables can remove the next-to-highest-order term of a polynomial. This is why for quadratic iteration we use $z^2 + c$, and for cubic iteration, $z^3 + az + b$.

Next, because the Mandelbrot set for $z^3 + az + b$ lives in complex 2-dimensional space, which is geometrically equivalent to real 4-dimensional space, we develop some familiarity with 4-dimensional space. Some people think that understanding 4-dimensional geometry requires challenging intellectual gymnastics. Certainly, some problems are very difficult, but gaining a basic understanding of 4-dimensional geometry is straightforward. To demonstrate this, we'll find how many cubes make up the boundary of a hypercube. After that, we sketch the Sierpinski hypertetrahedron, the 4-dimensional analog of the Sierpinski tetrahedron.

A.71 Removing the next-to-highest power

For the polynomial $a_2 z^2 + a_1 z + a_0$, we can begin to manipulate the polynomial, with the goal of removing the $a_1 z$ term, by changing the variable z to w, defined by $z = w - a_1/(2a_2)$. This gives

$$a_2 z^2 + a_1 z + a_0 = a_2(w - a_1/(2a_2))^2 + a_1(w - a_1/(2a_2)) + a_0$$
$$= a_2 w^2 + a_0 - a_1^2/(4a_2)$$

To translate the iteration formula

$$z_{k+1} = a_2 z_k^2 + a_1 z_k + a_0$$

in terms of w_k and w_{k+1}, we will substitute again, this time with $z_k = w_k - a_1/(2a_2)$ and $z_{k+1} = w_{k+1} - a_1/(2a_2)$. Now the iteration becomes

$$w_{k+1} - a_1/(2a_2) = a_2(w_k - a_1/(2a_2))^2 + a_1(w_k - a_1/(2a_2)) + a_0$$
$$= a_2 w_k^2 + a_0 - a_1^2/(4a_2)$$

We see the w_k term has disappeared from the right side of the equation. Isolating w_{k+1} on the left side, we obtain

$$w_{k+1} = a_2 w_k^2 + a_0 + (2a_1 - a_1^2)/(2a_2)$$

This is almost what we want. The only problem is that the coefficient of w_k^2 is a_2, but we want the coefficient to be 1. We achieve this with a final pair of substitutions, $u_k = a_2 w_k$ and $u_{k+1} = a_2 w_{k+1}$. The iteration becomes

$$u_{k+1} = u_k^2 + a_2(a_0 + (2a_1 - a_1^2)/(2a_2))$$
$$= u_k^2 + (a_2 a_0 + a_1 - a_1^2/2)$$

This argument can be generalized to polynomials with any exponent, n, specifically, by substituting

$$z_{k+1} = w_{k+1} - \frac{a_{n-1}}{na_n} \quad \text{and} \quad z_k = w_k - \frac{a_{n-1}}{na_n}$$

the iteration

$$z_{k+1} = a_n z_k^n + a_{n-1} z_k^{n-1} + a_{n-2} z_k^{n-2} + \cdots + a_1 z_k + a_0$$

is transformed into

$$w_{k+1} = b_n w_k^n + b_{n-2} w_k^{n-2} + \cdots + b_1 w_k + b_0$$

where the coefficients b_n, b_{n-2}, ..., b_1, and b_0 are determined by the coefficients a_n, a_{n-1}, ..., a_1, and a_0. One more step changes the coefficient b_n into 1.

A.72 Cubes in the hypercube boundary

We build from more familiar dimensions to less familiar. And what is more familiar than the unit cube? To be specific, by the unit cube we mean the set of all the points (x, y, z) for which $0 \leq x \leq 1$, $0 \leq y \leq 1$, and $0 \leq z \leq 1$. The edges of this cube, drawn in perspective, are shown in each of the six drawings of Fig. A.43.

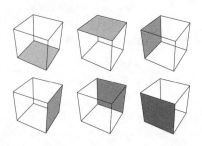

Figure A.43: The six square faces of a cube.

The boundary of this cube consists of six squares – if you're indoors, look at the surrounding walls, floor, and ceiling. Each square is the part of the cube where one of the three variables is set to one of its two extreme values, 0 or 1.

A similar argument can show that the unit hypercube consists of all points (x, y, z, w) with $0 \leq x, y, z, w \leq 1$. The 24 edges of this hypercube, drawn in perspective, are shown in the upper left image of Fig. A.44. The boundary of the hypercube is obtained by setting the four variables, one at a time, to 0 or 1. This gives the eight cubes shown in Fig. A.44.

Now if you're thinking, "Wait a minute. Some of these don't look like cubes," we have to point out that you didn't object to our using the word "square" to refer to the shaded shapes in Fig. A.43. Strictly speaking, as shapes in the plane, these are just quadrilaterals, not squares. Familiarity with perspective drawing helped us recognize these as representing squares. In the same way, the parallelepipeds of Fig. A.44 are perspective images of cubes.

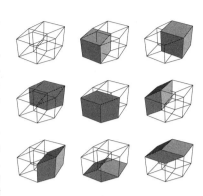

Figure A.44: The unit hypercube (first image) and the eight cubes constituting its boundary.

Once we find a clear way to approach the problem, there is nothing too tricky to understand. And we can keep going. How many hypercubes make up the boundary of the 5-dimensional cube? How many 5-dimensional cubes make up the boundary of the 6-dimensional cube?

A.73 The Sierpinski hypertetrahedron

Cubes bounding a hypercube are not so difficult, but fractals in 4 dimensions get a little more challenging, or so we might think. While this is indeed true for the cubic Mandelbrot set that lives in four real dimensions, some other fractals are much simpler. For example, consider the Sierpinski gasket. In the plane, the gasket consists of three copies, each scaled by 1/2. In 3-dimensional space, the gasket, now the Sierpinski tetrahedron, consists of four copies, each scaled by 1/2. Then in 4-dimensional space, the gasket analog – we'll call it the Sierpinski hypertetrahedron – consists of five copies, each scaled by 1/2. Where these five copies are placed depends on how we project 4-dimensional space into 3 dimensions – that is, on where we place the light to cast the shadow from 4 dimensions into 3.

In Fig. A.45 we show a few stages in constructing one projection of a Sierpinski hypertetrahedron. The five copies are arranged according to the vertices of the first shape: three around the middle, one above,

Figure A.45: Stages in constructing a projection of a Sierpinski hypertetrahedron.

and one below. That is, the second image consist of five copies of the first image, each scaled by 1/2, three placed around the middle, one above, and one below. The picture we see looks complicated because it's a projection into 2 dimensions of a projection into 3 dimensions of an object in 4 dimensions.

The third image is five copies of the second, all scaled by 1/2, three placed around the middle, one above, and one below. The fourth image is five copies of the third image, all scaled by 1/2, three placed around the middle, one above, and one below. Iterated, this will produce a projection into 3 dimensions of the Sierpinski hypertetrahedron.

All projections of the hypertetrahedron into 3 dimensions display five copies each scaled by 1/2. The placements of those copies is determined by the particular projection. That of Fig. A.45 is not the only one.

A.74 Parameters for the images of Fig. 5.26

For the magnifications constituting the second row of Fig. 5.26, the intervals are $-0.15 \leq c \leq 0.15$ and $0.9 \leq d \leq 1.24$, $-1 \leq a \leq -0.5$ and $-0.25 \leq b \leq 0.25$, $-0.46 \leq a \leq -0.36$ and $-0.78 \leq c \leq -0.68$, $-2.2 \leq a \leq -1$ and $-0.6 \leq d \leq 0.6$.

This note refers to Sect. 5.8, **Unanswered questions**.

A.75 Some variations in the $1/n^2$ rule

Benoit's n^2 conjecture, that the diameters of the n-cycle discs attached to the Main cardioid of the Mandelbrot set have radii of approximately $1/n^2$, was published in 1985, though an earlier lecture of his was the basis for this paper. Benoit reported that during this lecture, he no-

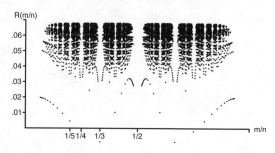

Figure A.46: Departures from the $1/n^2$ rule.

ticed two people in the audience writing furiously. At the end of the lecture, both men approached Benoit with sketches of proofs of this conjecture. These were Guckenheimer and McGehee, who published their proof in 1984.

Even brief inspection shows that the $1/n^2$ rule requires some modification: the principal sequence 5-cycle disc is not the same size as the Farey sequence 5-cycle disc between the principal 2- and 3-cycle discs. For this modification we need the "internal argument" m/n of an n-cycle disc attached to the Main cardioid. This is most easily understood by examples. The 2-cycle disc is attached at $m/n = 1/2$ and is halfway around the cardioid from the cusp. The 3-cycle discs are attached at $m/n = 1/3$ (upper disc) and $2/3$, the 4-cycle discs at $1/4$ and $3/4$. (For an internal argument m/n, m and n can have no common factors.) The refined n^2 rule is that the radius of an n-cycle disc attached at internal argument m/n is approximately

$$\gamma(m/n) = \sin(\pi m/n)/n^2$$

Programmer and artist Kerry Mitchell took a more geometrical approach, defining the radius $r(m/n)$ of an m/n disc as the distance between the point where the disc is attached to the cardioid and the center of the disc. (Recall the center of an n-cycle component is the value of c for which $z_0 = 0$ belongs to an n-cycle for $F_c(z) = z^2 + c$.) In Fig. A.46 we plot the ratio $r(m/n)/\gamma(m/n)$. When this ratio is not 1, the actual radius departs from the predicted $\gamma(m/n)$. But the point of including this figure is to show that the variations from the n^2 rule are fractally distributed. So far, we have no idea why.

Chapter 6. Quantifying fractals: What is fractal dimension?

The next three notes refer to Sect. 6.1, **Similarity dimension**. Here we give a proof that the Moran equation $r_1^d + \cdots + r_N^d = 1$ has a unique solution. Then we sketch the extension of the Moran equation to a fractal composed of pieces of size r, r^2, r^3, ..., and to a fractal composed of pieces of sizes given by some variations on this pattern.

A.76 Why the Moran equation has a unique solution

Recall the basic similarity dimension formula,

$$d_s = \frac{\log(N)}{\log(1/r)},$$

for fractals consisting of N pieces, each scaled by the factor r. If, instead, a fractal consists of N pieces having (possibly different) scaling factors $r_1, \ldots, r_N$ with each r_i satisfying $0 < r_i < 1$, then the similarity dimension d_s is the solution d of the Moran equation:

$$r_1^d + r_2^d + \cdots + r_N^d = 1$$

That this equation has a unique solution, or any solution at all, might not be obvious, so let's prove it. Define a function $f(d)$ by

$$f(d) = r_1^d + \cdots + r_N^d$$

In terms of this function f, the Moran equation is just $f(d) = 1$. So to prove the Moran equation has one and only one solution for every set of contraction factors $r_1, \ldots, r_N$, we'll show that the graph $y = f(d)$ crosses the line $y = 1$ for exactly one value of d. This follows from three observations.

1. $f(0) = r_1^0 + \cdots r_N^0 = N$. The only self-similar shape consisting of a single shrunken piece is a point, so every interesting fractal has $N \geq 2$.

2. Because each contraction factor $r_i < 1$, as $d \to \infty$, we see that $r_i^d \to 0$. Then $f(d) = r_1^d + \cdots + r_N^d$ is the sum of terms each going to 0 as $d \to \infty$, and so $f(d)$ goes to 0 as $d \to \infty$.

These two observations show that the graph of $y = f(d)$ lies above $y = 1$ for $d = 0$, and goes to 0 as d gets larger. Because this function f is continuous, we see the graph $y = f(d)$ must cross $y = 1$ at least once. How do we know that $y = f(d)$ doesn't wiggle up and down, crossing $y = 1$ several times? This is where the third observation comes in.

3. The graph $y = f(d)$ always decreases, so it can't wiggle up and down. To see why the graph always decreases, compute the derivative of $f(d)$.
$$f'(d) = r_1^d \ln(r_1) + \cdots + r_N^d \ln(r_N)$$
Because each scaling factor r_i satisfies $0 < r_i < 1$, each $\ln(r_i)$ is negative. This shows $f'(d) < 0$ for all d; that is, the graph of $f(d)$ decreases for all d.

Consequently, the graph $y = f(d)$ crosses the line $y = 1$ for exactly one value of d. That is, the Moran equation has one and only one solution.

To illustrate this result, Fig. A.47 shows $f(d)$ curves for two sets of contraction factors, $r_1 = r_2 = r_3 = 1/2$ and $r_4 = r_5 = 1/4$ (top curve), and $r_1 = r_2 = 1/2$ and $r_3 = r_4 = 1/4$ (bottom curve). Each curve crosses the $y = 1$ line at one point, whose d-coordinate is the dimension of the fractal having the contraction factors of that $f(d)$ curve.

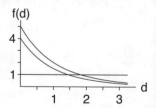

Figure A.47: Graphs of $f(d)$ curves for two sets of contraction factors.

A.77 Extending the Moran equation to infinite series

In preparation for extending the Moran equation to fractals having an infinite collection of scaling factors, $r_1, r_2, \ldots$, recall the formula for the

A.77. THE INFINITE MORAN EQUATION

sum of a geometric series, reviewed in Sect. A.2.

$$1 + r + r^2 + r^3 + \cdots = \frac{1}{1-r} \quad \text{if } |r| < 1.$$

The extension of the Moran equation to infinite families of contractions holds in great generality, but we needn't look at complicated bookkeeping to understand the main ideas. For that, seeing how and why the infinite Moran equation works for the fractal in Fig. A.48, which we saw earlier as fractal (c) in Fig. 6.3, suffices. Here we have outlined the larger pieces of the fractal.

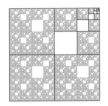

Figure A.48: Infinitely many scaling factors.

This fractal consists of 3 pieces scaled by $1/2$, 2 pieces scaled by $1/4$, 2 pieces scaled by $1/8$, and so on, 2 pieces scaled by scaled by $1/2^n$ for all $n > 1$. Fig. A.48 illustrates this decomposition. Call A_1 the part of the fractal consisting of 3 copies scaled by $1/2$ (certainly, no one is surprised that this is the gasket); A_2 the fractal made of 3 pieces scaled by $1/2$ and 2 scaled by $1/4$; A_3 the fractal made of 3 pieces scaled by $1/2$, 2 pieces scaled by $1/4$, and 2 pieces scaled by $1/8$; and so on. Continuing this notation, give the name A_∞ to the fractal shown in Fig. A.48. In Fig. A.49 we see A_1, A_2, and A_3, illustrating that as n increases, A_n converges to A_∞.

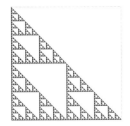 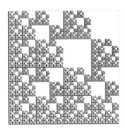 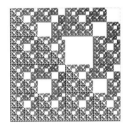

Figure A.49: The sets A_1, A_2, and A_3 for the fractal of Fig. A.48

For all $n \geq 1$, define a function $f_n(d)$ by

$$f_n(d) = 3 \cdot (1/2)^d + 2 \cdot (1/4)^d + 2 \cdot (1/8)^d + \cdots + 2 \cdot (1/2^n)^d$$

The three reasons any Moran equation has a unique solution ($f_n(0) > 1$, $\lim_{d \to \infty} f(d) = 0$, and $f'_n(d) < 0$ for all d) apply to every function $f_n(d)$, so for each $n \geq 1$ there is a unique number d_n, the dimension of A_n, satisfying $f_n(d_n) = 1$.

A simple consequence of the fact that each $f_n(d)$ is a decreasing function of d is that

$$f_n(d) > 1 \text{ for all } d < d_n, \text{ and } f_n(d) < 1 \text{ for all } d > d_n.$$

We single out this point because we'll use it twice in establishing the infinite Moran equation.

First, note that

$$f_{n+1}(d) = 3 \cdot (1/2)^d + 2 \cdot (1/4)^d + 2 \cdot (1/8)^d + \cdots + 2 \cdot (1/2^n)^d$$
$$+ 2 \cdot (1/2^{n+1})^d$$
$$= f_n(d) + 2 \cdot (1/2^{n+1})^d$$

Now use this to evaluate $f_{n+1}(d_n)$:

$$f_{n+1}(d_n) = f_n(d_n) + 2 \cdot (1/2^{n+1})^d = 1 + 2 \cdot (1/2^{n+1})^d > 1$$

From the observation mentioned above, this means that $d_n < d_{n+1}$. That is, the sequence $d_1, d_2, \ldots$ is increasing. Note that in passing from A_n to A_{n+1} we don't just add two more copies scaled by $1/2^{n+1}$, we add infinitely many smaller and smaller copies to A_n. Comparing the left, middle, and right images of Fig. A.49 illustrates this.

Now we'll define the infinite series that determines the dimension of A_∞. For all $d > 0$, let

$$f_\infty(d) = 3 \cdot (1/2)^d + 2 \cdot (1/4)^d + 2 \cdot (1/8)^d + 2 \cdot (1/16)^d + \cdots$$

To show that this defines a function, we demonstrate that the series converges.

$$f_\infty(d) = 3 \cdot (1/2)^d + 2 \cdot (1/2^2)^d + 2 \cdot (1/2^3)^d + 2 \cdot (1/2^4)^d + \cdots$$
$$= (1/2)^d + 2 \cdot (1/2)^d \cdot \left(1 + (1/2)^d + (1/2^d)^2 + (1/2^d)^3 + \cdots\right)$$

The bracketed terms are a geometric series, convergent because $|1/2^d| < 1$ so long as $d > 0$. Summing the series, we find a simpler expression for $f_\infty(d)$:

$$f_\infty(d) = (1/2^d) + \frac{2 \cdot (1/2^d)}{1 - (1/2^d)}$$

Now we'll show that there is a unique number d_∞ with $f_\infty(d_\infty) = 1$ by using the same three steps that showed each $f_n(d_n) = 1$ for a unique d_n, though for the middle step we'll use the simpler expression for $f_\infty(d)$.

First, as $d \to 0$,

$$f_\infty(d) \to 3 \cdot (1/2)^0 + 2 \cdot (1/4)^0 + 2 \cdot (1/8)^0 + 2 \cdot (1/16)^0 + \cdots$$
$$= 3 + 2 + 2 + 2 + \cdots$$

That is, $\lim_{d \to 0} f_\infty(d) = \infty$. So certainly $f_\infty(d) > 1$ for d near 0.

Second, $\lim_{d \to \infty} f_\infty(d) = \lim_{d \to \infty} \left((1/2^d) + \frac{2 \cdot (1/2^d)}{1 - (1/2^d)}\right) = 0.$

A.77. THE INFINITE MORAN EQUATION

And third, the graph of $f_\infty(d)$ is decreasing for all d because the derivative $f'_\infty(d)$ is negative:

$$f'_\infty(d) = 3 \cdot (1/2)^d \ln(1/2) + 2 \cdot (1/4)^d \ln(1/4) + 2 \cdot (1/8)^d \ln(1/8) + \cdots$$

Each term is negative, so $f'_\infty(d) < 0$ for all $d > 0$.

By a similar argument, we compute the second derivative, $f''_\infty(d) =$

$$3 \cdot (1/2)^d (\ln(1/2))^2 + 2 \cdot (1/4)^d (\ln(1/4))^2 + 2 \cdot (1/8)^d (\ln(1/8))^2 + \cdots$$

Every term is positive, so the graph of f_∞ is concave up. The corresponding calculation shows that the graphs of every f_n are concave up.

A consequence of the first three observations is that there is a unique number d_∞ satisfying $f_\infty(d_\infty) = 1$.

Now we know that for each n, d_n is the dimension of A_n, and by comparing Figs. A.48 and A.49, we see that as $n \to \infty$, the A_n converge to A_∞. So to show d_∞ is the dimension of A_∞, we show that $d_n \to d_\infty$ as $n \to \infty$. To do this, we begin by comparing $f_n(d)$ and $f_\infty(d)$.

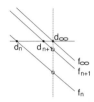

Figure A.50: $d_n \to d_\infty$.

$$\begin{aligned}
f_\infty(d) &= 3 \cdot (1/2)^d + 2 \cdot (1/4)^d + \cdots + 2 \cdot (1/2^n)^d + 2 \cdot (1/2^{n+1})^d + \cdots \\
&= f_n(d) + 2 \cdot (1/2^{n+1})^d + 2 \cdot (1/2^{n+2})^d + 2 \cdot (1/2^{n+3})^d + \cdots \\
&= f_n(d) + 2 \cdot (1/2^d)^{n+1} + 2 \cdot (1/2^d)^{n+2} + 2 \cdot (1/2^d)^{n+3} + \cdots \\
&= f_n(d) + 2 \cdot (1/2^d)^{n+1} \left(1 + (1/2^d) + (1/2^d)^2 + (1/2^d)^3 + \cdots \right) \\
&= f_n(d) + \frac{2 \cdot (1/2^d)^{n+1}}{1 - (1/2^d)}
\end{aligned}$$

Solving for $f_n(d)$, we find that for all $d > 0$,

$$f_n(d) = f_\infty(d) - \frac{2 \cdot (1/2^d)^{n+1}}{1 - (1/2^d)}$$

Then taking $d = d_\infty$,

$$\begin{aligned}
f_n(d_\infty) &= f_\infty(d_\infty) - \frac{2 \cdot (1/2^{d_\infty})^{n+1}}{1 - (1/2^{d_\infty})} \\
&= 1 - \frac{2 \cdot (1/2^{d_\infty})^{n+1}}{1 - (1/2^{d_\infty})}
\end{aligned}$$

Because $f_n(d_n) = 1$ and $f_n(d_\infty) < 1$, we see that each $d_n < d_\infty$.

For each n, $f_n(d_n) = 1$ and so $f_\infty(d_n)$ is greater than 1 because $f_\infty(d_n)$ contains every term of $f_n(d_n)$, plus more terms.

Finally, look again at

$$f_n(d) = f_\infty(d) - 2\frac{(1/2^d)^{n+1}}{1-(1/2^d)}$$

Because $f_\infty(d_\infty) = 1$, we see that as $n \to \infty$, $f_n(d_\infty) \to 1$. In Fig. A.50, the horizontal line is $f = 1$, the vertical dashed line is $d = d_\infty$. Circles are $f_n(d_\infty)$; the points where the graphs of $f_n(d)$, $f_{n+1}(d)$, and $f_\infty(d)$ cross the line $f = 1$ are the filled-in dots are d_n, d_{n+1}, and d_∞. Because the graphs of all the f_n and f_∞ are decreasing and concave up, and because $f_n(d_\infty) \to 1$, we see that $d_n \to d_\infty$, and we're done.

Completing the calculation for our example, as usual, we'll take $x = (1/2)^d$ and rewrite the infinite Moran equation as

$$3x + 2(x^2 + x^3 + x^4 + \cdots) = 1$$

Factoring an x^2 from all the bracketed terms gives us

$$3x + 2x^2(1 + x + x^2 + x^3 + \cdots) = 1$$

Summing the bracketed geometric series and solving for x, we find $x = 2 \pm \sqrt{3}$. Both these are positive, but we still can rule out one solution when we recall that in order to sum a geometric series, we must have $|x| < 1$. Then the solution is $x = 2 - \sqrt{3}$, and consequently

$$d(A_\infty) = \frac{\log(2-\sqrt{3})}{\log(1/2)} \approx 1.899$$

In order to calculate the exact value of the dimension using the infinite Moran equation, we must find the sum of the infinite series. The example we've just seen is a simple geometric series, something we already know how to sum. Now we'll give another example where the series is a bit trickier. The point of this example is to demonstrate another strategy for summing infinite series. There are many, many more such tricks.

A.78 Two more applications of the infinite Moran equation

Suppose a fractal is made of 2 pieces scaled by r, 3 pieces scaled by r^2, 4 pieces scaled by r^3, and so on. In this case, the infinite Moran equation becomes

$$2r^d + 3(r^d)^2 + 4(r^d)^3 + 5(r^d)^4 + \cdots = 1$$

A.78. MORE INFINITE MORAN EXAMPLES

In order to find d, first substitute $x = r^d$ and sum the series

$$2x + 3x^2 + 4x^3 + 5x^4 + \cdots$$

The relation between the coefficients and the exponents should remind you of something from a first-semester calculus class: derivatives of powers of x. So suppose $g(x) = x^2 + x^3 + x^4 + x^5 + \cdots$. Then $g'(x) = 2x + 3x^2 + 4x^3 + \cdots$, just the series we wanted. But what function is $g(x)$? That's easy.

$$1 + x + g(x) = 1 + x + x^2 + x^3 + x^4 + \cdots = \frac{1}{1-x} \text{ when } |x| < 1.$$

Because $r < 1$ and d is positive, $x = r^d$ is necessarily less than 1, and the series converges. So now we have $g(x) = 1/(1-x) - 1 - x$. Consequently, we can set the left-hand side of our infinite Moran equation equal to the derivative of the series for $g(x)$.

$$2x + 3x^2 + 4x^3 + 5x^4 + \cdots = g'(x) = \frac{1}{(1-x)^2} - 1 = \frac{2x - x^2}{(1-x)^2}$$

Then the infinite Moran equation becomes

$$\frac{2x - x^2}{(1-x)^2} = 1$$

which simplifies to the quadratic equation

$$2x^2 - 4x + 1 = 0$$

The solutions are $x = 1 \pm \sqrt{2}/2$, both positive, but recall the condition $|x| < 1$. Then we take the smalller value for r^d,

$$r^d = 1 - \sqrt{2}/2$$

so

$$d = \frac{\log(1 - \sqrt{2}/2)}{\log(r)}$$

With some experimentation, we can use the infinite Moran equation to find the dimensions of a substantial collection of fractals with infinitely many scaling factors.

We'll end with a simple example whose dimension we know already, and which can be calculated by another instance of the infinite Moran equation. When we computed the similarity dimension of the gasket, we decomposed it into three pieces scaled by $1/2$, or nine pieces scaled by $1/4$, or 27 pieces scaled by $1/8$, and so on.

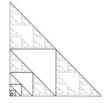

Figure A.51: Another gasket decomposition.

But suppose we break the gasket up into pieces of different sizes. For example, in Fig. A.51 we break the gasket into two pieces scaled by $1/2$, two pieces scaled by $1/4$, two pieces scaled by $1/8$, and so on. Then taking $x = (1/2)^d$, the infinite Moran equation for the gasket is
$$2x + 2x^2 + 2x^3 + \cdots = 1$$
Dividing by 2, adding 1 to both sides, and summing the resulting geometric series, we find $x = 1/3$ and recover the familiar result $d = \log(3)/\log(2)$.

The next three notes refer to Sect. 6.2, **Box-counting and mass dimension**. Here we'll derive the limit formula for d_b from the scaling hypothesis and show why the $\epsilon \to 0$ limit in the definition of d_b can be replaced by the sequential limit, $\epsilon_n \to 0$, and list some other kinds of boxes we can count to find this dimension.

A.79 The limit form of the box-counting dimension formula

Recall the definition of the box-counting dimension. For every $\epsilon > 0$, denote by $N(\epsilon)$ the minimum number of boxes of side length ϵ needed to cover the shape. Then the box-counting dimension, d_b is
$$d_b = \lim_{\epsilon \to 0} \frac{\log(N(\epsilon))}{\log(1/\epsilon)}$$
This is true only if the limit exists. If not, replace the limit by the limsup and liminf, obtaining the upper and lower box-counting dimensions. If the limit superior (limsup) and limit inferior (liminf) are unfamiliar, Google remains you true and trusted friend. We won't explore these subtleties here, but are pointing them out so anyone who asks, "And if the limit doesn't exist, then what?" will know there is an answer, but it's a bit complicated.

This box-counting dimension formula can be derived from the scaling hypothesis. We'll work through this quickly. First, for simplicity write the scaling hypothesis as an equality, rather than as an approximation,
$$N(\epsilon) = k(\epsilon) \cdot (1/\epsilon)^{d_b} \quad \text{rather than} \quad N(\epsilon) \approx k \cdot (1/\epsilon)^{d_b}$$
We can replace the $\approx$ with $=$ in the scaling hypothesis because k might vary with ϵ, but so long as it doesn't vary too rapidly, specifically, so that the *slow variation condition*
$$\lim_{\epsilon \to 0} \frac{\log(k(\epsilon))}{\log(1/\epsilon)} = 0$$
holds, we still obtain the limit definition of d_b from the scaling hypothesis.

A.80. BOX-COUNTING DIMENSION: SEQUENTIAL FORM

To derive the limit formula for d_b, take the log of both sides

$$\log(N(\epsilon)) = \log(k(\epsilon)) + d_b \log(1/\epsilon)$$

and solve for d_b

$$d_b = \frac{\log(N(\epsilon))}{\log(1/\epsilon)} + \frac{\log(k(\epsilon))}{\log(1/\epsilon)}$$

By the slow variation condition, we obtain the limit formula for d_b, provided, of course, that the limit exists.

A.80 The sequential form of the box-counting dimension formula

An apparent problem with applying this formula is that we must know the minimum number of boxes for *every* size ϵ. For any PhysicalWorld object, the best we can do is count $N(\epsilon_n)$ for a sequence of box sizes $\epsilon_1 > \epsilon_2 > \epsilon_3 > \dots$ Suppose the sequence of box sizes has these properties:

- $\epsilon_n \to 0$ as $n \to \infty$, and
- there is a constant c, $0 < c < 1$, with $\epsilon_n \geq c \cdot \epsilon_{n-1}$.

As long as these conditions are met, we'll show that if the *sequential limit*

$$\lim_{n \to \infty} \frac{\log(N(\epsilon_n))}{\log(1/\epsilon_n)}$$

exists, then so does

$$\lim_{\epsilon \to 0} \frac{\log(N(\epsilon))}{\log(1/\epsilon)}$$

and the limits are equal. This is why, for example, we can compute d_b of the gasket by calculating only $N(1/2^n)$.

Here's why the existence of the sequential limit implies the existence of the limit and why both limits are equal. Because $\epsilon_n \to 0$, for every $\epsilon > 0$, there is an n with $\epsilon_{n-1} > \epsilon \geq \epsilon_n$. For this n,

$$N(\epsilon_{n-1}) \leq N(\epsilon) \leq N(\epsilon_n)$$

We'll build the sequential limit on the left and right, trapping the $\epsilon \to 0$ limit between them. This will show the result.

The inequalities above need not be strict, because small changes in ϵ might not change the number of boxes. From this and the relation $1/\epsilon_{n-1} < 1/\epsilon \leq 1/\epsilon_n$, we see

$$\log(N(\epsilon_{n-1})) \leq \log(N(\epsilon)) \leq \log(N(\epsilon_n))$$

and
$$\log(1/\epsilon_{n-1}) < \log(1/\epsilon) \le \log(1/\epsilon_n)$$

Combining these gives
$$\frac{\log(N(\epsilon_{n-1}))}{\log(1/\epsilon_n)} \le \frac{\log(N(\epsilon))}{\log(1/\epsilon)} < \frac{\log(N(\epsilon_n))}{\log(1/\epsilon_{n-1})}$$

Adjusting the subscripts gives what we'll call the *squeeze relation*,
$$\frac{\log(N(\epsilon_{n-1}))}{\log(1/\epsilon_{n-1})} \cdot \frac{\log(1/\epsilon_{n-1})}{\log(1/\epsilon_n)} \le \frac{\log(N(\epsilon))}{\log(1/\epsilon)} < \frac{\log(N(\epsilon_n))}{\log(1/\epsilon_n)} \cdot \frac{\log(1/\epsilon_n)}{\log(1/\epsilon_{n-1})}$$

Now for $1 > \epsilon_{n-1}$, which is true for large-enough n, $\epsilon_{n-1} > \epsilon_n$ implies
$$\frac{\log(\epsilon_n)}{\log(\epsilon_{n-1})} > 1$$

and $\epsilon_n \ge c \cdot \epsilon_{n-1}$ implies
$$\frac{\log(c)}{\log(\epsilon_{n-1})} + 1 \ge \frac{\log(\epsilon_n)}{\log(\epsilon_{n-1})}$$

Combining these gives
$$\lim_{n \to \infty} \frac{\log(\epsilon_n)}{\log(\epsilon_{n-1})} = 1 \text{ and consequently } \lim_{n \to \infty} \frac{\log(\epsilon_{n-1})}{\log(\epsilon_n)} = 1$$

Then as $n \to \infty$, the left and right sides of the squeeze relation approach the same limit, so $\log(N(\epsilon))/\log(1/\epsilon)$, being trapped between these sides, also must approach the same value, d_b.

One consequence of this result is that d_b can be calculated as a sequential limit for *any* sequence $\epsilon_n \to 0$ with $\epsilon_n \ge c \cdot \epsilon_{n-1}$. However, unfortunate choices of ϵ_n can lead to very slow convergence to d_b.

A.81 Some other choices for $N(\epsilon)$

Finally, we mention that we have computed d_b using boxes of side length ϵ. In his 2003 book, Falconer shows that d_b of a shape A can be computed by taking $N(\epsilon)$ to be

- the minumum number of boxes of side length ϵ needed to cover A
- the minumum number of sets of diameter ϵ needed to cover A,
- the number of lattice boxes of side ϵ intersecting A, and
- the maximum number of disjoint ϵ-discs with centers in A.

A.82. RANDOM MORAN EQUATION

Misapplication of the third form of $N(\epsilon)$ is the source of many incorrect calculations of d_b. The point is this. The first form of $N(\epsilon)$ emphasizes that we must find the smallest number of boxes, the most efficient cover, therefore the cover that most closely reflects the shape of the set. But programming this requires some work and bit of skill. It is far more common to superimpose a grid, a lattice of squares, and count the number of squares that intersect the set. While this does give d_b in the $\epsilon \to 0$ limit, finding the limiting behavior may take very, very small ϵ. So be careful with lattice calculations.

The next four notes refer to Sect. 6.4, **Random, IFS with memory, and nonlinear fractals**. Here we extend the Moran equation to several additional types of fractals and, along the way, derive the formula for finding eigenvalues of matrices.

A.82 The random Moran equation for continuous probabilities

We have seen that the Moran equation $r_1^d + \cdots + r_n^d = 1$ can be modified to accommodate each scaling factor $r_1, \ldots, r_n$ being selected randomly:

$$1 = \mathbb{E}(r_1^d + \cdots + r_n^d) = \mathbb{E}(r_1^d) + \cdots + \mathbb{E}(r_n^d).$$

The last equality is called the *linearity* of the expected value. In Sect. 6.4 we worked through examples where the scaling factors are chosen randomly from a small collection of values. Here we show how to apply the random Moran equation when the scaling factors are continuously distributed.

Suppose the scaling factors r_1, r_2, r_3, and r_4 are uniformly distributed in $[1/4, 1/2]$, so for any value r_* of r satisfying $1/4 \leq r_* \leq 1/2$, the probability $Pr\{r \leq r_*\}$ is given by

$$Pr\{r \leq r_*\} = Pr\{1/4 \leq r \leq r_*\} = \int_{1/4}^{r_*} k\, dt$$

where k is called the normalization constant. To find the value of k, note that because the values of r always lie between $1/4$ and $1/2$, we can find the value of k by this calculation

$$1 = Pr\{1/4 \leq r \leq 1/2\} \quad \text{because } r \text{ lies between } 1/4 \text{ and } 1/2$$

$$= \int_{1/4}^{1/2} k\, dt \quad \text{probability distribution definition}$$

$$= kt \Big|_{1/4}^{1/2} \quad \text{evaluating the integral}$$

$$= k(1/2 - 1/4)$$

That is, $k = 4$.

Therefore the expected value of each scaling factor r in this example is

$$\mathbb{E}(r) = \int_{1/4}^{1/2} 4t\,dt = \frac{3}{8}$$

and so

$$\mathbb{E}(r^d) = \int_{1/4}^{1/2} 4t^d\,dt = \frac{4}{d+1}\left(\left(\frac{1}{2}\right)^{d+1} - \left(\frac{1}{4}\right)^{d+1}\right)$$

If we put together the expected value of r^d and the linearity of the expected value, in this example the random Moran equation simplifies to

$$\frac{16}{d+1}\left(\left(\frac{1}{2}\right)^{d+1} - \left(\frac{1}{4}\right)^{d+1}\right) = 1$$

Solving this numerically (our only option in this case) gives $d \approx 1.42481$.

We can accommodate nonuniform probability distributions (the probabilities near some r are higher than those near other values) by replacing the constant k by a function $k(t)$. This then becomes the beginning of another approach to compute the dimension of natural fractals: measuring structures across a range of scales gives approximations of the probability distribution of the scaling factors. Then we use these in computing $\mathbb{E}(r_i^d)$. This is difficult to implement; box-counting is simpler. However, this random Moran equation has been used in some applications.

The Moran equation for IFS with memory

For IFS with memory, the generalization of the Moran equation involves finding the eigenvalues of the transition matrix. We will sketch the minimum amount of linear algebra we need to understand the eigenvalue equation. Then we will show how to find eigenvalues using Mathematica. A good graphing calculator can do this calculation, too. Feel free to skip to A.85 if you've had a linear algebra course.

A.83 The basic steps for computing eigenvalues

Examples will be of 2×2 matrices, which are computationally simple, though these ideas generalize to any square matrix, with the note that for larger matrices solving the eigenvalue equation can become much trickier. We take this detour so eigenvalues, an essential tool here and later, will not be just an unexamined trick.

Although we are interested only in eigenvalues, understanding why the eigenvalue equation is what it is requires introducing eigenvectors.

A.83. COMPUTING EIGENVALUES

For a square matrix A a non-zero vector $\vec{v}$ is an *eigenvector* with *eigenvalue* λ if the *eigenvector equation* holds.

$$A \cdot \vec{v} = \lambda \vec{v}$$

If $\vec{v} = \vec{0}$, the eigenvector equation holds for all λ, which is not so useful. To avoid this vacuous case, the condition $\vec{v} \neq \vec{0}$ is imposed on eigenvectors.

How can we find for which λ the eigenvector equation has a solution? First, it helps to move all the terms to the left side of the equation.

$$A \cdot \vec{v} - \lambda \vec{v} = \vec{0}$$

We can't factor out $\vec{v}$, because A is a matrix and λ is a number. But remember that for the identity matrix I, $I \cdot \vec{v} = \vec{v}$, which lets us rewrite the previous equation as

$$A \cdot \vec{v} - \lambda I \cdot \vec{v} = \vec{0}$$

Because λI is a matrix, now we can factor out $\vec{v}$, obtaining

$$(A - \lambda I) \cdot \vec{v} = \vec{0}$$

If $\det(A - \lambda I) \neq 0$, the matrix $A - \lambda I$ is invertible. Then we can multiply both sides of this equation on the left by $(A - \lambda I)^{-1}$, obtaining

$$(A - \lambda I)^{-1} \cdot (A - \lambda I) \cdot \vec{v} = (A - \lambda I)^{-1} \cdot \vec{0}$$
$$\vec{v} = \vec{0}$$

That is, in order for the eigenvector equation to have a non-zero eigenvector solution, the matrix $A - \lambda I$ must *not* be invertible, so

$$\det(A - \lambda I) = 0$$

This is called the *eigenvalue equation*. The eigenvalues λ of a matrix M are the solutions of the eigenvalue equation.

For example, consider these matrices:

$$A = \begin{bmatrix} 1 & 2 \\ 0 & 4 \end{bmatrix}, \quad B = \begin{bmatrix} 1 & 2 \\ 3 & 4 \end{bmatrix}, \quad C = \begin{bmatrix} 1 & 2 \\ -3 & 4 \end{bmatrix}$$

To find the eigenvalues of A, solve

$$0 = \det\left(\begin{bmatrix} 1 & 2 \\ 0 & 4 \end{bmatrix} - \lambda \begin{bmatrix} 1 & 0 \\ 0 & 1 \end{bmatrix}\right) = \det\begin{bmatrix} 1-\lambda & 2 \\ 0 & 4-\lambda \end{bmatrix}$$
$$= (1-\lambda)(4-\lambda) - 0 \cdot 2$$

This is already factored, so the eigenvalues are

$$\lambda = 1 \quad \text{and} \quad \lambda = 4$$

To find the eigenvalues of B, begin the same way we did for A.

$$0 = \det\left(\begin{bmatrix} 1 & 2 \\ 3 & 4 \end{bmatrix} - \lambda \begin{bmatrix} 1 & 0 \\ 0 & 1 \end{bmatrix}\right) = \det\begin{bmatrix} 1-\lambda & 2 \\ 3 & 4-\lambda \end{bmatrix}$$

$$= (1-\lambda)(4-\lambda) - 3 \cdot 2 = \lambda^2 - 5\lambda - 2$$

That is, the eigenvalues are the solutions of the quadratic equation $\lambda^2 - 5\lambda - 2 = 0$, namely

$$\lambda = \frac{5 + \sqrt{33}}{2} \quad \text{and} \quad \lambda = \frac{5 - \sqrt{33}}{2}$$

Arguing similarly, we find that the eigenvalues of C are the solutions of the quadratic equation $\lambda^2 - 5\lambda + 10 = 0$. These are complex:

$$\lambda = \frac{5 + i\sqrt{15}}{2} \quad \text{and} \quad \lambda = \frac{5 - i\sqrt{15}}{2}$$

A.84 Finding eigenvalues with *Mathematica*

To find the eigenvalues using *Mathematica*, first we must know how to write a matrix in *Mathematica*. For example, the matrix

$$m = \begin{bmatrix} 0 & 1 & 1 & 0 \\ 1 & 0 & 1 & 1 \\ 1 & 1 & 0 & 1 \\ 1 & 1 & 1 & 1 \end{bmatrix}$$

is entered in *Mathematica* as

$$m = \{\{0,1,1,0\},\{1,0,1,1\},\{1,1,0,1\},\{1,1,1,1\}\}$$

That is, enter the matrix values by row, with the entries of each row enclosed in curly brackets (yes, "curly brackets" is the technical term). In *Mathematica*, defined functions and constants are indicated by capital letters, so we use a lowercase letter to denote the matrix. The command

$$\text{Eigenvalues}[m]$$

returns a list of the eigenvalues of the matrix m, in this case

$$-1, -1, 0, 3$$

four eigenvalues (-1 occurs twice) because the matrix is 4×4.

A.85 The memory Moran equation and infinite Moran equation

Once we know how to compute eigenvalues, we can solve the memory Moran equation,

$$\rho[m_{ij}r_i^d] = 1$$

As we've mentioned, $m_{ij} = 0$ or 1 depending on whether $T_i \circ T_j$ is forbidden or allowed, and $\rho[M]$ is the largest eigenvalue of M, also called the *spectral radius* of M. The fractal of Fig. A.48 (and Fig. 6.4), seen again in Fig. A.52, can be generated by an IFS with memory, with only the composition $T_4 \circ T_1$ forbidden. Because all the scaling factors are $r = 1/2$, the memory Moran equation simplifies:

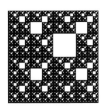

Figure A.52: The fractal of Fig. A.48.

$$1 = \rho[m_{ij}r_i^d] = \rho[m_{ij}r^d] = r^d\rho[m_{ij}]$$

The last equality comes from the fact that multiplying every entry of a matrix by a number multiplies all the eigenvalues of the matrix by this same number. Then for this particular fractal, the memory Moran equation becomes

$$r^d \rho \begin{bmatrix} 1 & 1 & 1 & 1 \\ 1 & 1 & 1 & 1 \\ 1 & 1 & 1 & 1 \\ 0 & 1 & 1 & 1 \end{bmatrix} = 1$$

The eigenvalues of this matrix are $2 \pm \sqrt{3}, 0$, and 0, so the spectral radius is $2 + \sqrt{3}$. But if you remember the dimension calculation using the infinite Moran equation, you'll recall that there we used $2 - \sqrt{3}$. How can we use $2 + \sqrt{3}$ here? We need to be a bit careful about how the solution of the quadratic equation is related to the scaling factor. Here the memory Moran equation is

$$(1/2)^d(2 + \sqrt{3}) = 1$$

Multiplying both sides by 2^d gives

$$2 + \sqrt{3} = 2^d$$

Taking logs and solving for d, we obtain this value for the dimension

$$d = \frac{\log(2 + \sqrt{3})}{\log(2)}$$

Compare this to our result from the infinite Moran equation:

$$d = \frac{\log(2 - \sqrt{3})}{\log(1/2)}$$

Simple algebra shows these are equal.

The next four notes refer to Sect. 6.5, **Some dimension rules**. Here we explain the open set condition, guaranteeing that the overlaps of pieces of a fractal are small enough that the Moran equation gives a sensible value of dimension. Rather than sketch the proofs, some of which are quite involved, we'll illustrate how to apply these results. For example, we show that the function $f(x) = x + x^2$ defined for x in the interval $0 \leq x \leq 1$ has distortion bounded above and below and consequently preserves dimensions. We'll list rigorous results for when the dimension rules are valid. Finally, we'll illustrate one of the very many subtle points involved in trying to compute the dimension of self-affine fractals.

A.86 The Open Set condition

Here is Hutchinson's *open set condition*, a principle which guarantees that the solution of the Moran equation is the dimension of a self-similar fractal.

Although we will only look at sets lying in the plane, these ideas can be applied with much greater generality.

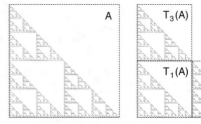

Figure A.53: An illustration that the gasket IFS rules satisfy the open set condition.

First, we must establish some terminology. A set A is *open* if for any point p of A, a small-enough disc centered at p lies entirely in A. For example, $A = \{(x, y) : x^2 + y^2 < 1\}$ is an open set because every point p of A lies a distance $d < 1$ from the origin $(0, 0)$, so the disc centered at p and having radius $(1 - d)/2$ lies entirely inside A. The set $B = \{(x, y) : x^2 + y^2 \leq 1\}$ is not an open set because, for example, the point $(1, 0)$ belongs to B and every disc, no matter how small, centered at $(1, 0)$ contains points not in B, and contains other points in B.

Now consider the fractal generated by the IFS $\{T_1, \ldots, T_n\}$, with each T_i a similarity, that is, $r_i = s_i$, so the transformation contracts by the same amount in the x- and y-directions. This IFS satisfies the open set condition if there is an open set A having these two properties

- A contains the union $T_1(A) \cup \cdots \cup T_n(A)$, and

- each $T_i(A)$ intersects none of the other $T_j(A)$.

For the right isosceles gasket, in Fig. A.53 we see that the set $A = \{(x, y) : 0 < x < 1, 0 < y < 1\}$ works to verify the open set condition. For

A.87 Functions that preserve fractal dimension

Certainly a similarity transformation preserves dimension: the transformed shape is just a copy of the original, magnified or shrunk by the same amount everywhere. But many more transformations preserve dimension. In particular, if there are numbers a and b, with $0 < a \leq b < \infty$, making

$$a|x - y| \leq |f(x) - f(y)| \leq b|x - y|$$

for all points x, y in the shape, then the original shape and the transformed shape have the same dimension. Such a function is called *bi-Lipschitz*.

To illustrate this concept, we'll show that the function $f(x) = x + x^2$ is bi-Lipschitz on the interval $[0, 1]$. That is, we must find numbers a and b, with $0 < a \leq b < \infty$, making

$$a|x - y| \leq |f(x) - f(y)| \leq b|x - y|$$

for all x, y in $[0, 1]$. First, note that $f'(x) = 1 + 2x$, so for all x in $[0, 1]$, $1 \leq f'(x) \leq 3$. Now apply the mean value theorem. For any points x and y in $[0, 1]$, there is a point z, between x and y, giving

$$(x - y)f'(z) = f(x) - f(y)$$

Taking absolute values, and noting that for all z in $[0, 1]$, $1 \leq f'(z) \leq 3$, we see

$$|x - y| \leq |f(x) - f(y)| \leq 3|x - y|$$

That is, on the interval $[0, 1]$, $f(x) = x + x^2$ is bi-Lipschitz with $a = 1$ and $b = 3$.

For example, at the top of Fig. A.54 we see the familiar Cantor middle-thirds set, which consists of 2 pieces each scaled by $1/3$ and so has dimension $\log(2)/\log(3)$. At the bottom of the figure is the image of that Cantor set under the function $f(x) = x + x^2$, and its dimension also is $\log(2)/\log(3)$. Try to think of another way to compute the dimension of this stretched and distorted Cantor set.

Figure A.54: Cantor set (top), and a nonlinear image (bottom)

A.88 Careful statements of dimension rules

Although the dimension rules stated in Sect. 6.5 of the text are valid for large classes of fractals, some do not hold universally. The precise results are these. Recall d_h is the Hausdorff dimension (Think similarity dimension, but generalized to apply to non-self-similar shapes.) and d_b is the box-counting dimension.

1. d_h is *stable*, that is, $d_h(A \cup B) = \max\{d_h(A), d_h(B)\}$; d_b is stable when it exists.

2. The upper box-counting dimension, $\overline{d_b}$, is stable.

3. The lower box-counting dimension, $\underline{d_b}$, is not stable.

4. d_h is *countably stable*, that is, $d_h(A_1 \cup A_2 \cup \cdots) = \sup\{d_h(A_1), d_h(A_2), \ldots\}$.

5. Neither $\underline{d_b}$ nor $\overline{d_b}$ is countably stable.

6. $d_h(A \times B) \geq d_h(A) + d_h(B)$.

7. $\overline{d_b}(A \times B) \leq \overline{d_b}(A) + \overline{d_b}(B)$.

8. $d_h(A \times B) = d_h(A) + d_h(B)$ if $d_h(A) = \overline{d_b}(A)$.

9. For A and B subsets of n-dimensional space,
$d_h(A \cap f(B)) \leq \max\{0, d_h(A) + d_h(B) - n\}$ for all translations f
$d_h(A \cap f(B)) \geq d_h(A) + d_h(B) - n$ for a "large" class of translations f

A.89 Dimensions of some self-affine fractals

Here we present some of McMullen's work on computing dimensions of self-affine fractals, fractals whose pieces are scaled by different factors in different directions. In general, there is little hope of finding a version of the Moran equation that can be applied to all self-affine fractals, but McMullen solved the problem for an important class of examples. McMullen's formulas have many applications. We'll focus on two. One is that we can construct a simple example for which the box-counting dimension is less than the Hausdorff dimension. The other is an example where, unlike the similarity dimension, the Hausdorff dimension depends on the locations, not just the sizes, of the pieces.

McMullen constructed a family of self-affine sets by dividing the unit square into an $m \times n$ collection of rectangles, each of width $1/n$ and height $1/m$. Keep some of the rectangles, remove the others. Subdivide each remaining rectangle into an $m \times n$ collection of still smaller rectangles and

A.89. DIMENSIONS OF SELF-AFFINE FRACTALS

keep the same pattern of smaller rectangles. Continue. See the second and third fractals of Fig. A.55 for two examples. The first example is simpler: the product of a Cantor middle-halves set and a Cantor middle-thirds set. From the product rule we see $d_b = d_h = \log(2)/\log(4) + \log(2)/\log(3)$, though of course McMullen's formulas give the same result. The second and third fractals are substantially more complicated.

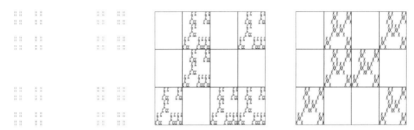

Figure A.55: Three self-affine fractals, with the first subdivision into rectangles indicated on the second and third fractals.

For a self-affine fractal A, made from this subdivided collection of $1/n \times 1/m$ rectangles, McMullen proved the Hausdorff dimension d_h is given by

$$d_h = \log_m\left(\sum_{j=1}^{m} t_j^{\log_n(m)}\right)$$

where t_j is the number of occupied rectangles in row j. Also, he proved the box-counting dimension d_b is given by

$$d_b = \log_m(r) + \log_n(N/r)$$

where r is the number of non-empty rows and N is the total number of rectangles occupied in the first step of the construction.

For the second fractal of Fig. A.55, we see $t_1 = 3$, $t_2 = 1$, and $t_3 = 2$, so

$$d_h = \log_3\left(3^{\log_4(3)} + 1^{\log_4(3)} + 2^{\log_4(3)}\right) \approx 1.48664$$

For computing the box-counting dimension, note $r = 3$, $N = 6$, so

$$d_b = \log_3(3) + \log_4(6/3) = 1.5$$

Then we see that for the second fractal of Fig. A.55, $d_h < d_b$.

For the third fractal of Fig. A.55, we see $t_1 = t_2 = t_3 = 2$, so the Hausdorff dimension is

$$d_h = \log_3\left(3 \cdot 2^{\log_4(3)}\right) = 1.5$$

So the second and third fractals of Fig. A.55 have different Hausdorff dimensions, even though both consist of six pieces, all with the same scaling.

Chapter 7. Further developments

The next three notes refer to Sect. 7.1, **Driven IFS and data analysis**. Here we'll note that in the IFS driven by the $r = 4$ logistic map with equal-size bins, every empty address contains an empty length-2 address. Also, we'll give an example of how to test if an empty length-3 address signals an exclusion in the data or just too little data.

A.90 Markov partitions and forbidden pairs

A driven IFS is determined by a time series and a choice of bins. In these examples, the time series are generated by iterating logistic maps,

$$x_{n+1} = L_r(x_n) = rx_n(1 - x_n).$$

That is, we pick a starting value x_0 between 0 and 1 and a value of r between 0 and 4. Then we produce the time series $x_0, x_1, x_2, \ldots$ by $x_1 = L_r(x_0)$, $x_2 = L_r(x_1)$, and so on. Here we'll divide the time series into equal-size bins.

In Fig. A.56 we see the graph of the $r = 4$ logistic map $L_4(x)$. Note that as x ranges over all of $[0, 1]$, $L_4(x)$ ranges over all of $[0, 1]$. The four bins are determined by dividing $[0, 1]$ into four equal-length intervals. On both the x- and y-axes, we'll label these bins B_1, B_2, B_3, and B_4, 1, 2, 3, and 4 for short. Taken together, these bins cover the whole interval $[0, 1]$, and no two bins have even one point in common. Specifically, $B_1 = [0, 1/4]$, $B_2 = [1/4, 1/2]$, $B_3 = [1/2, 3/4]$, and $B_4 = [3/4, 1]$. To find where points in each bin are sent, locate that bin on

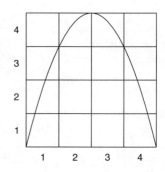

Figure A.56: A partition of the $r = 4$ logistic map.

the x-axis, look at the part of the graph of $L_4(x)$ above that bin, and notice through which bins along the y-axis that part of the graph passes. For example, above B_1 we see that the graph passes through B_1, B_2, and B_3. That is, for every point x in B_1, the point $L_4(x)$ must lie in B_1, in B_2, or in B_3. Moreover, the part of the graph above B_1 passes through all of B_1, B_2, and B_3. That is, for each point y in B_1 or B_2 or B_3, there is some x in B_1 for which $L_4(x) = y$.

Similarly analyzing the part of the graph above B_2, B_3, and B_4, we

A.90. MARKOV PARTITIONS AND FORBIDDEN PAIRS

see

$$L_4(B_1) = B_1 \cup B_2 \cup B_3$$
$$L_4(B_2) = B_4$$
$$L_4(B_3) = B_4$$
$$L_4(B_4) = B_1 \cup B_2 \cup B_3$$

Breaking an interval into disjoint bins is called a *partition* of the interval. The example we've been discussing is an instance of a *Markov partition*: the portion of the graph above each bin completely crosses every bin it enters.

One reason Markov partitions are important is that with addresses defined by the bins of a Markov partition, for all $n > 2$, every empty length-n address contains an empty length-2 address. Rather than write out a general proof, we'll show an example of a graph divided into a non-Markov partition and see what can go wrong. We'll find an empty length-3 address whose two length-2 subaddresses (the first two digits and the last two digits of the length-3 address) are non-empty, and see that the non-Markov nature of the partition is responsible for the empty length-3 address.

Before this, we must take care of one technical detail. The graph of the $r = 3.85$ logistic map does not fill the entire interval $[0, 1]$ of y-values. See Fig. A.57. The largest value of $L_{3.85}(x)$ is $L_{3.85}(1/2)$, and the smallest value of $L_{3.85}(x)$, after perhaps a few initial lower values, is $L_{3.85}(L_{3.85}(1/2))$. The interval $[L_{3.85}(L_{3.85}(1/2)), L_{3.85}(1/2)]$ is called the *trapping interval*, because once a point iterates into this interval, no later iterates will leave it. Consequrently, this is the interval that we'll divide into four equal-size bins. To see that this is not a Markov partition,

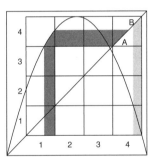

Figure A.57: A partition for the $r = 3.85$ logistic map.

note that while the part of the graph above B_1 crosses all of B_3, it crosses only parts of B_2 and B_4. Now, although both $1 \to 4$ and $4 \to 1$ are allowed, we see $1 \to 4 \to 1$ is forbidden. The function $L_{3.85}$ takes points from only a small part of B_1 into B_4, labeled A on the diagonal line in the graph. And $L_{3.85}$ takes points from only a small part of B_4, labeled B on the diagonal, into B_1. So long as A and B have no points in common, the composition $1 \to 4 \to 1$ will never be observed.

Think about these L_4 and $L_{3.85}$ examples, and you'll see why this problem cannot arise with a Markov partition.

A.91 Driven IFS and graph shapes

Here we present examples of what aspects of a driven IFS can be altered by changing the shape of a function's graph, but not the occupied bins.

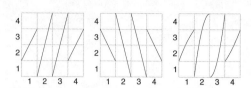

Figure A.58: Same partition and transitions.

A natural question is how much of the driven IFS depends on the specifics of the function. Suppose we have several functions with the same Markov partition, that cross the same bins. Fig. A.58 shows three such graphs.

For all three functions we see

$$B_1 \to B_2 \cup B_3, \quad B_2 \to B_1 \cup B_2 \cup B_3 \cup B_4,$$
$$B_3 \to B_1 \cup B_2 \cup B_3 \cup B_4, \quad B_4 \to B_2 \cup B_3$$

In the second image of Fig. A.59 we see the transition graph; the first image is the memory IFS plot for these allowed transitions. If the time series generated by iterating each of the three functions of Fig. A.58 drives an IFS, we'd expect the result to look like the first image of Fig. A.59. Certainly, every point of the driven IFS must lie on the memory IFS fractal. But if we generate enough points, will they come close to filling this fractal?

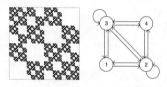

Figure A.59: IFS with memory and its transition graph.

In Fig. A.60 we find the answer, or at least a picture of the answer for these functions. In all three cases, the points of the driven IFS appear to fill the IFS of Fig. A.59, evidently to any address length we wish, provided

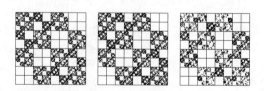

Figure A.60: The driven IFS for Fig. A.58.

we generate enough time series points. More on this in the next section. But notice that while the first and second driven IFS are filled (approximately) uniformly, the third is noticeably clumpy. The shape of the function can influence how the regions of the IFS are filled, but (usually) can't change the list of occupied addresses.

Usually this is true, but there are obvious exceptions. For example, if the function has a stable cycle, then the iterates of many points eventually settle down to this cycle, and so may not visit all of the memory IFS fractal. Fig. A.61 illustrates this. The function is a modification of the first graph of Fig. A.58. All x values where the graph is horizontal iterate immediately to the 2-cycle shown in the second graph of Fig. A.61.

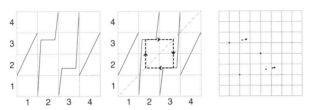

Figure A.61: Function, 2-cycle, and driven IFS.

These two horizontal intervals account for $1/3$ of the length of $[0, 1]$, but the 2-cycle attracts more still. All points that iterate to these intervals also converge to the 2-cycle. If we remove from $[0, 1]$ everything that iterates to the 2-cycle, what remains is a Cantor set. Unless we start with a point of this Cantor set, the driven IFS consists of a sequence of points that converge to the 2-cycle $\{(1/3, 2/3), (2/3, 1/3)\}$, points evenly spaced on the 2–3 diagonal. Although the bins B_1, B_2, B_3, and B_4 are a Markov partition for this function, here the driven IFS does not fill up the memory IFS determined by the forbidden pairs of the graph. To avoid this problem, we must be sure that the iterates of all, or at least most, points visit all the allowed addresses. Avoiding stable cycles is a good step. Of course, visiting all allowed addresses us guaranteed if the function is mixing .

A.92 Markov chains and detecting forbidden pairs

Markov chains (same mathematician, different math) can be used to estimate the likelihood that an address is unoccupied because of an exclusion in the dynamics.

Careful examination of the trapping interval partition in the $L_{3.85}$ example showed us why the transition $1 \to 4 \to 1$ is forbidden, which implies that the address 141 is empty in the driven IFS. But suppose we don't know the process that generated the time series, but we have a driven IFS with $10,000$ data points and notice that address 141 is unoccupied. Is this address empty because we don't have a long-enough time series, or is it empty because that combination is forbidden in the process that generated the time series? We'll develop a Markov chain model to approach this question.

First, we assemble some data from the time series. We'll use n_i to denote the number of time series points in bin B_i. Here are the counts

$$n_1 = 2871, \ n_2 = 1844, \ n_3 = 1674, \ n_4 = 3611$$

Also, note that n_i is the number of driven IFS points in length-1 address square i.

Next, let n_{ij} denote the number of driven IFS points in length-2 address square ij. Here are the non-zero counts.

$$n_{11} = 815, \ n_{14} = 2056, \ n_{21} = 1032, \ n_{24} = 812, \ n_{31} = 899,$$
$$n_{33} = 31, \ n_{34} = 243, \ n_{41} = 125, \ n_{42} = 1844, \ n_{43} = 44$$

With these counts we'll build a Markov chian to find the expected number of iterates before address 141 is occupied. This is similar to the Markov chain we built in A.8. Nevertheless, we'll review the main points of constructing a Markov chain.

Recall that each successive time series point lands in one of four bins, B_1, B_2, B_3, or B_4. If the next time series point lands in B_i, then in the driven IFS the next transformation applied is T_i. This number i is appended on the left of the address of the current driven IFS point. As we read through the time series, we'll keep track of the address of the current point and find the expected number of steps before the triple 141 occurs.

We'll see that we need four states for our Markov chain. The last state, which we'll label state D, corresponds to at least one driven IFS point being in address 141.

How do we get to state D? We must be in state C: address 141 is unoccupied and the two left-most entries of the address of the driven IFS point are 41. Then apply T_1 to enter state D.

To get to state C, we must be in state B: address 141 is unoccupied and the left-most entry of the address of the driven IFS point is 1. Then apply T_4 to enter state C.

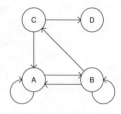

Figure A.62: The graph of the Markov chain to test the occupancy of address 141.

Then state A is everything else. Address 141 is unoccupied, and either the left-most entry of the driven IFS point address is 2 or 3, or the two left-most address entries are 42, 43, or 44.

In Fig. A.62 we see a graphical representation of the states of this Markov chain and the possible transitions between the states. The expected number of iterates to reach state D is calculated from the Markov chain transition matrix, so next we'll use the address occupancies to estimate the transition probabilities.

To estimate $Pr(A \to B)$, we have to count the number of points in A and the number of those points that go to B. Remember, we're using

A.92. MARKOV CHAINS AND FORBIDDEN PAIRS

only pair occupancy information in these calculations. First, to be in state A, the left-most addess digit can be 2 or 3, but with 4 we must be more careful. If the left-most pair is 41, we are in state C, so the complete description of state A is this: the left-most digit is 2 or 3, or the left-most pair is 42, 43, or 44. That is, the number of points in state A is

$$n_2 + n_3 + n_{42} + n_{43} + n_{44} = 1844 + 1674 + 1844 + 44 = 5406$$

Going from A to B is achieved by $2 \to 1$, $3 \to 1$, or $4 \to 1$. (To understand why it's $4 \to 1$ instead of $2 \to 4 \to 1$, for example, remember that only pair information can be used in these calculations. The transition $2 \to 4 \to 1$, which would correspond to hopping from address 42 to address 1, uses triple information, not pair information.) Then the number of points that go from A to B is

$$n_{12} + n_{13} + n_{14} = 0 + 0 + 2056 = 2056$$

and so we have
$$Pr(A \to B) = \frac{2056}{5406} \approx .380$$

The other transition from A is $A \to A$. We can take a shortcut, instead of counting all the ways to go from state A to A. From the graph in Fig. A.62, we know A must go to A or B. Then $Pr(A \to A) + Pr(A \to B) = 1$, and we've just calculated $Pr(A \to B)$, so

$$Pr(A \to A) = 1 - Pr(A \to B) \approx 1 - .380 = .620$$

To estimate $Pr(B \to B)$, observe that $B \to B$ means the left-most address digit is 1, and remains 1 in the next iteration. Any address with left-most entry 1 lies in state B (from the definition of B), so in this example n_1 is the number of points in state B, and n_{11} the number of points in B that stay in B. Then

$$Pr(B \to B) = \frac{n_{11}}{n_1} = \frac{815}{2871} \approx .284$$

Similarly,
$$Pr(B \to C) = \frac{n_{41}}{n_1} = \frac{125}{2871} \approx .044$$

and for $B \to A$, the final transition from B is

$$Pr(B \to A) = 1 - (Pr(B \to B) + Pr(B \to C)) \approx .672$$

Finally,
$$Pr(C \to D) = \frac{n_{14}}{n_4} = \frac{2056}{3611} \approx .569$$

and
$$Pr(C \to A) = 1 - Pr(C \to D) = 1 - .569 = .431$$

Assembling these numbers, we can see the transition matrix for this Markov chain:

$$M = \begin{bmatrix} .620 & .672 & .431 & 0 \\ .380 & .284 & 0 & 0 \\ 0 & .044 & 0 & 0 \\ 0 & 0 & .569 & 1 \end{bmatrix}$$

In A.8 we developed a Markov chain model to find the expected number of steps until the last state is entered. Adapting that approach to this transition matrix, define the matrix Q by removing the bottom row and right column of M.

$$Q = \begin{bmatrix} .620 & .672 & .431 \\ .380 & .284 & 0 \\ 0 & .044 & 0 \end{bmatrix} \quad \text{and} \quad (I-Q)^{-1} = \begin{bmatrix} 75.3 & 72.6 & 32.4 \\ 39.9 & 39.9 & 17.2 \\ 1.8 & 1.8 & 1.8 \end{bmatrix}$$

As before, the expected number of iterations needed to enter state D is the sum of the first column entries, that is, 117.

This means that if the dynamics of this system were determined by pair transition probabilities, then address 141 should be visited in about 117 iterations. That we do not visit this address in 10,000 iterations is a very strong indication that the address 141 corresponds to a combination forbidden by the dynamics of the system.

To support this interpretation, we compute the standard deviation of the number of iterations needed to enter state D, that is, until the address 141 first appears. Write $N = (I - Q)^{-1}$,

$$\vec{t} = \langle t_1, t_2, t_3 \rangle = \langle 1, 1, 1 \rangle N = \langle 117, 114.3, 51.4 \rangle$$

The variance of the number of steps before entering the absorbing state when starting from state i is the ith entry of

$$\vec{t}(2N - I) - \langle t_1^2, t_2^2, t_3^2 \rangle$$

(To find this formula, the easiest approach is to Google "absorbing Markov chain.") The standard deviation, the square root of the variance, of the number of iterations needed to enter state D starting from state 1 is about 114.5. Then 10,000 iterations is about 87 standard deviations from the mean, so we can be pretty secure that address 141 is a forbidden triple, and is not empty because of a paucity of data points.

consequence of a forbidden pair.

Of course, we cannot be completely sure of this conclusion, because we don't know how unlikely 87 standard deviations from the mean is. If times to see the first occurrence of address 141 followed the normal distribution, then not seeing 141 in 10,000 iterations would never have happened in the history of the universe. But we don't know whether or not these times are normally distributed. In fact, there's building evidence that the entry times follow a power law distribution. Still, this many standard deviations from the mean is reason enough to believe that we have detected a forbidden triple.

The next four notes refer to Sect. 7.2, **Driven IFS and synchronization**.

Robert May's 1976 paper introduced general scientific readers to intricate, chaotic dynamics obtained by iterating the logistic map, a simple parabola. How can there still be questions about a parabola that no one can answer? But there are such questions.

Jakobson's paper establishing that for the logistic map chaos is visible – roughly, for r values randomly selected in the range $[0, 4]$, there is a positive probability that $rx(1-x)$ is chaotic – was an important contribution to the study of chaotic dynamical systems that exploded in the 1980s. After all, with so many people studying chaos, we need to know that chaotic systems are not exceedingly rare.

The study of synchronized processes is a rapidly growing branch of science. Despite all that has been written in this field, the best introduction still is Steven Strogatz's book *Sync: The Emerging Science of Spontaneous Order*. The older author bought a copy at Snowbound Books in Marquette, Michigan, in the summer of 2003. This book is so compelling that he read it in two days, annoying his wife, who thought they had gone to the Upper Peninsula for a vacation.

A.93 Bifurcation diagrams of some other functions

It is true that every family of differentiable functions with a single maximum, not too flat (non-zero second derivative at the maximum), has a bifurcation diagram similar qualitatively, and in some sense quantitatively, to the logistic map bifurcation diagram. But if we give up differentiability, or even continuity, entirely new diagrams can arise. For examples, we'll use the tent map T_r (Fig. A.63, first) and a discontinuous hybrid H_r (second) of

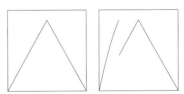

Figure A.63: The tent map and a discontinuous map

the logistic and tent maps. Here are the functions.

$$T_r(x) = \begin{cases} rx & \text{for } 0 \leq x \leq 1/2 \\ r - rx & \text{for } 1/2 \leq x \leq 1 \end{cases}$$

$$H_r(x) = \begin{cases} (8/3)rx(1-x) & \text{for } 0 \leq x \leq 1/4 \\ T_r(x) & \text{for } 1/4 \leq x \leq 1 \end{cases}$$

The factor of $8/3$ guarantees that the logistic and tent branches of H_r have the same maximum heights. For both T_r and H_r, the maximum height is $r/2$. Consequently, to keep both graphs in the unit square, r must lie in the range $0 \leq r \leq 2$.

In Fig. A.64 we see the bifurcation diagram for the tent map. For $0 \leq r < 1$, the graph of T_r lies below the diagonal line and so all inital values x_0 iterate to 0. In this r-range, the tent bifurcation diagram consists of the line of height 0. At $r = 1$ there is an abrupt change, not to a stable non-zero fixed point or cycle,

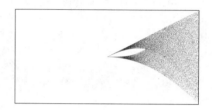

Figure A.64: T_r bifurcation diagram

but immediately to chaos. The diagram is not as simple as a view at this magnification suggests. Zooming in around $(1, 1/2)$ reveals some interesting fine structures.

In the first image of Fig. A.65 we see the bifurcation diagram for H_r. The right half of this diagram looks very much like the right half of the tent map diagram. But what about the left half? Magnified in the middle image, the largest feature is the 2-cycle signaled by two diagonal lines. Going left and right, we see 3-cycles, 4-cycles, 5-cycles, and so on. These might remind you of the principal sequence of discs attached around the Main cardioid of the Mandelbrot set. Next, note that between the 2- and 3-cycles we find a 5-cycle, between the 3- and 4-cycles we find a 7-cycle, and so on. Yes, the Farey sequence is here, too. But the function H_r is not related in any obvious way to the Mandelbrot function $z^2 + c$. Perhaps these patterns have different causes. Another good question.

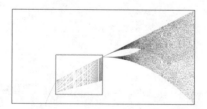 

Figure A.65: The bifurcation diagram for H_r and two magnifications.

A.94 Liapunov exponents

The Liapunov exponent is a way to quantify the degree of chaos, the time horizon over which some prediction is possible. The most familiar aspect of chaos is sensitivity to initial conditions. For time series generated by iterating many functions f, including the logistic map for some values of r, this means that iterates from nearby initial points eventually diverge from one another, and the series from each appear to be unrelated. We use the magnitude of the derivative of f to measure the divergence of nearby trajectories. More common is to measure the exponential divergence rate of nearby trajectories. For this, we average the log of the magnitude of the derivative along the iterates $f^i(x_0)$ of a point x_0. This average is λ, the Liapunov exponent.

$$\lambda = \lim_{N \to \infty} \frac{1}{N} \sum_{i=1}^{N} \ln\left(\left. \left| \frac{df}{dx} \right| \right|_{f^i(x_0)} \right)$$

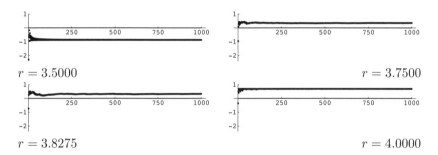

Figure A.66: Estimating the logistic map Liapunov exponent for several r values.

In Fig. A.66 we see graphs of approximations of λ for $N = 1, \ldots, 1000$, and for the indicated r values. That the graphs appear to fall along a horizontal line suggests that these numerical sequences have converged to the Liapunov exponents of these logistic maps. A bit more work is needed to establish the result conclusively, but for us, these graphs suffice.

A negative Liapunov exponent suggests that the iterates of nearby points converge to the same behavior. This is the opposite of sensitivity to initial conditions, and in fact we know that for the $r = 3.5$ logistic map, the iterates of almost every initial point converge to a 4-cycle.

A positive Liapunov exponent suggests sensitivity to initial conditions, and the magnitude of the exponent indicates how rapidly iterates of nearby trajectories diverge and, consequently, the time horizon of prediction of future behavior in terms of the uncertainty in the measurement of initial conditions.

A.95 Bifurcation diagrams of networks of logistic maps

Here we show the bifurcation diagrams of some coupled networks of logistic maps, varying the coupling strength. Each coupling-bifurcation diagram is a representation of how the dynamics of the iterate averages vary with the coupling strength. Global patterns are revealed, but reading the detailed behavior for any c-value may require some practice or augmentation by additional tools. Before getting to that, we'll find a range of c-values for which two coupled logistic maps

$$x_{n+1} = (1-c)rx_n(1-x_n) + cry_n(1-y_n)$$
$$y_{n+1} = crx_n(1-x_n) + (1-c)ry_n(1-y_n)$$

synchronize. To do this, we'll find a number $h < 1$ for which

$$|x_{n+1} - y_{n+1}| \leq h|x_n - y_n|$$

Then for all $k > 0$,

$$|x_{n+k} - y_{n+k}| \leq h^k |x_n - y_n|$$

So as $k \to \infty$, we see $|x_{n+1} - y_{n+1}| \to 0$ and the two logistic maps synchronize. Substituting in the expressions for x_{n+1} and y_{n+1} and simplifying, we find

$$\begin{aligned}|x_{n+1} - y_{n+1}| &= |(1-2c)rx_n(1-x_n) - (1-2c)ry_n(1-y_n)|\\ &= r|1-2c|\cdot|x_n - x_n^2 - (y_n - y_n^2)|\\ &= r|1-2c|\cdot|x_n - y_n - (x_n^2 - y_n^2)|\\ &= r|1-2c|\cdot|x_n - y_n|\cdot|1-(x_n+y_n)|\\ &\leq r|1-2c|\cdot|x_n - y_n|\end{aligned}$$

where the inequality $|1-(x_n+y_n)| \leq 1$ follows from $0 \leq x_n \leq 1$ and $0 \leq y_n \leq 1$. Then we see these two logistic maps synchronize if $|1-2c|r < 1$. Solving for c we find

$$\frac{1}{2} - \frac{1}{2r} < c < \frac{1}{2} + \frac{1}{2r}$$

Back to augmenting the coupling-bifurcation diagrams. For any c-value, we can plot the driven IFS and deduce bin transition probabilities, forbidden pairs, and such. In fact, the first time the older author noticed synchronization of two coupled logistic maps (*not* the first time this phenomenon was noticed), it was because the driven IFS of the average looked identical to the driven IFS of a single logistic map.

But we've looked at driven IFS a lot, so we'll take this opportunity to introduce another approach, the *return map*. Given a sequence of values $x_1, x_2, x_3, x_4, \ldots$, we plot the points

$$(x_1, x_2), (x_2, x_3), (x_3, x_4), \ldots$$

A.95. NETWORKS OF LOGISTIC MAPS

and also the points

$$(x_1, x_2, x_3), (x_2, x_3, x_4), (x_3, x_4, x_5), \ldots$$

These are called the 2- and 3-dimensional return maps. If the values x_n are generated by iterating a function, $x_{n+1} = f(x_n)$, then the points of the 2-dimensional return map fall on the graph of the function f. If this happens for values obtained by measuring some physical variable, then curve-fitting through the return map points lets us approximate the generating function f, and from this we can predict later values of the sequence.

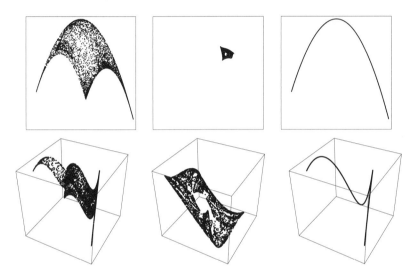

Figure A.67: Return maps for $c = .01$ (first), $c = .1$ (second), and $c = .25$ (third).

In Fig. A.67 we see the 2- and 3-dimensional return maps for two coupled logistic maps, both with $r = 3.9$. To interpret these, we recall the upper first coupling-bifurcation diagram of Fig. 7.22, shown again in Fig. A.68.

The first return maps in Fig. A.67 have $c = .01$. Certainly, the 2-dimensional return map is not the graph of a function. Given $z_n = (x_n + y_n)/2$, we cannot predict z_{n+1}. However, the 3-dimensional return map (the first image of the second row of Fig. A.67) unfolds on a surface, so given z_n and z_{n+1}, we can predict z_{n+2}. For this system, knowing two previous iterates is enough to predict the next.

The second return maps have $c = .1$. The 2-dimensional return map is a small blob with a hole. The 3-dimensional return map, magnified to reveal its structure, again lies on a surface, but some shapes have been cut out of the surface. We do not understand this picture yet.

The third return maps have $c = 0.25$, According to our calculation of the range of c values giving synchronization, these logistic maps fall outside the range of guaranteed synchronization, and yet they do synchronize. Our calculation gives sufficient, but evidently not necessary, conditions for synchronization. But how does the return map signal synchronization? The 2-dimensional return map consists of points lying on the parabola with maximum height .975, the maximum height of the $r = 3.9$ logistic map. Again, despite the sensitivity to initial conditions of chaotic logistic maps, these two are marching in lockstep.

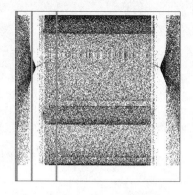

Figure A.68: A coupling-bifurcation diagram.

In Fig. A.68 we see the both $r = 3.9$ coupling-bifurcation diagram with $c = .01, .1$, and .25 marked with gray vertical lines. We can interpret the parts of the diagram in these gray lines by thinking about the top three return maps of Fig. A.67. Recall that for each c value, the corresponding vertical slice of the coupling-bifurcation diagram is a plot, along that single vertical line, of the iterates z_{n+1}, after the first few are dropped so the remaining iterates settle down to their eventual behavior. In each of these return maps, the shadow, or projection, to the z_{n+1}-axis is exactly the collection of points in the coupling-bifurcation diagram vertical line for that c-value.

The large region of the $c = .01$ return map projects to points apparently scattered densely in the interval $.1 \leq z_{n+1} \leq .975$. The much smaller $c = .1$ return map projects to a much smaller interval in the coupling-bifurcation diagram. The $c = .25$ return map projects to a large vertical interval in the coupling-bifurcation diagram. The large rectangular section of the middle of the coupling-bifurcation diagram consists of projections of synchronized return maps, similar to the $c = .25$ example.

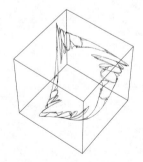

Figure A.69: Return map for $r = 3.44$ and 3.88 and $c = .9$.

These coupling-bifurcation diagrams are effective means to grasp how the distribution of z_{n+1}-values changes with c, but they may not be so useful for distinguishing detailed behaviors. For instance, the $c = .01$ and $c = .25$ coupling-bifurcation diagram sections appear to be almost identical, yet one corresponds to synchronized dynamics, the other not. The other coupling-bifurcation diagrams of Fig. 7.22 can be unpacked by looking at individual return maps.

In Fig. A.69 we see the return map for the average of the iterates of the $r = 3.44$ and $r = 3.88$ logistic maps, coupled with $c = .9$. We don't understand the structure of this picture, either.

A.96 Fractal map lattices and fuzzy synchronization

Finally, we mention what appears to be an unusual form of synchronized behavior for logistic maps arranged on the vertices of the third stage of building the Sierpinski gasket. In this example, the averages of the nearest neighbors are coupled in groups of three; the averages of these groups of three are coupled in nearest neighbor groups of three (that is, in groups of nine logistic maps), and the averages of those groups are coupled. For example, the rule for updating the first logistic map variable x_1 is

$$\begin{aligned}x_1^{n+1} &= \alpha L_1(x_1^n) + \beta(L_2(x_2^n) + L_3(x_3^n)) + \gamma \langle L_1(x_1^n), L_2(x_2^n), L_3(x_3^n) \rangle \\ &+ \delta(\langle L_4(x_4^n), L_5(x_5^n), L_6(x_6^n) \rangle + \langle L_7(x_7^n), L_8(x_8^n), L_9(x_9^n) \rangle) \\ &+ \epsilon \langle L_1(x_1^n), \ldots, L_9(x_9^n) \rangle + \zeta(\langle L_{10}(x_{10}^n), \ldots, L_{18}(x_{18}^n) \rangle \\ &+ \langle L_{19}(x_{19}^n), \ldots, L_{27}(x_{27}^n) \rangle)\end{aligned}$$

where $L_i(x) = r_i x(1-x)$ and pointy brackets denote the average of the bracketed terms. Take the coupling constants $\alpha = .01$, $\beta = .04$, $\gamma = .01$, $\delta = .23$, $\epsilon = .14$, and $\zeta = .15$. In Fig. A.70 we see the return maps of x_1 (first), of $\langle x_1, x_2, x_3 \rangle$ (second), of $\langle x_1, \ldots, x_9 \rangle$ (third), and of $\langle x_1, \ldots, x_{27} \rangle$ (fourth). The fifth image is a plot of the differences $x_1^n - \langle x_{10}^n, x_{11}^n, x_{12}^n \rangle$. In fact, x_1 is synchronized to the averages of the group of three and of the group of nine logistic maps that contain x_1, but not to the average of the whole network. The first, second, and third return maps are very close to identical, but the fourth is not. Note the first three return maps are *not* clean parabolas. They appear to be parabolas out of focus (though identically so – this was a surprise to us), so we call this "fuzzy synchronization."

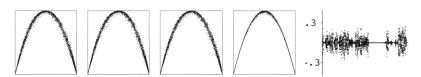

Figure A.70: Return maps and a difference plot of a fractal network.

Fractal networks can give synchronized chaotic processes, but as illustrated here, the return maps need not be simple parabolas; they can be more complex. At the moment, we do not understand this phenomenon.

More questions than answers. Not a bad place to be if you cherish curiosity, as we do.

The next three notes refer to Sect. 7.3, **Multifractals from IFS**. Introduced by Benoit in his 1974 *Journal of Fluid Mechanics* paper, multifractals were developed to study turbulence, a phenomenon so irregular that it cannot be described by a single power law.

A.97 Conditions that simplify drawing $f(\alpha)$ curves

Here we describe some conditions that simplify drawing $f(\alpha)$ curves. Specifically, we'll see why the minimum and maximum probabilities determine the extreme values of α and why, if all the r values are the same, the minimum probability corresponds to the maximum α. For illustration, we'll use the probabilities from Example 1 of Sect. 7.3: $p_1 = 0.1$, $p_2 = p_3 = p_3 = 0.3$, and $r_1 = r_2 = r_3 = r_4 = 0.5$. For any sequence of indices $i_1, \ldots, i_n$,

$$0.1^n \leq \text{Prob}(i_1 \ldots i_n) \leq 0.3^n$$

Then

$$\log(0.1^n) \leq \log(\text{Prob}(i_1 \ldots i_n)) \leq \log(0.3^n)$$

and because $\log(0.5^n) < 0$, for all n we have

$$\frac{\log(0.1)}{\log(0.5)} = \frac{\log(0.1^n)}{\log(0.5^n)} \geq \frac{\log(\text{Prob}(i_1 \ldots i_n))}{\log(0.5^n)} \geq \frac{\log(0.3^n)}{\log(0.5^n)} = \frac{\log(0.3)}{\log(0.5)}$$

So the minimum and maximum probabilities determine the range of α: if all the $r_i = r$,

$$\max(\alpha) = \frac{\log(\min(p_i))}{\log(r)} \qquad \min(\alpha) = \frac{\log(\max(p_i))}{\log(r)}$$

More generally, the max and min values of α can be found like this:

$$\max(\alpha) = \max\left(\frac{\log(p_i)}{\log(r)}\right) \qquad \min(\alpha) = \min\left(\frac{\log(p_i)}{\log(r)}\right)$$

Next, we show how to get α and $f(\alpha)$ from the generalized Moran equation. First we'll do the simpler case that all the $r_i = r$. In this case, the generalized Moran equation

$$p_1^q r_1^{\beta(q)} + \cdots + p_n^q r_n^{\beta(q)} = 1$$

can be solved explicitly for $\beta(q)$:

$$\beta(q) = -\frac{\log(p_1^q + \cdots + p_n^q)}{\log(r)}$$

Originally we defined the Hölder exponent, α, as a function of the address of a region. In order to exploit the relation between α and the generalized Moran equation, we can define α as a function of q. The connection between these approaches is revealed by noting that large positive q weight more heavily the addresses with the highest probabilities, and that large negative q weight more heavily the addresses with the lowest probabilities. In A.98 we give a sketch of why $\alpha = -d\beta/dq$. Then

$$\alpha = -\frac{d\beta}{dq} = \frac{p_1^q \ln(p_1) + \cdots + p_n^q \ln(p_n)}{(p_1^q + \cdots + p_n^q)\ln(r)}$$

So finding α in terms of r, q, and the p_i is relatively easy when all the r_i take the same value. With a bit more work, we can find an explicit expression for α in the more general situation of varied r_i. To do this, we begin by differentiating the generalized Moran equation with respect to q,

$$p_1^q \ln(p_1) r_1^{\beta(q)} + p_1^q r_1^{\beta(q)} \ln(r_1)\frac{d\beta}{dq} + \cdots + p_n^q \ln(p_n) r_n^{\beta(q)} + p_n^q r_n^{\beta(q)} \ln(r_n)\frac{d\beta}{dq} = 0$$

And then we can solve for $d\beta/dq$,

$$\frac{d\beta}{dq} = -\frac{p_1^q r_1^{\beta(q)} \ln(p_1) + \cdots + p_n^q r_n^{\beta(q)} \ln(p_n)}{p_1^q r_1^{\beta(q)} \ln(r_1) + \cdots + p_n^q r_n^{\beta(q)} \ln(r_n)}$$

In either case, identical r_i or varied r_i, we can calculate $\alpha = -d\beta/dt$. In A.99 we'll sketch why $f(\alpha)$ is given by

$$f(\alpha) = \alpha \cdot q + \beta(q)$$

For no, we'll just use this relation.

Next, we will show that the maximum point of the $f(\alpha)$ curve is the dimension of the set generated by the IFS. First, recall that the maximum point on a curve occurs where its derivative is 0, so

$$\begin{aligned}
0 = \frac{df}{d\alpha} &= q + \alpha\frac{dq}{d\alpha} + \frac{d\beta}{d\alpha} && \text{by the product rule} \\
&= q + \alpha\frac{dq}{d\alpha} + \frac{d\beta}{dq}\frac{dq}{d\alpha} && \text{by the chain rule} \\
&= q + \alpha\frac{dq}{d\alpha} - \alpha\frac{dq}{d\alpha} && \text{because } d\beta/dq = -\alpha \\
&= q
\end{aligned}$$

Next, by substituting $q = 0$ into $f(\alpha) = \alpha \cdot q + \beta(q)$, we see the maximum value of $f(\alpha)$ is $\beta(0)$. Finally, substituting $q = 0$ into the generalized Moran equation, we find it turns into the regular Moran equation for the IFS,

$$1 = p_1^0 r_1^{\beta(0)} + \cdots p_n^0 r_n^{\beta(0)}$$
$$= r_1^{\beta(0)} + \cdots + r_n^{\beta(0)}$$

That is, this is just the familiar Moran equation, with $\beta(0)$ equal to the dimension of the attractor of the IFS. Assembling these pieces, we see that the maximum value of $f(\alpha)$ is $\beta(0)$, and that $\beta(0)$ is the dimension of the IFS.

The generalized Moran equation $p_1^q r_1^{\beta(q)} + \cdots + p_n^q r_n^{\beta(q)}$ can be generalized even further by combining it with the memory Moran equation $\rho[m_{ij} r_i^d] = 1$. In this way we can find the $f(\alpha)$ curve for probability measures generated by IFS with memory. Recall p_{ij} is the probability that T_i can follow T_j. Then $\beta(q)$ is the solution of

$$\rho[p_{ij}^q r_i^{\beta(q)}] = 1$$

where $\rho[M]$ is the spectral radius of the matrix M, that is, the largest eigenvalue. Once we have $\beta(q)$, the values of α and $f(\alpha)$ can be found as before.

Here are some illustrations. The six images of Fig. A.71 show points on the $f(\alpha)$ curves for the corresponding transition probability matrices. The T_i are our usual square IFS rules.

$$(a) \begin{bmatrix} .40 & .33 & .50 & .25 \\ .20 & 0 & 0 & .25 \\ .20 & .33 & 0 & .25 \\ .20 & .33 & .50 & .25 \end{bmatrix} \quad (b) \begin{bmatrix} .25 & .33 & .50 & .25 \\ .25 & 0 & 0 & .25 \\ .25 & .33 & 0 & .25 \\ .25 & .33 & .50 & .25 \end{bmatrix}$$

$$(c) \begin{bmatrix} .50 & .25 & .50 & .25 \\ .50 & .25 & .50 & .25 \\ 0 & .25 & 0 & .25 \\ 0 & .25 & 0 & .25 \end{bmatrix} \quad (d) \begin{bmatrix} 0 & .50 & .33 & .33 \\ .50 & .25 & 0 & .33 \\ .25 & 0 & .33 & .34 \\ .25 & .25 & .34 & 0 \end{bmatrix}$$

$$(e) \begin{bmatrix} 0 & .30 & .30 & .30 \\ .30 & .30 & 0 & .30 \\ .30 & 0 & .30 & 0 \\ .40 & .40 & .40 & .40 \end{bmatrix} \quad (f) \begin{bmatrix} .10 & .30 & .30 & .30 \\ .30 & .30 & .10 & .30 \\ .30 & .10 & .30 & .10 \\ .30 & .30 & .30 & .30 \end{bmatrix}$$

One thing we see is that the highest point of the $f(\alpha)$ curve is the dimension of the IFS with memory, as determined by the transition matrix M when all the non-zero entries are changed to 1s. We'll call this the "unitized" transition matrix. Comparing (c) and (e), we can see that a

A.98. THE SLOPE OF THE $\beta(Q)$ CURVE IS $-\alpha$

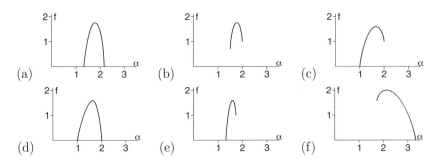

Figure A.71: The $f(\alpha)$ curves for the IFS with the memory given by these matrices.

narrower range in probabilities gives a narrower $f(\alpha)$ curve. A more interesting comparison is (a) and (b) with (d) and (e). The fractal generated by the unitized transition matrix for (a) and (b) can be produced without memory, while the fractal generated by the unitized transition matrix for (d) and (e) cannot. How are the effects of the probabilities disentangled from those of memory? We are just starting to explore these issues.

A.98 The slope of the $\beta(q)$ curve is $-\alpha$

We'll start with an illustrative calculation, then hint at the more general approach. From the generalized Moran equation

$$p_1^q r_1^{\beta(q)} + \cdots + p_n^q r_n^{\beta(q)} = 1$$

we see that the effect of the maximum p_i is revealed by the $q \to \infty$ limit: for smaller p_i, $p_i^q \to 0$ faster as $q \to \infty$.

We'll need a bit more detail. Differentiating the generalized Moran equation with respect to q we find

$$0 = p_1^q r_1^{\beta(q)} \left(\ln(p_1) + \ln(r_1) \frac{d\beta}{dq} \right) + \cdots + p_n^q r_n^{\beta(q)} \left(\ln(p_n) + \ln(r_n) \frac{d\beta}{dq} \right)$$

Because for each i, $0 < p_i < 1$ and $0 < r_i < 1$, each $\ln(p_i) < 0$ and each $\ln(r_i) < 0$, so we see $d\beta/dq < 0$: β is a decreasing function of q.

Differentiating the generalized Moran equation again and doing a bit of algebra,

$$0 = p_1^q r_1^{\beta(q)} \left(\left(\ln(p_1) + \ln(r_1) \frac{d\beta}{dq} \right)^2 + \ln(r_1) \frac{d^2\beta}{dq^2} \right)$$
$$+ \cdots + p_n^q r_n^{\beta(q)} \left(\left(\ln(p_n) + \ln(r_n) \frac{d\beta}{dq} \right)^2 + \ln(r_n) \frac{d^2\beta}{dq^2} \right)$$

Because each $\ln(r_i) < 0$, $d^2\beta/dq^2 \geq 0$; so long as not all $\ln(p_i)/\ln(r_i)$ are equal, $d^2\beta/dq^2 > 0$. So β is a decreasing, concave up function of q.

To simplify the final bits of the argument, suppose that $r_1 = \cdots = r_n = r$, and $p_1 < p_2 \leq \cdots \leq p_{n-1} < p_n$. Then from the simplified expression $\beta(q) = -\log(p_1^q + \cdots + p_n^q)/\log(r)$, to understand the $q \to \infty$ limit we separate out p_n, the maximum probability.

$$\lim_{q \to \infty} \beta(q) = \lim_{q \to \infty} -\frac{\log\left(p_n^q\left(\left(\frac{p_1}{p_n}\right)^q + \cdots + \left(\frac{p_{n-1}}{p_n}\right)^q + 1\right)\right)}{\log(r)}$$

$$= \lim_{q \to \infty} -\frac{\log(p_n^q)}{\log(r)} - \lim_{q \to \infty} \frac{\log\left(\left(\frac{p_1}{p_n}\right)^q + \cdots + \left(\frac{p_{n-1}}{p_n}\right)^q + 1\right)}{\log(r)}$$

$$= -q\frac{\log(p_n)}{\log(r)}$$

The last equality follows because each of $(p_1/p_n)^q, \ldots, (p_{n-1}/p_n)^q$ goes to 0 as $q \to \infty$. That is, for large q, $\beta(q)$ looks like

$$\beta = -q\log(p_n)/\log(r) = -q\log(\max(p_i))/\log(r)$$

a straight line through the origin with slope $-\log(\max(p_i))/\log(r)$.

A similar argument shows that for large negative q (that is, negative q with $|q|$ large), $\beta(q)$ looks like the straight line

$$\beta = -q\log(p_1)/\log(r) = -q\log(\min(p_i))/\log(r)$$

How to relate this back to the max and min values of α is straightforward. Recall that because all $r_i = r$, $\min(\alpha) = \log(\max(p_i))/\log(r)$. Then for large q,

$$\frac{d\beta}{dq} = -\log(\max(p_i))/\log(r) = -\min(\alpha)$$

and as $q \to -\infty$,

$$\frac{d\beta}{dq} = -\log(\min(p_i))/\log(r) = -\max(\alpha)$$

That $\alpha = -d\beta/dq$ for all α requires more work, but the basic idea is contained in this example.

The general argument uses the *Legendre transform* of β. For each α between $\min(\alpha)$ and $\max(\alpha)$, the Legendre transform of β is

$$f(\alpha) = \inf_{-\infty < q < \infty} \{\beta(q) + \alpha q\}$$

A.99. WHY $F(\alpha) = \alpha \cdot Q + \beta(Q)$

In Fig. A.72 we see a $\beta(q)$ curve and two lines of slope $-\alpha$, one passing through the point $(q, \beta(q))$ and tangent to the curve at that point, the other passing through another point $(q', \beta(q'))$ on the curve. The intersection of these lines with the β-axis is $\beta(q) + \alpha q$ and $\beta(q') + \alpha q'$: the lines have slope $-\alpha$ and pass through the points $(0, \beta)$ and $(q, \beta(q))$, and $(0, \beta)$ and $(q', \beta(q'))$, respectively. From the figure we see that the minimum value of the intersection with the β-axis occurs where the line of slope $-\alpha$ is tangent to the $\beta(q)$ curve. At the minimum of $\beta(q) + \alpha q$, the derivative is 0. That is,

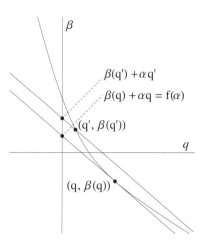

Figure A.72: A β curve and two lines in the family for the Legendre transform of $\beta(q)$.

$$0 = \frac{d}{dq}(\beta(q) + \alpha q) = \frac{d\beta}{dq} + \alpha$$

and so $\alpha = -d\beta/dq$.

This brief sketch doesn't answer some questions. Why is this α the Hölder exponent? What is the Legendre transform and why do we use it? But our sketch for the minimum and maximum values of α and the minimum and maximum values of $-d\beta/dq$ are enough of a hint.

A.99 A sketch of why $f(\alpha) = \alpha \cdot q + \beta(q)$

Recall that for an IFS with attractor A and transformations $T_1, \ldots, T_k$, where r_i is the contraction factor of T_i, then for any length-n address $\vec{i} = (i_1 \cdots i_n)$, the region with address $\vec{i}$ is $A_{\vec{i}} = T_{i_1} \circ \cdots \circ T_{i_n}(A)$. For each α between $\min(\alpha)$ and $\max(\alpha)$, define

$$A(\alpha) = \left\{ x \in A : \lim_{n \to \infty} \frac{\log(\mu(A_{\vec{i}}(x)))}{\log(|A_{\vec{i}}(x)|)} = \alpha \right\}$$

where $A_{\vec{i}}(x)$ is the length-n address region that contains x, $\mu(A_{\vec{i}}(x))$ is the probability of landing in this region, and $|A_{\vec{i}}(x)|$ is the side length of the region.

Next, we identify those addresses where the (region side length)$^\alpha$ is close to the probability of landing in the region:

$$Q_n(\alpha) = \{\vec{i} = (i_1 \cdots i_n) : |A_{\vec{i}}|^\alpha \approx \mu(A_{\vec{i}})\}$$

For reference, I_n denotes the list of all length-n addresses. Now

$$\sum_{\vec{i} \in Q_n(\alpha)} |A_{\vec{i}}|^{\alpha \cdot q + \beta} = \sum_{\vec{i} \in Q_n(\alpha)} (|A_{\vec{i}}|^{\alpha})^q |A_{\vec{i}}|^{\beta}$$

$$\approx \sum_{\vec{i} \in Q_n(\alpha)} \mu(A_{\vec{i}})^q |A_{\vec{i}}|^{\beta}$$

$$\leq \sum_{\vec{i} \in I_n} \mu(A_{\vec{i}})^q |A_{\vec{i}}|^{\beta}$$

$$= \sum_{i_1,\ldots,i_n=1}^{k} (p_{i_1} \cdots p_{i_n})^q (r_{i_1} \cdots r_{i_n})^{\beta} \quad \text{(a)}$$

$$= \sum_{i_1,\ldots,i_n=1}^{k} (p_{i_1}^q r_{i_1}^{\beta}) \cdots (p_{i_n}^q r_{i_n}^{\beta})$$

$$= \left(\sum_{i=1}^{k} p_i^q r_i^{\beta} \right)^n \quad \text{(b)}$$

$$= 1^n = 1 \quad \text{(c)}$$

Equality (a) follows because the probability of landing in $A_{\vec{i}}$ is $p_{i_1} \cdots p_{i_n}$ and the side length of $A_{\vec{i}}$ is $r_{i_1} \cdots r_{i_n}$, the product of the contraction factors of the transformations taking A to $A_{\vec{i}}$.

Equality (b) follows from the multinomial expansion, for example,

$$(x + y + z)^2 = xx + xy + xz + yx + yy + yz + zx + zy + zz:$$

squaring $x + y + z$ gives the sum of all pairs of x, y, and z. Changing the number of terms to n and the exponent to k, we have (b).

Equality (c) is the generalized Moran equation, $p_1^q r_1^{\beta} + \cdots p_k^q r_k^{\beta} = 1$.

The big leap is figuring out what $\sum_{\vec{i} \in Q_n(\alpha)} |A_{\vec{i}}|^{\alpha \cdot q + \beta} \leq 1$ tells us. The dimension used in multifractals is the Hausdorff dimension, a more delicate and complex notion than we have used elsewhere in this book. For the last step of the sketch, we must take a quick glance at Hausdorff dimension.

Hausdorff dimension is based on Hausdorff measure, a way to quantify a shape in any dimension, integer or not. Suppose A is the set we wish to measure. By a *cover* of A we mean a collection of sets $B_1, B_2, \ldots$ which cover A, that is, $A \subseteq \cup_i B_i$. For any number $\delta > 0$, we say $\{B_i\}$ is a δ-*cover* if the diameter of each B_i is $|B_i| \leq \delta$. The diameter of a set that is not a circle is just what you'd expect: the greatest distance between any pair of points in the set. Then for each $\delta > 0$ and for each dimension $d > 0$,

$$\mathcal{H}_{\delta}^d(A) = \inf \left\{ \sum_{i=1}^{\infty} |B_i|^d \ : \ \text{where } \{B_i\} \text{ is a } \delta\text{-cover of } A \right\}$$

A.99. WHY $F(\alpha) = \alpha \cdot Q + \beta(Q)$

Computing this can be tricky, because we must somehow figure out what happens with every δ-cover, and then take the greatest lower bound of all these sums. If we use a smaller δ, then the collection of δ-covers is smaller (Every set in a 1/4-cover also is a set in a 1/2-cover, but the larger sets in a 1/2-cover need not belong to any 1/4-cover.) and so the inf may be larger or could stay the same, but can't be smaller. From this we can see that the limit

$$\mathcal{H}^d(A) = \lim_{\delta \to 0} \mathcal{H}^d_\delta(A)$$

exists, though it may be ∞. This is the *d-dimensional Hausdorff measure*.

You can see that this is pretty involved, and we still haven't gotten to the Hausdorff dimension. A simple example will help us here. Consider a filled-in square. The square is 2-dimensional and area is the familiar measure of 2-dimensional shapes. But what's the volume of a square? A square doesn't enclose any volume at all; the only sensible answer is that the volume if a square is 0. What about the length of the square? We could slice the square into strips and place the strips end to end in a ribbon. Thinner strips give longer ribbons, and it's easy to see that no matter how long the ribbon, we always can make one longer by slicing the strips more thinly. The obvious answer is that the length of the square is infinite.

This example illustrates a fundamental principle: measuring in a dimension lower than that of a shape gives the value ∞; measuring in a dimension higher than that of the shape gives 0. This holds for Hausdorff measure. There is a number, $d_H(A)$, where the Hausdorff measure jumps from ∞ to 0. That is,

$$\mathcal{H}^d(A) = \infty \text{ for } d < d_H(A), \text{ and } \mathcal{H}^d(A) = 0 \text{ for } d > d_H(A).$$

Finally, we can go back to $\sum_{\vec{i} \in Q_n(\alpha)} |A_{\vec{i}}|^{\alpha \cdot q + \beta} \leq 1$. For every $\delta > 0$, there is a n large enough that $|A_{\vec{i}}| < \delta$. Then $\{A_{\vec{i}} : \vec{i} \in Q_n(\alpha)\}$ is a δ-cover of $A(\alpha)$. (A careful argument is a bit trickier, but this is the basic idea.) Then because

$$\sum_{\vec{i} \in Q_n(\alpha)} |A_{\vec{i}}|^{\alpha \cdot q + \beta} \leq 1$$

we see that

$$\inf \left\{ \sum_{i=1}^{\infty} |B_i|^{\alpha \cdot q + \beta} : \text{ where } \{B_i\} \text{ is a } \delta\text{-cover of } A(\alpha) \right\} \leq 1$$

That is,

$$\mathcal{H}^{\alpha \cdot q + \beta}(A(\alpha)) \leq 1$$

Recall that $f(\alpha) = d_H(A(\alpha))$ is the number where $\mathcal{H}^d(A(\alpha))$ jumps from ∞ to 0. Then $\mathcal{H}^{\alpha \cdot q + \beta}(A(\alpha)) \leq 1$ means that $d_H(A(\alpha)) \leq \alpha \cdot q + \beta$.

Generally, establishing an upper bound on Hausdorff dimension is easier than finding a lower bound. A cleverly chosen family of δ-covers works for an upper bound; for the lower bound we must identify some behavior common to all δ-covers. A complete argument that $f(\alpha) = \alpha \cdot q + \beta$ is given in Falconer's *Fractal Geometry. Mathematical Foundations and Applications*. Nevertheless, we hope this little sketch makes plausible how $f(\alpha)$ and $\alpha \cdot q + \beta$ are related.

This note refers to Sect. 7.4, **Applications of multifractals**

A.100 Multifractal finance cartoons and trading time

Here we give some details of Benoit's multifractal finance cartoons, and sketch how to apply the Trading Time Theorem.

We'll begin by illustrating how a multifractal cartoon can produce a time series of varying roughness, that is, of varying Hölder exponents. We'll end with a sketch of the proof of the

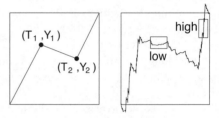

Figure A.73: First: A multifractal generator. Second: The seventh stage, noting regions of different roughnesses.

Trading Time Theorem, showing that if time is rescaled in a way that reflects changes in volatility – slowdown time during periods of high volatility and speed-up time during periods of low volatility – a multifractal cartoon becomes fractional Brownian motion.

In the first image of Fig. A.73 we see the first iteration of an IFS that produces a multifractal cartoon. We'll always assume the cartoon starts at $(0,0)$ and ends at $(1,1)$ (a reasonable assumption in the 1990s when Benoit was exploring these cartoons, but maybe a different choice would better match current market realities; nowadays we shouldn't assume that the general trend of the market is up). In the second image of the figure is a shuffled seventh stage of the construction. Notice that this graph has regions of low roughness and regions of high roughness, of low volatility and of high volatility.

Multifractal cartoons have memory – previous jumps influence current jumps – and the probability of jumps falls off according to a power law, not the exponential of a bell curve. Both these features are also observed in real data. Very large jumps can mask the memory effects, making them mostly vanish when the scale of the graph is set by the big jumps. Can

A.100. TRADING TIME

we disentangle the visual effects of memory and large jumps?

The Trading Time Theorem provides an answer. First, we'll establish some notation, using the first image of Fig. A.73 as a model. The points (T_1, Y_1) and (T_2, Y_2) are called *turning points* of the graph. The variable T stands for time, *clock time* to be precise. Our goal is to rescale time according to volatility. This will be *trading time*.

Volatility refers to vertical jumps in the graph, so we'll need to measure the sizes of these jumps. The variable Y stands for whatever we're measuring. For these applications, we may as well call it price. Then the right side of Fig. A.73 shows a time record of price. The price and clock time increments are

$$\Delta Y_1 = Y_1 - 0, \ \Delta Y_2 = Y_2 - Y_1, \ \Delta Y_3 = 1 - Y_2$$
$$\Delta T_1 = T_1 - 0, \ \Delta T_2 = T_2 - T_1, \ \Delta T_3 = 1 - T_2$$

The Hölder exponent of each branch (straight line segment in the first image of Fig. A.73) is defined by

$$H_i = \frac{\log |\Delta Y_i|}{\log(\Delta T_i)}$$

If at least some branches have different Hölder exponents, then the cartoon has several power laws, and so is a multifractal. If all the branches have the same Hölder exponent, then the cartoon is what Benoit called a *unifractal*, a simple fractal characterized by a single power law. In fact, unifractal cartoons are fractional Brownian motion, with the common value of H_i equal to the index α of fBm.

The Trading Time Theorem states that, expressed in multifractal trading time, every multifractal cartoon becomes fractional Brownian motion.

The main step of the proof is just the Moran equation, applied to $|\Delta Y_i|$. Because each $|\Delta Y_i|$ satisfies $0 < |\Delta Y_i| < 1$, the proof that the Moran equation has a unique solution can be adapted, more or less word for word, to show there is a unique solution D of the trading time equation

$$|\Delta Y_1|^D + |\Delta Y_n|^D + \cdots + |\Delta Y_n|^D = 1,$$

where n is the number of branches in the cartoon generator. Now the *trading time increments* $\Delta \tau_i$ are defined to be

$$\Delta \tau_i = |\Delta Y_i|^D$$

In the price vs. trading time cartoon, the first turning point is $(\Delta \tau_1, \Delta Y_1)$; the other turning points are adjusted similarly.

To verify that the price vs. trading time cartoon is unifractal, for each i we see

$$H_i = \frac{\log |\Delta Y_i|}{\log(\Delta \tau_i)} = \frac{\log |\Delta Y_i|}{\log(|\Delta Y_i|^D)} = \frac{1}{D}$$

In the first image of Fig. A.74 we see a shuffled cartoon with each $\Delta T_i = 1/4$, and with $\Delta Y_1 = \Delta Y_3 = 1/2$, $\Delta Y_2 = -1/4$, and $\Delta Y_4 = 1/4$. Because all the ΔT_i are the same, and some of the ΔY_i are different, this cartoon certainly is multifractal.

Usually, to find the trading time increments we must solve the Moran equation approximately by numerical methods, using the Solve button on a graphing calculator, for example. But every now and again, we can solve it exactly by algebraic methods. If you think about when we could solve the Moran equation exactly for the similarity dimension, you can figure this out without reading the rest of this paragraph. If you don't quite see it yet, give us a minute. For these $|\Delta Y_i|$, the Moran equation becomes

$$2 \cdot \left(\frac{1}{2}\right)^D + 2 \cdot \left(\frac{1}{4}\right)^D = 1$$

Substituting $x = (1/2)^D$, this becomes the quadratic $2x + 2x^2 = 1$. The positive solution is $x = (-1 + \sqrt{3})/2$ and so the exponent D is

$$D = \frac{\log((-1 + \sqrt{3})/2)}{\log(1/2)}$$

Then we can find the trading time increments:

$$\Delta \tau_1 = \Delta \tau_3 = \left(\frac{1}{2}\right)^D = \left(\frac{1}{2}\right)^{\log((-1+\sqrt{3})/2)/\log(1/2)}$$

$$\Delta \tau_2 = \Delta \tau_4 = \left(\frac{1}{4}\right)^D = \left(\left(\frac{1}{2}\right)^D\right)^2$$

Taking the log of both sides of these equations, after a little bit of algebra we see

$$\Delta \tau_1 = \Delta \tau_3 = \frac{-1 + \sqrt{3}}{2} \qquad \Delta \tau_2 = \Delta \tau_4 = \frac{2 - \sqrt{3}}{2}$$

One reason the Trading Time Theorem is important is that conversion to trading time absorbs the large price jumps into the rescaled time, allowing us to focus more on the dependence relations, on how the past influences the future.

This works for data produced by iterating a cartoon generator. For

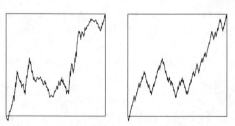

Figure A.74: First: A multifractal cartoon. Second: The equivalent fractional Brownian motion in multifractal trading time.

A.100. TRADING TIME

the trading time calculations, we need the ΔT_i and ΔY_i values, which we deduce from this generator. Finding a generator to mimic, even approximately, a real data time series is a much more difficult problem.

The theorem is useful for another reason: the time rescaling itself. When the volatility is low, a large amount of clock time is compressed into a small amount of trading time. When the volatility is high, a small amount of clock time is expanded into a large amount of trading time. An accurate clock time to trading time converter would estimate the times of high and of low volatility, or at least the fractions of clock time where we see high and also low volatility. This is something we imagine would be useful to know, but your authors have no idea why.

Appendix B
Solutions to the problems

Here we give the solutions to the exercises sprinkled throughout the text. Here is a map the solutions, organized by sections.

B.1	Solutions for Sect. 2.1 A simple way to grow fractals	407
B.2	Solutions for Sect. 2.5 IFS with memory	409
B.3	Solutions for Sect. 6.1 Similarity dimension	413
B.4	Solutions for Sect. 6.2 Box-counting dimension	414
B.5	Solutions for Sect. 6.4 Random, with memory, nonlinear	418
B.6	Solutions for Sect. 6.5 Dimension rules	428
B.7	Solutions for Sect. 7.1 Driven IFS	430
B.8	Solutions for Sect. 7.3 Multifractals from IFS	434

B.1 Solutions for the exercises of Sect. 2.1.

Here are IFS rules for the fractals of Fig. 2.6.

1. 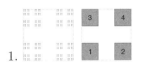 2. 3.

1.

	r	s	θ	φ	e	f
1	1/3	1/3	0	0	0	0
2	1/3	1/3	0	0	2/3	0
3	1/3	1/3	0	0	0	2/3
4	1/3	1/3	0	0	2/3	2/3

2.

	r	s	θ	φ	e	f
1	1/2	1/2	0	0	0	0
2	1/2	1/2	0	0	1/2	0
3	1/2	1/2	0	0	0	1/2
4	1/4	1/4	0	0	3/4	3/4

407

3.

	r	s	θ	φ	e	f
1	1/2	1/2	0	0	0	0
2	1/2	1/2	0	0	1/2	0
3	1/2	1/2	0	0	0	1/2
4	1/4	1/4	0	0	3/4	3/4
5	-1/4	-1/4	0	0	1	1

4. 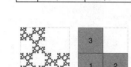 5. 6.

4.

	r	s	θ	φ	e	f
1	1/2	1/2	0	0	0	0
2	1/2	1/2	180	180	1	1/2
3	1/2	1/2	180	180	1/2	1

5.

	r	s	θ	φ	e	f
1	1/2	1/2	0	0	0	0
2	-1/2	1/2	90	90	1	1/2
3	1/2	1/2	0	0	0	1/2

6.

r	s	θ	φ	e	f
-1/2	1/2	0	0	1/2	0
-1/2	1/2	90	90	1	1/2
1/2	1/2	0	0	0	1/2

7. 8. 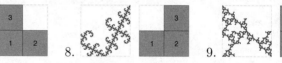 9.

7.

	r	s	θ	φ	e	f
1	1/2	-1/2	0	0	0	1/2
2	1/2	1/2	-90	-90	1/2	1/2
3	1/2	1/2	0	0	0	1/2

8.

	r	s	θ	φ	e	f
1	1/2	1/2	-90	-90	0	1/2
2	1/2	1/2	180	180	1	1/2
3	1/2	1/2	0	0	1/2	1/2

9.

	r	s	θ	φ	e	f
1	1/2	1/2	-90	-90	0	1/2
2	1/2	1/2	180	180	1	1/2
3	1/2	1/2	0	0	0	1/2

B.2 Solutions for the exercises of Sect. 2.5.

Exercise 1. Fig. B.1 is the same as Fig. 2.35, but here we superimpose the length-3 address squares.

(a) We see the forbidden triples are

$$111, 122, 133, 144, 211, 222, 233, 244, 311, 322, 333, 344, 411, 422, 433, 444$$

and there are no forbidden pairs, so manifestly this example is not the result of forbidden pairs.

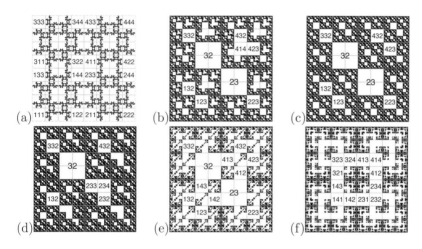

Figure B.1: Fractals for exercise 1 of Sect. 2.5.

(b) We see the forbidden pairs 23 and 32 and the forbidden triples 123, 132, 223, 332, 432, 423, and 414. This last forbidden triple does not contain a forbidden pair, so this IFS is not the result of forbidden pairs.

(c) We see the forbidden pairs 23 and 32 and the forbidden triples 123, 132, 223, 332, 432, and 423. Each forbidden triple contains a forbidden pair, so this IFS is determined by forbidden pairs.

(d) The only forbidden pair is 32; the forbidden triples are 132, 232, 233, 234, 332, and 432 The triples 233 and 234 do not contain the pair 32, so this IFS is not determined by forbidden pairs.

(e) The forbidden pairs are 23 and 32; the forbidden triples are 123, 132, 142, 143, 223, 332, 412, 413, 423, and 432. The triples 142, 143, 412, and 413 do not contain either pair 23 and 32, so this IFS is not determined by forbidden pairs.

(f) There are no forbidden pairs; consequently, none of the forbidden triples, 141, 142, 143, 231, 232, 234, 321, 323, 324, 412, 413, and 414, are determined by forbidden pairs.

410   APPENDIX B. SOLUTIONS

Exercise 2. For each of the fractals of Fig. 2.34 we show the empty length-2 address, draw the transition graph by removing from the 16 possible arrows those for the transitions forbidden by the empty length-2 address, and check the conditions for generating this fractal by an IFS without memory.

(a) The empty length-2 addesses are 14, 24, and 34, so the forbidden transitions are $4 \to 1$, $4 \to 2$, and $4 \to 3$. Because all four arrows go to state 4, we see that state 4 is a rome, the only one. Because there are no paths from the rome to states 1, 2, or 3, the non-romes, this forbidden pairs IFS cannot be generated by an IFS without memory.

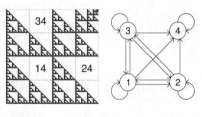

(b) The empty length-2 addresses are 22, 23, 24, 32, 33, and 34, so the forbidden transitions are $2 \to 2$, $3 \to 2$, $4 \to 2$, $2 \to 3$, $3 \to 3$, and $4 \to 3$. Because all four arrows go to states 1 and 4, these states are romes. In the transition graph there are paths $1 \to 2$ and $1 \to 3$ to the two non-

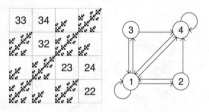

romes, and there are no loops passing through only non-romes. In fact, there are no paths at all passing through only non-romes. Consequently, this forbidden pairs IFS can be generated by an IFS without memory. This table describes an IFS that generates this fractal.

r	s	θ	φ	e	f
1/2	1/2	0	0	0	0
1/2	1/2	0	0	1/2	1/2
1/4	1/4	0	0	1/2	0
1/4	1/4	0	0	0	1/2

(c) The empty length-2 addresses are 22, 23, 24, 33, and 34, so the forbidden transitions are $2 \to 2$, $3 \to 2$, $4 \to 2$, $3 \to 3$, and $4 \to 3$. Because all four arrows go to states 1 and 4, these states are romes. In the transition graph there are paths $1 \to 2$ and $1 \to 3$ to the two non-romes, and

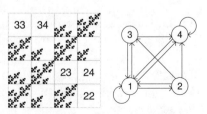

there are no loops passing through only non-romes. Consequently, this forbidden pairs IFS can be generated by an IFS without memory. The path $1 \to 2 \to 3$ produces the fifth row in this IFS table.

B.2. SOLUTIONS FOR SECT. 2.5

r	s	θ	φ	e	f
1/2	1/2	0	0	0	0
1/2	1/2	0	0	1/2	1/2
1/4	1/4	0	0	1/2	0
1/4	1/4	0	0	0	1/2
1/8	1/8	0	0	1/4	1/2

(d) The empty length-2 addresses are 22, 23, 24, 32, 33, 34, 42 and 43, so the forbidden transitions are $2 \to 2$, $3 \to 2$, $4 \to 2$, $2 \to 3$, $3 \to 3$, $4 \to 3$, $2 \to 4$, and $3 \to 4$. Because all four arrows go to state 1, that state is a rome,. There are paths $1 \to 2$, $1 \to 3$, and $1 \to 4$ to each non-rome.

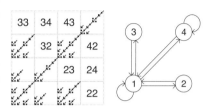

However, there is a loop, $4 \to 4$, so while this forbidden pairs IFS can be generated by an IFS without memory, the table would be infinitely long.

(e) The empty length-2 addresses are 14, 23, 32, and 41, so the forbidden transitions are $4 \to 1$, $3 \to 2$, $2 \to 3$, and $1 \to 4$. Each vertex has only three arrows going to it, so there are no romes. Consequently, this forbidden pairs IFS cannot be generated by an IFS without memory.

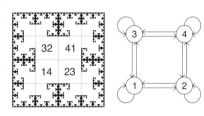

(f) The empty length-2 addresses are 11, 23, 32, and 44, so the forbidden transitions are $1 \to 1$, $3 \to 2$, $2 \to 3$, and $4 \to 4$. Each vertex has only three arrows going to it, so there are no romes. Consequently, this forbidden pairs IFS cannot be generated by an IFS without memory.

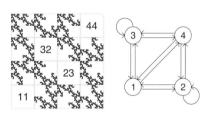

(g) The empty length-2 addresses are 12, 24, 31, and 43, so the forbidden transitions are $2 \to 1$, $4 \to 2$, $1 \to 3$, and $4 \to 3$. Each vertex has only three arrows going to it, so there are no romes. Consequently, this forbidden pairs IFS cannot be generated by an IFS without memory. Is this getting monotonous?

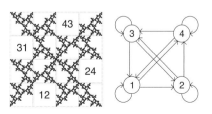

(h) The empty length-2 addresses are
11, 14, 41, 43, and 44, so the forbidden transitions are $1 \to 1$, $4 \to 1$, $1 \to 4$, $4 \to 3$, and $4 \to 4$. Because all four arrows go to states 2 and 3, these states are romes. The non-romes are states 1 and 4, and there are paths $2 \to 1$, $2 \to 4$, $3 \to 1$. There are no loops through non-romes; in fact, every arrow of the transition graph starts or ends with a rome. Consequently, this forbidden pairs IFS can be generated by an IFS without memory, with this table.

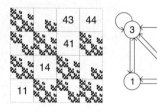

r	s	θ	φ	e	f
1/2	1/2	0	0	1/2	0
1/2	1/2	0	0	0	1/2
1/4	1/4	0	0	1/4	0
1/4	1/4	0	0	0	1/4
1/4	1/4	0	0	3/4	1/2

(i) The empty length-2 addresses are 13, 14, 23, and 24 (Why? See the next paragraph.), so the forbidden transitions are $1 \to 3$, $1 \to 4$, $2 \to 3$, and $4 \to 2$. Because all four arrows go to states 3 and 4, these states are romes, Because there is no path from either rome to either non-rome, this forbidden pairs IFS cannot be generated by an IFS without memory.

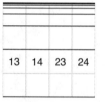

But wait, from the IFS how can we tell that addresses 13, 14, 23, and 24 are empty? Couldn't the horizontal line L bisecting the unit square be on the tops of addresses 13, 14, 23, and 24 rather than on the bottoms of addresses 31, 32, 41, and 42? The line L can be produced in several ways. For example, we can apply T_3 and T_4 to the 1–2 line, or we can apply T_1 and T_2 to the 3–4 line. (There are two other ways as well: apply T_1 to the 3–4 line and apply T_4 to the 1–2 line. Do you see the fourth way to generate the line L?) If we apply T_1 and T_2 to the 3–4 line, then we must allow the transitions $3 \to 1$, $3 \to 2$, $4 \to 1$ and $4 \to 2$. But this would map everything in addresses 3 and 4 into addresses 13, 14, 23 and 24. In other words, these addresses would be filled with a cascade of horizontal lines converging to the line L. We don't see this, so the transitions $3 \to 1$, $3 \to 2$, $4 \to 1$ and $4 \to 2$ must be forbidden.

The other length-2 addresses that look like they might be empty are 11, 12, 21, 22, 31, 32, 41, and 42. For 11, 12, 21, and 22, the corresponding transitions must be allowed in order to draw the 1–2 line. And 31, 32, 41, and 42, cannot be empty because the corresponding transitions must be allowed in order to map the 3–4 line to the line L.

B.3 Solutions for the exercises of Sect. 6.1.

To find the similarity dimension of the fractals in Fig. 6.6, shown again in Fig. B.2, determine the scaling factors r_i of the pieces of the fractal and then apply the basic similarity dimension formula or the Moran equation, as appropriate.

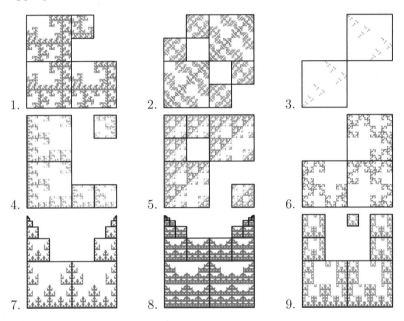

Figure B.2: Similarity dimension exercises.

Exercise 1. We see four copies, with scaling factors $r_1 = r_2 = r_3 = 1/2$ and $r_4 = 1/4$. Taking $x = (1/2)^d$, the Moran equation becomes $3x + x^2 = 1$. The solutions are $x = (-3 \pm \sqrt{13})/2$. Taking the positive solution for x, the dimension is $d_s = \log((-3 + \sqrt{13})/2)/\log(1/2)$.

Exercise 2. We see six copies, with scaling factors $r_1 = r_2 = 1/2$ and $r_3 = r_4 = r_5 = r_6 = 1/4$. Taking $x = (1/2)^d$, the Moran equation becomes $2x + 4x^2 = 1$. The solutions are $x = (-1 \pm \sqrt{5})/4$. Taking the positive solution for x, the dimension is $d_s = \log((-1 + \sqrt{5})/4)/\log(1/2)$.

Exercise 3. We see two copies, with scaling factors $r_1 = r_2 = 1/2$, so by the basic similarity dimension formula, $d_s = \log(2)/\log(2) = 1$. This fractal doesn't look anything like a line segment, but it and a line segment share a dimension.

Exercise 4. We see five copies, with scaling factors $r_1 = r_2 = 1/2$ and $r_3 = r_4 = r_5 = 1/4$. Taking $x = (1/2)^d$, the Moran equation becomes $2x + 3x^2 = 1$. The solutions are $x = -1$ and $x = 1/3$. Taking the positive solution for x, the dimension is $d_s = \log(1/3)/\log(1/2) = \log(3)/\log(2)$.

Exercise 5. We see six copies, with scaling factors $r_1 = r_2 = 1/2$ and $r_3 = r_4 = r_5 = r_6 = 1/4$. Taking $x = (1/2)^d$, the Moran equation becomes $2x + 4x^2 = 1$. The solutions are $x = (-1 \pm \sqrt{5})/4$. Taking the positive solution for x, the dimension is $d_s = \log((-1 + \sqrt{5})/4)/\log(1/2)$. Does this look familiar? It should.

Exercise 6. We see three copies, with scaling factors $r_1 = r_2 = r_3 = 1/2$, so by the basic similarity dimension formula, $d_s = \log(3)/\log(2)$.

Exercise 7. We see infinitely many copies, with scaling factors $r_1 = r_2 = 1/2$, $r_3 = r_4 = 1/4$, $r_5 = r_6 = 1/8$, $r_7 = r_8 = 1/16$, and so on. The infinite Moran equation, with $x = (1/2)^d$, is

$$1 = 2x + 2x^2 + 2x^3 + \cdots = 2x \cdot \frac{1}{1-x}$$

This equation has only one solution, $x = 1/3$, and so the similarity dimension $d_s = \log(3)/\log(2)$.

Exercise 8. This fractal has two copies scaled by $1/2$, four copies scaled by $1/4$, four copies scaled by $1/8$, and so on. The infinite Moran equation is

$$\begin{aligned} 1 &= 2x + 4x^2 + 4x^3 + 4x^4 + \cdots \\ &= 4x(1 + x + x^2 + x^3 + \cdots) - 2x \\ &= \frac{4x}{1-x} - 2x \end{aligned}$$

This becomes $2x^2 + 3x - 1 = 0$, with solutions $x = (-3 \pm \sqrt{17})/4$. Using the positive solution (which satisfies $x < 1$), we find the similarity dimension is $d_s = \log((-3 + \sqrt{17})/4)/\log(1/2)$.

Exercise 9. This fractal consists of two copies scaled by $1/2$, four copies scaled by $1/4$, and one copy scaled by $1/8$. Taking $x = (1/2)^d$, the Moran equation becomes $x^3 + 4x^2 + 2x - 1 = 0$. Dividing by $x + 1$ and $x - 1$, we find $x^3 + 4x^2 + 2x - 1 = (x + 1)(x^2 + 3x - 1)$. The solutions are $x = -1$ and $x = (-3 \pm \sqrt{13})/2$. The similarity dimension is $d_s = \log((-3 + \sqrt{13})/2)/\log(1/2)$.

B.4 Solutions for the exercises of Sect. 6.2.

Here's how to calculate the box-counting dimension for the shapes of Fig. B.3.

B.4. SOLUTIONS FOR SECT. 6.2

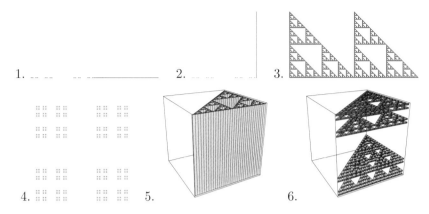

Figure B.3: Images for the box-counting dimension exercises.

Exercise 1. Scaling symmetry of the Cantor middle-thirds set suggests using squares of side length $1/3^n$. On the right we see coverings with boxes of side length $1/3$ and $1/9$. The covering by boxes of side $1/3$ requires $2+3$, 2 for the Cantor set and 3 for the line segment. The covering by boxes of side $1/9$ needs $4+9$ boxes. Then the general pattern is $N(1/3^n) = 2^n + 3^n$, the first term coming from the boxes to cover the Cantor set, the second from those to cover the line segment. Then the box-counting dimension is

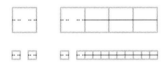

$$d_b = \lim_{n\to\infty} \frac{\log(N(1/3^n))}{\log(1/(1/3^n))} = \lim_{n\to\infty} \frac{\log(2^n + 3^n)}{\log(3^n)} = \lim_{n\to\infty} \frac{\log(3^n((2/3)^n + 1))}{n\log(3)}$$
$$= \lim_{n\to\infty} \frac{\log(3^n) + \log((2/3)^n + 1)}{n\log(3)} = \lim_{n\to\infty} 1 + \frac{\log((2/3)^n + 1)}{n\log(3)} = 1$$

Exercise 2. Scaling symmetry of the Cantor middle-thirds set suggests using squares of side length $1/3^n$. On the right we see coverings with boxes of side length $1/3$ and $1/9$. The covering by boxes of side $1/3$ requires $2 + 3 - 1$: 2 for the Cantor set, 3 for the line

segment, and -1 because the lower right box is counted twice, in the cover of the Cantor set and in the cover of the line segment. The covering by boxes of side $1/9$ needs $4+9-1$ boxes, again noting the lower right box is counted in both covers. Then the general pattern is $N(1/3^n) = 2^n + 3^n - 1$, the first term coming from the boxes to cover the Cantor set, the second from those to cover the line segment, and the -1 because the lower right box still is counted twice, once in the Cantor set, once in the line segment.

So the box-counting dimension is

$$d_b = \lim_{n\to\infty} \frac{\log(N(1/3^n))}{\log(1/(1/3^n))} = \lim_{n\to\infty} \frac{\log(2^n + 3^n - 1)}{\log(3^n)}$$

$$= \lim_{n\to\infty} \frac{\log(3^n((2/3)^n + 1 - 1/3^n))}{n\log(3)}$$

$$= \lim_{n\to\infty} \frac{\log(3^n) + \log((2/3)^n + 1 - 1/3^n)}{n\log(3)}$$

$$= \lim_{n\to\infty} 1 + \frac{\log((2/3)^n + 1 - 1/3^n)}{n\log(3)} = 1$$

Exercise 3. Scaling symmetry of the Sierpinski gasket suggests using squares of side length $1/2^n$. To see the pattern, consider the first few $N(1/2^n)$, in which we count separately the boxes that cover each copy of the gasket. On the top right we see the fractal is covered by $N(1/2) = 3 + 3$ boxes of side length $1/2$: three for the left gasket and three for the right. On the bottom right we see the fractal is covered by $N(1/4) = 9 + 9$ boxes of side length $1/4$: 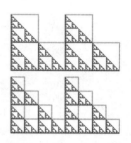 nine for the left gasket and nine for the right. The general pattern is $N(1/2^n) = 2 \cdot 3^n$. Then the box-counting dimension is

$$d_b = \lim_{n\to\infty} \frac{\log(N(1/2^n))}{\log(1/(1/2^n))} = \lim_{n\to\infty} \frac{\log(2 \cdot 3^n)}{\log(2^n)} = \lim_{n\to\infty} \frac{\log(2) + n\log(3)}{n\log(2)}$$

$$= \lim_{n\to\infty} \left(\frac{\log(2)}{n\log(2)} + \frac{\log(3)}{\log(2)}\right) = \frac{\log(3)}{\log(2)}$$

Exercise 4. Scaling symmetry of the Cantor set suggests using squares of side length $1/3^n$. To see the pattern, consider the first few $N(1/2^n)$. In the first image on the right we see this fractal is covered by 4 boxes of side $1/3$. In the second image we see the fractal is covered by $16 = 4^2$ boxes of side length $1/9$. Then the general pattern is $N(1/3^n) = 4^n$ and so the box-counting dimension is

$$d_b = \lim_{n\to\infty} \frac{\log(N(1/3^n))}{\log(1/(1/3^n))} = \lim_{n\to\infty} \frac{\log(4^n)}{\log(3^n)} = \lim_{n\to\infty} \frac{n\log(4)}{n\log(3)} = \frac{\log(4)}{\log(3)}$$

B.4. SOLUTIONS FOR SECT. 6.2

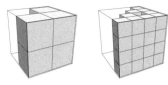

Exercise 5. Scaling symmetry of the Sierpinski gasket suggests using cubes of side length $1/2^n$. Any cube whose side length divides 1 evenly will work for the line segment. The top face of each cube is a square, itself a box in a cover of the gasket. The front (or side) face of each cube is a box in a cover of the line segment. In the first image on the right we see $3 \cdot 2 = 6$ boxes of side $1/2$ are needed, 3 to cover the gasket and 2 to cover the line segment. In the second image we see $9 \cdot 4 = 6^2$ boxes of side $1/4$ are needed for the cover. To see the general pattern, reall that covering the gasket uses 3^n boxes of side length $1/2^n$; covering a line segment uses 2^n boxes of side $1/2^n$. Then covering the product of the gasket and the line segment uses $N(1/2^n) = 3^n \cdot 2^n = 6^n$ boxes. The box-counting dimension is

$$d_b = \lim_{n \to \infty} \frac{\log(N(1/2^n))}{\log(1/(1/2^n))} = \lim_{n \to \infty} \frac{\log(6^n)}{\log(2^n)} = \lim_{n \to \infty} \frac{n \log(6)}{n \log(2)}$$
$$= \frac{\log(6)}{\log(2)} = \frac{\log(2) + \log(3)}{\log(2)} = 1 + \frac{\log(3)}{\log(2)}$$

We can obtain this value more directly once we recognize that this fractal is composed of six copies of itself, three on the top and three on the bottom, each scaled by $1/2$.

Exercise 6. Yet again, the scaling symmetry of the Sierpinski gasket suggests using cubes of side length $1/2^n$. Scaling symmetry of the Cantor middle-halves set suggests using cubes of side length $1/4^n$. These smaller cubes also will work for the gasket, so we use cubes of side length $1/4^n$.

The first picture on the right shows that we can cover the fractal by $9 \cdot 2$ boxes of side $1/4$; the second picture shows we need $81 \cdot 4$ boxes of side $1/16$. In general, the top face of each cube is a square, a box in a cover of the gasket. The front (or side) face of each cube is a box in a cover of the Cantor set. Covering the gasket uses 9^n boxes of side length $1/4^n$; covering the Cantor middle-halves set uses 2^n boxes of side length $1/4^n$. Then covering the product of the gasket and the Cantor set uses $N(1/4^n) = 9^n \cdot 2^n = 18^n$ boxes. The box-counting dimension is

$$d_b = \lim_{n \to \infty} \frac{\log(N(1/4^n))}{\log(1/(1/4^n))} = \lim_{n \to \infty} \frac{\log(18^n)}{\log(4^n)} = \lim_{n \to \infty} \frac{n \log(18)}{n \log(4)}$$
$$= \frac{\log(18)}{\log(4)} = \frac{\log(2) + \log(9)}{\log(4)} = \frac{\log(2)}{\log(4)} + \frac{\log(9)}{\log(4)} = \frac{1}{2} + \frac{\log(3)}{\log(2)}$$

This product is self-similar, consisting of 18 copies, 9 on the top and 9 on the bottom, each scaled by 1/4. Applying the basic similarity dimension formula, we find $d_s = \log(18)/\log(4)$, a more direct approach.

B.5 Solutions for the exercises of Sect. 6.4.

Exercise 1. Because of the random nature of the scaling factors,

$$r_1 = \begin{cases} 1/2 & \text{with } p = 1/2 \\ 1/4 & \text{with } p = 1/2 \end{cases} \quad \text{and} \quad r_2 = \begin{cases} 1/2 & \text{with } p = 1/4 \\ 1/4 & \text{with } p = 3/4 \end{cases}$$

we must use the random Moran equation, $\mathbb{E}(r_1)^d + \mathbb{E}(r_2)^d = 1$. First, compute the expected values of r_1^d and r_1^d:

$$\mathbb{E}(r_1)^d = \frac{1}{2} \cdot \left(\frac{1}{2}\right)^d + \frac{1}{2} \cdot \left(\frac{1}{4}\right)^d \quad \text{and} \quad \mathbb{E}(r_2)^d = \frac{1}{4} \cdot \left(\frac{1}{2}\right)^d + \frac{3}{4} \cdot \left(\frac{1}{4}\right)^d$$

Combining the $(1/2)^d$ terms and the $(1/4)^d$ terms in the random Moran equation gives

$$\frac{3}{4} \cdot \left(\frac{1}{2}\right)^d + \frac{5}{4} \cdot \left(\frac{1}{2}\right)^d = 1$$

Taking $x = (1/2)^d$ so $x^2 = (1/4)^d$, the random Moran equation simplifies to the quadratic $3x + 5x^2 = 4$. The solutions are $x = (-3 \pm \sqrt{89})/10$. Taking the positive solution and solving $x = (1/2)^d$ for d gives

$$d = \frac{\log((-3+\sqrt{89})/10)}{\log(1/2)} \approx 0.636$$

We have given a numerical approximation of the expected value of the dimension for use in part (d) of this problem.

(b) The maximum possible value of the dimension of this random fractal occurs when $r_1 = r_2 = 1/2$ for all choices. Then the basic similarity dimension formula gives

$$d = \frac{\log(N)}{\log(1/r)} = \frac{\log(2)}{\log(1/(1/2))} = 1$$

(c) The minimum possible value of the dimension of this random fractal occurs when $r_1 = r_2 = 1/4$ for all choices. Then the basic similarity dimension formula gives

$$d = \frac{\log(2)}{\log(1/(1/4))} = \frac{\log(2)}{\log(4)} = \frac{\log(2)}{\log(2^2)} = \frac{1}{2}$$

(d) In (a) we saw that a numerical approximation for the expected value of the dimension is 0.636, closer to the minimum value 0.5 than to the

B.5. SOLUTIONS FOR SECT. 6.4

maximum 1. From the probabilities of the scaling factors we see that r_1 is 1/2 and 1/4 about equally often, but r_2 is 1/4 more often – in fact, three times more often – than it is 1/2. Because the smaller scaling factor occurs more often, we expect the dimension will be lower.

Exercise 2. Here the scalings are

$$r_1 = 1/2 \quad \text{and} \quad r_2 = \begin{cases} 1/2 & \text{with } p = 1/2 \\ 1/4 & \text{with } p = 1/2 \end{cases}$$

The expected value of r_1^d is just $(1/2)^d$, because r_1 always is 1/2. The expected value of r_2^d is

$$\mathbb{E}(r_2^d) = \frac{1}{2} \cdot \left(\frac{1}{2}\right)^d + \frac{1}{2} \cdot \left(\frac{1}{4}\right)^d$$

Grouping together the $(1/2)^d$ terms, the random Moran equation becomes

$$\frac{3}{2} \cdot \left(\frac{1}{2}\right)^d + \frac{1}{2} \cdot \left(\frac{1}{4}\right)^d = 1$$

Taking $x = (1/2)^d$, the random Moran equation becomes a quadratic equation $3x + x^2 = 2$. The solutions are $x = (-3 \pm \sqrt{17})/2$. Taking the positive value of x and solving $x = (1/2)^d$ for d, we find that the expected value of the dimension is

$$d = \frac{(-3 + \sqrt{17})/2}{\log(1/2)} \approx 0.833$$

(b) The maximum possible dimension occurs when $r_2 = 1/2$ always. This gives $N = 2$ copies scaled by $r = 1/2$, so by the basic similarity dimension formula

$$d = \frac{\log(N)}{\log(1/r)} = \frac{\log(2)}{\log(1/(1/2))} = 1$$

(c) The minimum possible dimension occurs when $r_2 = 1/4$ always. Because then $r_1 = 1/2$ and $r_2 = 1/4$, we use the Moran equation, $(1/2)^d + (1/4)^d = 1$. Substituting $x = (1/2)^d$, the Moran equation becomes the quadratic $x + x^2 = 1$. The solutions are $x = (-1 + \sqrt{5})/2$. Taking the positive value and solving $x = (1/2)^d$ for the dimension d we find

$$d = \frac{\log((-1+\sqrt{5})/2)}{\log(1/2)} \approx 0.694$$

Exercise 3. Now the scaling factors are

$$r_1 = \begin{cases} 1/2 & \text{with prob } p \\ 1/4 & \text{with prob } 1-p \end{cases} \quad \text{and} \quad r_2 = \begin{cases} 1/2 & \text{with prob } p \\ 1/4 & \text{with prob } 1-p \end{cases}$$

(a) When $p = 1$, $r_1 = r_2 = 1/2$ and by the basic similarity dimension formula
$$d = \frac{\log(N)}{\log(1/r)} = \frac{\log(2)}{\log(1/(1/2))} = 1$$
Because these are the largest possible values of the scaling factors, this is the maximum value of the dimension.

When $p = 0$, $r_1 = r_2 = 1/4$ and by the basic similarity dimension formula
$$d = \frac{\log(2)}{\log(1/(1/4))} = \frac{\log(2)}{\log(4)} = \frac{1}{2}$$
Because these are the smallest possible values of the scaling factors, this is the minimum value of the dimension.

(b) With the values of the scaling factors we find
$$\mathbb{E}(r_1^d) = \mathbb{E}(r_2^d) = p \cdot \left(\frac{1}{2}\right)^d + (1-p) \cdot \left(\frac{1}{4}\right)^d$$
Then the random Moran equation becomes
$$2p \cdot \left(\frac{1}{2}\right)^d + 2(1-p) \cdot \left(\frac{1}{4}\right)^d = 1$$
Substituting $x = (1/2)^d$, the random Moran equation becomes the quadratic $px + (1-p)x^2 = 1/2$. Solving for p we obtain
$$p = \frac{1/2 - x^2}{x - x^2}$$
Because we want the dimension to be $d = 3/4$, the relation $x = (1/2)^d$ gives $x = (1/2)^{3/4}$. Substituting this into the equation for p, we find
$$p = \frac{1/2 - (1/2)^{3/2}}{(1/2)^{3/4} - (1/2)^{3/2}} \approx 0.606$$

Our expression for the probability giving $3/4$ for the expected value of the dimension can be extended in an obvious way:
$$p = \frac{1/2 - (1/2)^{2d}}{(1/2)^d - (1/2)^{2d}}$$

This is plotted in Fig. B.4, the probability ranging between 0 and 1 on the y-axis, the dimension ranging between $1/2$ and 1 on the x-axis. Note that the graph is not a straight line: the slope of the curve's tangent decreases as the dimension increases.

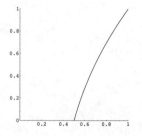

Figure B.4: Probability as a function of dimension.

B.5. SOLUTIONS FOR SECT. 6.4

Exercise 4. In order to find the similarity dimension of the fractals determined by each of these transition graphs, we'll write the corresponding transition matrix and find the spectral radius. For each, we'll write the appropriate version of the Moran equation, and if possible, solve it. Let's begin. Each transition graph arrow $i \to j$ corresponds to allowing the composition $T_j \circ T_i$ and to a 1 in the (j, i) entry (row j, column i) of the matrix. Here are the matrices for each transition graph.

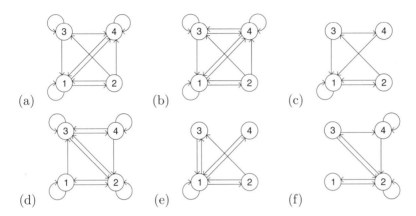

Figure B.5: Transition graphs for Exercise 4.

(a) $\begin{bmatrix} 1 & 1 & 1 & 1 \\ 1 & 0 & 0 & 0 \\ 0 & 1 & 1 & 0 \\ 1 & 1 & 1 & 1 \end{bmatrix}$ (b) $\begin{bmatrix} 1 & 1 & 1 & 1 \\ 1 & 0 & 0 & 0 \\ 0 & 1 & 1 & 1 \\ 1 & 1 & 1 & 1 \end{bmatrix}$ (c) $\begin{bmatrix} 1 & 1 & 1 & 1 \\ 1 & 0 & 0 & 0 \\ 0 & 1 & 0 & 0 \\ 0 & 0 & 1 & 0 \end{bmatrix}$

(d) $\begin{bmatrix} 1 & 1 & 0 & 0 \\ 1 & 1 & 1 & 1 \\ 1 & 1 & 1 & 1 \\ 0 & 0 & 1 & 1 \end{bmatrix}$ (e) $\begin{bmatrix} 1 & 1 & 1 & 1 \\ 1 & 0 & 0 & 0 \\ 1 & 1 & 0 & 0 \\ 1 & 0 & 0 & 0 \end{bmatrix}$ (f) $\begin{bmatrix} 0 & 1 & 0 & 0 \\ 1 & 1 & 1 & 1 \\ 0 & 1 & 0 & 0 \\ 0 & 0 & 1 & 1 \end{bmatrix}$

The eigenvalues are
(a) $(3 + \sqrt{5})/2$, $(3 - \sqrt{5})/2$, 0, 0
(b) 3, 0, 0, 0
(c) 1.92756, $-0.0763789 \pm 0.814704i$, -0.774804
(d) 3, 1, 0, 0
(e) $1 \pm \sqrt{2}$, -1, 0
(f) 2.24698, -0.801938, 0.554958, 0
(The exact solutions in (c) and (f) are far too complicated to write down, so we asked Mathematica for numerical approximations of the eigenvalues, N[Eigenvalues[M]].) In each case, the largest is the spectral radius $\rho[M]$ of the transition matrix M. Then because each scaling factor is $r = 1/2$, the

solution of the memory Moran equation $r^d \rho[m] = 1$ is

$$d = \frac{\log(\rho[M])}{\log(1/r)}$$

Then the dimensions of the fractals generated by these transition graphs are

(a) $\log((3+\sqrt{5})/2)/\log(2) \approx 1.38848$
(b) and (d) $\log(3)/\log(2) \approx 1.58496$
(c) $\log(1.92356)/\log(2) \approx 0.943779$
(e) $\log(1+\sqrt{2})/\log(2) \approx 1.27155$
(f) $\log(2.24698)/\log(2) \approx 1.16799$

Not surprisingly, there is a good correlation between the number of 1s in the transition graph and the dimension of the fractal.

Next, let's see which of these transition graphs can be produce fractals having dimensions that can be computed without the memory Moran equation. Recall a vertex is a rome if arrows from all four vertices end at the rome. This means the address of the rome contains a copy of the whole fractal scaled by $1/2$.

(a) From the transition graph we see that 1 and 4 are romes, signaling two copies scaled by $1/2$. Then $1 \to 2$ gives a copy scaled by $1/4$ in address 21, and $1 \to 2 \to 3$ gives a copy scaled by $1/8$ in address 321. (These two conditions guarantee that the fractal can be generated by an IFS without memory.) The path $1 \to 2 \to 3 \to 3$ gives a copy scaled by $1/16$ in address 3321, and so on. The infinite Moran equation is

$$2 \cdot (1/2)^d + (1/4)^d + (1/8)^d + (1/16)^d + \cdots = 1$$

Taking $x = (1/2)^d$, the infinite Moran equation becomes a geometric series equation, which we can sum and solve for x:

$$2x + x^2 + x^3 + x^4 + x^5 + \cdots = 1$$
$$x + (x + x^2 + x^3 + x^4 + x^5 + \cdots) = 1$$
$$x + \frac{x}{1-x} = 1 \quad \text{summing the bracketed series,}$$
$$\text{assuming } |x| < 1$$
$$x(1-x) + x = 1 - x$$
$$x = \frac{3 \pm \sqrt{5}}{2}$$

Both solutions are positive, but $(3+\sqrt{5})/2 > 1$, violating the $|x| < 1$ requirement for summing the geometric series. Consequently we take $x = (3-\sqrt{5})/2$. Solving $x = (1/2)^d$ for d gives

$$d = \log((3-\sqrt{5})/2)/\log(1/2).$$

B.5. SOLUTIONS FOR SECT. 6.4

But the memory Moran equation calculation gave

$$d = \log((3+\sqrt{5})/2)/\log(2).$$

How can we reconcile this difference?

$$\frac{\log((3-\sqrt{5})/2)}{\log(1/2)} = \frac{\log((3-\sqrt{5})/2)}{\log(2^{-1})} = \frac{\log((3-\sqrt{5})/2)}{-\log(2)}$$
$$= \frac{-\log((3-\sqrt{5})/2)}{\log(2)} = \frac{\log(2/(3-\sqrt{5}))}{\log(2)} = \frac{\log((3+\sqrt{5})/2)}{\log(2)}$$

There's no difference at all: the two methods just give different expressions of the same answer.

(b) From the transition graph we see that 1 and 4 are romes, so addresses 1 and 4 each contain a copy of the fractal scaled by 1/2. The arrows $1 \to 2$ and $4 \to 3$ give copies scaled by 1/4 in addresses 21 and 34. Then traversing the loop $3 \to 3$ gives copies scaled by 1/8 in address 321 and 334, two copies scaled by 1/16 in addresses 3321 and 3334, and so on. The infinite Moran equation is

$$2 \cdot (1/2)^d + 2 \cdot (1/4)^d + 2 \cdot (1/8)^d + 2 \cdot (1/16)^d + \cdots = 1$$

Taking $x = (1/2)^d$, the infinite Moran equation becomes a geometric series equation, which we can sum and solve for x:

$$2x + 2x^2 + 2x^3 + 2x^4 + 2x^5 + \cdots = 1$$
$$x + x^2 + x^3 + x^4 + x^5 + \cdots = \frac{1}{2}$$
$$\frac{x}{1-x} = \frac{1}{2}$$
$$2x = 1 - x$$

This gives $x = 1/3$. Solving $x = (1/2)^d$ for d we find

$$d = \frac{\log(1/3)}{\log(1/2)} = \frac{\log(3^{-1})}{\log(2^{-1})} = \frac{-\log(3)}{-\log(2)} = \frac{\log(3)}{\log(2)}$$

agreeing with the solution obtained by the memory Moran equation.

(c) From the transition graph we see 1 is a rome, so address 1 contains a copy of the fractal scaled by 1/2. The paths $1 \to 2$, $1 \to 2 \to 3$, and $1 \to 2 \to 3 \to 4$ give one copy each scaled by 1/4 in address 21, by 1/8 in address 321, and 1/16 in address 4321. The only transition from 4 returns to the rome 1, so these copies make up the entire fractal. The Moran equation is

$$(1/2)^d + (1/4)^d + (1/8)^d + (1/16)^d = 1$$

Taking $x = (1/2)^d$, The Moran equation becomes the polynomial equation

$$x + x^2 + x^3 + x^4 = 1$$

There is a quartic formula, a much larger cousin of the quadratic formula, but you really don't want to see it, or at least we really don't want to see it. Solving this equation numerically we obtain

$$0.51879, -1.29065, -0.114071 \pm 1.21675i$$

Because x must be positive, the only appropriate solution is $x = 0.51879$. Solving $x = (1/2)^d$ for d gives

$$d = \frac{\log(0.51879)}{\log(0.5)} \approx 0.946777$$

This is quite close to the value we got with the memory Moran equation. We'd like to believe the difference is numerical noise. How can we check this? Recall that with all the scaling factors equal to $1/2$, the memory Moran equation gives the dimension

$$d = \frac{\log(\rho)}{\log(2)}$$

where ρ is the largest eigenvalue of the transition matrix. On the other hand, applied to the transition graph, the Moran equation gives

$$d = \frac{\log(x)}{\log(1/2)}$$

where x is the positive solution of the Moran equation. Comparing the denominators for these expressions for d, we see the values of d are equal if $\rho = 1/x$. So are they equal? The eigenvalues are the solutions of the eigenvalue equation

$$0 = \det \begin{bmatrix} 1-\lambda & 1 & 1 & 1 \\ 1 & 0-\lambda & 0 & 0 \\ 0 & 1 & 0-\lambda & 0 \\ 0 & 0 & 1 & 1-\lambda \end{bmatrix} = \lambda^4 - \lambda^3 - \lambda^2 - \lambda - 1$$

From the transition graph and the Moran equation we see x is a solution of

$$x^4 + x^3 + x^2 + x = 1$$

Let's substitute $\lambda = 1/x$ into the eigenvalue equation:

$$(1/x)^4 - (1/x)^3 - (1/x)^2 - 1/x = 1$$
$$1 - x - x^2 - x^3 = x^4$$
$$1 = x^4 + x^3 + x^2 + x$$

B.5. SOLUTIONS FOR SECT. 6.4

That is, if λ satisfies the eigenvalue equation, then $x = 1/\lambda$ satisfies the Moran equation for the transition graph. A few details remain, but this is the main step in the proof that both methods give the same value for the dimension of this fractal.

(d) From the transition graph we see that 2 and 3 are romes, giving copies of the fractal scaled by $1/2$ in addresses 2 and 3. Then $2 \to 1$ and $3 \to 4$ give copies scaled by $1/4$ in addresses 12 and 43. And the (non-rome) loops $1 \to 1$ and $4 \to 4$ give copies scaled by $1/8$ in addresses 112 and 443, copies scaled by $1/16$ in addresses 1112 and 4443, and so on. The infinite Moran equation is

$$2 \cdot (1/2)^d + 2 \cdot (1/4)^d + 2 \cdot (1/8)^d + 2 \cdot (1/16)^d + \cdots = 1$$

This is the equation we found in example (b), so again we get the solution $d = \log(1/3)/\log(1/2)$. This equals $\log(3)/\log(2)$, the solution of the memory Moran equation.

(e) From the transition graph we see that 1 is a rome, giving a copy of the fractal scaled by $1/2$ in address 1. The paths $1 \to 2$, $1 \to 3$, and $1 \to 4$ give copies scaled by $1/4$ in addresses 21, 31, and 41. The path $1 \to 2 \to 3$ gives a copy scaled by $1/8$ in address 321. All other paths return to the rome, so these are all the copies making up this fractal. The Moran equation is

$$(1/2)^d + 3 \cdot (1/4)^d + (1/8)^d = 1$$

Substituting $x = (1/2)^d$, the Moran equation becomes

$$x + 3x^2 + x^3 = 1$$

Dividing by $x + 1$ we see that

$$x^3 + 3x^2 + 1 - 1 = (x+1)(x^2 + 2x - 1)$$

Applying the quadratic formula to the second factor, we see that the solutions are $x = -1$, $x = -1 - \sqrt{2}$, and $x = -1 + \sqrt{2}$. Only the last solution is positive, so solving $x = (1/2)^d$ for d we find

$$d = \frac{\log(-1+\sqrt{2})}{\log(1/2)} = \frac{\log(-1+\sqrt{2})}{-\log(2)} = \frac{-\log(-1+\sqrt{2})}{\log(2)}$$
$$= \frac{\log(1/(-1+\sqrt{2}))}{\log(2)} = \frac{\log(1+\sqrt{2})}{\log(2)}$$

The last expression is the value of d we got from the memory Moran equation.

(f) From the transition graph we see that 2 is a rome, so address 2 contains a copy of the fractal scaled by $1/2$. The paths $2 \to 1$ and $2 \to 3$ give copies in addresses 12 and 32 scaled by 1/4. The path $2 \to 3 \to 4$ gives a copy in address 432 scaled by 1/8. Then the loop $4 \to 4$ gives a copy scaled by 1/16 in address 4432, a copy scaled by 1/32 in address 44432, and so on. So here again we must use the infinite Moran equation :

$$(1/2)^d + 2 \cdot (1/4)^d + (1/8)^d + (1/16)^d + (1/32)^d + \cdots = 1$$

With $x = (1/2)^d$, this becomes a geometric series equation which we can sum and simplify a bit

$$x + 2x^2 + x^3 + x^4 + x^5 + \cdots = 1$$
$$x^2 + (x + x^2 + x^3 + x^4 + x^5 + \cdots) = 1$$
$$x^2 + \frac{x}{1-x} = 1 \quad \text{if } |x| < 1$$
$$(1-x)x^2 + x = 1 - x$$
$$0 = x^3 - x^2 - 2x + 1$$

This doesn't factor easily, and the exact solution by the cubic formula is very complicated, so we find approximate numerical solutions:

$$1.80194, 0.445042, -1.24698$$

Recalling that we must have $|x| < 1$ in order to sum the geometric series, we take $x \approx 0.445042$. Solving $x = (1/2)^d$ for d gives

$$d \approx \log(0.445042)/\log(.5) \approx 1.16799,$$

the answer we got, accurate to five places, with the memory Moran equation.

To check the method we developed in part (c), let's compare the Moran equation and the eigenvalue equation from the memory Moran equation. Recall that if x is the positive solution of the Moran equation, then $d = \log(x)/\log(1/2)$, and if ρ is the largest eigenvalue of the transition matrix, then $d = \log(\rho)/\log(2)$. Then we see these two expressions for d will give the same value if $\rho = 1/x$. We've seen that the value of x is a solution, in fact, the only positive solution, of the Moran equation

$$x^3 - x^2 - 2x + 1 = 0$$

and ρ is the largest solution of the eigenvalue equation

$$0 = \det \begin{bmatrix} 0-\lambda & 1 & 0 & 0 \\ 1 & 1-\lambda & 1 & 1 \\ 0 & 1 & 0-\lambda & 0 \\ 0 & 0 & 1 & 1-\lambda \end{bmatrix} = \lambda^4 - 2\lambda^3 - \lambda^2 + \lambda$$
$$= \lambda(\lambda^3 - 2\lambda^2 - \lambda + 1)$$

Substituting $\lambda = 1/x$ into the cubic factor ($\lambda = 0$ is not the solution we seek), we find

$$0 = (1/x)^3 - 2(1/x)^2 - 1/x + 1$$
$$= 1 - 2x - x^2 + x^3 \quad \text{multiplying both sides by } x^3$$

which is the Moran equation for this fractal.

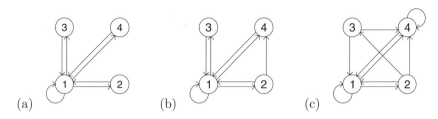

Figure B.6: Transition graphs for the Moran equations of Exercise 5.

Exercise 5. (a) The Moran equation $x + 3x^2 = 1$ says the fractal has one copy (indicated by x) scaled by $1/2$ and three copies scaled by by $1/4$. The copy scaled by $1/2$ corresponds to a rome in the transition graph; the three copes scaled by $1/4$ are generated by arrows from the rome to a non-rome. And that's all. The Moran equation gives no information about which vertex is a rome, so let's take vertex 1. Then in the transition graph there must be arrows $1 \to 1$, $2 \to 1$, $3 \to 1$, and $4 \to 1$. Because there is only one rome and three non-romes, the three copies scaled by $1/4$ are produced by three arrows $1 \to 2$, $1 \to 3$, and $1 \to 4$. The transition graph is the first graph of Fig. B.6.

(b) The Moran equation $x + 3x^2 + x^3 = 1$ says the fractal has one copy scaled by $1/2$, three scaled by $1/4$, and one scaled by $1/8$. The copy scaled by $1/2$ corresponds to a rome, the three copes scaled by $1/4$ are generated by arrows from the rome to a non-rome, the one copy scaled by $1/8$ is generated by a path from rome to non-rome to a different non-rome. The reason the second non-rome must be different from the first is that if there were a loop at the first non-rome, this would generate an infnite cascade of smaller and smaller copies and the Moran equation would include an infinite series. With the copy scaled by $1/2$ and three copies scaled by $1/4$ we can generate by the graph of (a). The copy scaled by $1/8$ is produced by an arrow between non-romes. In the second graph of Fig. B.6 we've taken $2 \to 4$, but any of $2 \to 3$, $3 \to 2$, $3 \to 4$, $4 \to 2$ or $4 \to 3$ would work.

(c) The Moran equation $2x + x^2 + x^3 = 1$ says the fractal has two copies scaled by $1/2$, one scaled by $1/4$, and one scaled by $1/8$. We'll take 1 and 4 to be romes, but of course any two vertices would do. This requires eight arrows in the transition graph, four going to 1 and four coing to 4. The one copy scaled by $1/2$ is produced by an arrow from a rome to a non-rome. In the third graph of Fig. B.6 we've taken $1 \to 2$, but $1 \to 3$, $4 \to 2$, or $4 \to 3$ would work. Finally, the copy scaled by 1/8 requires a path from rome to non-rome to the other non–rome, so $1 \to 2 \to 4$ in this graph. As in (b), the second non-rome must be different from the first to avoid an infinite cascade of copies.

B.6 Solutions for the exercises of Sect. 6.5.

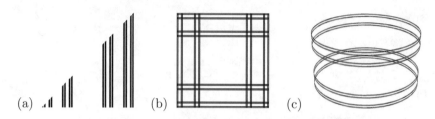

Figure B.7: Fractals for Exercises 1, 2, and 3.

In Fig. B.7 everything that looks like a Cantor set is a Cantor set. The Cantor middle-thirds set for (a) and (c), the Cantor middle-halves set for (b).

Exercise 1. Call X the fractal given in the problem and Y a copy of X rotated 180° and placed as indicated in Fig. B.8. Then X and Y have the same dimension by the invariance principle. By the product rule $X \cup Y = \mathcal{C} \times I$, which has dimension $\log(2)/\log(3) + 1$. Then by the stability rule

$$\log(2)/\log(3) = \dim(X \cup Y)$$
$$= \max\{\dim(X), \dim(Y)\}$$
$$= \dim(X)$$

where the last equality follows because we have $\dim(X) = \dim(Y)$.

Figure B.8: The solution of Exercise 1.

Exercise 2. In Fig. B.9 we see the given fractal X is the union of the fractals Y and Z. Both Y and Z are products of a Cantor middle-halves set $\mathcal{C}$ and a unit interval I, so by the product rule

$$\dim(Y) = \dim(Z) = \dim(\mathcal{C} \times I) = \dim(\mathcal{C}) + \dim(I)$$
$$= \log(2)/\log(4) + 1 = 3/2$$

Then by the stability rule,

$$\dim(X) = \dim(Y \cup Z) = \max\{\dim(Y), \dim(Z)\} = 3/2$$

B.6. SOLUTIONS FOR SECT. 6.5

Figure B.9: The fractal X of Exercise 2, and fractals Y and Z that make it up.

Exercise 3. Fractal (c) of Fig. B.7 is a product of a Cantor middle-thirds set $\mathcal{C}$ and a circle C. By the product rule

$$\dim(\mathcal{C} \times C) = \dim(\mathcal{C}) + \dim(C) = \log(2)/\log(3) + 1$$

Exercise 4. The dimension of a Cantor set $\mathcal{C}$ consisting of $N = 2$ pieces scaled by r is $\log(2)/\log(1/r)$. By the product rule

$$\dim(\mathcal{C} \times \mathcal{C} \times \mathcal{C}) = 3 \cdot \log(2)/\log(1/r) = \log(2^3)/\log(1/r)$$

Then $\dim(\mathcal{C} \times \mathcal{C} \times \mathcal{C}) = 1$ gives $r = 1/8$.

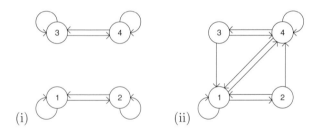

Figure B.10: The transition graphs of Exercise 5.

Exercise 5. (a) This IFS with memory generates two line segments, the 1–2 line and the 3–4 line. Both line segments have dimension 1, so by the stability rule, the attractor of this IFS has dimension 1.

(b) Addresses 1 and 4 are romes, so the squares with addresses 1 and 4 contain copies of the fractal scaled by $1/2$. The transitions $1 \to 2$ and $4 \to 3$ give two copies scaled by $1/4$. The Moran equation is

$$2(1/2)^d + 2(1/4)^d = 1$$

Substituting $x = (1/2)^d$, the Moran equation becomes the quadratic equation $2x^2 + 2x - 1 = 0$. Solving for x and then for d, we find the dimension is $d = \log((-1 + \sqrt{3})/2)/\log(1/2)$. Okay, (b) has nothing to do with the techniques of this section. We just wanted to give you some more practice with transition graphs and the Moran equation.

B.7 Solutions for the exercises of Sect. 7.1.

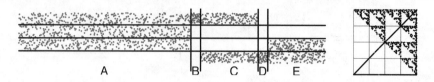

Figure B.11: The time series and driven IFS for exercise 1(a).

Exercise 1. (a) As seen in the first image of Fig. B.11, the time series splits into five regimes. In A the data points lie in bins 2, 3, and 4, scattered without apparent order. In the driven IFS, this corresponds to applying transformations T_2, T_3, and T_4 in (apparently) random order. That is, for data in regime A the driven IFS is just the random IFS algorithm for the IFS $\{T_2, T_3, T_4\}$, and so it generates points on the gasket with vertices $(1,0)$, $(0,1)$, and $(1,1)$.

In regime B, all the data points lie in bin 4, and successive application of T_4 produces a sequence of points lying on the gasket of regime A and going to $(1,1)$.

In regime C, all the data points lie in bins 4 and 1, so the driven IFS generates points on the 1–4 diagonal .

In regime D, all the data points lie in bin 1, and successive applications of T_1 produce a sequence of points going along the line of regime C to $(0,0)$.

In regime E, all the data points lie in bins 1 and 2, so the driven IFS generates points on the 1–2 line .

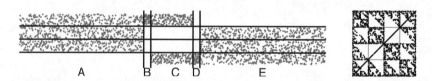

Figure B.12: The time series and driven IFS for exercise 1(b).

(b) As seen in the first image of Fig. B.12, the time series splits into five regimes. In A the data points lie in bins 2, 3, and 4, scattered without apparent order. In the driven IFS, this corresponds to applying transformations T_2, T_3, and T_4 in (apparently) random order. That is, for data in regime A the driven IFS is just the random IFS algorithm for the IFS $\{T_2, T_3, T_4\}$, and so it generates points on the gasket with vertices $(1,0)$, $(0,1)$, and $(1,1)$.

In regime B, all the data points lie in bin 4, and successive applications of T_4 produce a sequence of points lying on the gasket of regime A and going to $(1,1)$.

B.7. SOLUTIONS FOR SECT. 7.1   431

In regime C, all the data points lie in bins 4 and 1, so the driven IFS generates points on the 1–4 diagonal.

In regime D, all the data points lie in bin 1, and successive applications of T_1 produce a sequence of points going along the line of regime C to $(0,0)$.

In regime E, all the data points lie in bins 1, 2, and 3, so the driven IFS generates points on the gasket with vertices $(0,0)$, $(1,0)$, and $(0,1)$.

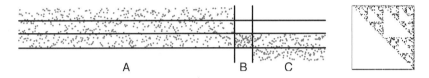

Figure B.13: The time series and driven IFS for exercise 2(a).

Exercise 2 (a) Having gone over 1 (a) and (b) in some detail, now it's time to hit the gas. Time series points in regime A produce driven IFS points on the gasket with vertices $(1,0)$, $(0,1)$, and $(1,1)$; points in regime B produce driven IFS points converging to the vertex $(1,0)$; and points in regime C produce driven IFS points along the line between vertices $(0,0)$ and $(1,0)$, the 1–3 line.

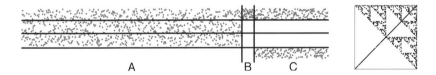

Figure B.14: The time series and driven IFS for exercise 2(b).

(b) Time series points in regime A produce driven IFS points on the gasket with vertices $(1,0)$, $(0,1)$, and $(1,1)$; points in regime B produce driven IFS points converging to the vertex $(1,1)$; and points in regime C produce driven IFS points along the 1–4 diagonal.

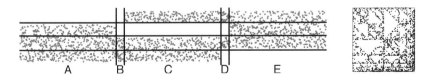

Figure B.15: The time series and driven IFS for exercise 3(a).

Exercise 3 (a) Time series points in regime A produce driven IFS points on the gasket with vertices $(0,0)$, $(1,0)$, and $(0,1)$; points in regime B produce driven IFS points converging to the 1–2 line; points in regime C

produce driven IFS points on the gasket with vertices $(0,0)$, $(1,0)$, and $(1,1)$; points in regime D produce driven IFS points converging to the line with vertices $(0,1)$ and $(1,1)$, the 3–4 line; and points in regime E produce driven IFS points on the gasket with vertices $(1,0)$, $(0,1)$, and $(1,1)$.

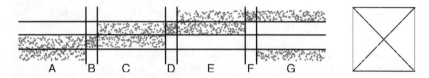

Figure B.16: The time series and driven IFS for exercise 3(b).

(b) Time series points in regime A produce driven IFS points on the 1–2 line; points in regime B produce driven IFS points converging to the point $(1,0)$; points in regime C produce driven IFS points on the 2–3 diagonal; points in regime D produce driven IFS points converging to the point $(0,1)$; points in regime E produce driven IFS points on the 3–4 line; points in regime F produce driven IFS points converging to the point $(1,1)$; and points in regime G produce driven IFS points on the 1–4 diagonal.

These driven IFS were computer-generated. Fig. B.17 shows hand-drawn sketches.

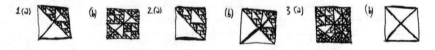

Figure B.17: Hand-drawn sketches of the solutions of problems 1, 2, and 3.

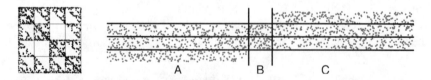

Figure B.18: The driven IFS of exercise 4(a) and a time series that could generate it.

Exercise 4(a) The driven IFS appears to be two gaskets sharing their hypotenuse. Which gasket the time series generates first doesn't matter. Time series points spread randomly through bins 1, 2, and 3 (regime A) produce driven IFS points on the gasket with vertices $(0,0)$, $(1,0)$, and $(0,1)$. The driven IFS exhibits no points not on these gaskets. To achieve a clean transition between gaskets, the time series must go through a regime

B.7. SOLUTIONS FOR SECT. 7.1   433

in bins common to both gaskets. This is the choice pictured in the time series, but the transitional regime also could consist of points only in bin 2, points only in bin 3, or (as pictured in regime B here) points randomly spread through bins 2 and 3. These produce driven IFS points on the 2–3 diagonal, the hypotenuse common to both gaskets. The third regime, C, contains points spread randomly through bins 2, 3, and 4, producing points on the gasket with vertices $(1,0)$, $(0,1)$ and $(1,1)$.

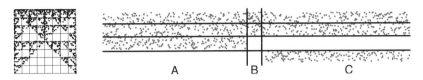

Figure B.19: The driven IFS of exercise 4(b) and a time series that could generate it.

(b) The driven IFS appears to be two gaskets sharing an edge, the top of the unit square. Regime A contains points spread randomly though bins 2, 3, and 4, producing driven IFS points on the gasket with vertices $(1,0)$, $(0,1)$, and $(1,1)$. To achieve a clean transition between the gaskets, regime B could consist of points in bins 3 and 4, producing driven IFS points on the 3–4 line, the edge common to these gaskets. This is the choice pictured in the time series, but the transitional regime also could consist only of points in bin 3 or of points only in bin 4. Regime C contains points spread randomly though bins 1, 3, and 4, producing driven IFS points on the gasket with vertices $(0,0)$, $(0,1)$, and $(1,1)$.

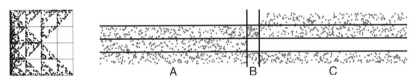

Figure B.20: The driven IFS of exercise 4(c) and a time series that could generate it.

(c) The driven IFS appears to be two gaskets sharing an edge, the left side of the unit square. Regime A contains points spread randomly though bins 1, 2, and 3, producing driven IFS points on the gasket with vertices $(0,0)$, $(0,1)$, and $(1,0)$. To achieve a clean transition between the gaskets, regime B could consist of points in bins 1 and 3, producing driven IFS points on the 1–3 line, the edge common to these gaskets. This is the choice pictured in the time series, but the transitional regime also could consist of points in bin 1 or of points in bin 3. Regime C contains points spread randomly though bins 1, 3, and 4, producing driven IFS points on the gasket with vertices $(0,0)$, $(0,1)$, and $(1,1)$.

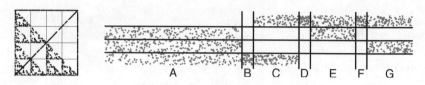

Figure B.21: The driven IFS of exercise 4(d) and a time series that could generate it.

(d) Here we see a gasket with vertices $(0,0)$, $(1,0)$, and $(0,1)$, so we start the time series (regime A) with points spread randomly between bins 1, 2, and 3. Regime B, points in bin 1, gives a clean transition between the gasket and the 1–4 diagonal, produced by time series points (regime C) spread randomly among bins 1 and 4. Points in bin D give a clean transition between this line and the 3–4 line (regime E). Points in bin 4 (regime F) give a clean transition between this line and the 2–4 line (regime G). Can you think of other orderings of these bins to achieve the same driven IFS?

These time series were computer-generated. Fig. B.22 shows hand-drawn sketches.

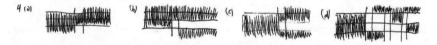

Figure B.22: Hand-drawn sketches of the solutions of problem 4.

B.8 Solutions for the exercises of Sect. 7.3.

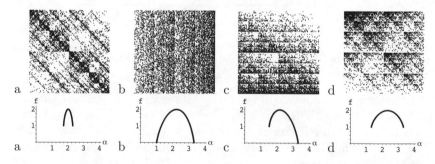

Figure B.23: Random IFS plots and $f(\alpha)$ curves.

IFS a appears to have about the same number of points in addresses 2 and 3 and about the same (smaller) number of points in addresses 1 and 4. This suggests $p_2 = p_3 > p_1 = p_4$. Consequently, $\min(\alpha)$ and $\max(\alpha)$

occur on line segments and so $f(\min(\alpha)) = f(\max(\alpha)) = 1$. This occurs in the $f(\alpha)$ curves a and d.

IFS b appears to have about the same number of points in addresses 2 and 4 and about the same (smaller) number of points in addresses 1 and 3. This suggests $p_2 = p_4 > p_1 = p_2$. Consequently, $\min(\alpha)$ and $\max(\alpha)$ occur on line segments and so $f(\min(\alpha)) = f(\max(\alpha)) = 1$. To distinguish IFS a from b, note that the points of IFS b are more uniformly spread, suggesting a smaller range of α values.

So we have IFS a corresponding to $f(\alpha)$ curve d and IFS b corresponding to $f(\alpha)$ curve a.

IFS c appears to have about the same number of points in addresses 1 and 2, which is more than those in address 3, which is in turn more than those in address 4. This suggests $p_1 = p_2 > p_3 > p_4$. Then the maximum α occurs at the point $(1,1)$, and the minimum α occurs on the 1–2 line. Consequently, for this IFS we have $f(\min(\alpha)) = 1$ and $f(\max(\alpha)) = 0$. That is, IFS c corresponds to $f(\alpha)$ curve c.

IFS d appears to have about the same number of points in addresses 2 and 4, which is more than in address 1 and is fewer than in address 3. This suggests $p_3 > p_1 = p_4 > p_1$. Then both the maximum and minimum values occur on single points, $(0,1)$ and $(0,0)$, respectively. This gives $f(\min(\alpha)) = f(\max(\alpha)) = 0$, and IFS d corresponds to $f(\alpha)$ curve b.

2. Because all the scaling factors are $r = 1/4$, the minimum α value corresponds to the maximum probability and the maximum α value corresponds to the minimum probability. That is,

$$\min(\alpha) = \frac{\log(\max \text{prob})}{\log(r)} = \frac{\log(0.2)}{\log(1/4)} \approx 1.161$$

$$\max(\alpha) = \frac{\log(\min \text{prob})}{\log(r)} = \frac{\log(0.05)}{\log(1/4)} \approx 2.161$$

The maximum probability occurs on $N = 2$ pieces scaled by $r = 1/4$, so by the basic similarity dimension formula

$$f(\min(\alpha)) = \frac{\log(N)}{\log(1/r)} = \frac{\log(2)}{\log(4)} = \frac{1}{2}$$

The minimum probability occurs on $N = 6$ pieces scaled by $r = 1/4$, so by the basic similarity dimension formula

$$f(\max(\alpha)) = \frac{\log(N)}{\log(1/r)} = \frac{\log(6)}{\log(4)} \approx 1.292$$

The maximum value of $f(\alpha)$ is the dimension of the entire fractal. This consists of $N = 11$ pieces scaled by $r = 1/4$, so once more we call upon the basic similarity dimension formula

$$\max(f(\alpha)) = \frac{\log(N)}{\log(1/r)} = \frac{\log(11)}{\log(4)} \approx 1.730$$

3. Because this fractal has different scaling factors ($r = 1/2$ and $r = 1/4$), to find the minimum and maximum values of α we must compute each $\log(\text{prob})/\log(r)$.

$$\frac{\log(\max \text{prob})}{\log(r)} = \frac{\log(0.25)}{\log(1/2)} = 2$$

$$\frac{\log(\min \text{prob})}{\log(r)} = \frac{\log(0.1)}{\log(1/4)} \approx 1.661$$

Then we see max prob corresponds to $\max(\alpha)$ and min prob corresponds to $\min(\alpha)$.

Because the minimum probability occurs on $N = 5$ pieces scaled by $r = 1/4$, we apply the basic similarity dimension formula to obtain

$$f(\min(\alpha)) = \frac{\log(N)}{\log(1/(1/4))} = \frac{\log(5)}{\log(4)} \approx 1.161$$

Similarly, because the maximum probability occurs on $N = 2$ pieces scaled by $r = 1/2$,

$$f(\max(\alpha)) = \frac{\log(N)}{\log(1/(1/2))} = \frac{\log(2)}{\log(2)} = 1$$

The maximum value of $f(\alpha)$ is the dimension of the whole fractal, which consists of 2 pieces scaled by $1/2$ and 5 pieces scaled by $1/4$. Because the scaling factors take more than one value, we use the Moran equation

$$2 \cdot (1/2)^d + 5 \cdot (1/4)^d = 1$$

Taking $x = (1/2)^d$ so $x^2 = (1/4)^d$, the Moran equation becomes a quadratic equation

$$2x + 5x^2 = 1$$

The solutions are

$$x = \frac{-1 \pm \sqrt{6}}{5}$$

Taking the positive value of x and solving $x = (1/2)^d$ for d by taking logs we find

$$\max(f(\alpha)) = d = \frac{\log((-1 + \sqrt{6})/5)}{\log(1/2)} \approx 1.786$$

4. As we saw in exercise 3, because this fractal has different scaling factors, we must compute each $\log(\text{prob})/\log(r)$ to find the minimum and

B.8. SOLUTIONS FOR SECT. 7.3

maximum values of α.

$$\frac{\log(0.3)}{\log(1/3)} \approx 1.096$$

$$\frac{\log(0.17)}{\log(1/3)} \approx 1.613$$

$$\frac{\log(0.1)}{\log(1/3)} \approx 2.096$$

$$\frac{\log(0.01)}{\log(1/9)} = \frac{\log(0.1^2)}{\log((1/3)^2)} = \frac{\log(0.1)}{\log(1/3)} \approx 2.096$$

The minimum value of α occurs for $N = 2$ pieces scaled by $r = 1/3$, so by the basic similarity dimension formula

$$f(\min(\alpha)) = \frac{\log(N)}{\log(1/r)} = \frac{\log(2)}{\log(1/(1/3))} = \frac{\log(2)}{\log(3)} \approx 0.631$$

The maximum value of α is a bit trickier: it occurs on those pieces with $r = 1/3$ and $p = 0.1$, and also on those pieces with $r = 1/9$ and $p = 0.01$. The value of $f(\max(\alpha))$ is the dimension of the fractal consisting of $N = 2$ pieces scaled by $r = 1/3$ and $N = 3$ pieces scaled by $r = 1/9$. Different scaling factors means we must use the Moran equation

$$2 \cdot (1/3)^d + 3 \cdot (1/9)^d = 1$$

Taking $x = (1/3)^d$ so $x^2 = (1/9)^d$, we obtain the quadratic equation

$$2x + 3x^2 = 1$$

The soutions are $x = -1$ and $x = 1/3$. Take the positive value for x and solving $x = (1/3)^d$ for d (You should be able to do this without taking logs.) we find

$$f(\max(\alpha)) = d = 1$$

The maximum value of $f(\alpha)$ is the is the dimension of the whole fractal: 5 pieces scaled by $1/3$ and 3 pieces scaled by $1/9$. Apply the Moran equation

$$5 \cdot (1/3)^d + 3 \cdot (1/9)^d = 1$$

Taking $x = (1/3)^d$ the Moran equation becomes the quadratic

$$5x + 3x^2 = 1$$

The solutions are $x = (-5 \pm \sqrt{37})/6$. Taking the positive value of x and solving $x = (1/3)^d$ for d, we find

$$\max(f(\alpha)) = d = \frac{\log((-5 + \sqrt{37})/6)}{\log(1/3)} \approx 1.589$$

References

Introduction

[1] B. Mandelbrot, *The Fractal Geometry of Nature*, Freeman, San Francisco, 1982.

Benoit's manifesto on fractal geometry , [1], appropriately the first item you see in our references, is an elaboration of [2], itself a translation and extension of [3]. Fifteen teachers report their experiences with fractals in the classroom in [4]. The four volumes of Benoit's papers, with extensive introductions, are [5]. His memoirs are [6]; [7] is the memorial volume edited by Nathan Cohen and the older author.

By now, fractals are described in many books spanning a wide range of technical levels. We'll mention a few, ordered approximately by the first publication date of the book, and grouping all the books of an author.

Kenneth Falconer's first three books on fractal geometry, [8, 9, 10] are masterpieces: far-reaching, mathematically rigorous, brilliantly clear. The older author has spent hundreds of delightful hours becoming familiar with these books. Falconer's book [11] is a superb, succinct introduction to the field. Francis Moon's book [12], first published in 1987 under the title *Chaotic Vibrations: An Introduction for Applied Scientists and Engineers* is an early collection of examples, both physical and dynamical, of fractals in complex mechanical and electrical systems. Jens Feder's book [13] gives a clear description of physical examples, including viscous fingering and percolation, and one of the earliest detailed expositions of multifractals. The first extensive presentation of IFS is Michael Barnsley's *Fractals Everywhere*, originally published in 1988. Barnsley updated and extended his approach in [15]. Bryan Kaye's book focuses on random processes generating fractals, with some applications to fracture. Gerald Edgar's books [17, 18] are careful mathematical studies approaching fractals through measure theory. First published in 1992, the book *Chaos and Fractals: New Frontiers in Science* [19] by Heinz-Otto Peitgen, Dietmar Saupe, and Hartmut Jürgens is an encyclopedia of the mathematics of fractals accessible at many levels. Denny Gulick's book [20], published first in 1992 under the title *Encounters with Chaos*, provides a clear mathematical development and some examples not found in other sources. Methods for calculating dimension, and a good collection of examples, including time series analysis, are main features of [21] by Harold Hastings and George Sugihara. First published in 1994, Steven Strogatz's book [22] includes a detailed analysis of differential equations, chaos, and fractals. And everything Strogatz writes is brilliant. In [23, 24] Guy David and Stephen Semmes introduce a more general class of fractals, BPI (Big Piece of Itself) spaces, with the goal of supporting some aspects of calculus on these fractals. They produce some examples different from those we'd seen before. Yakov Pesin's books are clear expositions of the role of dynamical systems in fractals. Jun Kigami writes on analysis on fractals [27]; Robert Strichartz focuses on differential equations on fractals.

There are other books, and scores, maybe hundreds, of conference proceedings. But for now, these are enough to mention.

[2] B. Mandelbrot, *Fractals: Form, Chance, and Dimension*, Freeman, San Francisco, 1977.

[3] B. Mandelbrot, *Les objets fractals: forme, hasard et dimension*, Flammarion, Paris, 1975.

[4] M. Frame, B. Mandelbrot, eds., *Fractals, Graphics, and Mathematics Education*, Mathematical Association of America, Washington, 2002.

[5] B. Mandelbrot, *Selecta*
Fractals and Scaling in Finance: Discontinuity, Concentration, Risk, Springer, New York, 1997.
Multifractals and $1/f$ Noise: Wild Self-Affinity in Physics, Springer, New York, 1999.
Gaussian Self-Affinity and Fractals, Springer, New York, 2002.
Fractals and Chaos. The Mandelbrot Set and Beyond, Springer, New York, 2004.

[6] B. Mandelbrot, *The Fractalist: Memoir of a Scientific Maverick*, Pantheon, New York, 2012.

[7] M. Frame, N. Cohen, eds., *Benoit Mandelbrot: A Life in Many Dimensions*, World Scientific, Singapore, 2015.

[8] K. Falconer, *The Geometry of Fractal Sets*, Cambridge Univ. Pr., Cambridge, 1985.

REFERENCES

[9] K. Falconer, *Fractal Geometry. Mathematical Foundations and Applications*, 3rd ed., Wiley, Chichester, 2014.

[10] K. Falconer, *Techniques in Fractal Geometry*, Wiley, Cinchester, 1997.

[11] K. Falconer, *Fractals: A Very Short Introduction*, Oxford Univ. Pr., Oxford, 2013.

[12] F. Moon, *Chaotic and Fractal Dynamics: An Introduction for Applied Scientists and Engineers*, Wiley, NY 1992.

[13] J. Feder, *Fractals*, Plenum, New York, 1988.

[14] M. Barnsley, *Fractals Everywhere*, 2nd ed., Academic Press, Boston, 1993.

[15] M. Barnsley, *SuperFractals: Patterns of Nature*, Cambridge Univ. Pr., Cambridge, 2006.

[16] B. Kaye, *A Random Walk through Fractal Dimensions* VCH, New York, 1989.

[17] G. Edgar, *Measure, Topology, and Fractal Geometry*, Springer, New York, 1990.

[18] G. Edgar, *Integral, Probability, and Fractal Measures*, Springer, New York, 1998.

[19] H.-O. Peitgen, H. Jürgens, D. Saupe, *Chaos and Fractals: New Frontiers in Science*, 2nd ed., Springer, NY, 2004. 1992

[20] D. Gulick, *Encounters with Chaos and Fractals* 2nd ed., CRC Press, Boca Raton, FL, 2012. 1992

[21] H. Hastings, G. Sugihara, *Fractals: A User's Guide for the Natural Sciences*, Oxford Univ. Pr. Oxford, 1993.

[22] S, Strogatz, *Nonlinear Dynamics and Chaos: With Applications to Physics, Biology, Chemistry, and Engineering* 2nd ed., Westview, Philadelphia, 2015. 1994

[23] G. David, S. Semmes, *Fractured Fractals and Broken Dreams: Self-Similar Geometry through Metric and Measure*, Oxford Univ. Pr., Oxford, 1997.

[24] S. Semmes, *Some Novel Types of Fractal Geometry*, Oxford Univ. Pr., Oxford, 2001.

[25] Y. Pesin, *Dimension Theory in Dynamical Systems: Contemporary Views and Applications*, Univ. Chicago Pr., Chicago, 1997.

[26] Y. Pesin, V. Climenhaga, *Lectures on Fractal Geometry and Dynamical Systems*, Amer. Math. Soc., Providence, RI, 2009.

[27] J. Kigami, *Analysis on Fractals*, Cambridge Univ. Pr., Cambridge, 2001.

[28] R. Strichartz, *Differential Equations on Fractals: A Tutorial* Princeton Univ. Pr., Princeton, 2006.

Chapter 1. What is the shape of the world around us?

Sect. 1.2. Symmetry under magnification

Many authors are keen observers of nature. Henry David Thoreau comes first to mind, and of course there are many others. But Olive Schreiner's descriptions in [29], first published in 1883, are rare. Almost a century before Benoit's *The Fractal Geometry of Nature*, she understood scaling in nature. Scaling has been staring us in the face for as long as we've had faces, but only recently have people begun to describe it.

The universality of fractals quote is on page 153 of [29]. An excellent, and very short, introduction to geometry based on symmetry under magnification is Kenneth Falconer's [11].

[29] O. Schreiner, *The Story of an African Farm*, Penguin Books, London, 1995.

Sect. 1.8. Some non-fractals

The quite detailed description of cats under hats under hats is on pages 32–33 of [30]. Philip Nel points out the non-fractality of the nested cats on pages 146–148 of [31]. The description of Officer MacCruiskeen's collection of nesting chests is on pages 61–64, near the end of chapter 5 of [32].

[30] Dr. Seuss, *The Cat in the Hat Comes Back*, Random House, New York, 1958.

[31] P. Nel, *The Annotated Cat: Under the Hats of Dr. Seuss and His Cats*, Random House, New York, 2007.

[32] F. O'Brien, *The Third Policeman*, Picador, London, 1967.

Chapter 2. Self-similarity in geometry

Sect. 2.1. A simple way to grow fractals

The mathematics of generating fractals by families of contraction maps was developed by Hutchinson [33]. The name IFS was coined by Barnsley and Steven Demko [34] and popularized as a method for compressing images [35]. Good general references for IFS are [36, 14, 15].

[33] J. Hutchinson, "Fractals and self-similarity," *Indiana Univ. J. Math.* **30** (1981), 713–747.

[34] M. Barnsley, S. Demko, "Iterated function systems and the global construction of fractals," *Proc. Roy. Soc. London A* **399** (1985), 243–275.

[35] M. Barnsley, A. Sloan, "A better way to compress images," *Byte* **13** (1988), 215–223.

[36] M. Barnsley, "Lecture notes on iterated function systems," pgs. 127–144 of [37].

[37] R. Devaney, L. Keen, eds., *Chaos and Fractals: the Mathematics Behind the Computer Graphics*, Amer. Math. Soc., Providence, RI, 1989.

Sect. 2.3. Fractal trees and ferns

The fractal fern, a key image in exciting early interest in IFS, appeared in [38].

[38] M. Barnsley, V. Ervin, D. Hardin, J. Lancaster, "Solution of an inverse problem for fractals and other sets," *Proc. Nat. Acad. Sci USA* **83** (1986), 1975–1977.

Sect. 2.2. Some classical fractals

Cantor's analysis of the Cantor set appeared in [39]. Serpinski presented the gasket as a curve with every point a branch point in [40] and the carpet as a universal plane curve in [41]. Menger introduced his sponge as a universal space curve in [42]. The Koch curve was introduced as a curve with no tangent anywhere in [43]. The papers of Menger and Koch are translated in [44].

[39] G. Cantor, "Über unendliche, lineare Punktmannigfaltigkeiten," *Math. Ann.* **21** (1883), 545–591.

[40] W. Sierpinski, "Sur une courbe dont tout point est un point de ramification," *Compte Rendu l'Acad. Sci., Paris* **160** (1915), 302–305.

[41] W. Sierpinski, "Sur une courbe cantorienne qui contient une image biunivoque et continue de toute courbe donnée," *Compte Rendu l'Acad. Sci., Paris* **162** (1916), 629–632.

[42] K. Menger, "Allgemeine Räume und Cartesische Räume," *Amsterdam Acad. Sci.* **29** (1926), 476–482.

[43] H. von Koch, "Sur une courbe continue sans tangente, obtenue par une construction géométrique élémentaire," *Arkiv för Mathematik* **1** (1904), 681–704.

[44] G. Edgar, ed., *Classics on Fractals*, Addison-Wesley, Reading, MA, 1993.

Sect. 2.5. IFS with memory

The conditions under which the attractor of an IFS with memory (determined by forbidden pairs) can be generated by an IFS without memory were presented in [45]. This is the same Jennifer Lanski who drew our cat sketches in Fig. 1.12. That the attractors of IFS with higher memory (determined by forbidden addresses of length > 2) are the attractors of IFS with forbidden pairs is presented in [46, 47]. The chart of gasket relatives is on pages 246 – 248 of [19]. Taylor's papers on the topological type of Sirepinski relatives are [48, 49, 50]. Topological types of IFS with 1-step memory are investigated in [51].

[45] M. Frame, J. Lanski, "When is a recurrent IFS attractor a standard IFS attractor?," *Fractals* **7** (1999) 257–266.

[46] R. Bedient, M. Frame, K. Gross, J. Lanski, B. Sullivan, "Higher block IFS 1: memory reduction and dimension computation", *Fractals*, **18** (2010), 145–155.

[47] R. Bedient, M. Frame, K. Gross, J. Lanski, B. Sullivan, "Higher block IFS 2: relations between IFS with different levels of memory" *Fractals*, **18** (2010), 399–409.

[48] T. Taylor, "Connectivity properties of Sierpiński relatives," *Fractals* **19** (2011), 481–506.

[49] T. Taylor, C. Hudson, A. Anderson "Examples of using binary Cantor sets to study the connectivity of Sierpiński relatives," *Fractals* **20** (2012), 61–75.

[50] T. Taylor, "Using epsilon hulls to characterize ans classify totally disconnected Sierpiński relatives," *Fractals* **23** (2015), DOI:10.1142/S0218348X15500152.

[51] K. Saxer-Taulbee, M. Frame, "Topological types of attractors of IFS with 1-step memory," in preparation.

Sect. 2.6. Random rendering of fractal images

As we see in Chapter 1 of [14], Barnsley gave the name "chaos game" to the random IFS algorithm. We have not adopted this name because the random IFS algorithm does not involve chaos in the technical sense. Mendivil's paper on generating de Bruijn sequences is [52]. Familiarity with some algebraic constructions – finite fields, for example – is assumed in this paper.

[52] F. Mendivil, "Fractals, graphs, and fields," *Amer. Math. Monthly* **110** (2003), 503–515.

Sect. 2.7. Circle inversion fractals

Mumford, Series, and Wright explain many, many examples of fractals generated by circle inversions and linear fractional transformations in their beautiful book [53]. Benoit's paper describing his fast algorithm for generating some circle inversion limit sets is [54]. The paper [55] includes the proof of Elton's ergodic theorem: that the random IFS algorithm generates fractals so long as on average the transformations are contractions.

[53] D. Mumford, C. Series, D. Wright, *Indra's Pearls: The Vision of Felix Klein*, Cambridge Univ. Pr., Cambridge, 2002.

[54] B. Mandelbrot, "Self-inverse fractals osculated by sigma-discs and the limit sets of inversion groups," *Math. Intelligencer* **5** (1983), 9–17.

[55] J. Elton, "An ergodic theorem for iterated maps," *Ergodic Th. Dyn. Syst.* **7** (1987), 481–488.

Sect. 2.8. Random fractals

Einstein's study of Brownian motion is [56]. A translation of Bachelier's thesis is printed in [57]. Fractional Brownian motion is introduced in [58], Lévy stable models in [59]. A good place to start studying fractal sums of pulses is [60].

[56] A. Einstein, *The Theory of Brownian Motion*, Dover, New York, 1956.

[57] P. Cootner, *The Random Character of Stock Market Prices*, MIT Press, Cambridge, 1964.

[58] B. Mandelbrot, J. Van Ness, "Fractional Brownian motions, fractional noises and applications," *SIAM Review* **10** (1968), 422–437.

[59] B. Mandelbrot, "The variation of certain speculative prices," *J. of Business* **36** (1963), 394–419.

[60] B. Mandelbrot, "Introduction to fractal sums of pulses," pgs. 110–123 of *Lévy Flights and Related Topics in Physics*, M. Shlesinger, G. Zaslavsky, U Frisch, eds., Springer, Berlin, 1995.

Sect. 2.9. And flavors stranger still

Mumford, Series, and Wright's book on fractals generated by linear fractional transformations is [53]. Thurston's program is described in [61, 62]. BPI spaces are presented in [23, 24].

[61] W. Thurston, "Three dimensional manifolds, Kleinian groups, and hyperbolic geometry," *Bull. Amer. Math. Soc.* **6** (1982), 357–381.

[62] W. Thurston, *Three-Dimensional Geometry and Topology*, Princeton Univ. Pr., 1997.

Chapter 3. Self-similarity in the wild

Sect. 3.3. Coastlines, Mountains, Rivers

Benoit's paper on coastlines is [63]; Richardson's measurements of coastline length scaling is [64]. Sapoval's papers on fractal drums are [65, 66]; his quotation about fractal coastlines is from [65], page 298.

One of Handelman's fractal mountain constructions is shown on page 271 of [1]. Some of Voss's fractal landscapes are shown on pages 264–265 and color plates C9, C11, and C13 of [1], and also color plates 1–13 of [67]. Benoit's jubilee volumes, where we find Musgrave's chapter on fractal forgeries of nature, are [68]. Fractal river networks are discussed in [69, 70, 71].

[63] B. Mandelbrot, "How long is the coast of Britain? Statistical self-similarity and fractional dimension," *Science* **156** (1967), 636–638.

[64] L. Richardson, "The problem of contiguity: an appendix of statistics of deadly quarrels," *Gen. Syst. Yearbook* **6** (1961), 139–187.

[65] B. Sapoval, "Experimental observation of local modes in fractal drums," *Physica D* **38** (1989), 296–298.

[66] B. Sapoval, T. Gobron, A. Margolina, "Vibrations of fractal drums," *Phys. Rev. Lett.* **67** (1991), 2974–2977.

[67] H.-O. Peitgen, D. Saupe, eds., *The Science of Fractal Images*, Springer-Verlag, New York, 1988.

[68] M. Lapidus, M. Frankenhuijsen, eds., *Fractal Geometry and Applications: A Jubilee of Benoit Mandelbrot*, Amer. Math. Soc., Providence, RI, 2004.

[69] M. Cieplak, A. Giacometti, A. Maritan, A. Rinaldo, I. Rodriguez-Iturbe, J. Banavar, "Models of fractal river basins," *J. Stat. Phys.* **91** (1998), 1–15.

[70] J. Banavar, A. Maritan, A. Rinaldo, "Size and form in efficient transportation networks," *Nature* **399** (1999), 130–132.

[71] A. Rinaldo, J. Banavar, A. Maritan, "Trees, networks, and hydrology," *Water Resources Research* **42** (2006), W06D07, doi:10.1029/2005WR004108.

Sect. 3.4. Fractal lungs

In [72]–[75] Weibel gives clear presentations of what we know about lung structure. One piece of this picture is the mother-daughter pulmonary branch diameter relation reported by Hess and Murray [76, 77]. The higher estimate of this relation is reported in [78]. Weibel and Gomez's transection method is described in [79]. The revised estimate of alveoli number by Ochs et al. is [80]. Our derivation of the lung surface area calculation is adapted from pgs. 30–33 of [81]. The more general case is treated on pages 70 – 76 of [82].

[72] E. Weibel, "Fractal geometry: a design principle for living organisms," *Amer. J. Physiol.* **261** (1991), L361–L369.

[73] E. Weibel, "Mandelbrot's fractals and the geometry of life: a tribute to Benoit Mandelbrot on his 80th birthday," pgs. 3–16 of *Fractals in Biology and Medicine, Vol. 4*, G. Losa, D. Merlini, T. Nonnenmacher, E. Weibel, eds., Birkhäuser, Basel, 2005.

[74] E. Weibel, "What makes a good lung? The morphometric basis of lung function," *Swiss Med. Wkly.* **139** (2009), 375–386.

[75] E. Weibel, "How Benoit Mandelbrot changed my thinking about biological form," pgs. 471–487 of [7].

[76] W. Hess, "Das Prinzip des kleinsten Kraftverbrauches im Dienste hämodynamischer Forschung," *Arch. Anat. Physiol.*, Physiologische Abteilung, 1914.

[77] C. Murray, "The physiological principle of minimum work. I. The vascular system and the cost of blood," *Proc. Nat. Acad. Sci. USA* **12** (1926), 207–214.

[78] B. Mauroy, M. Filoche, E. Weibel, B. Sapoval, "An optimal bronchial tree may be dangerous," *Nature* **427** (2004), 633–636.

[79] E. Weibel, D. Gomez, "A principle for counting tissue structures on random sections," *J. Appl. Physiol.* **17** (1962), 343–348.

[80] M. Ochs, J. Nyengaard, A. Jung, L. Knudsen, M. Voigt, T. Wahlers, J. Richter, H. Gundersen, "The number of alveoli in the human lung," *Am. J. Respir. Crit. Care Med.* **169** (2004), 120–124.

[81] E. Weibel, *Stereological Methods 1: Practical Methods for Biological Morphometry*, Academic Press, London, 1981.

[82] E. Weibel, *Stereological Methods 2: Theoretical Foundations*, Academic Press, London, 1981.

Sect. 3.6. Forests and trees

We first became aware of this result in the NOVA video [83]. More details can be found in [84, 85]. The work on arthropod populations appeared in [86].

[83] *Fractals. Hunting the Hidden Dimension*, 2008, a NOVA episode aired on PBS.

[84] G. West, B. Enquist, J. Brown, "A general quantitative theory of forest structure and dynamics," *Proc. Nat. Acad. Sci. USA* **106** (2009), 7040–7045.

[85] B. Enquist, G. West, J. Brown, "Extensions and evaluations of a general quantitative theory of forest structure and dynamics," *Proc. Nat. Acad. Sci. USA* **106** (2009), 7046–7051.

[86] D. Morse, J. Lawton, M. Dodson, M. Williamson, "Fractal dimension of vegetation and the distribution of arthropod body lengths," *Nature* **314** (1985), 731–733.

Sect. 3.7. Complex heartbeats

Some early papers reporting the presence of chaos in heartbeats are [87, 88].

In the late 1980s, highly visible papers reported putative discoveries of chaos in physiological systems. In 2009 the journal *Chaos* revisited the issue of heartbeat dynamics. In [89] Leon Glass, an early architect of the mathematical study of heartbeat patterns, summarized the study to date and introduced the current work. In this issue, possible sources of apparent chaos in cardiac signals are identified: [90] presents evidence that much of the observed fluctuation in heart rate is an effect of respiration, while [91] argues that fluctuations in the interbeat interval may result from variations in the acetylcholine levels around the sinoatrial node, the heart's pacemaker.

In this same issue of *Chaos*, multifractal characteristics of heartbeat are investigated in [92, 93]. While distinguishing chaos from certain kinds of noise appears to be challenging, detecting power law scalings that characterize fractals and multifractals has been achieved in some cases. The techniques needed to do this are quite a bit more involved than those we have studied here.

[87] R. Pool, "Is it healthy to be chaotic?" *Science* **243** (1989), 604–607.

[88] A. Goldberger, D. Rigney, B. West, "Chaos and fractals in human physiology," *Sci. Am.* **262** (1990), 43–49.

[89] L. Glass, "Introduction to controversial topics in nonlinear science: Is the normal heart rate chaotic?" *Chaos* **19** (2009), 028501.

[90] N. Wessel, M. Riedl, J. Kurths, "Is the normal heart rate 'chaotic' due to respiration?" *Chaos* **19** (2009), 028508.

[91] J. Zhang, A. Holden, O. Monfredi, M. Boyett, H. Zhang, "Stochastic vagal modulation of cardiac pacemaking may lead to erroneous indentification of cardiac 'chaos,'" *Chaos* **19** (2009), 028509.

[92] R. Baillie, A. Cecen, C. Erkal, "Normal heartbeat series are nonchaotic, nonlinear, and multifractal: new evidence from semiparametric and perimetric tests," *Chaos* **19** (2009), 028503.

[93] R. Sassi, M. Signorini, S. Cerutti, "Multifractality and heart rate variability," *Chaos* **19** (2009), 028507.

Sect. 3.8. Metabolic rates

Rubner's derivation of the $M^{-1/3}$ scaling of metabolic rate per unit mass was published in [94]. Kleiber's extensive empirical analysis of metabolic rate per unit mass, and his derivation of the $M^{-1/4}$ scaling, was published in [95]. West, Brown, and Enquist presented their explanation of the $M^{-1/4}$ scaling in [96, 97]. The other biological scalings – aortas, tree trunks, etc. – are found on page 1677 of [97]; the quotation about the genetic code and fractal distribution networks is on page 1679. The approach of Banavar and coworkers to explaining the metabolic rate per unit mass scaling is reported in [98, 99]. The Feynman quotation is from page 23 of [100]

[94] M. Rubner, "Über den Einfluss der Körpergrösse auf Stoff- und Kraftwechsel," *Zeitschrift für Biologie* **19** (1883), 535–562.

[95] M. Kleiber, "Body size and metabolism," *Hilgardia* **6** (1932), 315–353.

[96] G. West, J. Brown, B. Enquist, "A general model for the origin of allometric scaling laws in biology," *Science* **276** (1997), 122–126.

[97] G. West, J. Brown, B. Enquist, "The fourth dimension of life: fractal geometry and allometric scaling of organisms," *Science* **284** (1999), 1677–1679.

[98] J. Banavar, A. Maritan, A. Rinaldo, "Size and form in efficient transportation networks," *Nature* **399** (1999), 130–132.

[99] J. Banavar, M. Moses, J. Brown, J. Damuth, A. Rinaldo, R. Sibly, A. Maritan, "A general basis for quarter-power scaling in animals," *Proc. Nat. Acad. Sci. USA* **107** (2010), 15816–15820.

[100] R. Feynman, *The Pleasure of Finding Things Out*, Basic Books, New York, 1999.

Sect. 3.9. Fractals and DNA

The study of fractal correlations in DNA is an active topic; a thorough recounting of the work to date would fill another book. Richard Voss is a pioneer in a field now home to many voices and varied approaches. Voss reported his results in [101, 102].

Both the equilibrium globule model and the mathematical properties of the fractal globule model are described in Mirny's paper [103]. Originally called the crumpled globule model, the fractal DNA globule model was proposed by Grosberg and coworkers in [104] and suggested as a model for DNA folding in cell nuclei in [105]. Empirical evidence for fractal globules was reported in [106]. A summary of evidence for fractal DNA globules is given in [107]. Additional experiments to test the fractal globule hypothesis are suggested in [108].

Programmed DNA is a fascinating example of self-reference: DNA containing instructions for building itself into different shapes. A good description is [109]; DNA assembling Sierpinski gaskets is reported in [110].

DNA acting as fractal antennas is shown in [111].

[101] R. Voss, "Evolution of long-range fractal correlations and $1/f$ noise in DNA base sequences," *Phys. Rev. Lett.* **68** (1992), 3805–3808.

[102] R. Voss, "Long-range fractal correlations in DNA introns and exons," *Fractals* **2** (1994), 1–6.

[103] L. Mirny, "The fractal globule as a model of chromatin architecture in the cell," *Chromosome Res.* **19** (2011), 37–51.

[104] A. Grosberg, S. Nechaev, E. Shakhnovich, "The role of topological constraints in the kinetics of collapse of macromolecules," *J. Phys. France* **49** (1988), 2095–2100.

[105] A. Grosberg, Y. Rabin, S. Havlin, A. Neer, "Crumpled globule model of the three-dimensional structure of DNA," *Europhys. Lett.*, **23** (1993), 373–378.

[106] E. Lieberman-Aiden, N. van Berkum, L. Williams, M. Imakaev, T. Ragoczy, A. Telling, I. Amit, B. Lajoie, P. Sabo, M. Dorschner, R. Sandstrom, B. Bernstein, M. Bender, M. Groudine, A. Gnirke, J. Stamatoyannopoulos, L. Mirny, E. Lander, J. Dekker, "Comprehensive mapping of long-range interactions reveals folding principles of the human genome," *Science* **326** (2009), 289–293.

[107] A. Bancaud, S. Huet, N. Daigle, J. Mozziconacci, J. Beaudouin, J. Ellenberg, "Molecular crowding affects diffusion and binding of nuclear proteins in heterochromatin and reveals the fractal organization of chromatin," *EMBO J.* **28** (2009), 3785–3798.

[108] A. Bancaud, C. Lavelle, S. Huet, J. Ellenberg, "A fractal model for nuclear organization: current evidence and biological implications," *Nucleic Acids Res.* **40** (2012), 8783–8792.

[109] E. Winfree, *Algorithmic Self-Assembly of DNA*, Caltech Ph. D. thesis, 1988.

[110] P. Rothemund, N. Papadakis, E. Winfree, "Algorithmic self-assembly of DNA Sierpinski triangles," *PLOS Biology* **2** (2004), 2041–2053.

[111] M. Blank, R, Goodman, "DNA is a fractal antenna in electromagnetic fields," *Int. J. Radiation Biol.* **87** (2011), 409–415.

Sect. 3.10. How planets grow

A popular exposition of the fractal dust model of planet formation is [112]. The literature, both experimental and theoretical, on the role of fractal dust clumps in planet formation and in circumstellar material is substantial, and includes [113] – [125]. The quote from Dominik et al. is from pg. 780 of [114]. The observations of Peggy, Saturn's new and 63rd moon, were reported in [126].

[112] B. Daviss, "Dust demon: physicist David Peak has big plans for small dust balls," *Discover* **13** (March 1992), 114–116.

[113] J. Blum, G. Wurm, "Experiments on sticking, restructuring, and fragmentation of preplanetary dust aggregates," *Icarus* **143** (2000), 138–146.

[114] C. Dominik, J. Blum, J. Cuzzi, G. Wurm, "Growth of dust as an initial step toward planet formation," pgs. 783–800 of *Protostars and Planets. V* B. Reipurth, D. Jewitt, eds., Univ. of Arizona Pr., Tucson, 2007.

[115] B. Donn, P. Meakin, "The accumulation and structure of the cometary nucleus: the fluffy aggregate model," *Bull. Amer. Astron. Soc.* **20** (1988), 840.

[116] B. Donn, P. Meakin, "Aerodynamics of fractal grains: implications for the primordial solar nebula," *Bull. Amer. Astron. Soc.* **19** (1987), 847.

[117] B. Donn, P. Meakin, "Collisions of macroscopic fluffy aggregates in the primordial solar nebula and the formation of planetesimals," pgs 577–580 of *Proceedings of the 19th Lunar and Planetary Science Conference*, Lunar and Planetary Institute, Houston, 1989.

[118] B. Donn, P. Meakin, "Collisions of macroscopic fluffy aggregates in the primordial solar nebula," *Lunar Planetary Sci. Conf.* **19** (1989), 281–282.

[119] H. Kimura, H. Ishimoto, T. Mukai, "A study on solar dust ring formation based on fractal dust models," *Astron. Astrophys.* **326** (1997), 263–270.

[120] P. Meakin, D. Donn, "Aerodynamic properties of fractal grains: implications for the primordial solar nebula," *Astrophys. J.* **329** (1988), L39–L41.

[121] J. Nuth, S. Hallenbeck, F. Rietmeijer, "Laboratory studies of silicate smokes: Analog studies of circumstellar material," *J. Geophys. Res.* **105** (2000), 387–396.

[122] S. Weidenschilling, "Formation of planetesimals and accretion of the terrestrial planets," *Space Sci. Rev.* **92** (2000), 295–310.

[123] S. Weidenschilling, "Fractal aggregates and planetesimal formation," *Bull. Amer. Astron. Soc.* **20** (1988), 815.

[124] S. Weidenschilling, "Planetesimals from stardust," pgs. 281–293 of *From Stardust to Planetesimals*, Y. Pendleton, A. Tielens, eds. Astron. Soc. Pacific, San Francisco, CA, 1997.

[125] S. Weidenschilling, "When the dust settles: fractal aggregates and planetesimal formation," *Proc. Lunar and Planetary Sci. Conf.* **28** (1997), 1517–1518.

[126] C. Murray, N. Cooper, G. Williams, N. Altree, J. Boyer, "The discovery and dynamical evolution of an object at the outer edge of Saturn's A ring," *Icarus* **236** (2014), 165–168.

Sect. 3.11. Reversals of the Earth's magnetic field

To draw our graph of the timeline of the geomagnetic reversals data were assembled from the papers of Kent and Gradstein [127] and of Cande and Kent [128]. How these data were gathered is a fascinating story, involving magnetic alignments of rocks made of ancient lava, stripes of constant alignment on the ocean floor, and an estimation of the timescale of these events. Google will show you wonders. The Lévy flight analysis was done by Gaffin [129].

[127] D. Kent, F. Gradstein, "A Cretaceous and Jurassic geochronology," *Geol. Soc. Amer. Bull* **96** (1985), 1419–1427.

[128] S. Cande, D. Kent, "Revised calibration of the geomagnetic polarity timescale for the late Cretaceous and Cenozoic," *J. Geophys. Res.* **100** (1995), 6093–6095.

[129] S. Gaffin, "Analysis of scaling in the geomagnetic polarity reversal record," *Phys. of the Earth and Planetary Interiors* **57** (1989), 284–290.

Sect. 3.12. The distribution of galaxies

Hetherington's encyclopedia is [130]. A good exposition of Noether's theorem is the book by Lederman and Hill [131]. Those who want to get to some of the interesting properties of black holes, one of the most fascinating predictions of general relativity, using only calculus and linear algebra, can't do better than [132]. Many books give good introductions to Einstein's cosmology. Some we like are [133] – [144]. The calculations of Friedmann and Lemaître showing that Einstein's original gravity equations predict an expanding universe are [145, 146]. A thorough study of where and how fractals occur in cosmology is given in [147]. Swedenborg presented his hierarchical cosmology in [148]; see also Sects. 4.2 and 4.4 of [147]. A reprint of Lambert's *Cosmological Letters* is [149]. Harrison's account of Olbers' paradox is [150]; Poe's account is [151], the quotation is from pg. 26. Lemaître's speculations on the primordial atom are in [152]. The original Big Bang paper is [153]. The calculation of the current temperature of the cosmic background radiation was published in [154]. Popular accounts of the origin of the universe are [155, 156]. Pietronero's analysis is described in [147] and also in [157, 158]. Guth's ideas about chaotic inflation and fractal universes are in [159, 160, 161]. Descriptions of Linde's self-reproducing universe are in [162] – [165]. Connes' noncommutative geometry is presented in [166] and in his beautiful book [167]. Nottale's scale-invariant relativity is described in [168] – [171]. Peacock's quotes on fractal space-time are from [138], page 339 for the first, page 499 for the second.

[130] N. Hetherington, ed., *Encyclopedia of Cosmology*, Garland Publishing, New York, 1993.

[131] L. Lederman, C. Hill, *Symmetry and the Beautiful Universe*, Promethus Books, Amherst, 2008.

[132] E. Taylor, J. Wheeler, *Exploring Black Holes: Introduction to General Relativity*, Addison-Wesley, San Francisco, 2000.

[133] T. Frankel, *Gravitational Curvature: An Introduction to Einstein's Theory*, Freeman, San Francisco, 1979.

[134] J. Islam, *An Introduction to Mathematical Cosmology*, Cambridge Univ. Pr., Cambridge, 1992.

[135] S. Hawking, G. Ellis, *The Large-Scale Structure of Space-Time*, Cambridge Univ. Pr., Cambridge, 1973.

[136] D. Lawden, *An Introduction to Tensor Calculus, Relativity, and Cosmology*, 3rd ed., Wiley, New York, 1982.

[137] C. Misner, K. Thorne, J. Wheeler, *Gravitation*, Freeman, San Francisco, 1970.

[138] J. Peacock, *Cosmological Physics*, Cambridge Univ. Pr., Cambridge, 1999.

[139] P. J. E. Peebles, *The Large-Scale Structure of the Universe*, Princeton Univ. Pr., Princeton, 1980.

[140] P. J. E. Peebles, *Principles of Physical Cosmology*, Princeton Univ. Pr., Princeton, 1993.

[141] M. Ryan, L. Shepley, *Homogeneous Relativistic Cosmologies*, Princeton Univ. Pr., Princeton, 1975.

[142] R. Tolman, *Relativity, Thermodynamics, and Cosmology*, Oxford Univ. Pr., Oxford, 1934.

[143] R. Wald, *General Relativity*, Univ. of Chicago Pr., Chicago, 1984.

[144] S. Weinberg, *Gravitation and Cosmology: Principles and Applications of the General Theory of Relativity*, Wiley, New York, 1972.

[145] A. Friedmann, "Über die Krümmung des Raumes," *Z. Phys.* **10** (1922) ,377-386.

[146] G. Lemaître, "Un univers homogène de masse constante et de rayon croissant rendant compte de la vitesse radiale des nébuleuses extragalactiques," *Ann. Sci. Soc. Brussels* **47A** (1927) 49–59.

[147] Y. Baryshev, P. Teerikorpi, *Discovery of Cosmic Fractals*, World Scientific, Singapore, 2002.

[148] E. Swedenborg, *Principia: Opera Philosophica et Mineralia*, 1734.

[149] J. Lambert, *Cosmological Letters on the Arrangement of the World-Edifice*, Science History Publications, 1976.

[150] E. Harrison, *Darkness at Night: A Riddle of the Universe*, Harvard Univ. Pr., Cambridge, 1987.

[151] E. A. Poe, *Eureka: An Essay on the Material and Spiritual Universe*, Putnam, New York, 1848.

[152] G. Lemaitre, "The evolution of the universe: discussion," *Nature* **128** (1931), 699–701.

[153] R. Alpher, H. Bethe, G. Gamow, "The origin of chemical elements," *Phys. Rev.* **73** (1948), 803–804.

[154] R. Alpher, R. Herman, "Evolution of the universe," *Nature* **162** (1948), 774.

[155] S. Weinberg, *The First Three Minutes: A Modern View of the Origin of the Universe*, Basic Books, New York, 1977.

[156] G. Smoot, K. Davidson, *Wrinkles in Time: Witness to the Birth of the Universe*, HarperCollins, New York, 1993.

[157] P. Coleman, L. Pietronero, "The fractal structure of the universe," *Phys. Rep.* **213** (1992), 311–389.

[158] L. Pietronero, F. Labini, "Fractal universe," *Physica A* **280** (2000), 125–130.

[159] A. Guth, "Inflation and eternal inflation," *Physics Reports* **333** (2000), 555-574.

[160] J. Garriga, A. Guth, A. Vilenkin, "Eternal inflation, bubble collisions, and the persistence of memory," *Phys. Rev. D* **76** (2007), 123512-1–2.

[161] A. Guth, "Eternal inflation and its implications," *J. Phys. A* **40** (2007), 6811–6826.

[162] A. Linde, "The self-reproducing inflationary universe," *Sci. Am.*, November 1994, 32–39.

[163] A. Linde, "Eternally existing self-reproducing chaotic inflationary universe," *Phys. Lett. B* **175** (1986), 395–400.

[164] A. Linde, "Eternal chaotic inflation," *Mod. Phys. Lett. A* **1** (1986), 81–87.

[165] A. Linde, "A balloon producing balloons, producing balloons: a big fractal," a conversation with Andrei Linde, *Edge* April 27, 2013.

[166] A. Connes, C. Rovelli, "Von Neumann algebra automorphisms and time-thermodynamics relation in generally covariant quantum theories," *Classical and Quantum Gravity* **11** (1994), 2899–2923.

[167] A. Connes, *Noncommutative Geometry*, Academic Press, San Diego, 1994.

[168] L. Nottale, "Fractals and the quantum theory of spacetime," *Int. J. Mod. Phys. A* **4** (1989), 5047–5117.

[169] L. Nottale, "The theory of scale relativity," *Int. J. Mod. Phys. A* **7** (1992), 4899–4936.

[170] L. Nottale, *Fractal Space-Time and Micro-physics*, World Scientific, Singapore, 1993.

[171] L. Nottale, *Scale Relativity and Fractal Space-Time: A New Approach to Unifying Relativity and Quantum Mechanics*, World Scientific, Singapore, 2011.

Sect. 3.13. Is this all?

Cutting and Garvin's study relating dimension to perceived visual complexity is [172]. Hagerhall's correlation of the fractal dimension of perceived curves and EEG data are presented in [173]. Freeman and Skarda's work on olfaction and chaos is in [174, 175].

Moore's compendium of ammonite suture images is [176]. Calculations of ammonite suture dimensions is reported in [177, 178]. Gould's book that presents the left wall alternative to an evolutionary drive toward greater complexity is [179]. The quote about radical contingency as a fractal principle is from page 215.

Thorough descriptions of viscous fingering, electrodeposition, and dielectric breakdown are given in Vicsek's book [180], and in [181, 182, 183].

Eddington's eclipse expedition that verified Einstein's prediction of the gravitational deflection of starlight was reported in [184]. Zwicky's papers on gravitational lensing and dark matter are [185, 186]. Technical descriptions of the dark web of dark matter are in [187, 188]; a more popular exposition is in [189].

The *Bacillus subtilis* experiments are reported in [190, 191, 192]. Wild-type and mutant comparisons in other bacteria are studied in [193]. The communicating walkers model is described in [194, 195]. Fractal fungal growth is presented in [196, 197, 198].

The Gutenberg-Richter law can be found in [199]; Turcotte's relation of the Gutenberg-Richter exponent to the dimension of an earthquake fault is presented in [200]. Omori's law on the rate of aftershocks is presented in [201], Utsu's modification of Omori's law is in [202]. Okubo and Aki's computation and interpretation of the San Andreas fault dimension is found in [203]. Self-organized criticality is described in [204, 205]. In his book [205], Bak proposed SOC as an explanation for the presence of fractals in nature. In Section 8.2 of their book [21], Hastings and Sughara present an SOC explanation of earthquakes. The papers of Abe and Suzuki [206] – [214], and the paper of Madariaga [215], present applications of many aspects of modern network theory to the study of earthquakes.

A good reference on EEG data is [216]. The work of Babloyantz on chaotic and fractal aspects of EEG data is found in [217] – [220], for example. Watters and Martin study fractal structures and long-range correlations of EEG signals in [221, 222].

Galileo's work on sunspots is recorded in his book [223]. The papers [224, 225] report Milovanov and Zelenyi's fractal cluster of magnetic field lines model of sunspots. In [226], Shaikh et al. analyzed sunspot time series data and observed a persistent fractional Brownian motion character. The method of R/S analysis employed by Shaikh and others was developed by Hurst [227] to analyze the pattern of flooding of the Nile. Another analysis of the dimensionality of sunspot time series was done by Qin [228], based on Takens' delay embedding method [229].

Aschwanden and Aschwanden's papers [230, 231] investigate fractal aspects of solar flares and present some beautiful flare pictures. Mogilevsky and Shilova [232] propose a flare model based on Bak's self-organized criticality [205]. In [233] we find evidence for multifractality in flare brightness fluctuations. Multifractal characteristics of solar prominences are reported in [234, 235].

The 3-dimensional structure of the Cassiopeia A supernova remnant is reported in [236], and the radio astronomical observations of that remnant are discussed in [237].

The calculation suggesting Saturn's rings have a fat Cantor set cross-section is presented in [238].

The application of the area-perimeter relation to measure roughness of fractured metals was introduced in [239]. Feder's excellent book, [13], includes a description of the area-perimeter relation. Computer simulations of crack formation in various settings are reported in [240] – [243].

A box-counting calculation of soil pore dimension is reported in [244]; the slit island method is applied in [245]. The extensive review article by Perfect and Kay is [246]. The titles of [247, 248] speak for themselves.

Our reference for wildlife ecology is John Bissonette's book [249], especially the chapters by Milne [250] and Peak [251]. The Ott-Grebogi-Yorke (OGY) method of stabilizing unstable cycles using only sequential measurements of the system was first presented in [252]. This approach has been applied in a vast range of settings, from vibrating magnetic ribbons to cardiac arrhythmias. For other examples, Google is your true friend.

References to fractal aspects of clouds are found on pages 98–99 and 112 of [1] and in Section 12.1 of [13]. Lovejoy's study of clouds is [253].

Benoit's original papers on turbulence and multifractals are [254, 255]. Some of the investigations by Sreenivasan and his coworkers of multifractals and turbulence are [256] – [268]. Chhabara and Sreenivasan's papers on multifractals and negative dimensions are [257, 258]. Benoit's papers on negative dimensions include [269] – [274].

REFERENCES

Snow Crystals [275] is a gorgeous book, a work of patience and dedication. Early reports of simulated snowflake growth are [276, 277].

Animal foraging patterns are analyzed in [278, 279, 280].

The epidemiological and predator-prey differential equations models we discussed are studied in [281, 282]. Recall that the delay embedding method was proposed by Takens in [283]; elaborations and examples are presented in [284, 285, 286]. Some other appearances of fractals in epidemics are presented in [287, 288, 289]. Borges' parable "On exactitude in science" is included in the collection [290], among others.

The work of Clancy and coworkers on Markov models of ion channels is in [291] – [298]. Note especially that [291] uses Markov models of ion channel dynamics to identify the cardiac sodium channel mutation responsible for Long QT syndrome. The work of Liebovitch and coworkers on fractal models of ion channels is in [299] – [303]. Some variations on the fractal ion channel model are in [304, 305].

Wright's paper presenting the idea of fitness landscapes is [306]. Eigen and Schuster's fitness landscape variation with genotype base appears in [307]. Many examples of fitness landscapes are given in [308, 309, 310]. Sorkin's fractal fitness landscape paper is [311]. Chapter 8 of Adami's book [312] is "Fitness landscapes," section 8.3 is "Fractal landscapes," and in section 8.5 Adami presents evidence of the fractality of RNA landscapes. Some other examples of fractal fitness landscapes are in [313].

[172] J. Cutting, J. Garvin, "Fractal curves and complexity," *Perception & Psychophysics* **42** (1987), 365–370.

[173] C. Hagerhall, T. Laike, R. Taylor, M. Küller, R. Küller, T. Martin, "Investigations of human EEG response to viewing fractal patterns," *Perception* **37** (2008), 1488–1494.

[174] W. Freeman, "The physiology of perception," *Sci. Am.* **264** (February 1991), 78–85.

[175] C. Skarda, W. Freeman, "How brains make chaos in order to make sense of the world," *Behavioral and Brain Sci.* **10** (1987), 161–195.

[176] R. Moore, *Treatise on Invertebrate Paleontology, Part L: Mollusca*, Geological Society of America, Lawrence, KS, 1953.

[177] G. Boyajian, T. Lutz, "Evolution of biological complexity and its relation to taxonomic longevity in the Ammonoidea," *Geology* **20** (1992), 983–986.

[178] T. Lutz, G. Boyajian, "Fractal geometry of ammonoid sutures," *Paleobiology* **21** (1995), 329–342.

[179] S. Gould, *Full House: The Spread of Excellence from Plato to Darwin*, Harmony Books, New York, 1996.

[180] T. Vicsek, *Fractal Growth Phenomena*, 2nd ed., World Scientific, Singapore, 1992.

[181] K. Måløy, J. Feder, T. Jøssang, "Viscous fingering fractals in porous media," *Phys. Rev. Lett.* **55** (1985), 2688–2691.

[182] M. Matsushita, M. Sano, Y. Hayakawa, H. Honjo, Y. Sawada, "Fractal structures of zinc metal leaves grown by electrodeposition," *Phys. Rev. Lett.* **53** (1984), 286–289.

[183] L. Niemeyer, L. Pietronero, H. Weismann, "Fractal dimension of dielectric breakdown," *Phys. Rev. Lett.* **52** (1984), 1033–1036, and **57** (1986), 650.

[184] F. Dyson, A. Eddington, C. Davidson, "A determination of the deflection of light by the sun's gravitational field, from observations made at the total eclipse of May 29, 1919," *Philos. Trans. Roy. Soc. London, A* **220** (1920), 291–333.

[185] F. Zwicky, "Nebulae as gravitational lenses," *Phys. Rev.* **51** (1937), 290.

[186] F. Zwicky, "Die Rotverschiebung von extragalaktischen Nebeln," *Helv. Phys. Acta* **6** (1933), 110–127.

[187] N. Libeskind, C. Frenk, S. Cole, J. Helly, A. Jenkins, J. Navarro, C. Power, "The distribution of satellite galaxies: the great pancake," *Mon. Not. Roy. Astron. Soc.* **363** (2005), 146–152.

[188] N. Libeskind, A. Knebe, Y. Hoffman, S. Gottlöber, G. Yepes, M. Steinmetz, "The preferred direction of infalling satellite galaxies in the Local Group," *Mon. Not. Roy. Astron. Soc.* **411** (2011), 1525–1535.

[189] N. Libeskind, "Dwarf galaxies and the dark web," *Sci. Am.* **310** (March 2014), 46–51.

[190] H. Fujikawa, M. Matsushita, "Fractal growth of *Bacillus subtilis* on agar plates," *J. Phys. Soc. Japan* **58** (1989), 3875–3878.

[191] M. Matsushita, H. Fujikawa, "Diffusion-limited growth in bacterial colony formation," *Physica A* **168** (1990), 498–506.

REFERENCES

[192] H. Fujikawa, M. Matsushita, "Bacterial fractal growth in the concentration field of a nutrient," *J. Phys. Soc. Japan* **60** (1991), 88–94.

[193] T. Matsuyama, R. Harshey, M. Matsushita, "Self-similar colony morphogenesis by bacteria as the experimental model of fractal growth by a cell population," *Fractals* **1** (1993), 302–311.

[194] E. Ben-Jacob, O. Shochet, I. Cohen, A. Tenenbaum, A. Czirók, T. Vicsek, "Cooperative strategies in formation of complex bacterial patterns," *Fractals* **3** (1995), 849–868.

[195] E. Ben-Jacob, I. Cohen, O. Shochet, A. Tenenbaum, A. Czirók, T. Vicsek, "Cooperative formation of chiral patterns during growth of bacterial colonies," *Phys. Rev. Lett.* **75** (1995), 2899–2902.

[196] S. Matsuura, S. Miyazima, "Self-affine fractal growth front of *Aspergillus oryzae*," *Physica A* **191** (1992), 30–34.

[197] S. Matsuura, S. Miyazima, "Colony of the fungus *Aspergillus oryzae* and self-affine fractal geometry of growth fronts," *Fractals* **1** (1993), 11–19.

[198] S. Matsuura, S. Miyazima, "Formation of ramified colony of *Aspergillus oryzae* on agar media," *Fractals* **1** (1993), 336–345.

[199] B. Gutenberg, C, Richter, "Magnitude and energy of earthquakes," *Ann. Geophysics* **9** (1956), 1–15.

[200] D. Turcotte, "Fractals in geology and geophysics," pgs. 171–196 of *Fractals in Geophysics*, C. Scholz, B. Mandelbrot, eds., Birkhäuser, Basel, 1989.

[201] F. Omori, "On the aftershocks of earthquakes," *J. Coll. Sci. Imperial Univ. Tokyo* **7** (1894), 111–200.

[202] T. Utsu, "A statistical study of the occurrence of aftershocks," *Geophys. Mag.* **30** (1961), 521–605.

[203] P. Okubo, K. Aki, "Fractal geometry in the San Andreas fault system," *J. Geophys. Res.* **92** (1987), 345–355.

[204] P. Bak, C. Tang, K. Wiesenfeld, "Self-organized criticality: an explanation of $1/f$ noise," *Phys. Rev. Lett.* **59** (1987), 381–384.

[205] P. Bak, *How Nature Works: The Science of Self-Organized Criticality*, Springer, New York, 1996.

[206] S. Abe, N. Suzuki, "Earthquake networks, complex," pgs. 2530–2538, vol. 3 of *Encyclopedia of Complexity and Systems Science*, R. Meyers, ed., Springer, New York, 2009.

[207] S. Abe, N. Suzuki, "Zipf-Mandelbrot law for time intervals between earthquakes," arXiv:cond-mat/0208344

[208] S. Abe, N. Suzuki, "Complex network of earthquakes," pgs 1046–1053 of *Computational Science ICCS*, M. Bubak, ed., Springer, Berlin, 2004.

[209] S. Abe, N. Suzuki, "Scale-invariant statistics of the degrees of separation in directed earthquake network," *Physica A* **350** (2005), 588–596.

[210] S. Abe, N. Suzuki, "Aftershocks in modern perspectives: complex earthquake network, aging, and non-Markovianity," *Acta Geophysica* **60** (2012), 547–561.

[211] S. Abe, N. Suzuki, "Violation of the scaling relation and non-Markovian nature of earthquake aftershocks," *Physica A* **388** (2009), 1917–1920.

[212] S. Abe, N. Suzuki, "Scaling relation for earthquake networks," *Physica A* **388** (2009), 2511–2514.

[213] S. Abe, N. Suzuki, "Complex network of seismicity," *Progr. Theoret. Phys.* **162** (2006), 138–146.

[214] S. Abe, N. Suzuki, "Finite data-size scaling of clustering in earthquake networks," *Physica A* **390** (2011), 1343–1349.

[215] R. Madariaga, "Earthquake scaling laws," pgs. 2581–2599, vol. 3 of *Encyclopedia of Complexity and Systems Science*, R. Meyers, ed., Springer, New York, 2009.

[216] W. Freeman, R. Quiroga, *Imaging Brain Function with EEG*, Springer, New York, 2013.

[217] A. Babloyantz, "Some remarks on nonlinear data analysis of physiological time series," pgs. 51–62 of *Measures of Complexity and Chaos*, N. Abraham, A. Albano, A. Passamante, P. Rapp, eds., Plenum, New York, 1990.

[218] A. Babloyantz, "Probing dynamics of the cerebral cortex," pgs. 31–40 of *Complexity, Chaos, and Biological Evolution*, E. Mosekilde, L. Moseklide, eds., Plenum, New York, 1991.

[219] A. Babloyantz, A. Destexhe, "Strange attractors in the human cortex," pgs. 48–56 of *Temporal Disorder in Human Oscillatory Systems*, L. Rensing, ed., Springer, Berlin, 1987.

[220] A. Babloyantz, J. Salazar, C. Nicolis, "Evidence of chaotic dynamics of brain activity during the sleep cycle," *Phys. Lett. A* **111** (1985), 152–156.

[221] P. Watters, F. Martin, "A method for estimating long-range power law correlations from the electroencephalogram," *Biol. Psych.* **66** (2004), 79–89.

[222] P. Watters, "Time-invariant long-range correlations in electroencephalogram dynamics," *Int. J. Sys. Sci.* **31** (2000), 819–825.

[223] G. Galilei, C. Scheiner, *On Sunspots*, Univ. of Chicago Pr., Chicago, 2010.

[224] A. Milovanov, L. Zelenyi, "Fractal model for sunspot evolution," *Geophys. Res. Lett.* **19** (1992), 1419–1422.

[225] A. Milovanov, L. Zelenyi, "Applications of fractal geometry to dynamical evolution of sunspots," *Phys. Fluids B* **5** (1993), 2609–2615.

[226] Y. Shaikh, A. Khan, M. Iqbal, S. Behere, S. Bagare, "Sunspot data analysis using time series," *Fractals* **16** (2008), 259–265.

[227] H. Hurst, "Long-range storage capacity of reservoirs," *Trans. Am. Soc. Civil Eng.* **116** (1951), 770–808.

[228] Z. Qin, "A fractal study on sunspot relative number," *Chinese Astron. and Astrophys.* **18** (1994), 313–318.

[229] F. Takens, "Detecting strange attractors in turbulence," pgs. 366–381 of *Dynamical Systems and Turbulence, Warwick 1980*, D. Rand, L.-S. Young, eds., Springer, New York, 1981.

[230] M. Aschwanden, P. Aschwanden, "Solar flare geometries. I. The area fractal dimension," *Astrophys. J.* **674** (2008), 530–543.

[231] M. Aschwanden, P. Aschwanden, "Solar flare geometries. II. The volume fractal dimension," *Astrophys. J.*, **674** (2008), 544–553.

[232] E. Mogilevsky, N. Shilova, "Fractal-cluster analysis and small-scale structures of solar flares," *Geomagnetism and Aeronomy* **46** (2006), 303–308.

[233] Z. Yu, V. Anh, R. Eastes, D. Wang, "Multifractal analysis of solar flare indices and their horizontal visibility spectrum," *Nonlin. Processes Geophys.* **19** (2012), 657–665.

[234] E. Leonardis, S. Chapman, C. Foullon, "Turbulent characteristics in the intensity fluctuations of a solar quiescent prominence observed by the *Hinode* solar optical telescope," *Ap. J.* **745** (2012), 185.

[235] E. Leonardis, S. Chapman, W. Daughton, V. Roytershteyn, H. Karimabadi, "Identification of intermittent turbulence in fully kinetic simulations of magnetic reconnection," *Phys. Rev. Lett.* **119** (2013): 205002.

[236] Y. Kim, G. Rieke, O. Krause, K. Misselt, R. Indebetouw, K. Johnson, "Structure of the interstellar medium around Cas A," *Astrophys. J.* **678** (2008), 287–296.

[237] N. Roy, J. Chengalur, P. Dutta, S. Bharadwaj, "H 21 cm opacity fluctuation power spectra toward Cassiopeia A," *Month. Not. Roy. Astron. Soc.* **404** (2010), L45–L49.

[238] J. Avron, B. Simon, "Almost periodic Hill's equation and the rings of Saturn," *Phys. Rev. Lett.* **46** (1981), 1166–1168.

[239] B. Mandelbrot, D. Passoja, A. Paullay, "Fractal character of fracture surfaces of metals," *Nature* **308** (1984), 721–722.

[240] H. Herrmann, "Fractal deterministic cracks," *Physica D* **38** (1989), 192–197.

[241] H. Herrmann, "Fractures," pgs 174–205 of *Fractals and Disordered Systems*, A. Bunde, S. Havlin, eds., Springer-Verlag, Berlin, 1991.

[242] E. Louis, F. Guinea, F. Flores, "The fractal nature of fracture," pgs. 177–180 of *Fractals in Physics*, L. Pietronero, E. Tosatti, eds., North-Holland, Amsterdam, 1986.

[243] C. Lung, "Fractals and the fracture of cracked metals," pgs. 189–192 of *Fractals in Physics*, L. Pietronero, E. Tosatti, eds., North-Holland, Amsterdam, 1986.

[244] R. Uthayakumar, G. Prabakar, S. Aziz, "Fractal analysis of soil pore variability with two dimensional binary images," *Fractals* **19** (2011), 401–406.

[245] A. Papadopoulos, N. Bird, S. Mooney, A. Whitmore, "Fractal analysis of pore roughness in images of soil using the slit island method," *Vadose Zone J.* **7** (2007), 456–460.

[246] E. Perfect, B. Kay, "Applications of fractals in soil and tillage research: a review," *Soil & Tillage Res.* **36** (1995), 1–20.

[247] P. Baveye, J.-Y. Parlange, B. Stewart, *Fractals in Soil Science*, CRC Press, New York, 1998.

[248] Y. Pachepsky, J. Crawford, W. Rawls, *Fractals in Soil Science*, Elsevier, New York, 2000.

REFERENCES

[249] J. Bissonette, ed., *Wildlife and Landscape Ecology: Effects of Pattern and Scale*, Springer, New York, 1997.

[250] B. Milne, "Applications of fractal geometry in wildlife biology," pgs. 32–69 of [249].

[251] D. Peak, "Taming chaos in the wild: a model-free technique for wildlife population control," pgs. 70–100 of [249].

[252] E. Ott, C. Grebogi, J. Yorke, "Controlling chaos," *Phys. Rev. Lett.* **64** (1990), 1196–1199.

[253] S. Lovejoy, "Area-perimeter relation for rain and cloud areas," *Science* **216** (1982), 185–187.

[254] B. Mandelbrot, "Intermittent turbulence in self-similar cascades: divergence of high moments and the dimension of the carrier," *J. Fluid Mech.* **62** (1974), 331–358.

[255] B. Mandelbrot, "On the geometry of homogeneous turbulence, with stress on the fractal dimension of the iso-surface of scalars," *J. Fluid Mech.* **72** (1975), 401–416.

[256] A. Chhabara, R. Jensen, K. Sreenivasan, "Extraction of underlying multiplicative processes from multifractals via the thermodynamic formalism," *Phys. Rev. A* **40** (1989), 4593–4611.

[257] A. Chhabara, K. Sreenivasan, "Negative dimensions, theory, computation, and experiment," *Phys. Rev. A* **43** (1991), 1114–1117.

[258] A. Chhabara, K. Sreenivasan, "Probabilistic multifractals and negative dimensions," pgs. 271–288 of *New Perspectives in Turbulence*, L. Sirovich, ed., Springer, New York, 1991.

[259] C. Meneveau, R. Jensen, K. Sreenivasan, "Direct determination of the $f(\alpha)$ singularity spectrum and its application to fully developed turbulence," *Phys. Rev. A* **40** (1989), 5284–5294.

[260] C. Meneveau, K. Sreenivasan, "Measurement of $f(\alpha)$ from the scaling of histograms, and applications to dynamical systems and fully developed turbulence," *Phys. Lett. A* **137** (1989), 103–112.

[261] C. Meneveau, K. Sreenivasan, "The multifractal nature of turbulent energy dissipation," *J. Fluid Mech.* **224** (1991), 429–484.

[262] C. Meneveau, K. Sreenivasan, "Simple multifractal cascade model for fully developed turbulence," *Phys. Rev. Lett.* **59** (1987), 1424–1427.

[263] C. Meneveau, K. Sreenivasan, P. Kallasnath, M. Fan, "Joint multifractal measures: theory and applications to turbulence," *Phys. Rev. A* **41** (1990), 894–913.

[264] R. Prasad, C. Meneveau, K. Sreenivasan, "Multifractal nature of the dissipation field of passive scalars in fully turbulent flows," *Phys. Rev. Lett.* **61** (1988), 74–77.

[265] K. Sreenivasan, "Fractals and multifractals in fluid turbulence," *Ann. Rev. Fluid Mech.* **23** (1991), 539–604.

[266] K. Sreenivasan, R. Prasad, R. Ramshankar, "The fractal geometry of interfaces and the multifractal distribution of dissipation in fully turbulent flows," *Pure Appl. Geophysics* **131** (1989), 43–80.

[267] K. Sreenivasan, G. Stolovitzky, "Turbulent cascades," *J. Stat. Phys.* **78** (1995), 311–320.

[268] V. Yakhot, K. Sreenivasan, "Towards a dynamical theory of multifractals in turbulence," *Physica A* **343** (2004), 147–155.

[269] B. Mandelbrot, "A class of multifractal measures with negative (latent) values for the 'dimension' $f(\alpha)$," pgs 3–29 of *Fractals' Physical Origin and Properties*, L. Pietronero, ed., Plenum, Erice, 1988.

[270] B. Mandelbrot, "Two meanings of multifractality, and the notion of negative fractal dimension," pgs 79–90 of *Chaos: Soviet-American Perspectives on Nonlinear Science*, D. Campbell, ed., American Institute of Physics, New York, 1990.

[271] B. Mandelbrot, "Random multifractals: negative dimensions and the resulting limitations of the thermodynamic formalism," *Proc. Royal Soc. London A*, **434** (1991), 79–88.

[272] B. Mandelbrot, "Negative dimensions and Hölders, mulifractals and their Hölder spectra, and the role of lateral preasymptotics in science," pgs 409–432 of *J. Fourier Analysis and Applications, Special Issue*, J.-P. Kahane, et al. eds., (1995).

[273] B. Mandelbrot, "Multifractal power-law distributions, other 'anomalies,' and critical dimensions, explained by simple example," *J. Stat. Phys.* **110** (2003), 739–777.

[274] B. Mandelbrot, M. Frame, "A primer of negative test dimensions and degrees of emptiness for latent sets," *Fractals* **17** (2009), 1–14.

[275] W. Bentley, W. Humphreys, *Snow Crystals*, Dover, New York, 1962.

[276] J. Nittmann, H. Stanley, "Non-deterministic approach to anisotropic growth patterns with continuously tunable morphology: the fractal properties of some real snowflakes," *J. Phys. A* **20** (1987), L1185–L1191.

[277] J. Nittmann, H. Stanley, "Tip splitting without interfacial tension and dendritic growth patterns arising from molecular anisotropy," *Nature* **321** (1986), 663–668.

[278] M. Etzenhauser, M. Owens, D. Spalinger, S. Murden, "Foraging behavior of browsing ruminants in a heterogeneous landscape," *Landscape Ecology* **13** (1998), 55–64.

[279] J. Bascompte, C. Vilà, "Fractals and search paths in mammals," *Landscape Ecology* **12** (1997), 213–221.

[280] L. Seuront, "Behavioral fractality in marine copepods: endogeneous rhythms versus exogenous stressors," *Physica A* **390** (2011), 250–256.

[281] W. Schaffer, M. Kot, "Differential systems in ecology and epidemiology," pgs. 158–178 of *Chaos*, A. Holden, ed., Princeton Univ. Pr., Princeton, 1986.

[282] M. Gilpin, "Spiral chaos in a predator-prey model," *Amer. Naturalist* **113** (1979), 306–308.

[283] F. Takens, "Detecting strange attractors in turbulence," pgs. 366–381 of *Dynamical Systems and Turbulence, Warwick 1980*, D. Rand, L.-S. Young, eds., Springer, New York, 1981.

[284] E. Ott, T. Sauer, J. Yorke, eds., *Coping with Chaos: Analysis of Chaotic Data and the Exploitation of Chaotic Systems*, Wiley, New York, 1994.

[285] E. Ott, *Chaos in Dynamical Systems*, Cambridge Univ. Pr., Cambridge, 1993.

[286] K. Alligood, T. Sauer, J. Yorke, *Chaos: An Introduction to Dynamical Systems*, Springer, New York, 1997.

[287] W. Kendal, "Evidence for a fractal stochastic process underlying measles epidemics in Britain," *Fractals* **8** (2000), 29–34.

[288] R. Crandall, "The fractal character of space-time epidemics," *Proc. 2007 Int. Symp. Nonlinear Theory and Its Appl.*, (2007), 268–271.

[289] R. Wallace, "A fractal model of HIV transmission on complex sociogeographic networks: towards analysis of large data sets," *Environment and Planning A* **25** (1993), 137–148.

[290] J. L. Borges, *A Universal History of Infamy*, Dutton, New York, 1979.

[291] C. Clancy, Y. Rudy, "Linking a genetic defect to its cellular phenotype in a cardiac arrhythmia," *Nature* **400** (1999), 566–569.

[292] C. Clancy, R. Kass, "Defective cardiac ion channels: from mutations to clinical syndromes," *J. Clinical Invest.* **110** (2002), 1075–1077.

[293] C. Clancy, M. Tateyama, R. Kass, "Insights into the molecular mechanisms of bradycardia-triggered arrhythmias in long QT-3 syndrome," *J. Clinical Invest.* **110** (2002), 1251–1262.

[294] I. Splawski, K. Timothy, M. Tateyama, C. Clancy, A. Malhotra, A. Beggs, F. Cappuccio, G. Sagnella, R. Kass, M. Keating, "Variant of SCN5A sodium channel implicated in risk of cardiac arrhythmia," *Science* **297** (2002), 1333–1336.

[295] C. Clancy, Y. Rudy, "Na^+ channel mutation that causes both Brugada and long-QT syndrome phenotypes: a simulation study of mechanism," *Circulation* **105** (2002), 1208–1213.

[296] C. Clancy, M. Tateyama, H. Liu, X. Wehrens, R. Kass, "Nonequilibrium gating in cardiac Na^+ channels: an original mechanism of arrhythmia," *Circulation* **107** (2003), 2233–2237.

[297] C. Terrenoire, C. Clancy, J. Cormier, K. Sampson, R. Kass, "Autonomic control of cardiac action potentials: role of potassium channel kinetics in response to sympathetic stimulation," *Circulation Res.* **96** (2005), e25–e34.

[298] C. Clancy, R. Kass, "Theoretical investigation of the neuronal Na^+ channel SCN1A: abnormal gating and epilepsy," *Biophysical J.* **86** (2004), 2606–2614.

[299] L. Liebovitch, "Analysis of fractal ion channel gating kinetics: kinetic rates, energy levels, and activation energies," *Math. BioSci.* **93** (1989), 97–115.

[300] L. Liebovitch, "Testing fractal and Markov models of ion channel kinetics," *Biophys. J.* **55** (1989), 373–377.

[301] L. Liebovitch, J. Fischbarg, J. Koniarek, "Ion channel kinetics: a model based on fractal scaling rather than multistate Markov processes," *Math. BioSci.* **84** (1987), 37–68.

[302] L. Liebovitch, J. Fischbarg, J. Koniarek, I. Todorova, M. Wang, "Fractal model of ion-channel kinetics," *Biochimica et Biophysica Acta* **896** (1987), 173–180.

[303] L. Liebovitch, T. Tóth, "Using fractals to understand the opening and closing of ion channels," *Ann. Biomed. Eng.* **18** (1990), 177–194.

[304] A. French, L. Stockbridge, "Fractal and Markov behavior in ion channel kinetics," *Canad. J. Physiol. Pharmacol.* **66** (1988), 967–970.

[305] S. Cavalcanti, F. Fontanazzi, "Deterministic model of ion channel flipping with fractal scaling of kinetic rates," *Ann. Biomed. Eng.* **27** (1999), 682–695.

[306] S. Wright, "Evolution in Mendelian populations," *Genetics* **16** (1931), 97–159.

[307] M. Eigen, P. Schuster, *The Hypercycle: A Principle of Natural Self-Organization*, Springer, New York, 1979.

[308] S. Gavrilets, *Fitness Landscapes and the Origin of Species*, Princeton Univ. Pr., Princeton, 2004.

[309] M. Nowak, *Evolutionary Dynamics: Exploring the Equations of Life*, Harvard Univ. Pr, Cambridge, 2006.

[310] R. Soulé, J. Bascompte, *Self-Organization in Complex Ecosystems*, Princeton Univ. Pr., Princeton, 2006.

[311] G. Sorkin, "Efficient simulated annealing on fractal energy landscapes," *Algorithmica* **6** (1991), 367–418.

[312] C. Adami, *Introduction to Artificial Life*, Springer, New York, 1997.

[313] E. Weinberger, P. Stadler, "Why *some* fitness landscapes are fractal," *J. Theoret. Biol* **163** (1993), 255–275.

Chapter 4. Manufactured fractals

Most of the applications mentioned in the first paragraph are explored in this chapter. For those which aren't, we suggest these references: [314] for fractal fiberoptics, [315, 316] for fractal image compression, and [317] for fractal camouflage.

[314] L. Cook, "Fractal fiberoptics," *Applied Optics* **36** (1991), 5220-5222.

[315] Y Fisher, ed., *Fractal Image Compression: Theory and Applications*, Springer, New York, 1995.

[316] M. Barnsley, L. Hurd, *Fractal Image Compression*, A. K. Peters, Wellesley, MA., 1993.

[317] V. Billock, D. Cunningham, B. Tsou, "What visual discrimination of fractal textures can tell us about discrimination of camouflage tragest," Human Factors Issues in Combat Identification Workshop, Gold Canyon, AZ, May 13, 2005.

Sect. 4.1. Chemical mixers

The work of Coppens and his group on fractal chemical mixers is described in [318, 319, 320]. Simulations showing the advantages of hierarchical pore distributions in fuel cells are given in [321]. The design of materials modeled on bacteria colonies is in [322]. An application of engineering design principles to analyze the efficiency of the lungs is presented in [323].

[318] M.-O. Coppens, "Scaling-up and -down in a nature-inspired way," *Ind. Eng. Chem. Res.* **44** (2005), 5011–5019.

[319] M.-O. Coppens, "A nature-inspired approach to reactor and catalysis engineering," *Current Opinion in Chem. Eng.* **1** (2012), 281–289.

[320] S. Kjejstrup, M.-O. Coppens, J. Pharoah, P. Pfeifer, "Nature-inspired energy- and material-efficient design of a polymer electrolyte membrane fuel cell," *Energy & Fuels* **24** (2010), 5097–5108.

[321] J. Marquis, M.-O. Coppens, "Enhanced performance of low temperature PEM fuel cells by introducing hierarchically structured macroporosity to the cathode catalyst layer," *diffusion-fundamentals.org* **16** (2011), 1–2.

[322] Y. Huang, I. Krumanocker, M.-O. Coppens, "Fractal self-organization of bacteria-inspired agents," *Fractals* **20** (2012), 179–195.

[323] C. Hou, S. Gheorghiu, M.-O. Coppens, V. Huxley, P. Pfeifer, "Gas diffusion through the fractal landscape of the lung: how deep does oxygen enter the alveolar system?" pgs. 17–30 of *Fractals in Biology and Medicine, Vol. IV*, G. Losa, D. Merlini, T. Nonnenmacher, E. Weibel, eds., Birkhäuser, Basel, 2005.

Sect. 4.2. Fractal capacitors

Information on fractal capacitors is presented in [324] and on Microelectromechanicalsystems (MEMS) fractal capacitors in [325].

[324] H. Samavati, A. Hajimiri, A. Shahani, G. Nasserbahkt, T. Lee, "Fractal capacitors," *IEEE J. Solid-State Circuits* **33** (1998), 2035–2041.

[325] A. Elshurafa, A. Radwan, K. Salama, "RF MEMS fractal capacitors with high self-resonant frequencies," *J. Microelectromechanical Systems* **21** (2012), 10–12.

Sect. 4.3. Fractal antennas

The scaling symmetry of Maxwell's equations is presented in [326], a survey of the advantages of fractal antennas in [327], software-defined radio in [328], and military applications in [329]. A good history and introduction to the field is [330]. Puente's work is reported in [331]

[326] R. Hohlfeld, N. Cohen, "Self-similarity and the geometric requirements for frequency independence in antennae," *Fractals* **7** (1999), 79–84.

[327] N. Cohen, "Fractal antenna applications in wireless telecommunications," *Electronics Industries Forum of New England* (1997), 43–49.

[328] N. Cohen, R. Hohlfeld, D. Moschella, P. Salkind, "Fractal wideband antennas for software defined radio, UWB, and multiple platform applications," *Radio and Wireless Conf. Proc.* (2003), 99–102.

[329] N. Cohen, "Fractals' new era in military antenna design," *RF Design* (August 2005), 12–16.

[330] N. Cohen, "Fractal antenna and fractal resonator primer," pgs. 207–228 of [7].

[331] C. Puente, J. Romeu, R. Pous, A. Cardama, "On the behavior of the Sierpinski multiband fractal antenna," *IEEE Trans. Ant. Prop.* *46* (1998), 517–524.

Sect. 4.4. Invisibility cloaks

The body-sized microwave invisibility cloak is described in [332]; more details are given in [333].

[332] N. Cohen, "Body-sized wideband high fidelity invisibility cloak," *Fractals* **20** (2012), 227–232.

[333] N. Cohen, O. Okoro, D. Earle, P. Salkind, B. Unger, S. Yen, D. McHugh, S. Polterzycki, A. Shelman-Cohen, "Fractal-based wideband invisibility cloak," pgs. 229–238 of [7].

Sect. 4.5. Fractal materials

For general background on superconductors, in our opinion Chapter 21, "The Schrödinger Equation in a Classical Context: a Seminar on Superconductivity" of Volume 3 of Feynman's brilliant introduction to physics [334] can't be beat. These volumes were the older author's constant companion during his freshman year, so many decades ago. The most up-to-date source of information about fractal metamaterials is the Fractal Antenna Systems website:
www.fractenna.com

References for fractal superconductors are [335] and [336]. A general reference for defect migration is [337]; the original paper by Shlesinger and Montroll is [338].

[334] R. Feynman, R. Leighton, M. Sands, *The Feynman Lectures on Physics*, Addison-Wesley, Reading, MA, 1963 (volume 1), 1964 (volume 2), and 1965 (volume 3).

[335] J. Zaanen, "The benefit of fractal dirt," *Nature* **466** (2010), 825–827.

[336] M. Fratini, N. Poccia, A. Ricci, G. Campi, M. Burghammer, G. Aeppli, A. Bianconi, "Scale-free structural organization of oxygen interstitials in La_2CuO_{4+y}," *Nature* **466** (2010), 841–844.

[337] I. Peterson, "Time to Relax," *Sci. News* **135** (2013), 157–159.

[338] M. Shlesinger, E. Montroll, "On the Williams-Watts function of dielectric relaxation," *Proc. Nat. Acad. Sci. USA* **81** (1084), 1280–1283.

Sect. 4.6. Web infrastructure

The survey article [339] is a good place to start; more details and other examples are found in [340]. An early fractal model of Internet traffic is [341]. Flake and Pennock's analysis of the web as an ecosystem is [342]. A detailed study of the fractality of the Internet is presented in [343]. For instance, the self-similarity of Internet traffic is described on pages 214–217, the absence of a characteristic hierarchical level of the Internet on page 64. Finally, a network dynamics analysis of the Internet is in [344].

[339] W. Willinger, V. Paxson, "Where mathematics meets the Internet," *Not. Amer. Math. Soc.* **45** (1998), 961–970.

[340] K. Park, W. Willinger, *Self-Similar Network Traffic and Performance Evaluation*, Wiley, New York, 2000.

[341] W. Leland, M. Taqqu, W. Willinger, D. Wilson, "On the self-similar nature of ethernet traffic," *Proc. ACM SIGCOMM '93* (1993), 183–193.

[342] G. Flake, D. Pennock, "Self-organization, self-regulation, and self-similarity on the fractal web," pgs 88–119 of *The Colors of Infinity. The Beauty and Power of Fractals*, Springer, New York, 2010.

[343] R. Pastor-Satorras, A. Vespignani, *Evolution and Structure of the Internet: A Statistical Physics Approach*, Cambridge Univ. Pr., Cambridge, 2004.

[344] A. Barrat, M. Barthélemy, A. Vespignani, *Dynamical Processes on Complex Networks* Cambridge Univ. Pr., Cambridge, 2008.

Sect. 4.7. Fractal music

Harlan Brothers' papers on fractals and music are [345, 346, 347]. György Ligeti's quotation is from Rockwell's [348] *New York Times* story. Greenberg's comment on Prokofiev is from page 393 of [349]. In *Aspects of Schenkerian Theory*, David Beach's chapter is [350]; Charles Burkhart's is [351]. Schenker's opus is [352]. Gardner's remark about Voss's work is from [353]. The papers by Voss and Clarke on $1/f$ noise are [354, 355]; Voss's other paper on $1/f$ noise is [356]. Michael Steinberg's description of Wuorinen is from page 319 of [357]. Wuorinen's essay is [358].

[345] H. Brothers, "The nature of fractal music," pgs. 181–205 of [7].

[346] H. Brothers, "Structural scaling in Bach's Cello Suite no. 3," *Fractals* **15** (2007), 89–95.

[347] H. Brothers, "Intervallic scaling in the Bach cello suites," *Fractals* **17** (2009), 537–545.

[348] J. Rockwell, "Laurels at an auspicious time for Gyorgy Ligeti," November 11, 1986, *New York Times*.

[349] M. Gardner, *The Night Is Large*, St. Martin's, New York, 1996.

[350] D. Beach, "Schenker's theories: a pedagogical view," pgs. 1–38 of *Aspects of Schenkerian Theory*, D. Beach, ed., Yale Univ. Pr., New Haven, 1983.

[351] C. Burkhart, "Schenker's theory of levels and musical performance," pgs 95–112 of *Aspects of Schenkerian Theory*, D. Beach, ed., Yale Univ. Pr., New Haven, 1983.

[352] H. Schenker, *Neue Musikalische Theorien und Phantasien:*
Harmonielehre, Universal edition, Vienna, 1906.
Kontrapunkt 1, Universal edition, Vienna, 1910.
Kontrapunkt 2, Universal edition, Vienna, 1922.
Der freie Satz, Universal edition, Vienna, 1935.

English translations:
New Musical Theories and Fantasies
Harmony, transl. E. Borgese, Univ. of Chicago Pr., Chicago, 1954.
Counterpoint 1 and 2, transl. J. Rothgeb, Jürgen Thym, Schirmer Books, New York, 1987.
Free Composition, transl. E. Oster, Longman, New York, 1979.

[353] M. Gardner, "Mathematical recreations," *Sci. Am.*, April, 1978.

[354] R. Voss, J. Clarke, "$1/f$ noise in music and speech," *Nature* **258** (1975), 317–318.

[355] R. Voss, J. Clarke, "$1/f$ noise in music: music from $1/f$ noise," *J. Acoust. Soc. Amer.* **63** (1978), 258–263.

[356] R. Voss, "Random fractals: self-affinity in noise, music, mountains, and clouds," *Physica D* **38** (1989), 362–371.

[357] M. Steinberg, *Choral Masterworks: A Listener's Guide*, Oxford Univ. Pr., Oxford, 208.

[358] C. Wuorinen, "Music and Fractals" pgs. 501–506 of [7].

Sect. 4.8. Fractal literature

First, our references for *Fractals as subject*. The quotes from Stéphane Audeguy's *The Theory of Clouds* are from pages 44 and 245 of [359]. Jorge Luis Borges' story "The Garden of Forking Paths" is on pages 19–29 of [360] and on pages 119–128 of [361]. Bloch's comment about Borges' review of *Mathematics and the Imagination* [364] is on page 144 of [362]. Borges' review is reprinted on pages 249–250 of [363]. Arthur C. Clarke's *The Ghost from the Grand Banks* is [365]. The film *The Colors of Infinity* is [366]. Clarke's *The Exploration of Space* is [367]. The quote from Mark Cohen's *The Fractal Murders* is from page 5 of [368]. The quotations from Høeg's *Smilla's Sense of Snow*, are on pages 348 and 300-301 of [369]. Leibniz's *Monadology* is on pages 249–272 of [370]. Borges' "The Aleph" is on pages 138–154 of [371], on pages 15–30 of [372], and on pages 274–286 of [361]. The quote of Flann O'Brien is from page 73 of [32]. The quotations from Richard Powers' *Gold Bug Variations* are on pages 579 and 627 of [373]. The quotation from *Plowing the Dark* is on page 37 of

REFERENCES   475

[374]. The quote from Schreiner's *The Story of an African Farm* is on page 153 of [29]. The quotes from Stoppard's *Arcadia* [375] are from page 43 ("the New Geometry of Irregular Forms ..."), page 84 ("Mountains are not pyramids ..."), and page 76 ("In an ocean of ashes, ..."). But do read the play: it holds many wonders beyond those we cite. Benoit's "Clouds are not spheres" quote is on page 1 of [1]. The quote from John Updike's *Roger's Version* is on page 236 of [376]. The Kate Wilhelm quote is from pages 163–164 of [377].

Next, our references for *Structural fractals*. The paper of James Cutting and his students is [378]. Demetri Martin's palindrome is Chapter 14 of [7]. The sources cited for the analysis of scaling in Wallace Stevens' poems are [379, 380].

Finally, our references for *Metastructural fractals*. The David Mitchell quote is from Amanda Buckingham's article [381], the letter of Charles Dickens appears on pages 301–302 of [382], and the García Márquez quote is from page 11 of Peter Stone's interview [383]. The comparison of large, historic events to small, personal ones was used by John Lewis Gaddis in Chapter 7 of [384] to support the claim that history has fractal aspects. Unhappily, his example was of Stalin's acts of terror across many scales. We seek out less heart-breaking examples.

[359] S. Audeguy, *The Theory of Clouds*, Harcourt, Orlando, 2007.

[360] J. Borges, *Labyrinths: Selected Stories & Other Writings*, D. Yates, J. Irby, eds., New Directions Publ. Co., New York, 1964.

[361] J. Borges, *Jorge Luis Borges: Collected Fictions*, transl. A. Hurley, Penguin Books, New York, 1998.

[362] W. Bloch, *The Unimaginable Mathematics of Borges' Library of Babel*, Oxford Univ. Pr., Oxford, 2008.

[363] J. Borges, *Jorge Luis Borges. Selected Non-Fictions*, E. Weinberger, ed., Penguin Books, New York, 1999.

[364] E. Kasner, J. Newman, *Mathematics and the Imagination*, Simon & Schuster, New York, 1940.

[365] A. Clarke, *The Ghost from the Grand Banks*, Bantam, New York, 1990.

[366] N. Lesmoir-Gordon, A. Clarke, B. Mandelbrot, et. al., *The Colors of Infinity*, Films for the Humanities and Sciences, 1995.

[367] A. Clarke, *The Exploration of Space*, Harper & Brothers, New York, 1959.

[368] M. Cohen, *The Fractal Murders*, Moody Gap Press, Boulder, 2002.

[369] P. Høeg, *Smilla's Sense of Snow*, Dell, New York, 1993.

[370] G. Leibniz, *Basic Writings: Discourse on Metaphysics, Correspondence with Arnauld, Monadology*, Open Court Publ. Co., La Salle, 1968.

[371] J. Borges, *A Personal Anthology*, Grove Press, New York, 1967.

[372] J. Borges, *The Aleph and Other Stories, 1933–1969*, E. P. Dutton, New York, 1968.

[373] R. Powers, *The Gold Bug Variations*, Harper, New Your, 1991.

[374] R. Powers, *Plowing the Dark*, Farrar, Straus, and Giroux, New York, 2000.

[375] T. Stoppard, *Arcadia*, Faber and Faber, 1993.

[376] J. Updike, *Roger's Version*, Knopf, New York, 1986.

[377] K. Wilhelm, *Death Qualified: A Mystery of Chaos*, St. Martin's, New York, 1991.

[378] J. Cutting, J. DeLong, C. Nothelfer, "Attention and the evolution of Hollywood film," *Psychological Science* (2010), **21**, 432–439.

[379] L. Pollard-Gott, "Fractal repetition in the poetry of Wallace Stevens," *Language and Style* **19** (1989), 233–249.

[380] F. Doggett, *Wallace Stevens: The Making of the Poem*, Johns Hopkins Univ. Pr., Baltimore, 1980.

[381] A. Buckingham, "*Cloud Atlas* author talks originality," *Yale Daily News*, (April 9, 2014), 3.

[382] J. Hartley, ed., *The Selected Letters of Charles Dickens*, Oxford Univ. Pr., Oxford, 2012.

[383] G. García Márquez, "The art of fiction, no. 69," *The Paris Review*, **82** (1981).

[384] J. Gaddis, *The Landscape of History: How Historians Map the Past*, Oxford Univ. Pr., Oxford, 2002.

Sect. 4.9. Visual art

The T. E. Breitenbach quotation is from a December 11, 2012 email to the older author. One of our favorite general art references is [385], the source of our information on Cozens and the quotations, page 625 for the statements about Cozens, page 738 for Pollock. Dalí's *Visage of War* is on page 97 of [386]; his study for *Visage of War* is on page 96. Many references to decalcomania are given in [387]. Ernst's *Europe after the Rain* is on pages 242–243 of [388]. Natalie Eve Garrett's decalcomania project is [389], Tanja Geis's is [390], Claire Miller and Cara Norris's is [391]. These are among the most inventive, and visually striking, projects the older author saw during his decades teaching at Yale. Hokusai images and commentaries are in [392]. Pollock's drip paintings are reproduced in [393]. A record of the controversy, claims and counter-claims, are in [394] – [404]. Edgerton's analysis of the relations between Renaissance art and mathematics is in [405, 406]. Henderson's analysis of the influence of geometry on art is in [407]. Rhonda Roland Shearer's thesis that revolutions in geometry catalyze revolutions in art is in [408].

[385] H. Janson, *History of Art*, 4th ed, Abrams, New York, 1991.

[386] R. Descharnes, *Dalí*, Harry Abrams, New York, 1985.

[387] R. Passeron, *Phaidon Encyclopedia of Surrealism*, E. P. Dutton, New York, 1978.

[388] W. Spies, ed., *Max Ernst: A Retrospective*, Prestel, Munich, 1991.

[389] N. E. Garrett, "Aesthetics in visual art and fractals," Yale fractal geometry project, 1998.

[390] T. Geis, "Decalcomania," Yale fractal geometry project, 2001.

[391] C. Miller, C. Norris, "Fractal painting au natural," Yale fractal geometry project, 1999.

[392] H. Smith, *Hokusai: One Hundred Views of Mt. Fuji*, George Braziller, New York, 1988.

[393] E. Frank, *Pollock*, Abbeville Press, New York, 1991.

[394] R. Taylor, A. Micolich, D. Jonas, "Fractal analysis of Pollock's drip paintings," *Nature* **399** (1999), 422.

[395] R. Taylor, R. Guzman, T. Martin, G. Hall, A. Micolich, D. Jonas, B. Scannell, M. Fairbanks, C. Marlow, "Authenticting Pollock paintings using fractal geometry," *Pattern Recognition Lett.* **28** (2007), 695–702.

[396] K. Jones-Smith, H. Mathur, "Revisiting Pollock's drip paintings," *Nature* **44** (2006), E9–10.

[397] R. Taylor, A. Micolich, D. Jonas, "Taylor et al. reply," *Nature* **444** (2006), E10–11.

[398] K. Jones-Smith, H. Mathur, L. Krauss, "Drip paintings and fractal analysis," *Phys. Rev. E* **79** (2009), 046111-1, 12.

[399] A. Micolich, B. Scannell, M. Fairbanks, T. Martin, R. Taylor, "Comment on 'Drip paintings and fractal analysis' by K. Jones-Smith, H. Mathur, L. M. Krauss," arXiv:0712.1652v1 [cond-mat.stat-mech], December 11, 2007.

[400] K. Jones-Smith, H. Mathur, L. Krauss, "Reply to comment on 'Drip paintings and fractal analysis' by Micolich et al.," arXiv:0803.0530v1 [cond-mat.stat-mech], March 4, 2008.

[401] R. Taylor, A. Micolich, D. Jonas, "The construction of Jackson Pollock's fractal drip paintings," *Leonardo* **35** (2002), 205–7.

[402] J. Alvarez-Ramirez, C. Ibarra-Valdez, E. Rodriguez, L. Dagdug, "$1/f$-Noise structures in Pollock's drip paintings," *Physica A* **387** (2008), 281–295.

[403] B. Spehar, C. Clifford, B. Newell, R. Taylor, "Universal aesthetic of fractals," *Computers & Graphics* **27** (2003), 813–820.

[404] R. Taylor, B. Spehar, J. Wise, C. Clifford, B. Newell, C. Hagerhall, T. Purcell, T. Martin, "Perceptual and physiological responses to the visual complexity of fractal patterns," *Nonlinear Dynamics, Psychology, and Life Sciences* **9** (2005), 89–114.

[405] S. Edgerton, "Alberti's perspective: A new discovery and a new evaluation," *Art Bulletin* **48** (1966), 367–378.

[406] S. Edgerton, *The Renaissance Rediscovery of Linear Perspective*, Basic Books, New York, 1975.

[407] L. Henderson, *The Fourth Dimension and Non-Euclidean Geometry in Modern Art*, MIT Press, Cambridge, 2013.

[408] R. Shearer, "Chaos theory and fractal geometry: their potential impact on the future of art," *Leonardo* **25** (1992), 143–152.

Chapter 5. The Mandelbrot set

Sect. 5.1. Some pictures

Douady and Hubbard proved the Mandelbrot set is connected in [409]. Shishikura proved the boundary of the Mandelbrot set is 2-dimensional in [410].

[409] A. Douady, J. Hubbard, "Iteration des pôlynomes quadratiques complexes," *C. R. Acad. Sci. Paris* **294** (1982) 123–126.

[410] M. Shishikura, "The Hausdorff dimension of the boundary of the Mandelbrot set and Julia sets," *Ann. of Math.* **147** (1998) 225–267.

Sect. 5.3. Julia sets

An excellent introduction to chaos and to some of the geometry of Julia sets is [411].

[411] R. Devaney, *An Introduction to Chaotic Dynamical Systems*, 2nd ed., Addison-Wesley, Redwood City, 1989.

Sect. 5.4. The Mandelbrot set

Benoit's first paper on the Mandelbrot set is [412]. Lavaurs' algorithm was presented in [413]; Tan Lei's theorem in [414]. A good survey of our understanding of the Mandelbrot set at the start of the current century is [415]. Benoit's conjecture that the Mandelbrot set boundary has dimension 2 was presented in [416]; John Milnor's work on the complexity of the Mandelbrot set boundary is in [417]. As mentioned earlier, Shishikura's proof that the Mandelbrot set boundary is 2-dimensional was published in [410].

A derivation of Fermat's Little Theorem from counting the cycles of doubling map relatives is in [418].

[412] B. Mandelbrot, "Fractal aspects of the iteration of $z \to \lambda z(1-z)$ for complex λ and z," pgs. 249–259 of *Non-Linear Dynamics: Annals of the New York Academy of Sciences* **357** (1979), R. Helleman, ed., New York, 1979.

[413] P. Lavaurs, "Une description combinatoire de l'involution définie par M sur les rationnels à dénominateur impair," *C. R. Acad. Sci. Paris* **303** (1986), 143–146.

[414] L. Tan, "Similarity between the Mandelbrot set and Julia sets," *Commun. Math. Phys.* **134** (1990), 587–617.

[415] L. Tan, ed., *The Mandelbrot Set, Theme and Variations*, Cambridge Univ. Pr., Cambridge, 2000.

[416] B. Mandelbrot, "On the dynamics of iterated maps, V: conjecture that the boundary of the M-set has a fractal dimension equal to 2," pgs. 235–238 of *Chaos, Fractals and Dynamics*, P. Fischer, W. Smith, eds., Marcel Dekker, New York, 1985.

[417] J. Milnor, "Self-similarity and hairiness in the Mandelbrot set," pgs. 211–257 of *Computers in Geometry and Topology*, M. Tangora, ed., Marcel Dekker, New York, 1989.

[418] M. Frame, B. Johnson, J. Sauerberg, "Fixed points and Fermat: a dynamical systems approach to number theory," *Amer. Math. Monthly* **107** (2000), 422–428.

Sect. 5.6. Universality of the Mandelbrot set

Arthur Cayley's study of the basins of attraction for Newton's method was published in [419]. As recounted in his preface to [415], Hubbard began his computer experiments on the basins of attraction of Newton's method for $z^3 - 1$ in the spring of 1977. The Lakes of Wada construction appeared in [420]. It is in [421] that Curry, Garnett, and Sullivan report their discovery that for a particular family of complex cubic polynomials, Newton's method converges to stable cycles for points determining an intricate web of Mandelbrot sets in the parameter plane. Douady and Hubbard develop an explanation of the universality of the Mandelbrot set in [422]. McMullen writes about the universality of the Mandelbrot set in [423].

[419] A. Cayley, "The Newton-Fourier imaginary problem," *Amer. J. Math.* **2** (1879), 97.

[420] Y]K. Yoneyama, "Theory of continuous sets of points," *Tohoku Math. J.* **11–12** (1917) 43–158.

[421] J. Curry, L. Garnett, D. Sullivan, "On the iteration of a rational function: computer experiments with Newton's method," *Commun. Math. Phys.* **91** (1983), 267–277.

[422] A. Douady, J. Hubbard, "On the dynamics of polynomial-like maps," *Ann. Scient. Ec. Norm. Sup.* **18** (1985), 287–343.

[423] C. McMullen, "The Mandelbrot set is universal," pgs. 1–17 of [415].

Sect. 5.8. Unanswered questions

Douady and Hubbard conjectured the Mandelbrot set is locally connected in [409].

On two earlier occasions we mentioned that Shishikura's proof that the boundary of the Mandelbrot set has dimension 2 is presented in [410]. Tan Lei's proof of the similarity of some Julia sets to some regions of the Mandelbrot set boundary is in [426]. Benoit's statement of the $1/n^2$ conjecture was published in [424]; Guckenheimer and McGehee's proof of the $1/n^2$ conjecture appeared in [425]. Mitchell's graphs of the departures from the $1/n^2$ rule are shown on pages 96–97 of the *Fractals and Chaos* volume of [5].

A excellent early survey of what we do and don't know about the Mandelbrot set, including the relation between the conjecture that the Mandelbrot set is locally connected (MLC) and the hyperbolicity conjecture, was given by Bodil Branner in 1989, published in [428]. The older author was in the audience of Branner's lecture and was fascinated by how much of the Mandelbrot set structure she was able to explain. The intervening quarter-century has not dimmed this fascination at all. Mandelbrot set pictures are amazing, but even more amazing is how much of these pictures we can understand.

[424] B. Mandelbrot, "On the dynamics of iterated maps. III: The individual molecules of the M-set self-similarity properties, the empirical n^2 rule, and the n^2 conjecture," pgs. 213–224 of *Chaos, Fractals and Dynamics*, P. Fischer, W. Smith, eds., Marcel Dekker, New York, 1985.

[425] J. Guckenheimer, R. McGehee, "A proof of the Mandelbrot n^2 conjecture," Institute Mittag-Leffler, Report 15, 1984.

[426] L. Tan, "Ressemblance entre l'ensemble de Mandelbrot et l'ensemble de Julia au voisinage d'un point de Misiurewicz," pgs. 139–152 of [427].

[427] A. Douady, J. Hubbard, *Etude dynamique des polynomes complexes II*, Publ. Math. Orsay 1985.

[428] B. Branner, "The Mandelbrot set," pgs. 75–105 of [37].

Chapter 6. Fractal Dimension

Sect. 6.1. Similarity dimension

The Moran equation appears in [429].

[429] P. Moran, "Additive functions of intervals and Hausdorff measure," *Proc. Camb. Phil. Soc.* **42** (1946), 15–23.

Sect. 6.2. Box-counting dimension

Different approaches to computing box-counting are presented in Section 2.1 of [9].

Sect. 6.4. Random, IFS with memory, and nonlinear fractals

The random Moran equation $\mathbb{E}(r_1^d + \cdots + r_n^d) = 1$ is presented in Section 15.1 of [9]. The extension of the Moran equation to IFS with memory is derived in [430]. The nonlinear Moran equation is presented in Sect. 5.2 of [10]. The approach of Minkowski and Bouligand to computing box-counting dimension is presented at the end of Section 2.1 of [9]. A tiny glimpse of Hausdorff dimension is given in A.99.

[430] R. Mauldin, S. Williams, "Hausdorff dimension in graph-directed constructions," *Trans. Amer. Math. Soc.* **309** (1988), 811–829.

Sect. 6.5. Algebra of dimensions

The open set condition is presented in [33]. Proofs of the algebra of dimensions results can be found in Chapters 2, 3, 7, and 8 of [9], for example. McMullen's computations of the dimensions of some self-affine fractals is presented in [431]. Dimension calculations for other classes of self-affine fractals are given by Falconer and Miao [432] – [436], Bedford and Urbanski [437, 438], Gatzouras and Lalley [439, 440], Hueter and Peres [441], and Edgar [442, 443], among others.

[431] C. McMullen, "The Hausdorff dimension of general Sierpinski carpets," *Nagoya Math. J.* **96** (1984), 1–9.

[432] K. Falconer, "The Hausdorff dimension of self-affine fractals," *Math. Proc. Camb. Phil. Soc.* **103** (1988), 330–350.

[433] K. Falconer, "The Hausdorff dimension of self-affine fractals II," *Math. Proc. Camb. Phil. Soc.* **111** (1992), 169–179.

[434] K. Falconer, J. Miao, "Dimensions of self-affine fractals and multifractals generated by upper triangular matrices," *Fractals* **15** (2007), 289–299.

[435] K. Falconer, J. Miao, "Exceptional sets for self-affine fractals," *Math. Proc. Camb. Phil. Soc.* **145** (2008), 669–684.

[436] K. Falconer, J. Miao, "Random subsets of self-affine fractals," *Mathematika* **56** (2010), 61–76.

[437] T. Bedford, "On Weierstrass-like functions and random recurrent sets," *Math. Proc. Camb. Phil. Soc.* **106** (1989), 325–342.

[438] T. Bedford, M. Urbanski, "The box and Hausdorff dimension of self-affine sets," *Ergod. Th. Dyn. Syst.* **10** (1990), 627–644.

[439] D. Gatzouras, S. Lalley, "Statistically self-affine sets: Hausdorff and box dimensions," *J. Theoret. Prob.* **7** (1994), 437–468.

[440] S. Lalley, D. Gatzouras, "Hausdorff and box dimensions of certain self-affine fractals," *Indiana U. Math. J.* **41** (1992), 533–568.

[441] I. Hueter, Y. Peres, "Self-affine carpets on the square lattice," *Comb. Prob, Comput.* **6** (1997), 197–204.

[442] G. Edgar, "Fractal dimension of self-affine sets: some examples," *Measure Theory, Oberwolfach 1990, Suppl. ai Rendiconti del Circolo Matematico di Palermo* **28** (1992), 341–358.

[443] G. Edgar, "Kisswetter's fractal has Hausdorff dimension 3/2," *Real Analysis Exchange* **14** (1988), 215–223.

Sect. 6.6. Uses of dimensions

Richardson's work on coastline length measurements is given in [64]. This reference is a bit difficult to find; a more accessible source for the quotation about positive correlation between exponents and irregularity is page 214 of [444], which reprints a section of Richardson's paper. In [63] Richardson's coastline work was extended by Benoit to quantify coastlines by their dimensions rather than by their lengths. Gould's book *Full House* is [179]. Weibel has written about his work on the fractal geometry of lungs in many places. A good summary is [73]; the quotation of Weibel is from pages 5–6 of [73]. The dimension computations of Huang, Yen, McLaurine, and Bledsoe is reported in [445]. Jens Feder gives a clear description of DLA, including (page 33) the dimension calculations for DLA grown in 2- and in 3- dimensions, in [13]. The first paper on diffusion-limited aggregation was [446]. Vicsek gives a thorough description of clustering mechanisms, including cluster-cluster collisions and (page 216) the dimension of BCCA clusters, in [180]. Complications with large DLA clusters were noted in [447, 448].

[444] T. Körner, *The Pleasures of Counting*, Cambridge Univ. Pr., Cambridge, 1996.

[445] W. Huang, R. Yen, M. McLaurine, G. Bledsoe, "Morphometry of the human pulmonary vasculature," *J. Appl. Physiol.* **81** (1996), 2123–2133.

[446] T. Witten, L. Sander, "Diffusion-limited aggregation," *Phys. Rev. B* **27** (1983), 5686–5697.

[447] B. Mandelbrot, H. Kaufman, A. Vespignani, I, Yekutieli, C.-H. Lam, "Deviations from self-similarity in plane DLA and the 'infinite drift' scenario," *Europhys. Lett.* **29** (1995), 599–604.

[448] B. Mandelbrot, H. Kaufman, A. Vespignani, "Crosscut analysis of large radial DLA: departures from self-similarity and lacunarity effects," *Europhys. Lett.* **32** (1995), 199–204.

Sect. 6.7 A speculation about dimensions

The Lightman quote is from page 16 of [449]. Here are a few others relevant to the subject of our book. "Patterns within patterns within patterns, all perfect as the number π and precisely determined by the few quantum rules I had given," page 62. "the displacement of a single atom in any of those brains might change the outcome of a long sequence of events, ending in a different decision or action," page 108, a nice statement of sensitivity to initial conditions. Smolin's book is [450]; Vrobel's is [451]. Do not confuse Vrobel's excellent book with another *Fractal Time* book that claims fractal patterns in history prove the world will end in 2012. (At least, that's what the older author recalls from perusing the book, when, attracted by the title, he began reading, gradually became aware of its real topic, then set it back on the table with a long string of intricate muttered curses and an accusatory glare at the bookstore owner.) Look at the calendar and make your own assessment of this book.

As mentioned earlier, Nottale's fascinating work is described in [168, 169, 170, 171].

[449] A. Lightman, *Mr g*, Pantheon, New York, 2012.

[450] L. Smolin, *Three Roads to Quantum Gravity*, Basic Books, New York, 2002.

[451] S. Vrobel, *Fractal Time. Why a Watched Kettle Never Boils*, World Scientific, Singapore, 2011.

REFERENCES 485

Chapter 7. Further developments

Sect. 7.1. Driven IFS and data analysis

Stewart's paper introducing non-random versions of the random IFS algorithm is [452]. Driving IFS with DNA sequences was introduced by H. Joel Jeffrey [453, 454]. Markov partitions were introduced in [455, 456]; more information is in [411] and [457]. Excellent general references on symbolic dynamics are [458, 459]. A study of the relations between choice of Markov partition and symbolic dynamics is in [460]. A classical reference for Markov processes is [461]; a readable introduction is Chapter 15 of [462].

[452] I. Stewart, "Order within the chaos game," *Dynamics Newsletter* **3** (1989), 4–9.

[453] H. Jeffrey, "Chaos game representation of gene structure," *Nucl. Acid Res.* **18** (1990), 2163–2170.

[454] H. Jeffrey, "Chaos game visualization of sequences," *Comput. Graph.* **16** (1992), 25–33.

[455] Y. Sinai, "Markov partitions and C-diffeomorphisms," *Funct. Analy. Appl.* **2** (1968), 61–82.

[456] Y. Sinai, "Construction of Markov partitions," *Funct. Analy. Appl.* **2** (1968), 245–253.

[457] C. Beck, F. Schlögl, *Thermodynamics of Chaotic Systems: An Introduction*, Cambridge Univ. Pr., Cambridge, 1993.

[458] D. Lind, B. Marcus, *An Introduction to Symbolic Dynamics and Coding*, Cambridge Univ. Pr., Cambridge, 1995.

[459] B. Kitchens, *Symbolic Dynamics: One-Sided, Two-Sided, and Countable State Markov Shifts*, Springer, Berlin, 1998.

[460] H. Teramoto, T. Komatsuzaki, "How does the choice of Markov partition affect the resultant symbolic dynamics?" *Chaos* **20** (2010), 037113.

[461] J. Kemeny, L. Snell, *Finite Markov Chains*, Van Nostrand, Princeton, 1960.

[462] H. Tijms, *Understanding Probability: Chance Rules in Everyday Life*, 2nd ed., Cambridge Univ. Pr., Cambridge, 2007.

Sect. 7.2. Driven IFS and synchronization

Robert May introduced the chaos of the logistic map in [463]. A good discussion of the logistic map bifurcation diagram is in Chapter 1 of [411]. In [464], Jakobson proved that the chaos occurs for a significant collection of logistic map r-values. Steve Strogatz's wonderful book on synchronized behavior throughout science and nature, including stories about fireflies, is [465]. Coupled map lattices is a large topic, now absorbed into the more general *Network Science*. An early reference for CML is [466]; see also [467]. The dictionary of coupled maps was developed by C. Noelle Thew in her 2014 Yale applied math senior thesis [468]. Synchronization studies on fractal lattices were done by Elena Malloy [469] as an independent study at Yale.

In her 1996 Union College applied math senior thesis [470], Kathy Walter studied the bifurcation diagram of the discontinuous function (See A.93) made of part logistic map and part tent map , and she did electronics experiments (requiring a rare visit to PhysicalWorld by the older author) to investigate bifurcation properties of Chua's circuit. Liapunov exponents (See A.94) are a common tool for measuring the "degree of chaos" in a signal. A good reference with details about computation is [471].

[463] R. May, "Simple mathematical models with very complicated dynamics," *Nature* **261** (1976), 459–467.

[464] M. Jakobson, "Absolutely continuous invariant measures for one-parameter families of one-dimensional maps," *Commun. Math. Phys.* **81** (1981), 39–88.

[465] S. Strogatz, *Sync: The Emerging Science of Spontaneous Order*, Hyperion Books, New York, 2003.

[466] K. Kaneko, "Period-doubling of kink-antikink patterns, quasiperiodicity in antiferro-like structures and spatial intermittency in coupled logistic lattice: towards a prelude of a 'field theory of chaos,'" *Prog. Theoret. Phys.* **72** (1984), 480–486.

[467] K. Kaneko, *Theory and Applications of Coupled Map Lattices*, Wiley, Chichester, 1993.

[468] C. N. Thew, *A Dictionary of Driven Iterated Function Systems to Characterize Chaotic Time Series*, Yale University applied mathematics senior thesis, 2014.

[469] E. Malloy, C. N. Thew, M. Frame, "Synchronization and fuzzy synchronization in fractal lattices," in preparation.

[470] K. Walter, *Fanning New Worlds: Fractals of Discontinuous Functions*, Union College applied mathematics senior thesis, 1996.

[471] A. Wolf, "Quantifying chaos with Lyapunov exponents,", pgs. 273–290 of *Chaos*, A. Holden, ed., Princeton Univ. Pr., Princeton, 1986.

Sect. 7.3. Multifractals from IFS

Benoit introduced multifractals in [254]. Many of his papers on multifractals are collected, and given lengthy introductions, in the second volume of his *Selecta* [5]. Good mathematical introductions to multifractals are given by Falconer in Chapter 17 of his book [9] (In Sect. 17.3 you'll find a proof that $f(\alpha) = \alpha \cdot q + \beta$) and in Chapters 10 and 11 of his book [10]. An introduction to multifractals from the perspective of a physicist is in Chapter 6 of [13]; an approach for geophysicists is given in [472]. The generalization of the Moran equation to multifractals, $\rho[p_{ij}^q r_i^{\beta(q)}] = 1$, is derived in [473], equation (1.3.1).

[472] B. Mandelbrot, "Multifractal measures, especially for the geophysicist," *Pure and Applied Geophysics* **131** (1989), 5–42.

[473] L. Olsen, *Random Geometrically Graph Directed Self-Similar Multifractals*, Wiley, New York, 1994.

Sect. 7.4. Applications of multifractals

Kolmogorov's original paper on turbulence is [474]. Richardson's verse about "Big whorls" appears in [475]. Benoit's original work on turbulence, where some of the main ideas of multifractals were developed, is [254]. Meneveau and Sreenivasan did important early work on multifractals. Their paper on simple multifractal cascade models is [262]; their paper on turbulent energy dissipation is [261]. Sreenivasan also worked with Ashvin Chhabra [476, 257]. Their work included using multifractals to measure negative dimensnions. Sreenivasan and others have written many more papers on multifractals and turbulence. In January 2015, a Google Scholar search for multifractals and turbulence returned over 14,800 matches. Benoit continued his work on negative dimensions in [477, 478].

A good description of applications of multifractals in geophysics is Benoit's paper [472]. Some studies of multifractal aspects of earthquakes are [480, 481, 482]. Turcotte's paper [200] is a good general reference for fractals in geology. Earthquake magnitude examples come from the Wikipedia entry on the Richter magnitude scale. Fractal aspects of seismic fault lines are discussed in [479] and in [203].

Many of our ideas about fractals, multifractals, and the Internet are from the papers [339, 483], and from the book [340]. Information about the data capacity of the genome, the brain, the Internet came from page 104 of the November 2011 issue of *Scientific American*.

Popular descriptions of Benoit's use of multifractals to understand financial data are given in his *Scientific American* article [484], and also in his and Richard Hudson's book [485]. Much more mathematical detail is in the first volume of [5].

Hurst's paper on the annual flooding of the Nile is [227]. Benoit's and Wallis's papers on fBm and rainfall are [486, 487]. Lovejoy and Schertzer's book on weather, climate, and multifractals is [488]. [489, 490] are papers on multifractals and climate. There are, of course, many more.

[474] A. Kolmogorov, "The local structure of turbulence in incompressible viscous fluid for very large Reynolds numbers," *Proc. USSR Acad. Sci.* **30** (1941), 299-303.

[475] L. Richardson, *Weather Prediction by Numerical Process*, Cambridge Univ. Pr., Cambridge, 1922.

[476] A. Chhabra, K. Sreenivasan, "Scale-invariant multiplier distributions in turbulence," *Phys. Rev. Lett.* **68** (1992), 2762-2765.

[477] B. Mandelbrot, "Random multifractals: negative dimensions and the resulting limitations of the thermodynamic formalism," *Proc. Roy. Soc. A* **434** (1991), 79-88.

[478] B. Mandelbrot, M. Frame, "A primer of negative test dimensions and degrees of emptiness for latent sets," *Fractals* **17** (2009), 1-14.

[479] A. Öncel, I. Main, Ö. Alptekin, P. Cowie, "Spatial variations of the fractal properties of seismicity in the Anatolian fault zones," *Tectonophysics* **257** (1996), 189-202.

[480] T. Hirabayashi, K. Ito, T. Yoshi, "Multifractal analysis of earthquakes," *Pure Appl. Geophysics* **138** (1992), 591-610.

[481] D. Li, Z. Zheng, B. Wang, "Research into the multifractal of earthquake spatial distribution," *Tectonophysics* **233** (1994), 91-97.

[482] L. Telesca, V. Lapenna, M. Macchiato, "Mono- and multi-fractal investigation of scaling properties in temporal patterns of seismic sequences," *Chaos Solitons & Fractals* **19** (2004), 1-15.

[483] A. Feldmann, A. Gilbert, W. Willinger, "Data networks as cascades: investigating the multifractal nature of Internet WAN traffic," *Computer Communication Review* **28** (1998), 42-55.

[484] B. Mandelbrot, "A multifractal walk down Wall street," *Sci. Am.*, February 1999, 50–53.

[485] B. Mandelbrot, R. Hudson, *The (Mis)Behavior of Markets: A Fractal View of Risk, Ruin, and Reward*, Basic Books, New York, 2004.

[486] B. Mandelbrot, J. Wallis, "Noah, Joseph, and operational hydrology," *Water Resources Res.* **4** (1968), 909–918.

[487] B. Mandelbrot, J. Wallis, "Computer experiments with fractional gaussian noises, averages and variances," *Water Resources Res.* **5** (1969), 228–241.

[488] S. Lovejoy, D. Schertzer, *The Weather and Climate: Emergent Laws and Multifractal Cascades*, Cambridge Univ. Pr., Cambridge, 2013.

[489] J. Kantelhardt, E. Koscielny-Bunde, D. Rybski, P. Braun, A. Bunde, S. Havlin, "Long-term persistence and multifractality of precipitation and river runoff records," *J. Geophys. Res.* **111** (2006), D01106.

[490] Z.-G. Shao, P. Ditlevesen, "Contrasting scaling properties of interglacial and glacial climates," *Nature Commun.* **7** (2016), doi:10.1038/ncomms10951.

Sect. 7.5. Fractals and stories

About mutlifractal distributions of sentence lengths, in [491] the authors analyze sentence length variability in 113 long literary works, in several languages. Especially in stream of consciousness novels, the intricate pattern of sentence lengths gives rise to a multifractal distribution. The complexity of literature seems far too rich to be quantified by a single number such as a dimension. The added complexity of multifractals may make them a more promising tools for this task.

Cortázar's novel is [492]. The Lightman quote is from page 196 of [449].

[491] S. Drożdż, P. Oświęcimka, A. Kulig, J. Kwapień, K. Bazarnik, I. Grabska-Gradzińska, J. Rybicki, M. Stanuszek, "Quantifying origin and character of long-range correlations in narrative texts," *Information Sci.*, **331** (2016), 32–44.

[492] J. Cortázar, *Hopscotch*, Pantheon, New York, 1966.

Acknowledgments

First, we must thank Benoit Mandelbrot, the visionary who imaged fractals from an odd collection of rough, irregular shapes and behaviors. More personally, he was the older author's mentor, coauthor, partner in hundreds of exhilarating discussions, and dear friend. The friendship of Benoit's widow Aliette remains a treasure in the older author's life.

In a sense this book is a reboot of *Chaos Under Control*, the 1994 text written by physicist David Peak and the older author. In the 20 years since that book, the field has developed so many fresh ideas that there seemed to be a place for another book. Also, a poet and a mathematician bring a different focus than did a physicist and (the same) mathematician.

The Yale mathematics department has been, and continues to be, a comfortable home for the fractal geometry course that was the proving ground for many of these topics, and introduced the older author to hundreds upon hundreds of Yale students, including the younger author. Through their questions and interest, many of these students have helped us construct the net of ideas in this book. In particular, the older author benefitted from discussions with Katherine Davis, Natalie Eve Garrett, Tanja Geis, Eva Hoffman, Shoshana Iliaich, Mason Ji, Simo Kalla, Caroline Kanner, Rachel Kurchin, Sarah Larsson, Rachel Lawrence, Peter Lewis, Elena Malloy, Demetri Martin, William Martino, Clara Norris, Maureen O'Hanlon, Madeline Oliver, Hannah Otis, Josie Rodberg, Nader Sobhan, Katherine Stewart, Arielle Susu-Mago, Caroline Sydney, Kira Tebbe, Jackson Thea, Joseph Thornton, Enoch Wu, and especially Jennifer Lanski and C. Noelle Thew. Before these, the older author learned much working with his Union College students, including Maureen Angers, Barbara Bemis, Colleen Clancy, Tatiana Cogevina, Lynne Erdman, Barbara Fulton, Heather Kanser, Albert Kern, Martin Logan, Jeremy Lynch, Shontel Meachem, Kathleen Meloney, Michael Monarchi, Brianna Murratti, Melanie Rinaldi, Allison Pacelli, Tricia Pacelli, James Robertson, Adam Robucci, Steve Szydlik, and Kathy Walter. And certainly many others.

The younger author is grateful for the many friends and mentors who listened to her talk about fractals with genuine interest, especially Konrad Coutinho, Michael Holkesvik, Sumaya Ibraheem, James Giammona, Anna

Nasonova, Autumn Von Plinsky, and Alex Werrell. John Loge, longtime Dean of Timothy Dwight College and a role model to all, provided further inspiration on how to work mindfully, with integrity and curiosity, and, most importantly, how to know when just to go to sleep.

We also thank friends and colleagues: Ginger Booth for writing the software used by students in the older author's classes. Sarah Campbell for inviting the older author to consult on her book *Mysterious Patterns, Finding Fractals in Nature* (Boyds Mills Press, 2014). Benoit and I discussed writing a children's book on fractals. We didn't, but Sarah did. Chip Cohen for sharing his work on fractal antennas and invisibility cloaks, and for much other help besides. Amanda Folsom for many discussions of the mathematics of fractals. Brenda Johnson and Jim Sauerberg for exploring connections between number theory and dynamics. Mary Laine and Donna Laine for their photography of skull sutures and yarrow plants, and for many discussions about plants. Kerry Mitchell for sharing his work on subtle variations in the sizes of pieces of the Mandelbrot set. Nial Neger and Harlan Brothers for substantial contributions to our summer teacher training workshops, which in turn influenced our presentation of some topics in this book. William Segraves, former Yale College dean of science education, for support of the continued development of this course, and in general of everything involving curiosity about the natural world. Christine Waldron and Jeff Sorbo for sharing their valuable insights into pedagogy and coding.

The interest and encouragement of both our families is more important to us than they can know. In particular, the younger author thanks her parents, for their support of her writing from the beginning, as well as for instilling in her a healthy love of the wonders that science can reveal to us. She also would like to thank everyone who put up with her endlessly editing these chapters on various beautiful family vacations. Careful proofreading by Andy Szymkowiak and Jean Maatta caught typos, inelegant constructions, and logical gaps that passed through both our sieves. Andy's Linux skill was invaluable for constructing the index.

From the beginning, Joseph Calamia, our editor at Yale University Press, was an enthusiastic advocate of this project. Especially, we must thank him for his remarkable patience with delays caused by medical problems and increased geographical distance between your authors. Mary Pasti's very careful copyediting improved the text in ways far beyond catching the remaining typos and stylistic inconsistencies.

Finally, we owe a debt of gratitude to evolution, which gave us brains with complexity sufficient to understand fractal and other beautiful patterns in nature, and which also gave us cats, including Bopper.

Figure credits

Except for those listed here, the images were generated by the older author, or taken from the ClipArt collection *Art Explosion 750000*, Nova Development, 1995.

Frontispiece. The first image is by Amelia Urry, the second by Dr. Nathan Cohen.

Introduction. The second image of Fig. 1 is a photo of a lung cast made by Dr. Robert Henry, DVM.

The photo of Benoit was taken by Harlan Brothers.

Shape of the World. The first image of Fig. 1.1 is from [275].
The second image of Fig. 1.2 is a NASA photo, the third image was provided by Prof. Ewald Weibel.
The second and third images of Fig. 1.12 were drawn by Jennifer Lanski.
The second image of Fig. 1.13 was provided by Nial Neger.
The second image of Fig. 1.16 was generated by Dr. Ken Musgrave.

Self-similar geometry. The photos of Fig. 2.20 were taken by Harlan Brothers.

Wild self-similarity. The image of Fig. 3.3 is a NASA photo.
The images of Fig. 3.5 are from Google Maps.
The images of Figs. 3.7 and 3.8 were provided by Aliette Mandelbrot.
The image of Fig. 3.9 was generated by Dr. Ken Musgrave.
The images of Figs. 3.10 and 3.12 are from Google Maps.
The images of Figs. 3.13 and 3.14 were provided by Prof. Ewald Weibel.
The images of Fig. 3.21 were provided by Prof. Leonid Mirny.
The second image of Fig. 3.22 was provided by Aliette Mandelbrot.
The image of Fig. 3.23 is a NASA photo.

Manufactured fractals. The images of Figs. 4.6, 4.7, 4.8, and 4.9 are from Dr. Nathan Cohen.
The images of Fig. 4.12 are provided by CAIDA, of the University of California, San Diego. © 2002 The Regents of the University of California.
Fig. 4.13 was provided by Harlan Brothers.
Fig. 4.16 was provided by Dr. Josie Rodberg.
Fig. 4.17: © Salvador Dalí, Fundació Gala-Salvador Dalí, Artists Rights Society (ARS), New York 2014.

The images of Fig. 4.19 were produced by Natalie Eve Garrett (first), Tanja Geis (second), and Claire Miller and Cara Norris (third).
Figs. 4.20 and 4.21 were made by Katsushika Hokusai.

The Mandelbrot set. The third image of Fig. 5.24 was provided by Harlan Brothers.

Dimensions. The photos of Figs. 6.10 and 6.14 were taken by Donna Laine.
The pictures of Fig. 6.11 were provided by Harlan Brothers.
The images of Fig. 6.34 are from Google Maps.
The first image of Fig. 6.35 was provided by Aliette Mandelbrot.

Technical Appendix. The images of Fig. A.23 are from Google Maps. The first image of Fig. A.27 is Fig. 1 of [182], the second is Fig. 1 of [183], both reprinted with permission; © 1984 by the American Physical Society. The third is Fig. 1(d) of [194], from World Scientific Publishing.
The images of Fig. A.28 and A.29 and the first image of Fig. A.30 are from NASA.
The images of Fig. A.33 are from [275].
Fig. A.37 was provided by Dr. Josie Rodberg.

Index

1-2 line, 211, 230, 412, 429–432, 435
1-3 line, 431, 433
1-4 diagonal, 206, 430–432, 434
$1/f$
    distribution, 103, 114
    noise, 70, 473
    power spectrum, 72, 77, 104
2-3 diagonal, 206, 207, 211, 215, 383, 432, 433
2-4 line, 434
2-address correlation, 225
3-4 line, 211, 230, 412, 429, 432–434
3D-printing, 93

Abe, Sumiyoshi, 306, 457
Adami, Christopher, 325, 459
address
    correlation, 225
    length-1, 5, 35, 44, 207, 225, 384
    length-2, 5, 6, 35–38, 44, 183, 208, 209, 213, 215, 224, 277, 278, 380, 381, 384, 410–412
    length-3, 6, 35–38, 208, 209, 224, 380, 381, 409
    length-4, 224
    length-5, 224
    length-n, 44, 186, 275, 381, 399
    occupancy, 212, 224, 225, 384, 385
Aesthetics in Visual Art and Fractals, 123

air exchange zone, 64, 65
Aki, Keiiti, 235, 305, 457
Aleph, The, 106, 107, 474
Algorithmic Self-Assembly of DNA, 451
allometry, xviii, 73
α, 227–234, 236, 241, 284–286, 394–396, 398–403, 434–437, 487
Alpher, Ralph, 299, 300
Alptekin, Ömer, 235
alveoli, 64–66, 290, 291, 447
ammonite, 86, 197, 198
    suture, 86, 197, 198, 302, 457
amylase, 76, 206, 207
"Mandelbrot's Fractals and the Geometry of Life", 198
angle-doubling map, 337, 338, 347, 348, 350
antenna
    Euclidean, 92, 93
    fractal, *see* fractal, antenna
Apollonius, 46
Arcadia, 106, 111, 112, 116, 329
area-perimeter relation, 312, 313, 316, 458
arthropod population, 71, 314, 448
Artificial Life, 325
Aschwanden, Marcus, 309, 458
Aschwanden, Pascal, 309, 458
Aspects of Schenkerian Theory, 102, 473
asthma, 65, 290
Audeguy, Stéphane, 105, 106, 474
Avatar, 12
Avron, Joseph, 311, 312

Babloyantz, Agnes, 307, 457
Bach, J. S., 100, 105, 110
Bachelier, Louis, 282, 445
Bacillus subtilis, 87, 303, 457
backward Z, 211
bacteria, 77, 87, 198, 303, 304, 457
    communicating walkers model, 304, 457
    diffusion-limited growth, 304
    fractal growth, 87, 303
    growth on nutrient-depleted agar, 303, 304
Bak, Per, 306, 309, 457, 458
ballistic aggregation (BA), 80, 81
ballistic cluster-cluster aggregation (BCCA), 81, 82, 199, 483
Banavar, Jayanth, 75, 290, 296, 450
Bancaud, Aurélien, 78, 79
Barnsley, Michael, 20, 31, 439, 443, 445
Barrallo, Javier, 127
Baryshev, Yurij, 84, 299
Bascompte, J., 319
base-2 expansions, 259
base-3 expansions, 259, 260
basic similarity dimension formula, 162, 163, 166, 170, 313, 361, 413, 414, 418–420, 435–437
basin of attraction, 150, 321
Beach, David, 102, 473
Ben-Jacob, Eshel, 304
Bentley, Wilson, 317, 318
β (multifractal), 231, 232, 395–402, 487
β (scaling signal), 69, 70, 77
Bethe, Hans, 299
bi-Lipschitz function, 377
bifurcation diagram
    coupling, 222–224, 390–393
    hybrid map, 388, 486
    logistic map, 217, 218, 222, 223, 387, 486
    tent map, 388
bifurcation, period-doubling, 217, 218
bin
    boundaries, 209, 210, 212, 213, 225
    equal-size, 208–210, 212, 218, 219, 225, 380, 381
    equal-weight, 208–214, 219
    median-centered, 208, 210, 212
    occupancy, 213
biological area, 74, 294–296
biological volume, 74, 295
Bissonette, John, 458
Blank, Martin, 80
Bloch, William, 329, 474
Bopper, xv–xvii, 131, 492
Borges, Jorge Luis, xvii, xx, 106, 107, 322, 329, 459
Boyajian, G., 86, 302
BPI space, 53, 54, 288, 440, 446
branch point, 24, 137, 143, 244, 443
branching, xi, 2, 12, 56, 63–66, 73, 74, 85, 106, 107, 122–124, 242–244, 303, 315, 318, 325
Branner, Bodil, 481
Breitenbach, T. E., 120, 127, 477
broccoli, 3
Brothers, Harlan, xiv, 100, 101, 329, 473, 493, 494
Brown, James, 71, 74, 294–297, 450
Brown, Robert, 281
Brownian
    motion, xviii, 52, 81, 238, 241, 281–285, 445
    independent increments, 282, 284, 285
    normally distributed increments, 282, 284
    scaling, 238, 282–284
    stationary increments, 282
noise, 70

Buckingham, Amanda, 475
Buffon needle problem, 66, 291
Buffon, George-Louis, 66
Burkhart, Charles, 102, 473
butter, branching, 12

Calvin and Hobbes, 113
Cantor cube, 21, 23, 24
Cantor set, 4, 9, 21–24, 49, 51–53, 56, 82, 100, 103, 107, 116, 135, 136, 138, 149, 162, 169, 170, 175, 177, 194, 195, 246, 247, 256, 311, 321, 333, 334, 340, 377, 383, 415–417, 429, 443
  fat, 87, 311, 312, 458
  middle-halves, 175, 177, 192, 193, 195, 311, 379, 417, 428
  middle-thirds, 135, 169, 176, 177, 188, 191–195, 256–260, 274, 288, 311, 377, 379, 415, 428, 429
    every point is a limit point, 22, 23, 256, 260
    length 0, 22, 23, 256, 257, 311
    uncountably infinite, 22, 23, 256, 259, 274
Cantor's diagonal argument, 258, 259
Cantor, Georg, 21, 257–259, 443
capacitor, 91
  Euclidean, 91, 92
  fractal, *see* fractal, capacitor
  lateral, 91
Cardano, Gerolamo, 131
cardiac time series, *see* time series, cardiac
Cat in the Hat Comes Back, The, 9
cathedral of Anagni, 24
Cayley, Arthur, 150, 151, 353, 354, 480

Cello Suite no. 3 (J. S, Bach), 100
Center for Applied Internet Data Analysis (CAIDA), 99
Chaos, 72, 449
chaos, ix, xviii, 63, 72, 105, 111–113, 116, 117, 137, 216–219, 223, 302, 307, 332, 334, 337, 339, 387–389, 445, 449, 457, 479, 486
  mixing, 335, 339, 383
  periodic points are dense, 335, 339
  sensitivity to initial conditions, 216, 218, 334, 338, 389
"Chaos Theory and Fractal Geometry: Their Potential Impact on the future of Art", 127
Charlier, Carl, 299
chemical mixers, 90, 128, 235, 470
Chessie, xvi, 43, 131
Chhabara, Ashvin, 458
chromatin, 77–79
  euchromatin, 77, 78
  heterochromatin, 77–79
circle inversion, 46, 48–50, 187, 280, 281, 445
  attractor, 48, 49
  fractals, *see* fractal, circle inversion
  Mandelbrot's method, 280, 445
  radius property, 46
  random IFS algorithm, 49, 280
  ray property, 46
circuit switching, 97
circulatory system, 2, 65, 66, 74, 199
Clancy, Colleen, 323, 459
Clarke, Arthur C., 107, 113, 474
Clarke, John, 103, 104, 473
climate, 240
Cloud Atlas, 118
clouds, xiii, 2, 11, 56, 87, 105, 106, 112, 316, 318, 458,

475
area-perimeter relationship, 316
fractality, 11, 105, 125, 317, 458
coarse-graining data, 208, 226
coastline, xi, xiii, 2, 3, 11, 12, 58–61, 64, 67, 105, 106, 112, 171, 176, 196–198, 234, 242, 251, 289, 446, 483
COBE satellite, 300
Cohen, Mark, 107, 474
Cohen, Nathan, xv, 93, 94, 115, 326, 327, 439, 492, 493
Colors of Infinity, The, 107, 474
complex numbers, xviii, 130–132, 158, 331, 344
argument, 335, 337, 342
conjugate, 354
imaginary part, 132, 133
modulus, 335, 337, 342, 344, 354
multiplication, 132, 335, 336
polar representation, xviii, 335, 336, 342
real part, 132, 133
conducting airways, 64, 65
connected set, 135, 136, 138, 145, 147, 148, 156, 333, 334, 340, 479
Connes, Alain, 85, 454
Constable, John, 316
contingency, 117, 118, 302, 457
contraction, 17, 102, 186, 227, 281, 362, 363, 399, 400, 443, 445
Cootner, Paul, 282
Coppens, Marc-Olivier, 90, 325, 470
correlation, 68, 69, 75, 103, 206, 235, 286, 300, 327, 483
fractal, 68, 77, 450
function, xviii, 68, 69, 76, 292–294, 300, 324

length, 300
long-range, 76, 457
Cortázar, Julio, 243, 489
cosmic background radiation, 300, 454
"Cosmography of the Universe, On the", 299
cosmological
constant, 298, 299
principle, 298
Cosmological Letters, 299, 454
cosmology
hierarchical, 299, 454
homogeneous, 83, 297–300
isotropic, 297, 298
Newtonian, 297
coupled map lattice (CML), 220, 223, 486
synchronization, 223, 390, 392, 486
coupling
nearest neighbor, 220, 393
next nearest neighbor, 220
random, 220
coupling constant, 220, 221, 224, 393
Coverly set, 112, 117
Cowie, Patience, 235
Cozens, Alexander, 120, 121, 477
Crandall, Richard, 322
crepes, xix, 12
Crick, Francis, 75
critical point, 147, 148, 332, 333, 340, 351, 356
for $z^2 + c$, 148, 333, 334
for $z^n + c$, 352
for cubic polynomials, 352, 356
cross-product, 266–268
Crumples, xvi, 131
Curry, James, 152, 153, 356, 480
cut-out sets, 186
Cutting, James, 114, 301, 457, 475
cycle

2-, 139–142, 146, 152, 153, 217, 218, 223, 341–346, 349, 350, 360, 361, 383, 388
3-, 139–145, 216–218, 345, 346, 360, 361, 388
4-, 141–144, 216–219, 223, 345, 346, 349, 361, 388, 389
5-, 141, 142, 144, 145, 346, 360, 388
6-, 143, 345, 346
7-, 141, 224, 346, 388
m- (m divides n), 345
n-, 156, 341, 344, 345, 360, 361
intermittent, 217
stable, xviii, 139, 152, 153, 335, 341, 342, 344, 345, 350, 353, 356, 383, 388, 480
unstable, 152, 316, 335, 350, 351, 458

da Vinci, Leonardo, 121, 316
Dahl, Jonah, 316
Dalí, Salvador, 121, 127, 477
Dalí, 121
"Dammit, I'm mad", 115
dark matter, 83, 86, 87, 297, 457
dark night sky paradox, 201, 299
dark web (cosmological, not internet), 87, 457
Darkness at Night, 299
David, Guy, 54, 440
Davidson, Keay, 300
de Bruijn
graph, 278, 279
sequence, xviii, 278–280, 445
prefer 1 algorithm, 278, 279
de Bruijn, Nicolaas, 278
Death Qualified, 113
Decalcomania, 123
decalcomania, 122–124, 127, 477
deer skull suture, 171, 172, 174
Dekker, Job, 78

DeLong, Jordan, 114
δ-cover, 400–402
Demko, Seven, 443
Descharnes, Robert, 121
Devaney, Robert, 334
Dewdney, A. K., 129
Dickens, Charles, 118, 330, 475
dictionary of coupled maps, 224–226, 486
dielectric breakdown, 53, 86, 302–304, 457
diffusion-limited aggregation (DLA), 52, 80–82, 87, 199–201, 303, 304, 318, 483
dimension, xii, xviii, 58, 85, 126, 131, 146, 159, 175
as exponent, xviii, 58, 160, 167
box-counting, xix, 86, 126, 167–169, 171–175, 177, 187, 190, 191, 195, 197–199, 233, 286, 302, 303, 305, 368, 372, 377–379, 415–417, 457, 458, 482, 483
limit formula, 168, 369
log-log approach, 172
other things to count, 370
sequential form, 370
sequential formula, 169–171
drip, 126
Euclidean, xviii, 21, 23–25, 27, 28, 57, 74, 75, 78, 79, 90, 99, 154–157, 244, 246, 247, 260, 266, 282, 287, 296, 309, 310, 313, 321, 325, 357, 359, 360, 391, 392, 458
fractal, 6, 51, 56, 58, 78, 91, 92, 101, 105, 125, 159, 181, 184–186, 188–190, 195, 201, 203, 226, 227, 235, 236, 301, 305–307, 312–316, 320, 328, 418–420, 457, 458, 483

Hausdorff, 188, 190, 191, 195, 227–229, 232, 311, 312, 351, 376–379, 396, 400–402, 479, 481, 482
Lévy, 126
mass, 167, 178, 179, 199–201
  clusters of peas, 179
  crumpled paper, 178
negative, 235, 317, 458, 487
of memory IFS, 183–185, 189, 421–426, 482
of nonlinear fractals, 186
of random fractals, 180–182, 188, 189, 418–420, 482
rules
  intersection, 193, 200, 201
  invariance, 191, 195, 428
  monotonicity, 190, 194, 195
  product, 192–194, 428, 429
  stability, 190, 428, 429
similarity, 162–164, 166, 167, 169, 173, 175, 182, 190, 229, 232, 313, 361–367, 375, 395, 396, 404, 413, 414, 418, 419, 421, 422, 429
Dinky, xvi, 131
Discovery of Cosmic Fractals, The, 84, 299
displacement field, 313
Ditlevsen, Peter, 241
DNA, 67, 68, 75–77, 79, 110, 176, 206, 207, 324, 485
  equilibrium globule, 77, 78, 450
  extraterrestrial, 80
  fractal antenna, 79, 451
  fractal globules, *see* fractal, DNA globules
  long-range correlations, 77, 450
  programming, 451
  Sierpinski gasket, 79, 451
Doggett, Frank, 116
Dominik, Carsten, 81, 452

"Don't Get Around Much Anymore" (D. Ellington), 101
Donn, Bertran, 81
Douady, Adrien, 131, 153, 156, 157, 479–481
doubling map, 144, 479
drainage
  basin, 290
  pattern
    rectangular, 290
    trellis, 290
drip paintings, 125, 126, 477
dust (mathematical), *see* fractal, dust
dust (physical), *see* fractal, dust clump
Dusty, xvi, 131
dwell, 134, 138, 147
dynamical plane (z-plane), 138–140, 145

Earth's magnetic field, 82, 308
  reversals, 82, 83, 453
earthquake, 224, 233, 236, 237, 304, 308
  aftershock, 305, 457
  distribution, 87, 235, 236
  fault lines, 176, 235, 305, 306, 457, 487
  magnitude, 87, 235, 236, 304, 305, 487
  multifractal, 236, 487
  network theory, 306, 307, 457
  self-organized criticality, 306, 457
Edgar, Gerald, 439
Edgerton, Samuel, 127, 477
Eigen, Manfred, 324, 459
eigenvalue, xviii, xix, 183–185, 371–375, 396, 421, 424, 426
  equation, 372, 373, 424–426
  Mathematica command for, 374
eigenvector, 372, 373
  equation, 373

INDEX 501

Einstein, Albert, 83, 86, 250, 282, 297–299, 445, 454, 457
electrocardiogram (EKG), 87
electrodeposition, 86, 199, 302–304, 457
electroencephalogram (EEG), 85–87, 301, 302, 307, 457
Elshurafa, Amro, 326
EMBO Journal, 78
Encyclopedia of Cosmology, The, 297
Enquist, Brian, 71, 74, 294–297, 450
epidemic distribution, xviii, 87, 216, 319, 459
    fractal basin boundaries, 321
    fractal front propagation, 322
    fractal islands of survivors, 322
    fractal patterns in time, 320
    HIV, 322
    measles, 321, 322
    underlying random fractal process, 322
equilibrium globules, *see* DNA, equilibrium globules
Ernst, Max, 122, 477
escape criterion, xviii, 331, 351, 352
Etzenhauser, J., 319
Euclidean geometry, xi, 1, 2, 28, 63, 128
Eulerian cycle, 278, 279
Eureka, 84, 299
evolution, xii, 55, 65, 66, 71, 78, 79, 89, 197–199, 296, 302, 324, 325, 457, 492
expected value, 180, 284, 371, 372, 418
    formula, 180, 181
Exploration of Space, The, 107, 474
"Expressions about the Time and the Weather", 115

$f(\alpha)$, 227–234, 236, 241, 394–398, 401, 402, 435–437, 487
Falconer, Kenneth, 188, 370, 402, 439, 442, 482, 487
Fatou, Pierre, 136, 138, 147, 148, 332, 333, 340, 356
Feder, Jens, 199, 302, 312, 316, 439, 458, 483
Fermat's Little Theorem, 143, 348–350, 479
fern, xi, xii, 3, 56
    fractal, 29–33, 443
Feynman, Richard, 75, 450, 472
filaments, 83, 87, 249, 310
film (movie, not a thin layer of soap), 12, 102, 107, 114, 474
Filoche, M., 290
First Three Minutes, The, 300
fitness landscape, 324, 459
    fractal, 87, 324, 325, 459
    RNA landscape, 325, 459
    space of genotypes, 324, 459
fixed point, xviii, 137, 139, 140, 217, 320, 321, 332, 334, 340–344, 346–351, 356, 388
    stable, 340, 341
fixed point equation, 340
Flake, Gary, 328, 472
Flamsteed, John, 310
"Flock", 115
foraging path, 87, 126, 322
forbidden pair, 37, 38, 41, 42, 209, 212, 219, 383, 386, 390, 409–412, 444
forbidden triple, 37, 41, 42, 213, 219, 386, 387, 409
forest, 70, 71, 74, 249, 316
    power law, 71
Fourier cosine series, xviii, 292
fractal, ix, xii–xiv, xvi–xx, 1, 3–6, 8–10, 12, 13, 15–21, 23, 25, 26, 29–31, 36–44, 46, 48–56, 58, 60–

64, 67–70, 72, 73, 75, 77–79, 81–87, 89–91, 94–128, 155, 158, 159, 161–169, 172, 173, 175–184, 186–188, 190–196, 200, 201, 203, 205, 214, 215, 219, 221, 226, 227, 230, 235–237, 240, 242–244, 246, 247, 249–253, 256, 272–274, 280, 288–290, 298, 301–304, 306, 307, 309–311, 313–330, 350, 359, 361–363, 366, 367, 371, 372, 375–379, 397, 403, 409, 410, 413, 414, 416, 417, 421–423, 425, 427–429, 439, 440, 442, 443, 445, 446, 449, 450, 454, 457–459, 469, 470, 473, 475, 484, 486–488

antenna, 79, 80, 89, 92–96, 128, 326, 327, 471, 492
capacitor, 89, 91, 92, 249, 326, 470
circle inversion, 6, 7, 50, 114, 186, 187, 445
cracks and fractures, 87, 176, 312–315, 458
DNA globules, 77, 78, 249, 450
dust, 22, 131, 135, 136, 138, 139, 146, 148, 149, 333, 340
dust clump, 55, 80–83, 87, 177, 179, 196, 199, 281, 310, 452
fern, *see* fern, fractal
fitness landscape, *see* fitness landscape, fractal
fluid injector, 90
forgery, 12, 62, 110, 207, 239, 289, 318, 446
geometry, ix, xii–xvi, xix, 12, 17, 20, 58, 62, 63, 79, 92, 93, 110, 111, 115, 117– 119, 122, 127, 129, 158, 159, 200, 224, 232, 243, 249–251, 315, 316, 439, 483, 491
invariance property, 16, 48
invisibility cloak, 89, 94–96, 327, 471, 492
metamaterial, 95, 96, 472
metastructural, 118, 119, 330, 475
nonlinear, 4, 6, 7, 11, 50, 180, 185–188, 191, 377
palindrome, 115
random, xix, 8, 11, 12, 51, 52, 126, 179–182, 188, 189, 281, 284, 298, 322, 418
self-affine, 4, 6, 10, 162, 188, 193, 195, 376, 378, 379, 482
space-time, 203, 454
spiral, *see* spiral, fractal
statistically self-similar, 4, 7, 8, 52
structural, 114–118, 475
time, 67, 68, 95, 96, 120, 203, 484
tree, *see* tree, fractal
with memory, 4, 5, 10, 38, 183, 189, 382
Fractal Antenna Systems, 93, 94, 96, 472
"Fractal forgeries of Nature", 289
Fractal Geometry and Applications, 289
Fractal Geometry of Nature, The, xiii, 199, 312, 316, 317, 442
Fractal Growth Phenomena, 199, 302
Fractal Murders, The, 107, 474
Fractal Painting, au Natural, 123
Fractal Time, 203
Fractals, 199, 312, 316
Fractals in Soil Science, 315

INDEX 503

FractalWorld, xiii, xix, xx, 8, 149
fractional Brownian motion (fBm), xviii, 239, 240, 242, 243, 284–287, 402–404, 445, 488
    anti-persistent, 285, 286
    index α, 284, 286, 403
    persistent, 285, 286, 307, 458
Fratini, Michela, 95
Free Composition, 103
Freeman, Walter, 86, 302, 307, 457
Friedmann, Alexander, 298, 454
fuel cell, 89, 90, 325, 470
Fuji from the Seashore, 124
Fuji in a Thunderstorm, 125
Fujikawa, Hiroshi, 303
Full House, 197, 302, 483
Fuzzy, xvi, 131

Gaddis, John Lewis, 119, 475
galaxies, 83, 84, 86, 87, 297–299, 309
    distribution, 55, 83, 86, 179, 232, 237, 297, 300
        fractal, 55, 83, 201, 300, 301, 454
Galileo, 307, 457
Gamow, George, 299
García Márquez, Gabriel, 119, 475
"Garden of Forking Paths, The", xvii, 106, 474
Gardner, Martin, 102, 103, 328, 473
Garnett, Lucy, 152, 153, 356, 480
Garrett, Natalie Eve, 123, 477, 494
Garvin, Jeffrey, 301, 457
gasket, *see* Sierpinski, gasket
gasket relative, 31, 203, 272, 444
Geis, Tanja, 123, 477, 494
gene
    activation, 78
    expression, 77
    inactivation, 78
    regulation, 78, 224
    repression, 78
    transcription, 77
geometric series, xix, 165, 256, 277, 364, 366, 422
    summing, xviii, 165, 166, 256, 257, 259, 277, 311, 363, 366, 368, 422, 423, 426
Ghost from the Grand Banks, The, 107, 474
Giacometti, Augusto, 124
giant hogweed, 55
Gilpin, Michael, 321
Glass, Leon, 449
Glass, Philip, 100
Gold Bug Variations, The, 110, 474
Goldblum, Jeff, 113
goldenrod, 55
Gomez, Domingo, 291, 447
Goodman, Reba, 80
Google, 39, 98, 99, 277, 297, 326, 334, 368, 386, 453, 458, 487
Google Maps, 62, 63, 171, 197, 290, 493, 494
Gould, Stephen Jay, 86, 197, 198, 302, 457, 483
Grebogi, Celso, 316
Greenberg, Frank, 102, 473
Grosberg, Alexander, 78, 450
Guckenheimer, John, 156, 360, 481
Gutenberg, Beno, 304
Gutenberg-Richter law, 235, 236, 304, 305, 457
Guth, Alan, 85, 454
Guy, David, 288

Høeg, Peter, 108, 474
Hagerhall, Caroline, 301, 457
Handelman, Sig, 62, 289, 446
Harrison, Edward, 299, 454
Harshey, Rasika, 304
Hastings, Harold, 306, 457
Hausdorff

dimension, *see* dimension, Hausdorff
measure, 400, 401
heartbeats, 71, 72, 212, 232, 449
  intervals, 68, 71, 72, 210
heartbreak, 72
heavy tails, 328
Hele-Shaw cell, 302
Henderson, Linda Dalrymple, 127, 477
Henry, Robert, 493
Herman, Robert, 299
Herrmann, Hans, 313
Hess, W. R., 64, 65, 290, 447
Hetherington, Norriss, 297, 300, 453
hierarchical clustering
  of earthquake graphs, 306
  of galaxies, 83, 84, 201
  of logistic maps, 220
  of seismic activity, 235
Hilbert, David, 28, 297
Hill's equation, 311
Hill, Christopher, 297, 453
Hinode Solar Optical Telescope, 309
Hirabayashi, Tadashi, 235
histones, 77, 78
History of Art, 120, 125
Hokusai, Katsushika, 124, 125, 316, 477
Hölder exponent, 227, 228, 230, 231, 233, 234, 241, 286, 308, 395, 399, 402, 403
Hong, Young, 99
Hopscotch, 243
"How long is coast of Britain?", 58
Hubbard, John, 131, 151, 153, 156, 157, 479–481
Hubble, Edwin, 299
Humason, Milton, 299
Hurst exponent, 308
Hurst, Harold, 240, 308, 488
Hutchinson, John, 20, 190, 376, 443
hypercube, 74, 127, 154, 279, 358, 359
  cubes in its boundary, 357, 359

IFS, 20, 439, 443
  attractor, 17, 41, 48, 229, 230, 232, 262, 396, 399, 429
  deterministic algorithm, 17, 42, 44, 49, 273, 288
  driven, xix, 205–216, 218, 219, 221, 222, 224–226, 228, 229, 275, 339, 380, 382–384, 390, 430–434
  matrix formulation, 265
  random algorithm, 43, 44, 275, 277, 278, 430, 445
    probability an address is unoccupied, 383–386
    visiting addresses efficiently, 278
  rules, 18, 19, 25, 33, 34, 36, 40, 52, 67, 256, 265, 271, 376, 407
    reflection, 15, 17, 18, 267, 268
    rotation, 15, 18, 33, 262, 269
    scaling, 15–19, 25, 26, 30, 33, 67, 162, 261, 268
    translation, 15, 17, 18, 25, 33, 34, 262
  with memory, 34–39, 41, 182, 183, 185, 196, 219, 372, 375, 382, 383, 396, 429, 444, 482
  without memory, 38–41, 397, 410, 411, 422, 444
Imaging Brain Function with EEG, 307
Imakaev, Maxim, 77
Indra, 50
Indra's Pearls, 288
Indra's pearls, 50

INDEX

infinite set
    countable, 23, 26, 257, 258, 274
    uncountable, 22, 23, 26, 259, 260, 274
inflating universe, 84
Internet, 10, 62, 85, 89, 97, 99, 205, 237, 242, 327, 328, 488
    characteristic hierarchical level, 328, 472
    fractal, 98, 472
    traffic
        congestion, 327
        distribution, 97
        fractal, 327, 472
        inter-arrival times, 328
        multifractal, 237, 327
        queueing times, 328
        round-trip times, 328
        self-similarity, 98, 237, 327, 472
intersection (of shapes), 156, 190, 193, 194, 348, 349
Introduction to Chaotic Dynamical Systems, An, 334
ion channel kinetics, 87, 323
    fractal, 323
    Markov model, 323
IR echoes, 310
"Itchy & Scratchy Land", 113
Ito, Keisuke, 235

Jøssang, Torstein, 303
Jakobson, M., 218, 387, 486
Janson, H. W., 120, 121, 125
Jeffrey, H. Joel, 485
Jeopardy, 15
Jonas, David, 125, 126
Jones-Smith, Katherine, 126
Journal of Fluid Mechanics, 394
Julia set, xviii, 129, 132–134, 136, 138, 143, 145–148, 156, 332, 333, 335, 350, 351, 479, 481
    boundary, 134
    Cantor set (dust), 135, 136, 138, 139, 147, 148, 333, 340
    chaotic dynamics, 137, 332, 337–339
    connected, 134–138, 145, 147, 148, 333, 340
    filled-in, 125, 134, 135, 139, 142
    Misiurewicz points belong to their Julia sets, 145, 350
    neither connected nor dust, 148, 149
    number of lobes, 143
Julia, Gaston, 134, 136, 138, 147, 148, 332, 333, 340, 356
Jurassic Park, 113
Jürgens, Hartmut, 440

Kanner, Caroline, 115
Kant, Immanuel, 299
Kasner, Edward, 329
Kaufman, Henry, 200
Kay, B., 315, 458
Kaye, Bryan, 439
Kendal, Wayne, 322
Kigami, Jun, 440
Kim, Yeunjin, 310
Kleiber, Max, 73, 296, 449
Kleinian group limit sets, 53
Koch
    curve, xiii, 4, 26, 54, 103, 112, 160, 173, 174, 329, 443
    no tangents, 26, 27, 54, 227, 443
    randomized, 11, 51–53
    relatives, 166, 167
    pyramid, 28, 29
    snowflake, 21, 27
    tetrahedron, 21, 28, 29
Kolmogorov, Andrey, 234, 487
Kot, Mark, 319, 321, 322

labyrinth, 106

Lamé equation, 314
Lambert, Johann, 299, 454
Landscape of History, The, 119
Lanski, Jennifer, 9, 444, 493
Lapenna, Vincenzo, 236
Large-Scale Structure of the Universe, The, 300
lateral flux, 91
Lavaurs' algorithm, *see* Mandelbrot set, Lavaurs' algorithm
Lederman, Leon, 297, 453
left wall, 86, 198, 302, 457
Legendre transform, 398, 399
Leibniz, Gottfried Wilhelm, 107–109, 474
Leland, Will, 328
Lemaître, Georges, 454
Leo, xvi, 131
Leonardis, Ersilia, 309
lettuce leaves, 12
Levy flight (stable motion, processes), 52, 82, 126, 239, 286, 287, 445, 453
Li, Dongsheng, 236
Liapunov exponent, xix, 218, 389, 486
  calculation, 389
  interpretation, 218
Libeskind, Noam, 86
Lieberman-Aiden, Erez, 78
Liebovitch, Larry, 323, 459
Ligeti, György, 101, 105, 473
Lightman, Alan, 201, 247, 484, 489
lightning, 56, 86, 112, 125
limit point, 22, 23, 256, 260, 274
limit set, 274, 280, 281, 445
Lind, Doug, 39
Linde, Andrei, 84, 454
linear fractional transformations, 288, 445, 446
literature, 9, 89, 97, 105–120, 203, 252, 329, 330, 474, 475
"Little Girl in the Big Ten", 113

locally connected, 156, 157, 481
logarithm, xii, xviii, 58, 159, 161, 164, 169, 171, 172, 197, 228, 304, 330, 436
logistic map, 216–221, 223–225, 380, 381, 387–393, 486
  bifurcation diagram, *see* bifurcation diagram, logistic map
loop, 39, 41, 306, 410–412
Lovejoy, Shaun, 240, 241, 316, 458, 488
lung, xi, 3, 64, 65, 74, 90, 196, 198, 447
  area, 64, 295
    measurement, 66, 67, 290–292, 447
  fractality, 64, 67, 73, 198, 325
  volume, 64
Lung, C., 314
Lutz, T., 86, 302

Måløy, Knut, 302
Macchiato, Maria, 236
Madariaga, Raúl, 306, 457
Main, Ian, 235
Malloy, Elena, 486
Mandelbrojt, Szolem, 138
Mandelbrot formula, 132–134, 136, 138, 139, 333
  computational time, 332
  cubic, 157
  escape criterion for $z^2 + c$, xviii, 133, 331, 332, 352
  escape criterion for $z^n + c$, 351, 352
Mandelbrot sequence, 139, 140, 142
  testing stability of fixed points and cycles, 341
  where the 2-cycle is stable, 139, 343, 344
  where the fixed point is stable, 139, 342, 343

INDEX 507

why the fixed point at infinity is stable, 347
why the iterations start with 0, 340
Mandelbrot set, xii, xviii, 20, 104, 107, 108, 112, 113, 117, 127, 129–133, 136–154, 156–158, 250, 330–332, 340–348, 350–353, 356, 360, 361, 481
$1/n^2$ rule, 156, 360, 361, 481
abstract, 144
boundary, 104, 131, 145, 146, 156, 158, 351, 479, 481
center-locating equation, 345
centers of components, 344, 345, 361
connected, 145, 156, 479
counting discs and cardioids, 344–346
cubic (4-dimensional), xviii, 154, 155, 157, 359
 removing the next-to-highest power of a polynomial, 357, 358
Farey sequence, 140–143, 153, 346, 360, 388
hyperbolicity conjecture, 158, 345, 481
Lavaurs' algorithm, 143–145, 347, 348, 479
locally connected, 156, 157, 481
Main cardioid, 112, 139–144, 146, 153, 344–346, 361, 388
MLC, 481
multiplier rule, 141–144, 153, 346
other functions, 147–149
principal sequence, 140–143, 153, 345, 346, 388
universality, 149–153, 353, 355, 356, 480
Mandelbrot, Aliette, 491, 493, 494

Mandelbrot, Benoit, ix, xii–xvi, xx, 5, 12, 21, 22, 29, 49–51, 56, 58, 60–62, 67, 89, 100, 104–109, 112, 115, 117, 118, 121, 122, 124, 125, 127, 129, 131, 138, 145, 146, 153, 158, 159, 174, 176, 193, 196–201, 213, 234, 235, 237, 239, 240, 249–252, 280, 281, 284, 287, 289, 298, 301, 309, 312, 316, 317, 329, 360, 394, 402, 403, 439, 440, 442, 445–448, 458, 475, 479, 481, 483, 487, 488, 491–493
Maritan, Amos, 75, 290, 296
Markov
 chain, xix, 275, 383, 384, 386
  absorbing state, 277, 386
  probability of unoccupied addresses, 384–386
 models, 323, 459
 partition, 380–383, 485
Martin, Demetri, 115
Martino, William, 242
mass action, 315, 320, 321
Mathematical Intelligencer, 280
Mathematics and the Imagination, 329, 474
Mathur, Harsh, 126
MathWorld, xix, xx, 8, 10, 56, 129, 156, 274
matrix
 determinant, 263
 inverse, 264
 multiplication, 263, 264
 transition, 386
  IFS with memory, 182–185, 372, 396, 397, 421, 424, 426
  Markov chain, 276, 384, 386
Matsushita, Mitsugu, 303, 304
Matsuyama, Tohey, 304
Mauroy, B., 290

Maxwell's equations, 93, 326, 327
    scaling solutions, 326, 471
May, Robert, 216, 387, 486
McGehee, Richard, 156, 360, 481
McMullen, Curtis, 153, 156, 157, 188, 195, 378, 379, 480, 482
Mendivil, Franklin, 278, 445
Meneveau, Charles, 234, 487
Menger sponge, 21, 25, 443
Menger, Karl, 25, 443
metabolic rate, 73–75, 295, 296
    per unit mass, 55, 73, 74, 296, 449, 450
metamaterials, *see* fractal, metamaterials
metric, 301
    scale-invariant fluctuations, 301
Micolich, Adam, 125, 126
microelectromechanical systems, 326, 470
Miller, Claire, 123, 477, 494
Milne, Bruce, 315, 458
Milovanov, Alexander, 307, 457
Mirny, Leonid, 77, 450, 493
Misiurewicz point, 145, 146, 350, 351
Mitchell, David, 118, 330, 475
Mitchell, Kerry, 361, 481
mitochondria, 74, 294, 295
mixing, *see* chaos, mixing
Miyazaki, Hayao, 281
Mogilevsky, E., 309, 458
monad, 107–109
Monadology, The, 109, 474
Monet, Claude, 316
Montroll, Elliott, 96, 472
Moon, Francis, 439
Moore, Raymond, 86, 302, 457
Moran equation, xix, 162, 163, 165, 166, 184, 190, 228, 376, 377, 403, 404, 413, 414, 419, 423–427, 429, 436, 437, 481, 487
    converting to a cubic, 165, 414, 425
    converting to a quadratic, 164, 181, 404, 418–420, 429, 436, 437
    generalized, 231, 394–397, 400
    infinite, 165, 184, 362–368, 375, 414, 422, 423, 425, 426
    memory, 183–185, 372, 375, 396, 422–426, 482
    compared with the infinite Moran equation, 376, 423
    nonlinear, 186, 482
    random, 180, 181, 371, 372, 418–420, 482
    unique solution, 164, 361, 362
motivic parallelism, 102
mountains, 12, 61, 62, 102, 111, 112, 249, 289, 316, 446, 493
Mr g, 201
multifractals, xix, 72, 176, 205, 227, 232–235, 240, 241, 243, 309, 315, 317, 327, 394, 439, 449, 458, 487–489
    earthquakes, 235, 236
    $f(\alpha)$, see $f(\alpha)$
    $f(\alpha)$ curve
        basic geometry, 394, 395
    $f(\alpha) = \alpha \cdot q + \beta(q)$, 399–402
    finance cartoons, 107, 237–239, 402, 403
    from IFS, 228–232
    from IFS with memory, 396
    generalized Moran equation, 231, 232, 394, 396
    Internet, 232, 237
    Legendre transform, 398, 399
    slope of the $\beta(q)$ curve is $-\alpha$, 397–399
    turbulence, 234
Mumford, David, 445, 446
Munroe, Randall, 114
Murray, C. D., 64, 65, 290, 447

INDEX 509

Murray, Carl, 82
Musgrave, Ken, 12, 62, 289, 446, 493
music, 89, 97, 99–105, 110, 114, 119, 120, 127, 328, 329, 473
"Music and Fractals", 104

Nature Inspired Chemical Engineering (NICE), 90
Neger, Nial, xiv, 329
Nel, Philip, 442
networks, 328, 457
    circulatory, 2, 66, 74, 111, 199, 296
    fault line, 235
    fractal, 74, 226, 294, 322, 393, 446, 450
    gene regulatory, 224
    graph, 306
        clustering coefficient, 306
        connectivity distribution, 306
        scale-free, 306
        small worlds, 306
    hierarchical, 74, 90, 296
    information, 97–99, 237, 242, 327, 472
    logistic map, 216, 220, 221, 390, 393
    metabolic, 74, 224, 295
    river, 56, 62, 63, 290, 446
    sociogeographic, 322
    spanning, 75, 296
New Musical Theories and Fantasies, 102, 474
New York Notes: Music and Fractals, 104
Newman, James, 329
Newton's method, xviii, 149, 150, 152, 321
    basins of attraction, 150, 151
    for $z^2 - 1$, 151, 353, 354, 480
    for $z^3 - 1$, 151, 480
    formula (function), 150, 153, 353
    stable cycle, 152, 153, 356, 480
    unstable cycle, 152
Newton, Isaac, 83, 149, 250, 297, 353
Niemeyer, L., 303
Nittman, Johann, 318
Noether's theorem, 297, 298, 453
Noether, Emmy, 297
non-fractals, 8, 9, 33, 442
non-rome, 39, 41, 410–412
Norris, Cara, 123, 477, 494
Nothelfer, Christine, 114
Nottale, Laurent, 85, 203, 454, 484

O'Brien, Flann, 9, 109, 110, 474
OGY method, 316, 335, 458
Okubo, Paul, 235, 305, 457
Olbers' paradox (dark night sky paradox), 84, 299, 454
Olbers, Heinrich, 299
Omori's law, 305, 457
Omori, Fusakichi, 305, 457
"On exactitude in science", 322, 459
"On the self-similar nature of ethernet traffic", 327
Öncel, Ali, 235
One Hundred Years of Solitude, 119
Onnes, Heike, 95
open set condition, 190, 376, 377, 482
optical gasket, 11, 50
"Origins of Primordial Nucleosynthesis and Prediction of Cosmic Background Radiation", 300
Ott, Edward, 316

packet switching, 98
palindrome, 115
Papadopoulos, Apostolos, 314, 315
parameter plane (c-plane), 138, 140, 145, 153, 343, 480

parasitic capacitance, 326
Passoja, Dann, 312
Pastor-Satorras, Romualdo, 328
patch clamp, 323
Paullay, Alvin, 312
Paxson, Vern, 327
Peacock, John, 301, 454
Peak, David, xiii, 81, 315, 316, 458, 491
Peebles, P. J. E., 84, 300
Peggy, 82, 452
Peitgen, Heinz-Otto, 440
Pennock, David, 328, 472
Penzias, Arno, 300
Perfect, Edmund, 315, 458
periodic points are dense, *see* chaos, periodic points are dense
Perrin, Jean, 282
Pesin, Yakov, 440
PhilosophiæNaturalis Principia Mathematica, 297
PhysicalWorld, xix, xx, 8, 10, 56, 152, 174, 233, 298, 369, 486
Pietronero, Luciano, 84, 300, 454
Planck satellite, 300
planetesimal, 81
Plato, 159
Plowing the Dark, 110, 474
Poe, Edgar Allan, 84, 299, 454
Poincaré, Henri, 216
point at infinity, 47, 48, 334, 506
Pollard-Gott, Lucy, 115, 116, 118
Pollock, Jackson, 125, 126, 477
power law, 58, 59, 67–73, 77, 82–84, 98, 101, 103, 114, 117, 126, 167, 168, 177–179, 190, 197, 198, 205, 227, 233, 234, 236–239, 287, 289, 306, 308, 310, 316, 319, 323, 327–329, 387, 394, 402, 403, 449
power spectrum, 69, 70, 72, 77, 104, 293, 294
Powers, Richard, 110, 474

Principia: Opera Philosophica et Mineralia, 299
Principles of Physical Cosmology, 300
product (of shapes), 23, 51, 52, 169, 175, 177, 192–195, 311, 312, 379, 417, 428, 429
product of Cantor sets, 52, 53
 randomized, 52, 53
Prokofiev, Sergei, 102, 473
Puente, Carlos, 93, 471

Qin, Zhang, 307, 308, 458
quasispecies, 324
Queen Anne's lace, 3, 55
Quinnipiac River, 290
Quiroga, Rodrigo, 307

R/S analysis, 307, 308, 458
radius property, *see* circle inversion, radius property
Random Character of Stock Market Prices, The, 282
random porous medium, 303
ray property, *see* circle inversion, ray property
regimes, 126, 213, 214, 430
relativity
 general, 202, 298, 453, 457
 special, 127, 282
return map, 390–393
Richardson, Lewis Fry, 59, 60, 105, 106, 196–198, 234, 289, 446, 483, 487
Richter, Charles, 304
Riemann sphere, 334
Rinaldo, Andrea, 75, 290, 296
river, 12, 63, 171, 242, 249, 251, 290
 drainage basin, *see* drainage, basin
 drainage pattern
  rectangular, *see* drainage, pattern, rectangular

INDEX

trellis, *see* drainage, pattern, trellis
network, *see* network, river
Rockwell, John, 101, 473
Rodberg, Josie, 117, 329, 493, 494
Roger's Version, 112, 475
rome, 38–41, 184, 410–412, 422, 423, 425, 427, 429
root, 149–153, 345, 346, 353, 355, 356
roughness, xi, xii, 6, 31, 51, 56–58, 60–64, 117, 159, 176, 196, 205, 226, 227, 230, 232, 238, 242, 312, 314, 402, 458
Rovelli, Carlo, 85
Rubner, Max, 73, 295, 449
Rudy, Yoram, 323
Russian dolls, 9

Sainte-Marie, Camille, 278
Salmonella typhimurium, 304
Samavati, Hirad, 92
San Andreas fault, 235, 305
sand dunes, 56, 63
Sapoval, Bernard, 60, 61, 289, 290, 446
Saturn's rings, 87, 311, 312, 458
Saupe, Dietmar, 440
Saxer-Taulbee, Kacie, 273
scale ambiguity, 61, 121, 245, 246, 317
scale invariant, 59
Scale Relativity and Fractal Space-Time, 203
scale relativity theory, 85, 203, 454
scaling, xviii, 6, 7, 10, 15–20, 25, 26, 30, 33, 55, 61, 64, 67, 68, 73–75, 77, 83, 89, 100–105, 112, 114, 116–118, 124, 126, 128, 161, 162, 168, 169, 186, 188, 191, 193, 198, 226, 227, 234–238, 244, 245, 247,
282–284, 304–306, 309, 310, 323, 327–329, 379, 442, 446, 449, 450, 475
duration, 100
during, 101
factor, xix, 6, 8, 10, 136, 163–166, 180–183, 186, 193, 195, 240, 241, 275, 311, 312, 326, 361–363, 367, 371, 372, 375, 413, 414, 418–421, 424
fractal, 67, 87, 100, 116, 309
harmonic interval, 100
hypothesis, 168, 171, 172, 227, 368
melodic interval, 100
melodic moment, 100
metabolic, 73–75, 294–297, 449, 450
motivic, 100
multifractal, 309
pitch, 100
range, 96, 172, 178
relation, 240, 245
rule, 34, 262, 271
signal, 69, 70, 77
structural, 100
symmetry, 6, 169, 170, 243, 298, 415–417, 471
Schaffer, William, 319, 321, 322
Schenker, Heinrich, 102, 103, 473
Schertzer, Daniel, 241, 488
Schreiner, Olive, 2, 111, 442, 475
Schuster, Peter, 324, 459
Science, 58, 60, 72, 78
Scientific American, xiii, 72, 97, 102, 129, 488
secant line, 26, 27
SEIR model, 319–321
self-affine, *see* fractals, self-affine
self-organized criticality (SOC), 306, 309, 457, 458
self-similar, xi, xii, 1, 4–6, 8, 10, 12, 15, 23, 32, 51, 56, 63, 64, 67, 68, 83, 92,

93, 98, 100, 104, 109, 113, 114, 117, 121, 122, 127, 149, 159, 161, 162, 167–170, 173, 181, 186, 188, 190, 191, 193, 235, 237, 243, 251, 301, 324, 327, 328, 362, 376, 377, 417, 472
  about a point, 8, 9
  nonlinear, 1, 4, 6, 7, 11, 48, 50, 185, 186, 188, 191, 377, 482
  statistically, 1, 4, 7, 8, 52, 58, 188
  with memory, 1, 4–6, 10, 34–41, 46, 179, 182, 184, 196, 219, 372, 375, 376, 382, 383, 396, 410, 429, 444, 482
Semmes, Stephen, 54, 288, 440
sensitivity to initial conditions, *see* chaos, sensitivity to initial conditions
sequence notation, 338, 339
Series, Caroline, 445, 446
Serratia marcescens, 304
Seuront, L, 319
Shaikh, Yusuf, 307, 308, 457, 458
Shao, Zhi-Gang, 241
Shearer, Rhonda Roland, 127, 477
shift map, 338, 339
Shilova, N., 309, 458
Shishikura, Mitsuhiro, 131, 145, 146, 156, 351, 479, 481
Shlesinger, Michael, 96, 472
Sierpinski
  carpet, 21, 24, 25, 93, 443
  gasket, 4, 7, 8, 16, 21, 24, 33, 51, 79, 114, 121, 155, 161, 162, 220, 242, 246, 261, 262, 272, 326, 359, 393, 416, 417, 443, 451
    random, 51, 52
  hypertetrahedron, 154, 357, 359, 360

  tetrahedron, 21, 24, 25, 162, 203, 357, 359
Sierpiński, Wacław, 24, 443
Simon, Barry, 311, 312
Simpsons, The, 113, 216
Skarda, Christine, 302, 457
Slipher, Vesto, 298
slit island method, 312, 314, 315, 458
Smilla's Sense of Snow, 108, 474
Smolin, Lee, 202, 484
Smoot, George, 300
Snow Crystals, 317, 459
Snowbound Books, 387
snowflakes, 87, 317, 318
  branching pattern, 318
  symmetry, 2, 318
soil pores, 314, 315
  dimension, 314
solar flares, 308, 458
  fractal scaling, 309, 458
  multifractal fluctuations, 309
solar prominences, 309, 458
  multifractal scaling, 309, 458
Sonata in D Major (K. 284, W. A. Mozart), 101
Sorkin, Gregory, 324, 325, 459
space-filling, 93, 162, 198, 199
species-area relationship, 316
spectral radius, 183, 375, 396, 421
spiral, xi, 2
  fractal, 33, 34, 271
    IFS rules, 34, 272
  shell, 8, 9, 33
square IFS rules, 35, 36, 44, 182, 195, 206, 228, 231, 396
Sreenivasan, Katepalli, 234, 317, 458, 487
Steinberg, Michael, 104, 473
Stereological Methods, 292
Stevens, Wallace, 115, 116, 118, 475
Stewart, Ian, 205, 206, 485
stock market, 282, 287
Stone, Peter, 475

INDEX 513

Stoppard, Tom, 106, 111, 112, 116–118, 329, 475
stories, xvi, 97, 106, 110, 118, 119, 135, 149, 240, 242, 243, 245–247, 251, 252, 289, 318, 486
Story of an African Farm, The, 2, 111, 442, 475
stretched exponential relaxation, 96
Strichartz, Robert, 440
Strogatz, Steven, ix, x, 387, 486
Sughara, George, 306, 457
Sullivan, Dennis, 152, 153, 356, 480
sunspots, 307, 308, 457
    11-year cycle, 307
    fractal cluster of magnetic field lines, 307, 457
superconductivity, 95, 472
    high-temperature, 95, 96
supernova, 309
    remnant, 310, 311
        Cassiopeia A, 310, 458
        Crab Nebula, 310
Suzuki, Norikazu, 306, 457
Swedenborg, Emanuel, 83, 299, 454
Swift, Jonathan, 234
Sydney, Caroline, 115
symbolic dynamics, 339, 485
symmetry, 2, 6, 93, 297
    magnification, 1, 3, 6, 8, 9, 20, 124, 128, 170, 243, 298, 415–417, 471
    reflection, 1, 2
    rotation, 1, 2, 298, 318
    time translation, 298
    translation, 1, 2, 124, 297
Symmetry and the Beautiful Universe, 297
Sync: The Emerging Science of Spontaneous Order, 387
synchronization, xix, 222, 224, 486
    coupled maps, 223, 390, 392, 486
    fuzzy, 393
synchronized
    fireflies, 222
    logistic maps, 223, 393

Takens' delay embedding method, 302, 307, 321, 458, 459
Takens, Floris, 322, 458, 459
Tan Lei's theorem, 145, 146, 351, 479
    magnification factor, 350
Tang, Chao, 306
tangent line, xviii, 26, 27, 33, 47, 59, 149, 150, 152, 227, 231, 353, 399
Taqqu, Murad, 328
Taylor, Richard, 125, 126
Taylor, Tara, 273, 444
Teerikorpi, Pekka, 84, 299
Telesca, Luciano, 236
tent map, 387, 388, 486
    bifurcation diagram, *see* bifurcation diagram, tent map
"The Sail of Ulysses (Canto I)", 116
The Weather and Climate: Emergent Laws and Multifractal Cascades, 240
Théorie de la spéculation, 282
Thew, C. Noelle, 224, 486
Third Policeman, The, 9, 109, 110
Three Roads to Quantum Gravity, 202
Thurston, William, 54, 446
tidal volume, 64
time
    clock, 403, 405
    trading, 403–405
"Time and Punishment", 113, 216
time series, 68–70, 72, 207–211, 213–219, 225, 239, 242, 293, 306–308, 316, 380, 382–384, 389, 402, 405,

430–434, 440, 458
    cardiac, 210, 213, 214
time-like geodesics, 298
Timeaus, 159
Tombeau de Couperin, Le (M. Ravel), 101
topology, 50, 99, 272
Trading Time, 242
    increments, 403, 404
    Theorem, 107, 242, 402–404
transection method, 291, 447
transition graph
    IFS with memory, 38–42, 182, 183, 185, 189, 196, 274, 382, 421–425, 427, 429
    Markov chain, 275
transition matrix, *see* matrix, transition
trapping interval, 381, 383
Treatise on Invertebrate Paleontology, 86, 302
Treatise on Painting, 121
tree, xi, 3
    fractal, 29–31, 85
    power law, 70, 71
turbulence, 87, 234–237, 241, 310, 317, 394, 458, 487
Turcotte, Donald, 305, 457, 487
turning point, 239, 403

uncountable, *see* infinite set, uncountable
unifractal, 241, 403
union (of shapes), 170, 171, 176, 177, 190, 191, 376, 428
unit step function, 287
United Mine Workers, 196
universal curve, 24, 25
Updike, John, 112, 475
Uthayakumar, R., 314
Utsu, Tokuji, 305, 457

van Aardenne-Ehrenfest, Tatyana, 278
Van Ness, J. W., 284

vertex point, 26, 27
Vespignani, Alessandro, 200, 328
Vicsek, Tamás, 199, 302, 457, 483
video feedback, 10
Visage of War, 121, 122, 477
viscous fingering, 52, 86, 302, 457
visual art, 119–127, 477
voice traffic, 97, 98, 327
    Poisson distribution, 97
    Poisson model, 327
von Koch, Helge, 26, 443
Voss, Richard, 62, 75–77, 103, 104, 114, 289, 328, 446, 450, 473
Vrobel, Susie, 203, 484

Wada
    Lakes of, 353, 355, 356, 480
    property, 152
Wallace Stevens: The Making of the Poem, 116
Wallace, R., 322
Wallis, James, 240, 488
Walrus, 99
Walter, Kathy, 486
Wang, Binghong, 236
waterfall, 63
Watson, James, 75
Weber, William, 96
Weibel, Ewald, 64–67, 198, 290, 292, 447, 483, 493
Weinberg, Steven, 300
West, Geoffrey, 71, 74, 294–297, 450
white noise, 69, 70, 77
wide-band antennas, 92–94
Wiener, Norbert, 282
Wiesenfeld, Kurt, 306
Wildlife and Landscape Ecology, 315
Wilhelm, Kate, 113, 475
Willinger, Walter, 327, 328
Wilson, Daniel, 328
Wilson, Robert, 300
Winfree, Erik, 79

INDEX

WMAP satellite, 300
Wright, David, 445, 446
Wright, Sewall, 87, 324, 459
Wrinkles in Time, 300
Wuorinen, Charles, 104, 105, 473
Wurm, Gerhard, 82

X-ray brightness, 309
xkcd, 114

Yale Daily News, xvi, 330
yelm, 299
Yergin, Paul, 298
Yoneyama, Kunizo, 355
Yorke, James, 316
Yoshi, Toshikatsu, 235
Yu, Z., 309

Zelenyi, Lev, 307, 457
Zheng, Zhaobi, 236
Zwicky, Fritz, 86, 457